A Lexicon of

Medicinal Plants in India

DEDICATION

This work is dedicated to the memory of
Barindra Mitra,
the imaginative and courageous leader of
Naya Prokash
who inspired us to write these volumes.

A LEXICON *of* MEDICINAL PLANTS IN INDIA

D. N. GUHA BAKSHI P. SENSARMA D. C. PAL

NAYA PROKASH
206 Bidhan Sarani
Calcutta 700 006
India

A LEXICON OF MEDICINAL PLANTS IN INDIA (VOL. I)

First published : April 15, 1999

Published by

NAYA PROKASH
206 Bidhan Sarani
Calcutta 700 006
India

Printed by :
NEW SARADA PRESS
9C, Shib Narayan Das Lane,
Kolkata-700 006.

Distributor

N P SALES PVT LTD
206 Bidhan Sarani
Calcutta 700 006

Price : Rs 1950.00
US $ 150.00

ISBN 81-85421-39-0

PREAMBLE

Man has been using vegetal materials for medicinal purposes since hoary past. It may be difficult to determine how and where this aspect of man-plant relationship started, but it is undeniable that on the basis of empirical experience of generations, human societies developed certain systems of herbal medicines. In India also evolved some such systems, namely the Ayurvedic and and Siddha systems. Even before emergence of the Ayurvedic system, plants of medicinal value were referred to in the *Rigveda* and other ancient scriptures written in different languages. Besides these, the ethnic communities, numbering more than 400, and settled in different ecological niche, developed their respective systems of vegetal medicines, keeping harmony with their ambient flora. The Unani system of medicine, though exotic in origin, came to India in the medieval period, and soon got rooted firmly in the Indian milieu. In a very short period of time, India became an important centre of Unani research and development. With the Europeans came the Homoeopathic system of medicine. All these systems have their own respective pharmacopoeias – oral or written. The attitude of these systems towards plants and their pharmacopoeias are different. But the cycle(s) of biochemical reaction(s) caused by these medicines in cells of the ailing human body is not known. Naturally, the information contained in these systems requires to be verified by modern scientific methods before accepting/modifying or rejecting the same. Some efforts have been made towards this goal, and results are encouraging. A number of modern medicines have been developed from the data collected from the earlier-mentioned systems of medicines, which may be collectively named as *Traditional Medicine*. Chemical and pharmaceutical verifications of/and experimentations with the traditional medicines are being carried out in different centres of India and abroad. Quite a number of books in English and other languages have already been published. So, it will not be wrong to state that there is no dearth of literature on medicinal botany of plants found in India. But, we felt that there is not a single book which contains all—botanical, ecological, chemical and medicinal—information regarding the same plant species. We have endeavoured to fill up this blank by bringing out this work entitled *A Lexicon of Medicinal Plants in India.* We have tried to collect the following data regarding each species : (i) Botanical name of the species, its family, common synonyms and chromosome number as available; (ii) English and vernacular names of the species; (iii) Trade name, if any; (iv) Traditional uses; (v) Modern uses, (vi) Brief phytography for easy identification; (vii) Phenology; (viii) Distribution in India and neighbouring countries; (ix) Ecology and cultivation; (x) Chemical contents of the species; (xi) Adulterant, if any; and (xii) Additional relevant information, including medicinal/poisonous uses of the species in other countries.

Regarding botanical nomenclature of the species, every effort has been taken to update them, but there might be some errors, as we perhaps could not always keep appropriate record of changes caused by splitting and regrouping of plant families, genera and species. But as we have given the common synonyms, we believe, the purpose of identifying the plants will not be jeopardised by our omissions and commissions. For the same purpose, we have also tried to give outline sketches of the plants. In 'Traditional uses', we have endeavoured to record the uses made by the different ethnic communities of India, medicinal references in non-medicinal texts in Sanskrit language, uses mentioned in Ayurveda, Siddha, Unani and Homoeopathy systems of medicines, while under 'Modern use', we have put only the uses accepted in the Allopathic system of medicine. We must admit that the job of completing all these data from authentic publications is vast, and we could not touch all the texts; and at the same time, it should be recognised that all data regarding each species are still NOT available. Hence, in spite of our best efforts, some lacunae are there. These gaps will help to identify the areas of research, and we hope that with the works of future scientists, these blank spaces will be filled in near future. Further, for the sake of brevity, we have resorted to the taxonomic style of expression and have avoided writing full sentences. We firmly believe this will not be a hindrance in communication.

Our efforts will be amply rewarded if this *Lexicon* helps the scientists of all connected disciplines to determine their immediate programmes of research, and the industrialists and pharmaceutical concerns of various sectors to manufacture medicines and phytochemical products with indigenous materials and technologies.

FOREWORD

The herbal wealth of India and the knowledge of their medicinal properties have a long tradition, as referred to in *Rigveda* and other ancient literature. The topography of India in the tropical belt with its varied climatic zones make it a vast storehouse of medicinal plants. In this part of the biosphere, several ethnic groups lived in a world of their own, depending on plants for the cure of their ailments and wellbeing of their health.

There is no dearth of references of herbal medicines in this subcontinent initially embodied in treatises of *Ayurveda* and *Unani* and later of Allopathy. Commendable works have been published by stalwarts like Watt, Kirtikar, Basu, Chopra and Biswas, and lately in publications of Council of Scientific and Industrial Research.

Despite such comprehensive treatises, nowhere there is an up-to-date list of medicinal species of India,as referred to in different ancient literature including epics and different pharmacopoeias, with their latin and vernacular names, traditional uses, distinct biological features and the chemical contents so far explored.

The present attempt which aims to fill up this lacuna is indeed highly welcome. It is intended for common use to all interested in the study of plants and their medical properties. The long experience of the authors who are reputed Botanists, with expertise in exploration, collection, preservation and documentation of plants along with their deep knowledge of our *Puranas*, *Vedas*, classical literature and modern pharmacognosy, have led to a remarkable lexicon of medicinal plants. I congratulate them heartily and feel confident that this series will be of constant use to all lovers of plants, interested in their identification and medicinal properties, distribution, chemical characteristics, and medicinal potential.

Date : 11.11.98

Professor
CENTRE OF ADVANCED STUDIES
(CELL AND CHROMOSOME RESEARCH)
DEPARTMENT OF BOTANY
UNIVERSITY OF CALCUTTA

LIST OF ABBREVIATIONS

Abbreviation	Language	Abbreviation	Language
And	Andamani	Man	Manipuri
Asm	Assamese	Manb	Manbhumia
Aru	Arunachali	Mar	Marathi
Ben	Bengali	Meg	Meghalaya
Bhi	Bhil	Mik	Mikir
Bhu	Bhutanese	Miri	Miris (Mishing)
Gar	Garhwali	Mon	Monpa
Garo	Garo	Mun	Mundari
Gon	Gond	Nag	Naga
Guj	Gujarati	Nay	Neyadi (North Kerala)
Him	Himachal Pradesh	Nep	Nepali
Hin	Hindi	Ori	Oriya
Jam	Jammu	Orn	Oraon
Jau	Jaunsari	Pun	Punjabi
Kal	Kalat	Pus	Pushtu
Kan	Kannad	Raj	Rajasthani
Kas	Kashmiri	Sad	Sadri
Kha	Khasia	San	Santali
Kon	Konkani	Sik	Sikkimese
Kor	Korku	Sin	Sindhi
Kum	Kumaoni	Tam	Tamil
Lad	Ladakhi	Tel	Telegu
Lep	Lepcha	Tot	Toto
Lod	Lodha	Tri	Tripuri
Lus	Lusai	Urd	Urdu
Mal	Malayali		

We are extremely sorry to record that the imaginative and courageous publisher, Barindra Mitra, who inspired us to prepare this book, has suddenly passed away even before the manuscript of the first volume could be readied by us. Being Hindus, we believe, his soul will be pleased to see from the Heavenly Abode, the fulfillment of his idea. We are thankful to the successors of Late Barindra Mitra—Mrs Arati Mitra and Miss Bhubani Mitra. We are also thankful to Mr Manik Choudhury, Mr Samir Ghosh and other staff members of Naya Prokash and also to Mrs Krishna Neyogi and Mr Chandan Neyogi for supervision and surveillance from the very beginning of the work and successfully piloting the publication of this monumental work within a very short span of time. We are obliged to Dr P K Hajra, ex-Director, Botanical Survey of India, for allowing us to use the libraries and herbaria of Botanical Survey of India, whenever we required. Our gratitude are due to the members and staff of G C Bose Biological Research Unit, Bangabasi College, Calcutta, and the Department of Botany, University of Calcutta for their unstinted support and help. We are thankful to Dr Prabir Ranjan Sur of Economic Botany Section, Botanical Survey of India, and the staff of the library of the Industrial Section, Indian Museum, Botanical Survey of India, for their cooperation.

Shri Swaraj Mukherjee deserves warm thanks for typing out the entire manuscript.

We will fail in our duties if we do not record our gratitude to Mrs Chitra Bakshi, Mrs Ira Sensarma and Mrs Kalyani Pal, who have quietly extended their full support to us in this venture.

D. N. Guha Bakshi **P. Sensarma** **D. C. Pal**

Abelmoschus esculentus (L.) Moench. (Malvaceae)

(2n = 66,132)

Syn : *Hibiscus esculentus* L., *H. longifolius* Roxb.

English names : Lady's finger, Okro, Gumbo.

Vernacular names : Ben : Dhneras, Bhindi; **Guj :** Bhinda; **Hin :** Bhindi, Bhindi tori, Ram turai; **Kan :** Bhende kayi; **Kon :** Bhendi; **Mal :** Venda, Bendai; **Mar :** Bhendi; **Tam :** Vendai, Vendakka; **Tel :** Benda, Bendi.

Trade name : Bhindi.

Traditional use : *AYURVEDA* : seeds with bitter and sharp taste, cooling, tonic, carminative, aphrodisiac, cure diseases caused by *kapha* and *vata*, instestinal complaints, stomatitis, effective in heart diseases, allay thirst and check vomiting.

UNANI : it is considered as a good medicine for dyspepsia, gonorrhoea, urinary discharges, leucoderma and itch.

Modern use : *Immature capsule* (*decoction*) : emollient, demulcent, diuretic, in catarrhal affections, dysuria, gonorrhoea; *Mucilage from fruits and seeds* : emollient, demulcent, useful in gonorrhoea.

Phytography : Tall (0.3–2.1 m) annual herb, covered with rough hairs; leaves simple, cordate, palmately 3- to 5-lobed, coarsely toothed, petiolate, stipulate—stipules subulate; flowers—axillary, pedunculate, bracteate—bracts linear and subulate, large, yellow with crimson centre; capsules pyramidal or oblong (12 to 26 cm) with longitudinal ridges, green or creamy green, bursting from top to base, smooth or hairy; seeds many, rounded, hairy.

Phenology : *Flowering* and *Fruiting* : almost throughout the year.

Distribution : Native of tropical Africa; now cultivated throughout India, Bangladesh, Sri Lanka, Myanmar and Pakistan for its edible fruits.

Ecology and cultivation : Can grow on any type of soil, but thrives best in well-manured loamy soil; being essentially a warm-season crop, cannot endure continuous cold spell.

Chemical contents : *Whole plant* : aromatic, emitting clover-like odour; *Stem* and *Leaf* : iodine; *Pods* : rich in pectin, mucilage, iron and calcium; *Fresh pods* : vitamin A, thiamine, riboflavin, ascorbic acid and niacin; *Ripe seeds* : 16–22% edible oil, pigments, gossypetin and quercetin.

Adulterants : Ripe seeds substitute for coffee; stem gums substitute for gum tragacanth, gum arabic and gum karya.

Abelmoschus moschatus Medic. (Malvaceae)

Syn : *Hibiscus abelmoschus* L.

English names : Ambrettee, Musk mallow.

Vernacular names : Ben : Kalkasturi, Mushak dana; **Guj** and **Mar :** Kasturi bhindi; **Hin :** Kasturidana, Muskdana.

Trade name : Kasturidana.

Traditional use : *Seeds*: diuretic, demulcent, stimulant, antispasmodic, stomachic, tonic, carminative and aphrodisiac; *Seed paste* with *milk* : used to cure itch and in snake bite.

Modern use : *Juice of fresh plant* : febrifuge and expectorant; *Poultice made of whole plant* : applied to chest in bronchitis; *Mucilage from root and leaves* : in venereal diseases, urinary discharges; *Fatty fraction of fresh aqueous extraction of seeds* : antitumour.

Phytography : Annual, herbaceous, 60 cm to 1 m high, hispid; leaves usually palmately 5- to 7-lobed, coarsely toothed, acuminate, stipulate; flowers yellow with crimson centre, 10 cm in diameter; capsules 2.5–7.5 cm; fulvous hairy, oblong lanceolate, acute; seeds subreniform, blacking with a few raised striae.

Phenology : *Flowering* and *Fruiting* : Cold season.

Distribution : Warmer parts of India; cultivated in most tropical countries.

Ecology and cultivation : Like *Abelmoschus esculentus.*

Chemical contents : *Leaves* : β-sitosterol and its glucosides; *Petals* : β-sitosterol, myricetin and its glucosides; *Fruit-husk* : β-sitosterol; *Seeds* : macrocyclic musk compounds, phospholipids.

Remark : Oil obtained from seeds is used, in small quantities, as an exalting agent in fine perfumes.

Abies spectabilis (D. Don) Spach (Pinaceae)

Syn : *Abies webbiana* Lindl., *A. densa* Griff., *Picea webbiana* Loud., *Pinus webbiana* Wall., *P. seriata* Ham., *P. spectabilis* Lamb.

English names : Indian silver fir, East Himalayan fir, West Himalayan high level fir.

Vernacular names : Bhut : Dumsing; **Nep :** Gobra salla.

Trade name : Silver fir.

Traditional use : *TRIBES OF BOTH THE EASTERN AND WESTERN HIMALAYA, PARTICULARLY LEPCHAS, BHUTIAS AND NEPALIS* : use dried leaf-powder with leaf-juice of *Justicia adhatoda* L. and honey in haemoptysis and in catarrh of urinary bladder; give leaf-infusion to the infants in fever due to chest infection. *LODHA* : give diluted resin in tuberculosis. In West Bengal, leaf-decoction is given after parturition.
AYURVEDA : considered as a source of incense; powdered leaves form an ingredient of a number of formulations, viz., *Thalisyaadichoornam* for pulmonary TB and bronchitis,

Yograaja guggulu for rheumatism and chronic nervous diseases, *Nowshadhara* and *Gorojanam* pills; also used as a contraceptive.

Modern use : *Leaves* : carminative, expectorant, stomachic, tonic, astringent, used in asthma, bronchitis, etc.; *Juice of leaves* : antiperiodic.

Phytography : Leafy evergreen tree attaining 50 m in height, crown cylindric, branches flat, horizontal; leaves petioled, spirally arranged but more or less bifarious, very variable in length, very dark green and shining, tip very variable, midrib raised beneath, white beneath on either side of the midrib, persisting for 8 to 10 years; male cones sessile, solitary or clustered, female cones 10–15 cm long, top and base rounded, ripens in the same year; scales much longer than the bracts, deciduous, winged, wings broadly obovate, truncate.

Phenology : *Flowering* and *Fruiting* : Autumn and Winter, respectively.

Distribution : Sikkim between 3000 and 4300 m ; Bhutan.

Ecology and cultivation : Wild in temperate and subtemperate climate.

Chemical contents : *Tree* : yields a toxic resin which is mixed with oil of roses and applied in neuralgia; *Leaves* : an alkaloid, an essential oil.

Abroma augusta (L.) L.f. (Sterculiaceae)

(now known as *Ambroma* L. f.)

Syn : *Abroma augulata* Lam., *A. wheeleri* Retz., *A. fastuosum* Gaertn., *Theobroma augusta* L.

English name : Devil's cotton.

Vernacular names : Asm : Bon kopashi; **Hin** and **Ben :** Ulat kambal; **Tam :** Sivapputtutti.

Trade name : Ulat kambal.

Traditional use : *Seed* : yields a fatty oil, rich in linoleic acid and lowers cholesterol level in blood.

HOMOEOPATHY : used to control mind, different ailments of head, heart, eyes, ears, nose, face, mouth, throat, stomach, abdomen, urinary organs, male and female sexual organs, respiratory organs, neck, back and limbs, skin diseases, fever; to regulate appetite and sleep.

Modern use : *Fresh* and *dried root-bark* : used as uterine tonic and emmenagogue; regulates menstrual flow; *Fresh juice from root- bark* : useful in congestive and neuralgic forms of dysmenorrhoea, amenorrhoea, urinary trouble, bronchitis, broncho-pneumonia, carbuncles and poisonous boils; *Leaf* : used in diabetes, rheumatic pain and sinusitis.

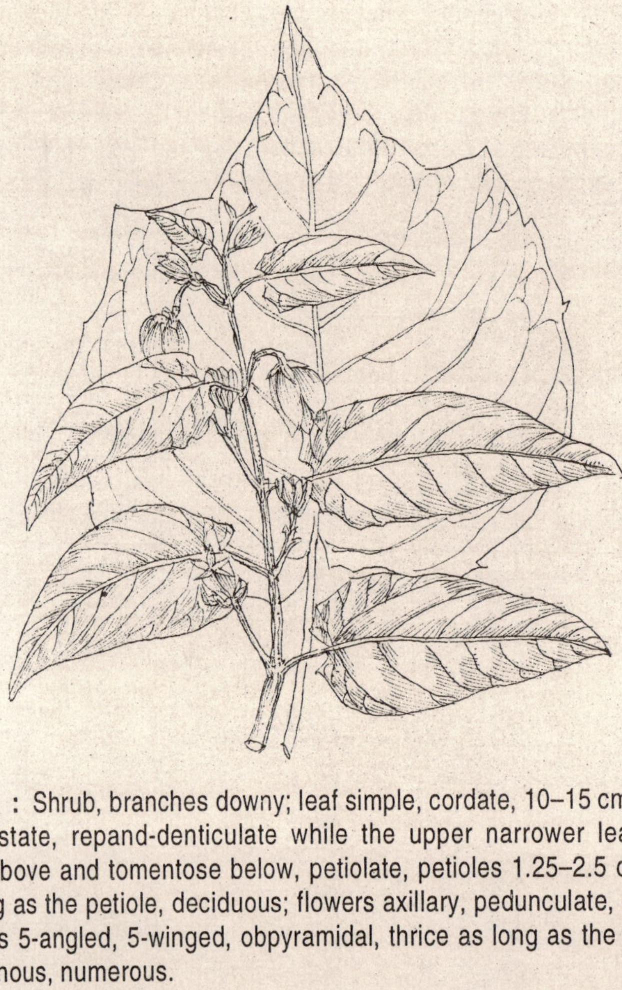

Phytography : Shrub, branches downy; leaf simple, cordate, 10–15 cm by 10–12.5 cm, base 3–7 costate, repand-denticulate while the upper narrower leaves are entire, glabrescent above and tomentose below, petiolate, petioles 1.25–2.5 cm long, stipules linear, as long as the petiole, deciduous; flowers axillary, pedunculate, peduncle 3.5 cm long; capsules 5-angled, 5-winged, obpyramidal, thrice as long as the persistent calyx; seeds albuminous, numerous.

Distribution : Widely distributed in hotter parts of India up to 1500 m; Bangladesh, Pakistan and the adjacent countries.

Ecology and cultivation : Plant prefers hot and moist climate; sometimes planted, occasionally found as an escape.

Chemical contents : *Plant* : fixed oil, a little resinous matter, alkaloid (minute quantity) and water-soluble bases.

Abrus precatorius L. (Fabaceae)
(2n = 22)

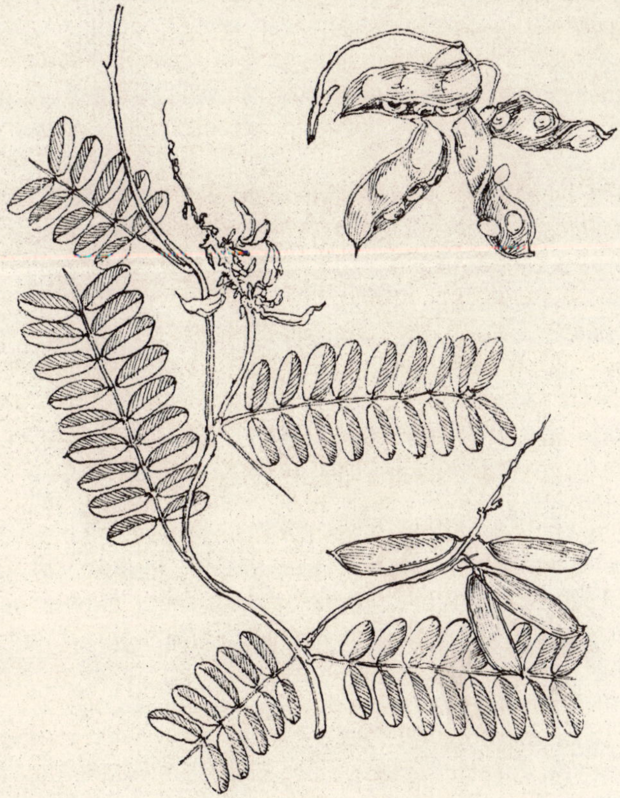

Syn : *Abrus minor* and *A. pauciflorus* Desv.

English names : Indian liquorice, Crab's eye.

Sanskrit name : Krishna gunja.

Vernacular names : Asm : Latuwani; **Ben :** Rati, Kunch; **Guj :** Gumchi; **Hin** and **Pun :** Rati; **Kan :** Gurgunn, Gulaganji; **Ori :** Kaincha, Gunja; **Mal :** Kunnikkura; **Tam :** Kunthamani; **Tel :** Gumginja.

Trade names : Rati, Kunch.

Traditional use : SANTAL: (i) grind the roots, make small pills, encase the pills in molasses and eat the same to treat night-blindedness; (ii) make a plaster by grinding the roots of white-fruited variety and apply the plaster on the painful part of inflammated sections of the gum; (iii) to treat white-coloured urine they drink a mixture made by grinding roots of the following : (a) white-fruited *A. precatorius*, (b) *Indigofera pulchella*, (c) *Panicum repens* and (d) *Spatholobus roxburghii*; (iv) to treat gravel they drink a mixture made of the following : (a) roots of *A. precatorius*, (b) the refuse of molasses, (c) exudation from a sapling of *Diospyros tomentosa*, (d) exudation from *Acacia catechu*, (e) little saltpeter, and (f) pinch of sulphur; (v) to treat the variety of childbed complaints (usually caused by anaemia) characterized by profuse diarrhoea, roots of *A. precatorius* are used in preparing two different varieties of mixtures; the ingredients of the mixtures are given below : (a) first variety : roots of *A. precatorius*, *Elaeodendron roxburghii*, *Coix lachryma-jobi*, *Piper longum*, *Ruellia suffruticosa*, white onion, rhizome of *Zingiber officinale*; (b) second variety : roots of *A. precatorius*, *Coix lachryma-jobi*, *Embelia robusta*, *Piper longum*, bark of *Casearea tomentosa*, *Elaeodendron roxburghii*, *Gmelina arborea*, *Emblica officinalis*, white onion, leaves of *Ocimum sanctum*, rhizome of *Curcuma angustifolia* and *Zingiber officinale* — all these are ground together, boiled and mixed with the refuse of molasses; (vi)roots as abortifacient and in paralysis; (vii) apply leaf-paste with lime-water (2:1) on swelling of glands; (viii) grind the leaves of white-flowered *A. precatorius*, warm slightly and plaster on the loins to kill pain there; (ix) grind leaves of *A. precatorius* along with leaves of *Lawsonia alba* and *Tamarindus indica* (1:1:1), add a little salt, boil a little and apply the plaster on the whole body to get relief from muscular pain caused by over-exhaustion; (x) make a paste of leaves of *A. precatorius* along with roots of *Carissa carandas* and *Gossypium arboreum*, warm the paste slightly and plaster the same over the whole body of the patient suffering from stealth convulsions; (xi) leaf-paste in leucoderma; (xii) seed-paste in skin diseases; (xiii) seeds after some processing as contraceptive. *MUNDA* : Root-paste in gonorrhoea. *ORAON* : dried root-powder as mild purgative.

AGNI PURANA : (i) husks of *A. precatorius* along with the same of *Vitis vinifera* and the decoction of *Polyalthia longifolia*, *Moringa pterigosperma*, *payomuca* and *triphala* (fruits of *Terminalia belerica*, *Terminalia chebula* and *Emblica officinalis*) destroys all intestinal worms; (ii) the mixture of powder of *A. precatorius*, marine salt and *pathya* in warm water removes all fevers; (iii) consumption of the seeds of *A. precatorius* along with the fruits of *Melia azadiracta*, *Holarrhena antidysenterica* (leaves) *Acorus calamus* (young leaves) and *Glycyrrhiza glabra* (powder of stem) causes vomiting; (iv) regular drinking of *A. precatorius* along with *Acorus calamus*, *Gloriosa superba*, *vasa, nisagada*, *Zingiber officinalis*, *Glycyrrhiza glabra* and marine salt daily in the morning enhances memory of young boys; (v) *A. precatorius* can enhance the span of a man's life, if it is eaten with marine salt and some other plants (*Tinospora cordifolia*, *pathya*, *citraka*, dried rhizome of *Zingiber officinalis*).

Modern use : *Roots* : emetic and alexiteric; *Decoction of roots* and *leaves* : for cough, cold and colic; *Seeds* : purgative, emetic, tonic, aphrodisiac, used in nervous disorder and cattle poisoning; *Poultice of seeds* : as suppository to bring about abortion; *Paste of seeds* : applied locally in sciatica, stiffness of shoulder joints and in paralysis.

Phytography : Copiously branched climber with slender branches; leaves alternate, pinnately compound with numerous deciduous leaflets; flowers small, in dense racemes on axillary peduncles or short branches; pods 2.5–3.7 by 1.0–1.25 cm; seeds bright scarlet and black or whitish or black or mixed black and white, large like pea.

Phenology : *Flowering* : August and September; *Fruiting* : January to March (even up to May).

Distribution : Occurring throughout greater parts of India, ascending the outer Himalaya up to 1200 m, occasionally planted in gardens.

Ecology and cultivation : Naturalised in tropical countries.

Chemical contents : *Root* and *Leaf* : glycyrrhizin, isoflavanquinones, abrusquinone A, B & C; leaves taste sweet and roots less so; roots also contain precol, abrol, abrasine and precasine. *Seed* : poisonous, principal constituents being 'abrin'; a fat-splitting enzyme, haemaagglutinin, urease; alkaloid (abrin), a glucoside (abralin) and a small quantity of fatty oil have also been isolated from seed. Pharmacologically, abrin is considered to be intensely poisonous. Besides abrin, a seed contains hypaphorine, two steroids—one oily and the other crystalline— β-sitosterol, stigmasterol, 5 B-cholanic acid, abricin, abridin, cholesterol, lectins and toxic proteins.

Remark : *Root* is used as an adulterant of the root of *Glycyrrhiza glabra* (Fam. *Fabaceae*).

Abutilon indicum (L.) Sweet (Malvaceae)

(2n = 26,42)

Syn : *Abutilon asiaticum* G. Don, *Sida indica* L.

English name : Indian mallow.

Sanskrit name : Kankatikta.

Vernacular names : Asm : Japapetari; **Ben :** Petari; **Guj :** Dabali, Hettuti; **Hin :** Jhampi, Kanghi; **Kan :** Thurubi gida, Shree-mudre gida; **Kon :** Mudro; **Mal :** Tuthy, Perinthotty; **Mar :** Mudra; **Ori :** Pediapedika; **Pun :** Pili-buti, Simbol; **Tam :** Perinthotty, Tuthal; **Tel :** Tuthiribenda.

Trade names : Jhampi, Petari.

Traditional use : SANTAL : plant in menorrhagia; eat boiled leaves in bleeding piles; TRIBES OF MADHYA PRADESH : *Leaf-paste* : in gonorrhoea; TRIBES OF GARHWAL : *Seed-paste* : on leprotic sores; BIRHOR : *Leaf-decoction* : as wash for vaginal infection; *Dried seed-powder* : as abortifacient.

AYURVEDA and *UNANI* : as febrifuge, anthelmintic and anti-inflammatory agent of genital organs, piles and spermatorrhaea; *Bark* : astringent, diuretic; *Infusion of root* : for fever; *Leaf* and *Seed* : as demulcent and laxative.

Modern use : *Infusion of root* : for fever; *Bark* : astringent, diuretic; *Leaf* and *Seed* : demulcent and laxative.

Phytography : Shrub; leaves simple, petiolate, cordate, nearly entire or irregularly toothed, surface rough with closely felted white hairs, stipules deflexed; flowers axillary, pedunculate, epicalyx absent, yellow, 2.5 cm in diameter, opening in the evening; fruit of ripe carpels separating from the axis.

Phenology : *Flowering* : Rainy season, but may continue up to December; *Fruiting* : October to January.

Distribution : Throughout the hotter parts of India; Pakistan, Bangladesh and the adjoining areas.

Ecology and cultivation : Plant prefers hot and humid climate; wild.

Chemical contents : *Plant* : Mucilage, aspargin.

Abutilon hirtum (Lamk.) Sweet [Malvaceae]

(2n = 14)

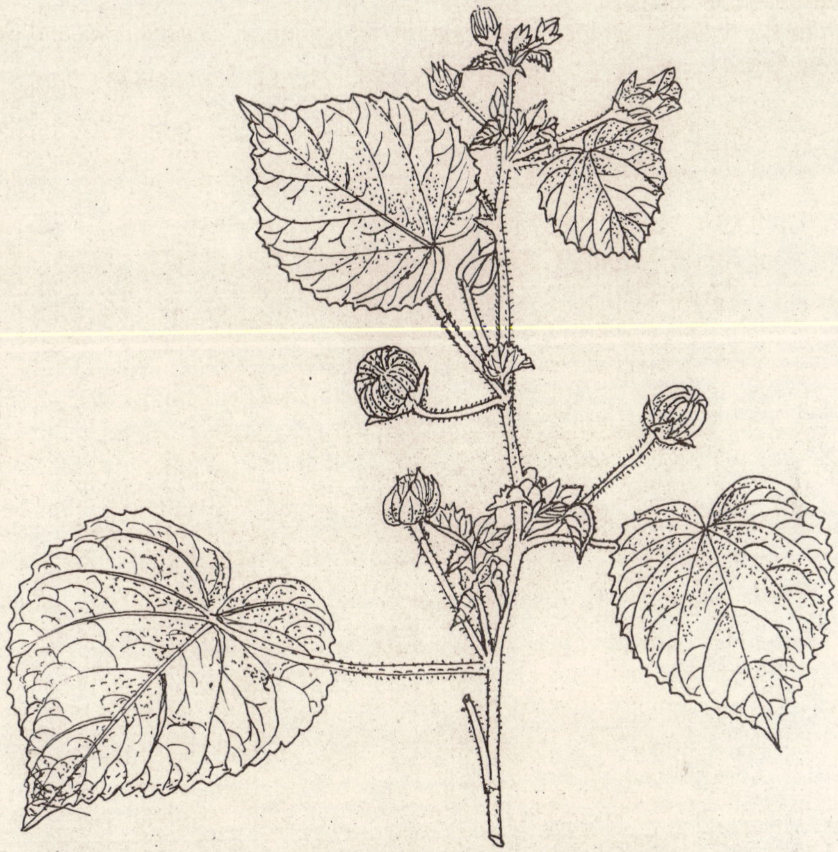

Syn : *Abutilon graveolens* W. & A., *Sida hirta* Lam, *Sida pilosa* L'Her.

Vernacular names : Ben : Barkanghi; **Hin :** Barkanghi; **Mar :** Tulti.

Trade name : Barkhangi.

Traditional use : *SIDDHA SYSTEM* : used in piles and as rejuvenative tonic; *Decoction* for anaemia, fever, hiccup, hoarseness of voice, urinary calculus, gout.

Modern use : *Root*: in fever; *Bark*: astringent, diuretic; *Leaf*: demulcent; *Seed*: laxative and demulcent.

Phytography : Wild shrub; branches covered with hairs, brownish; leaves simple, petiolate, stipulate, stipules linear-subulate, lamina orbicular-cordate, abruptly acuminate; subscrabid above; flowers axillary, pedunculate, orange-coloured with a darker centre, epicalyx absent.

Phenology : *Flowering*: August to December; *Fruiting*: October to January.

Distribution : Uttar Pradesh, Madhya Pradesh to South India; Pakistan (Baluchistan).

Ecology and cultivation : In tropical and subtropical regions.

Abutilon theophrastii Medic. (Malvaceae)

(2n = 42)

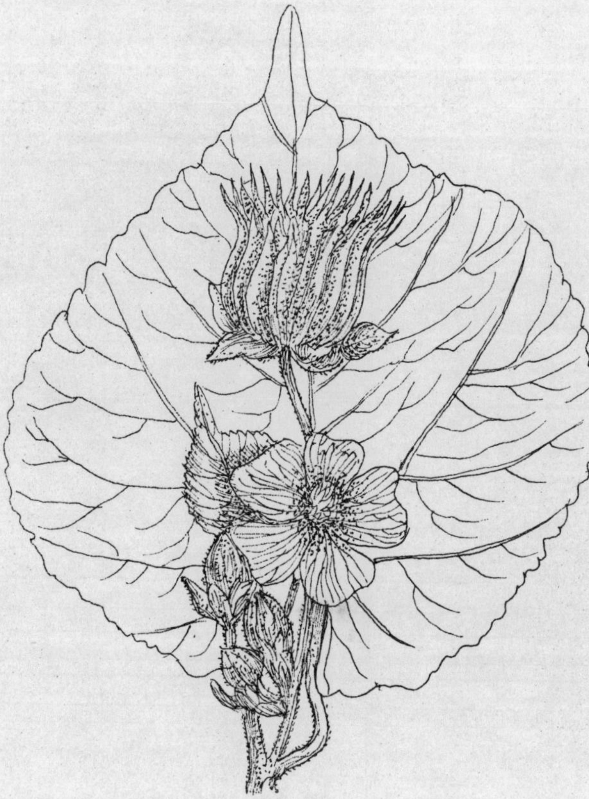

Syn : *Abutilon avicennae* Gaertn., *Sida abutilon* L.

English names : American jute, Indian mallow, China jute.

Sanskrit name : Jaya.

Vernacular name : Mar : Nahani khapat.

Traditional use : *Leaf* : demulcent; *Bark* : astringent, diuretic; *Root* : in fever; *Seed* : laxative, demulcent.

Modern use : *Decoction of seed* : in dysentery, fistula and sores of eye.

Phytography : Annual, herbaceous bushy shrub; leaves 7.5–10 .0 cm, orbicular-cordate, petiole 7.5 cm; flower axillary, solitary, peduncle 2.5 cm, yellow, staminal tube very short, carpels 15–20, oblong, truncate, pubescent, with 2 long horizontal spreading ciliate awns; seeds pilose.

Distribution : Kashmir, northwestern India, West Bengal; Bangladesh and Pakistan (Sind).

Ecology and cultivation : Wasteland; cultivation not known.

Chemical contents : *Leaf* : rutin (0.01%); *Seed* : yield 19% semidrying oil.

Remarks : It is also known as *Manchurian jute.* Fibre is very strong and stable.

Acacia caesia (L.) Willd. (Mimosaceae)

Syn : *Acacia intsia* Willd., *Mimosa caesia* L.

Vernacular names : Hin : Aila; **Kan :** Antarika; **Mal :** Attu, Incha, Inna; **Mar :** Chilari; **Ori :** Dentari; **Sant :** Kondrojanum; **Tam :** Karijindu; **Tel :** Kondakarinda.

Trade name : Aila.

Traditional use : *SANTAL WOMEN* : in deranged courses.

Modern use : *Soft beaten bark* : has cleansing properties and protect the skin against micro-organisms.

Phytography : Climber with minute, hooked prickles, branchlets and leaf-rachises finely downy or glabrous; leaves bipinnate, pinnae 5.0–7.5 cm long, leaflets 0.3–0.5 cm, minutely cuspidate, both surfaces glabrous or nearly so, the lower gray-green; heads yellow, less than 1.25 cm in width; pods straight, 10–15 cm long, strap-shaped, glabrous, dehiscent, 8- to 12-seeded.

Phenology : *Flowering* : September–October; *Fruiting* : April.

Distribution : Sub-Himalayan tract from Chenab eastwards, ascending to about 1200 m and 1700 m in Sikkim; found throughout India; Bangladesh and Pakistan to some extent.

Ecology and cultivation : Wasteland in the vicinity of jungles; not in cultivation.

Chemical contents : *Bark* : 17% tannin, stigmasterol, an aliphatic hydroxy compound and an amorphous powder obtained with all characteristics of saponin.

Remark : Bark used for washing hair.

Acacia catechu (L.f.) Willd. (Mimosaceae)

(2n = 26)

Syn : *Acacia catechuoides* Benth., *Mimosa catechuoides* Roxb.

English name : Cutch tree.

Sanskrit name : Khadira.

Vernacular names : Ben : Khayer; **Guj :** Kher, Kherabaval, Vodalia; **Hin :** Katha, Khair; **Kan :** Kaggali, Kachu; **Kon :** Kairazhadd, Kairarukh; **Mal :** Kadaram; **Ori :** Khaira; **Pun :** Katha, Khair; **Tam :** Baga, Kadiram, Karangalli, Othalei; **Tel :** Kadiramu, Kanchu, Sandra, Nalla sandra.

Trade names : Katha, Khair.

Traditional use : *SANTAL* : *Plant* : in cancer and bronchitis; *Bark-decoction* : as astringent, in dysentery and diarrhoea; *Gum* : in asthma; *Leaf-decoction* in dysentery; *TRIBES OF RAJASTHAN* : *Bark decoction* : in leucorrhoea and menstrual complaints; *Leaf decoction* in gonorrhoea; *RURAL FOLKS OF INDIA* : *Various parts of the plant* : in sore mouth, pain in chest, cancer, colic pain, gravel, dysentery, phthysis, bronchitis, consumption and strangulation of intestine; *Bark-juice* along with *asafoetida* : haemoptysis; *Bark* : abortifacient; *Mixture of flower tops* and *cumins* : in gonorrhoea. *AYURVEDA* : in leprosy.

Modern use : *Kheersal* (a white crystalline deposit of catechin) : for cough and sore throat; *Bark* : astringent; *EtOH* (50%) *extract of stem* : antivirus, spasmolytic; *Heartwood extract* : antifungus.

Phytography : Tree, up to 10 m in height and 1.5 m in girth; bark dark grayish brown, 2.25 cm thick; spikes axillary, cylindrical; flowers small, sessile, pale creamy white; fruits flat brown pods, shiny with a triangular beak at the apex and narrowed at the base; seeds 3–10 per pod.

Phenology : *Flowering* : Autumn; *Fruiting* : Winter.

Distribution : Assam, Bihar, Uttar Pradesh, Punjab, Himachal Pradesh, Madhya Pradesh, Maharashtra, Gujarat, Karnataka; Nepal, Pakistan and Bangladesh. It is also common in the dry plains and lower hill forests of upper and lower Myanmar.

Ecology and cultivation : Grows in abundance between 700 and 1700 m along the Terai belt, in drier regions.

Chemical contents : *Heartwood* : 8-flavonoids; *Gum* : D-galactose, L-arabinose, δ-rhamnose, L-glucuronic acid; on hydrolysis yields aldobiuronic acid.

Adulterants : Similar extracts are also prepared from other substances and called *cutch*; thus mangrove cutch is obtained from the bark of *Rhizophora mangle* L. and *Ceriops candolleana* Arnold (family *Rhizophoraceae*).

Remarks : Cutch or catechu, popularly known as Katha, an extract obtained from the heartwood, is used as an ingredient of *paan* (betel leaf masticatory) and is used for treating asthma.

Acacia farnesiana Willd. (Mimosaceae)

(2n = 52)

Syn : *Acacia indica* Des., *Mimosa farnesiana* L., *Vachellia farnesiana* W. & A.

English name : Cassie flower.

Sanskrit name : Arimaudah.

Vernacular names : **Asm :** Tarwa kadam; **Ben :** Guga babul; **Hin :** Gand babul; **Mal :** Ari velam; **Mar :** Deo babul; **Ori :** Kapur; **San :** Grabur; **Tam :** Kasturivel; **Tel :** Kasturiturmma.

Trade name : Gand babul.

Traditional use : SANTAL : *Various parts of the plant* : in cholera, convulsion, delirium, epilepsy, madness, sores and sterility in women; *Bark* : in malaria.

Modern use : *Bark* : astringent, demulcent, used in antifertility; *inflorescence* : in venereal diseases.

Phytography : Erect shrub with zigzag branches marked with minute raised gray dots; leaves alternate, bipinnate with minute leaflets, stipulate, stipules spiny, leaf rachis downy with minute petiolar gland, leaflets small, green, opposite, subglabrous, rigidly coriaceous;

flowers borne on axillary cymose heads, small, fragrant, bright yellow, bracteate, sessile; pods 5.0–7.5 cm long, 1.25 cm thick, dull brown, marked with 10–12 fine close horizontal constrictions; seeds exalbuminous.

Phenology : *Flowering* : chiefly September to February; *Fruiting* : almost throughout the year.

Distribution : Native of West Indies; now planted throughout India and adjacent countries.

Ecology and cultivation : Foothills, scrub-jungles; now raised by the Department of Forests.

Chemical contents : *Leaf* : tannins, alkaloids, rutin, apigenin-6, 8-bis-c-glucoside, cyanogens, linamarin, lotaustralin, and some amount of cyanide have been separated and identified, but some are still unidentified; *Flower* : contains 38 new constituents including three C_{11} compounds with characteristic fragrance of cassie oil, lactone, dihydroactinidiolide and δ-pinitol; *Pigments* : have isorhamnetin-3 ,7-glucorhamnoside; *Pod* : phenolics.

Acacia ferruginea DC (Mimosaceae)

Syn : *Mimosa ferruginea* Roxb.

Vernacular names : Hin : Sonkhair; **Guj :** Kaigu; **Kan :** Banni; **Mar :** Pandhra khair; **Tam :** Velvelam, Parambai; **Tel :** Ansandra.

Trade name : Sonkhair.

Traditional use : *Wood-oil* : used in scabies.

Modern use : *Bark* and *Pod* : astringent; *Gum* : demulcent, emollient and nutrient.

Phytography : Medium-sized tree with brown bark, branchlets slightly downy and glabrous, prickled; leaf alternate, compound, one gland present on the petiole and the other between the uppermost pinnae, leaflets vary from 0.6–0.8 cm in length, pale glaucous-green, rigidly sub coriaceous; spikes dense, 7.5–10.0 cm long; flowers small, yellowish; pods distinctly stalked, 7.5–10.0 cm in length and up to 1.87 cm in breadth, dark brown, glabrous, 4- to 6-seeded, upper suture narrowly winged.

Phenology : *Flowering* : March and April; *Fruiting* : November to January.

Distribution : Gujarat, Maharashtra, Madhya Pradesh and South India; Sri Lanka.

Ecology and cultivation : Foothills, scrub jungles; wild.

Acacia leucophlaea Willd. (Mimosaceae)

(2n = 26)

Syn : *Acacia alba* Willd., *A. arcuata* Decaisne, *Mimosa leucophlaea* Roxb., *M. alba* Rott.

Sanskrit name : Shveta barbura.

Vernacular names : Ben : Safed babul, Sveta babul; **Hin :** Peru, Safed kikar, Ronj, Hrinj; **Kan :** Bilijali; **Mar :** Huvar; **Tam :** Velvelum; **Tel :** Tellatumma.

Trade name : Safed kikar.

Traditional use : *Root* : abortifacient; *Bark* : astringent, also used in ulcer; *Ash (prepared with ghee)* : applied as antidote against any poison.
AYURVEDA : applied to cure *Raktapitta* (internal haemorrhage).
UNANI : constituernt of the drug 'Kath Safaid'.

Modern use : *Bark* : astringent, cooling, alexeteric, anthelmintic, antipyretic, cures inflammatory diseases like bronchitis, leprosy, biliousness, thirst, vomiting ,etc.; used in spermaturia and diarrhoea.

Phytography : Deciduous tree; leaves alternate, compound, stipules spiny, leaf-rachises downy, with several glands, leaflets crowded, rigidly coriaceous, small—length varies from 0.31 to 0.41 cm; panicles reaching 30 cm in length, with long compound densely tomentose ascending branches, the lower only from leafy nodes, heads yellow, small, not more than 0.62 cm in diameter, frequently affected by insect galls; flowers very small, not exceeding 0.1 cm in length; pods sessile, 10–20 cm in length, subindehiscent, clothed with pale brown or gray persistent tomentum.

Phenology : *Flowering* : August–October; *Fruiting* : December.

Distribution : Found in the dry regions of the country, especially in Rajasthan, Punjab, Madhya Pradesh and throughout the dry forest tracts of the peninsula; Pakistan (Sind, Punjab and NW Frontier Province) and Myanmar.

Ecology and cultivation : Grows in arid, desertic and semi desertic areas with typical xerophytic characters; wild.

Chemical contents : *Root-bark* : leucophleol, leucophleoxol, leucoxol; *Stem-bark* : n-hexacosanol, β-amyrin and β-sitosterol; *Heartwood* : n-octacosanol, β-sitosterol; *Leaf*, *Bud*, *Flower* and *Pod* : a deadly substance HCN; *Fresh flower* : myricetin, a flavonoid pigment.

Adulterants : *Bark* : substitute for Avarun bark (*Cassia auriculata* L.) and Konam bark (*Cassia fistula* L.).

Acacia modesta Wall. (Mimosaceae)

(2n = 26)

Syn : *Mimosa dumosa* Roxb., *M. obovata* Roxb.

Vernacular names : Hin : Phulai; **Mar :** Kantosariyo; **Pun :** Phulahi.

Trade name : Phulai.

Traditional use : *Young branches* : used as tooth brush.

Modern use : *Gum* : restorative.

Phytography : Middle-sized tree, branches grey and glabrous, prickled, prickles dark brown and polished; leaves alternate, compound, leaf-rachis very slender, finely downy, one small gland present at the base and the other between two uppermost pinnae, leaflets thin, 0.3–0.9 cm long, shortly stalked, subcoriaceous, glabrous, oblique at the base, obtuse or minutely cuspidate at the apex, pale green; spikes 5.0–7.5 cm long, not very dense; flowers small with greenish corolla; pods 5.0–7.5 cm long, narrowed gradually into a short stalk, glossy, drab, glabrous, 6- to 8-seeded.

Distribution : Sub-Himalayan tract of the Punjab from the Jamuna eastwards ascending up to 1300 m; Pakistan (Salt Range, elevated portions of Sind-Sagar Doab, lower hills of Hazara).

Ecology and cultivation : In deciduous forests and open thorny scrub jungles; wild.

Chemical contents : *Stem-bark* : octacosanol, α-amyrin, betulin and ψ-sitosterol; *Heartwood* : γ-sitosterol and pinotol; *Leaves* : octacosane and hentriacontane, octacosanol and hentriacontanol.

Acacia nilotica (L.) Willd. ex Del. sub. sp. *indica* (Benth.) Brenan. (Mimosaceae)

(2n = 52, 102)

Syn : *Acacia arabica* (Lamk.) Willd. var. *indica* Benth.; *Mimosa arabica* Lamk.

English names : Babul, Babul acacia, Gum tree.

Sanskrit name : Vaboula.

Vernacular names : Asm : Torua kadam; **Ben :** Babla, Babul; **Guj :** Kaloabaval; **Hin :** Babul, Kikar; **Kan :** Kari jali; **Kon :** Babulzhadd; **Mal & Tam :** Karuvelam; **Mar :** Babhul(a); Kikar; **Ori :** Baburi; **Pun :** Kikar; **Tel :** Nellatumma, Suma,Tuma.

Trade names : Babul, Gum arabic, Babla.

Traditional use : Different parts are used in hairfall, earache, syphilis, cholera, dysentery, leprosy and rinderpest. *Bark* : cooling, astringent, demulcent, anthelmintic, cures skin diseases, bleeding piles, used in asthma and bronchitis, decoction in leucorrhoea; *Gum* : in sore throat, given with latex of *Calotropis procera* to cure asthma, stops bleeding, urinary and vaginal discharges, useful in diabetes; *Leaf* : bruised leaves to eye sores in children; eaten in throat infection; *Paste of burnt leaf* : on itch; *Flower, pod* and *gum resin* : used in diarrhoea and dysentery; *Pod* : in impotence and urino-genital disorders.
UNANI : Constituent of the drug called 'Keekar' ('Mughilan').

Modern use : In pharmacy, it is used in preparing emulsions, tablets, pills, etc., while *decoction of bark* is used in gurgle.

Phytography : Shrub or tree, branchlets straight, finely grey downy, prickles 0.6–5.0 cm long, white, ascending, pungent when old; leaves bipinnate, leaf-rachises downy with several glands, pinnae 1.25–3.7 cm long, leaflets membranous, green, 0.3–0.6 cm long, glabrous; peduncles short, densely grey downy, profusely bracteate; heads yellow, 1.25 cm in diameter; pods rigidly coriaceous, subindehiscent, densely persistently grey-downy, 7.5–5.0 cm long, distinctively stalked, 8- to 12-seeded.

Phenology : *Flowering* : June–July; *Fruiting* : almost throughout the year.

Distribution : From Punjab to Bihar, western peninsula; Pakistan, Bangladesh and Sri Lanka.

Ecology and cultivation : Xerophyte, grown scatteredly in crop-fields in the drier regions of India, Pakistan, Bangladesh and Sri Lanka, and along roads, avenues, etc.

Chemical contents : *Gum* : galactose, l-arabinose, l-rhamnose, 4-aldobiouronic acids and arabinobioses; *Bark* : several polyphenols, tannin; *Root-bark* : octacosanol, betulin, β-amyrin, β-sitosterol; *Pod* : gallic acid and 2 condensed tannins.

Adulterants : Amritsar gum (*Acacia modesta*), Cape gum (*Acacia farnesiana* Willd. and *Acacia leucophlaea* Willd.) are used.

Remarks : Wood used for charcoal; timber used for agricultural implements; leaves are used in India for smoking opium; gum is used commercially as gum arabic, the gum is extensively used by calico-printers and dyers; the bark is used in tanning. In India, the lac insect is found on this tree; the pods yield a useful black dye. It is also a useful plant for afforestation in dry and arid zones.

Acacia pennata (L.) Willd. (Mimosaceae)

(2n = 26)

Syn : *Acacia megaladena* Desv., *A. prensans* Lowe., *A. paludosa* Miq., *A. pinnata* Dalz. & Gibs., *Mimosa pennata* Roxb.

Vernacular names : Ben : Kuchui; **Hin :** Biswul; **Kum :** Agla; **Tam :** Indu; **Tel :** Karusikaya; **Nep :** Arfu.

Trade name : Biswul.

Traditional use : TRIBES OF BASTAR : *Decoction of young leaves* : in body pain, fever and headache; TOTO : *Dried bark powder* : to relieve dandruff and on the wounds of snake-bite.

AYURVEDA : *bark* : acrid, bitter, sour and hot, cures diseases of blood biliousness, *tridosha*, bronchitis and asthma.

Modern use : *Leaf* : chewed with sugar and cumin in bleeding gums; *Juice of bark* : antidote to snake poison.

Phytography : Climber with copious scattered prickles, branchlets few and finely downy; leaf alternate, compound, leaf-rachis finely downy with a large plate-shaped gland at the base of petiole and several between the upper pinnae, leaflets firm in texture, very small, dark green above and grey-green below, glabrous; flowers in copiously panicled globose heads, heads 1.25 cm in diameter, flowers small; pods stalked, straight, strap-shaped, thin, long, 15 to 20 cm in length, glabrous, 8- to 12-seeded, sutures rather raised.

Phenology : *Flowering* : Rainy season; *Fruiting* : Winter.

Distribution : Central and eastern Himalaya up to 1500 m, Uttar Pradesh, West Bengal, Bihar, Madhya Pradesh, Maharashtra, Gujarat, South India; Bangladesh.

Ecology and cultivation : Occurring throughout India, chiefly along rivers and streams ; wild .

Chemical contents : *Bark* : lupeol and α-spinasterol; *Stem* : β-sitosterol.

Remarks : Bark is a substitute of soap. Plant is a host for lac insects.

Acacia senegal Willd. (Mimosaceae)

(2n = 26)

Syn : *Acacia verek* Guill. & Per., *A. rupestris* Stocks ex. Boiss., *Mimosa senegal* L.

English name : Gum arabic tree.

Sanskrit name : Sveta khadira.

Vernacular names : Guj : Goradio babul; **Hin :** Kumta; **Mar :** Khor; **Raj :** Kumta.

Trade names : Kordofan gum, Kumta.

Traditional use : *Gum arabic* : externally applied to nodules in leprosy.

Phytography : Low erect tree, branches flexuose, glaucous, grey; leaf alternate, compound, stipules spiny and hooked, leaf-rachis shorter than 2.5 cm, with gland at the base and one between the upper pair of pinnae, leaflets small, glabrous, pale glaucous green, rigidly subcoriaceous; spikes 5.0–7.5 cm long, not very dense; flowers yellowish, small, calyx united and deeply toothed; pods stalked, usually 7.5 cm long and 2.0 cm broad, grey, firm, indehiscent, 5- to 6-seeded.

Phenology : *Flowering* and *Fruiting* : November–March generally.

Distribution : Punjab, Rajasthan; Pakistan (Sind and Baluchistan).

Ecology and cultivation : Xerophytic plant, wild.

Chemical contents : *Root* : ceryl alcohol, betulin, sitosterol; *Bark* : uvaol, octacosanol, β-amyrin, sitosterol, β-sitosterol, erythrodiol; *Heartwood* : sitosterol; *Gum* : an acid polysaccharide containing sugars, δ-galactose, l-arabinose, γ-rhamnose and δ-glucuronic acid; *Leaf* : nonacosane, hentriacontane, hentriacontanol, δ-pinitol and di-Me-tryptamine; *Pod* : polyphenols, kaempferol, quercetin, β-amyrin and β-sitosterol; *Flower* : n-octacosanol, sitosterol, ceryl alcohol and β-sitosterol glucoside.

Adulterants : *Gummi indicum*, also called *Gatti gum* or *Indian gum* is an exudation from the stem of *Anogeissus latifolia* Wall. (Fam. Combretaceae); *Mesquite gum* is obtained from *Prosopis chilensis*, artificial gum made from starch and dextrin and taika gum from *Acacia seyal*.

Remark : The bulk of the *Acacia* is used in the textiles, confectionery, mucilage, paste and polishing industries.

Acalypha fruticosa Forssk. (Euphorbiaceae)

Syn : *Acalypha betulina* Retz., *A. amentcea* Roxb., *A. capitata* Wall.

English name : Birch-leaved acalypha.

Vernacular names : Kan : Chinni, Chinnimara; **Tam :** Sinni marum, Kuppamen; **Tel :** Chinniaka, Chinni.

Trade name : Chinni.

Traditional use : *Leaf-juice* : in ophthalmia; *Leaf-infusion* : to wash pustules; *Root* : in gonorrhoea.

Modern use : In digestive trouble and vulnerary; also in alterative and attenuant.

Phytography : A strong-smelling shrub, stem pubescent, covered with yellow waxy glands; leaves alternate, petiole short, lamina 2.5–7.5 cm in length, crenate, obtuse or subacute; spikes axillary, solitary, short, usually male with a few female flowers at the base; flowers minute, unisexual, bracteate, bracts of female flowers many toothed and enlarged in fruit; sepals membranous, 4, united, in male flowers pubescent; capsules tomentose; seeds smooth, testa crustaceous, albumen fleshy, cotyledons broad, flat.

Phenology : *Flowering* : Peak during July–September; *Fruiting* : Almost throughout the year.

Distribution : Tamil Nadu, Karnataka and Kerala mainly; Myanmar.

Ecology and cultivation : Foothill scrub jungles up to 900 m; aggressive coloniser.

Acalypha hispida Burm. f. (Euphorbiaceae)

Syn : *Acalypha fallax* Muell., *A. corchorifolia* Vahl, *A. boehmerioides* Miq.

English names : Chenilic plant, Red hot cactail.

Vernacular names : Ben : Muktajhuri; **Mal :** Vattattali; **Tel :** Moorukonda.

Traditional use : RURAL FOLK : (i) *Leaf* (beaten up with green tobacco leaf and infusion of rice) : applied to ulcers; (ii) *Flower* : in diarrhoea and dysentery; other uses like *Acalypha indica.*

Modern use : Similar to *Acalypha indica.*

Phytography : Annual herb, 15–45 cm high, branched from base, flaccid or rigid, pubescent, sometimes deciduous, hairs present on stem; leaves alterate, petiolate, simple, lamina membranous, shape very variable, 2.5–10.0 cm long; spikes 2.5–12.5 cm long; flowers small, unisexual, heavily bracteate; capsules hispid; seeds globosely ovoid.

Phenology : *Flowering* and *Fruiting* : Cold season.

Distribution : Cultivated in gardens from Karnataka southwards; Myanmar, Sri Lanka.

Ecology and cultivation : Prefers humid and hot climate, not in cultivation.

Chemical contents : *Plant* : acalyphine and cyanoginetic glucoside.

Acalypha indica L. (Euphorbiaceae)

Syn : *Acalypha ciliata* Wall., *A. canescens* Wall., *A. spicata* Forssk.

English name : Indian acalypha.

Sanskrit name : Aritamanjari.

Vernacular names : Ben : Muktajhuri; **Guj :** Vanchikanto; **Hin :** Kuppi, Kokli; **Kan :** Kuppigida, Tuppakeere; **Kon :** Makdda xhempddi, Kolea xhempddi; **Mal :** Kuppaimani; **Mar :** Khokali, Khajoti; **Ori :** Indra marisa; **Tam :** Kuppaimani; **Tel :** Kuppinta, Muripindi.

Trade names : Muktajhuri, Kokli.

Traditional use : *TRIBES OF KERALA, RAJASTHAN AND MADHYA PRADESH* : *Fruits* : in asthma, cough, bronchitis and earache; *Plant* and *Fruit* : as expectorant, laxative and in pneumonia and rheumatism; *Leaf* : in skin diseases like scabies.
BHAVA PRAKASH NIGHANTA : five parts of the plant (viz., *leaf, flower, fruit, root* and *stem*) can be used in the diseases caused by *kapha* (phlegm).
AYURVEDA : *Fresh leaf extract* with common salt applied in eczema; *Plant* used in gastrointestinal and respiratory affections; eating of fresh plant makes fatal to human beings.
UNANI : *Decoction of entire plant* (1/2 to 1 oz) as safe and speedy laxative; *Bruised leaf* : in constipation in children as suppository; *Fresh leaf* : emetic and in croup; *Leaf-paste* with common salt and lime : parasiticide, useful in ringworm, applied on burns; *Fresh leaf-juice* : in rheumatism, arthritis and skin infections; *Powdered leaf* : for bed sores and maggot-infected wounds.
HOMOEOPATHY : severe cough associated with bleeding from lungs, haemoptysis, incipient phthisis, gastrointestinal and respiratory affections; also as emetic to children.

Modern use : According to BPC, *Fresh* or *dry plant* : a gastrointestinal irritant; a substitute for ipecacuanha; *Leaf* : laxative, anthelmintic, externally effective in scabies; *Fresh leaf* : ground into paste, made into ball (marble-size) and introduced into the rectum of children to cure acute constipation; emetic, cures earache, acute mania, diuretic, remedy for bronchitis, asthma and pneumonia.

Phytography : Annual herb, 30 cm–1 m high, pubescent; leaves simple, alternate, long-petioled, 2.5–5.0 cm, rhombic-ovate; spikes axillary, 2.5–7.5 cm long, slender, erect; flowers unisexual, monoecious, bracteate; capsules 1-seeded, concealed by the bract; seeds ovoid, smooth.

Phenology : *Flowering* and *Fruiting* : Winter.

Distribution : Hotter parts of India; Pakistan, Bangladesh and Sri Lanka.

Ecology and cultivation : A troublesome weed in gardens and roadsides; not cultivated, prefers humid and hot climate.

Chemical contents : *Plant* : kaempferol, HCN (27 mg/100 g); *Twigs* and *Leaves* : acalyphamide and other amides, quinone, sterols, cyanogenic glucosides.

Adulterant : *Acalypha paniculata* Miq. is a substitute for *Acalypha indica* L.

Acalypha racemosa Wall. ex Baill. (Euphorbiaceae)

(2n = 20)

Syn : *Acalypha paniculata* Miq., *A. wallichiana* Thw., *A. filiformis* Heyne.

Vernacular name : Mal : Valiakuppameni.

Traditional use : *Juice* : in erysipelas and haemorrhage.

Phytography : Undershrub or herb, finely pubescent; leaves simple, alternate, 2.5–10.0 cm, petioles 5–15 cm, coarsely unequally crenate-serrate; male spikes axillary, female spikes axillary or terminal; flowers unisexual, small, bracteate; capsules trilobed, glandular; seeds reticulately pitted.

Phenology : *Flowering* and *Fruiting* : almost throughout the year.

Distribution : From Malabar southwards; Sri Lanka.

Ecology and cultivation : Grows in moist and shady places, hills up to 1200 m, also in village wastelands; wild.

Acampe papillosa (Lindl.) Lindl. (Orchidaceae)

Syn : *Saccolobium papillosum* Lindl., *Sarcochilus praemorsus* Spreng.

Sanskrit name : Rasna.

Vernacular names : Mal : Rasna ; **Mar :** Nakuli.

Trade name : Rasna.

Modern use : *Root* : bitter tonic, used in rheumatism.

Phytography : Stem erect, grows up to 40 cm; leaves densely alternate, thick recurved, fleshy, may be 12 cm in length; racemes very short, not more than 4 cm in length, rarely branched, many- flowered, dense; flowers small, about 1 cm across, dark yellow or spotted brown, lip upcurved and white, adnate to the column, fragrant, long-lasting.

Phenology : *Flowering* : peak in Autumn; *Fruiting* : Winter.

Distribution : West Bengal and lower Himalayan mountain from Sikkim eastwards, Assam, Ganga delta, Maharashtra and Karnataka; Bangladesh.

Ecology and cultivation : Epiphytic; wild.

Chemical contents : Alkaline, bitter resin.

Acampe praemorsa Blat. and McC. (Orchidaceae)

Syn : *Acampe wightiana* Lindl., *Saccolobium papillosum* Dalz. & Gibs., *Vanda wightiana* Lindl.

Vernacular names : Kan : Marabale; **Mal :** Taliyamaravada; **Mar :** Kanbher.

Traditional use : *Plant* : credited with antityphoid properties; *Paste of leaves* : applied to fractures.

Modern use : *Plant* : bitter tonic used in rheumatism.

Phytography : Stem woody, 30–50 cm tall, longitudinally striated, sheathed; leaves alternate, petiolate, thick, fleshy, 15–30 cm long; scapes very stout, racemes up to 6 cm in length, rarely branched, many-flowered, compact; flowers small, 1 cm in diameter, bracteate, creamy yellow with dark brownish stripes, lip creamy yellow, barred with crimson.

Phenology : *Flowering* : September–October; *Fruiting* : Winter.

Distribution : West Bengal, Maharashtra, Gujarat; Sri Lanka.

Ecology and cultivation : Epiphyte.

Acanthus ilicifolius L. (Acanthaceae)

(2n = 44,48)

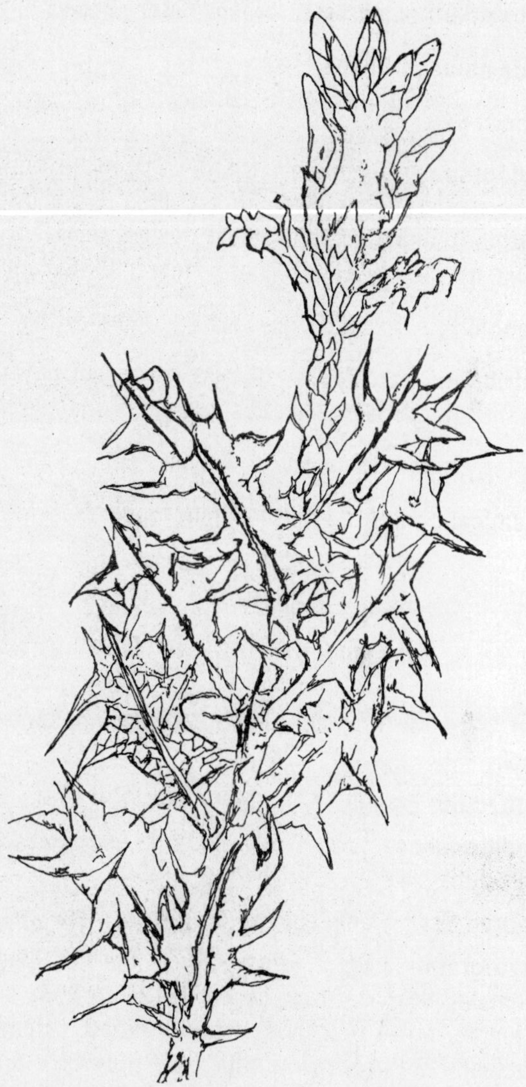

Syn : *Acanthus doloarius* Blanco, *Dilivaria ilicifolia* Nees

Sanskrit name : Harikasa.

Vernacular names : Ben : Hargoza, Harkachkanta; **Kan :** Holeculli; **Mal :** Payinaculli; **Mar :** Maranda, Maraneli; **Ori :** Harkamcli; **Tam :** Kalutai mulli; **Tel :** Alei.

Trade name : Harkasa.

Traditional use : TRIBES OF SUNDARBANS : *Root* (boiled in mustard oil): in paralysis of limbs; FOLKS OF GOA : *Leaf* : as fomentation in rheumatism and neuralgia.

Modern use : *Plant* : in asthma ; *Decoction of plant* : in dyspepsia; *Leaf* and *tender shoot* : in snake bite; *Root* : in asthma, paralysis, leucorrhoea and debility; *Leaf* : as fomentation in rheumatism, neuralgia and in snake bite.

Phytography : Erect herb; stems up to 1.5 m, in clumps, little divided, glabrous; leaves shortly petioled, oblong or elliptic, base usually spinous, toothed or pinnatifid, rigid, glabrous; spikes 10–40 cm, terminal, commonly solitary; flowers mostly opposite, bract and bracteoles present, sepals 4, outer 2 elliptic rounded, inner 2 broadly lanceolate, subacute, petals 5, blue, united, 2-lipped, corolla tube short, pubescent within, stamens 4, didynamous, shorter than corolla lip, filaments stout, anthers 1-lobed, bearded, carpels 2, united, ovary 2-chambered having 2 ovules in chamber, style short, bifid; capsules shining chestnut-brown, ellipsoid, compressed, 0.6–0.8 cm long; testa white, very lax.

Phenology : *Flowering* and *Fruiting* : almost throughout the year.

Distribution : Mangroves of Indian peninsula; Sri Lanka, Bangladesh, Pakistan and the adjoining areas.

Ecology and cultivation : Commonly grown on the river banks, tidal canal sides, low swampy areas in the mangrove forests and its vicinity; wild.

Chemical contents : *Plant* : acanthicifoline, oleanolic acid, β-sitosterol, lupeol, quercetin and its glucopyranoside, trigonellin; *Root* : saponin, glycoside of 3β-OH-lup-20(29)-ene.

Acer mono Maxim (Aceraceae)

Syn : *Acer pictum* Thunb., *A. laetum* C. A. Mey, *A. truncatum* Bunge., *A. cultratum* Wall., *A. sterculiaceum* Griff.

English name : Maple.

Vernacular names : Gar : Potli; **Hin :** Kanch; **Him :** Dhadonjra; **Pun :** Kanzal.

Trade name : Kanzal.

Traditional use : Local application of leaves causes blistering of skin.

Modern use : *Bark* : astringent.

Phytography : Tree of small or moderate size; leaves simple, 5- to 7-lobed, opposite or subopposite, petioles glabrous, 2.5–11.25 cm in length, lamina glabrescent except hairy tufts on the axils of 5–7 basal nerves, lanceolate or caudate, entire, often have latex or resin cells; cymes corymbose, each cyme 2.5–15.0 cm in length; flowers small, on slender pedicels; fruits dry, indehiscent, winged and are blown for long distances by strong winds.

Distribution : From Indus to Assam, ascending from 1300 to 2500 m in northwest Himalaya, common in mixed forests above 3000 m. Also found in Bhutan, China, Japan, Persia and the Caucasus.

Ecology and cultivation : Prefers rocky areas in alpine zones.

Achillea millefolium L. (Asteraceae)

(2n = 18)

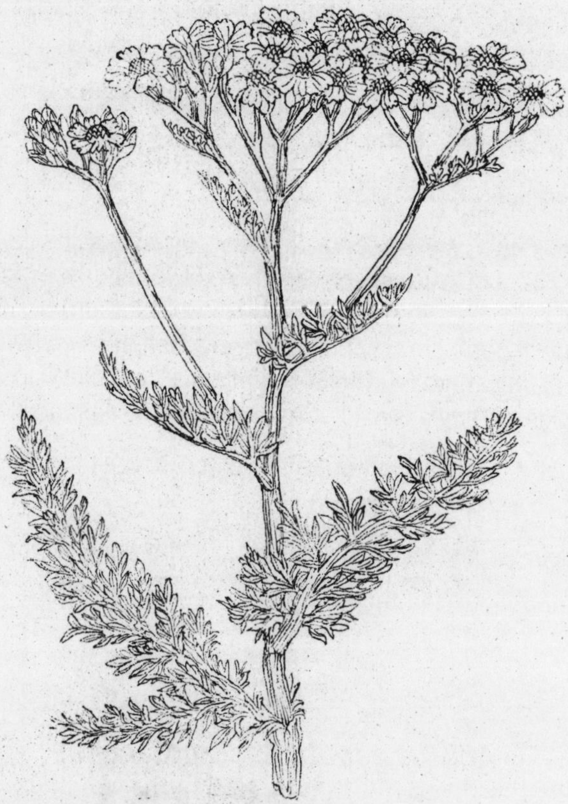

Syn : *Achillea cuspidata* DC., *A. lanulosa* Nutt.

English names : Milfoil, Yarrow.

Vernacular names : Hin : Gandana; **Kas :** Inomadnu, Akarkhara; **Mar :** Rajmari; **Urd :** Biranjasifa.

Trade name : Biranjasif.

Traditional use : *TRIBAL* : *Plant*: in fever, vaginal haemorrhages, diaphoretic, in epilepsy; *Leaf* and *Flower*: bitter aromatic, tonic, astringent and diaphoretic.
UNANI: Ingredient of 'Basanjasif'.

Modern use : In many parts of Europe, a decoction of the leaves and flower heads is employed as a carminative tonic and aromatic stimulant. It is antispasmodic, emmenagogue, vulnerary, useful in colds, colic, in commencement of fever, flatulence, heart burn, hysteria, intermittent fever, kidney disorders and obstructed perspiration.

Phytography : Herb; stem up to 1 m in height; leaves strap-shaped, bipinnate, segments linear, 3- to 5-cleft; flower heads white, ovoid, in flattened corymbs, each flower head possesses a flat chaffy receptacle surrounded by an ovoid-cylindric involucre of imbricated and keeled oblong bracts, inserted into the receptacle are 4–6, white or rarely pink or purple-tinged pistillate ray florets and numerous yellow perfect disc florets.

Distribution : Common in many parts of western Himalaya from Kashmir to Kumaon (1200–4000 m); indigenous to North Asia, Europe and America.

Ecology and cultivation : Cultivated in gardens.

Chemical contents : *Herb*: achillin, betaine, coline, trigonelline, betonicine, stachydrine, hydrocarbons, ceryl alcohol, β-sitosterol, leukodin. 'Milfoil oil', the essential oil from aerial parts of flowering herb, has azulene and bitter principles in colones; Achilleine—an alkaloid present in living herb but absent in dry-plants; *Inflorescence*: maximum essential oil, millefin, de-Ac-martricarine, 5-Off-tetra-MeO-flavone, artemesin, casticin, proazulene, achillicin; *Leaf*: apigenin, cosmasin, luteolin and its glucosides, rutin.

Adulterant : Used as an adulterant of older flowers of *Sambucus nigra* L. (Caprifoliaceae).

Remark : The odour is characteristically aromatic, the taste aromatic, astringent and bitter.

Achras zapota L. (Sapotaceae)

(2n = 26)

Syn : *Achras sapota* L., *Mimusops manilkara* Don, *Mimusops indica* A. DC.

English names : Gum chicle, Mexican chicle, Sapota, Sapodilla plum, Tree potato.

Vernacular names : Ben : Safeda; **Hin :** Chiku, Sapota; **Mal :** Shima ellulopai; **Mar :** Chikali, Kirni, Rayan.

Trade name : Safeda.

Modern use : *Bark* : tonic, febrifuge; *Fruit* : preventive against biliousness and febrile attacks; *Seed* : diuretic.

Phytography : Tree grows up to 30 m in height, young parts often rusty tomentose, rows of laticiferous sacs in the bark, pith, cortex and leaves; leaves alternate, simple, entire, leathery; flowers small, solitary, bisexual, actinomorphic, sepals 4, united, petals 4, united, funnel-shaped, 1.25–1.80 cm in diameter, scales present in the corolla tube; stamens many, epipetalous; carpels united; berry with sclerenchymatous outer layer.

Phenology : *Flowering* : September–November; *Fruiting* : almost throughout the year.

Distribution : Evergreen tree of the West Indies and Central America, cultivated more or less throughout India; Bangladesh, Pakistan and Sri Lanka.

Ecology and cultivation : Common in the dry forests; cultivated in the shady places.

Chemical contents : *Bark* : a water-soluble tuberculostatic principle, glucocide, tannin, sapinin alkaloid; *Seed-kernel* : 20% liquid fat, sapotinin.

Adulterants : Substitute for 'Gutta percha', but less gutta and more resin. Balata is a dried latex obtained from *Manilkara bidentata* (A. DC) Chevalier (*Mimusops balata* Miq.). It possesses the general properties of gutta percha.

Remark : The fruit is eaten by the poor people at times of scarcity.

Achyranthes aspera L. (Amaranthaceae)

Syn : *Aerva aspera* Spreng.

English name : Prickly chaff flower.

Sanskrit name : Apamarga.

Vernacular names : **Asm** : Ubtisath; **Ben** : Apang; **Guj** : Kutri, Safed aghado; **Hin** : Chirchita, Latijira; **Kan** : Uttarani; **Kon** : Dove aghado; **Mal** : Katalady; **Mar** : Aghada; **Ori** : Apamaranga; **Pun** : Kutri, Puth kanda; **Tam** : Nahiroorvy; **Tel** : Attareni.

Trade names : Apang, Latijira.

Traditional use : TRIBES OF CHOTANAGPUR : *Root* : as abortifacient; SANTAL : *Plant* : in diabetes, dropsy; *Fresh plant-paste with that of black pepper* : to cure effect of bite of mad dog; *Plant-decoction* : in cough, kidney stone, pneumonia and as an ingredient of 'cystone'—used in urinary tract infection; *Root* : in abdominal tumour, cough, snake bite, toothache, whooping cough; *Leaf* : in leucoderma; *Leaf-infusion* : in cold and cough; *Leaf-extract* : in eczema and leprosy in the name of 'Achyrol'; *Seed-powder* : to reduce hunger; TRIBES OF ARUNACHAL PRADESH AND UTTAR PRADESH : *Root* : in asthma, to women for antifertility purpose, bleeding in delivery.
ATHARVA VEDA : an important medicinal plant having wormicidal properties, it cures piles, troubles of thigh, fever and oral ailments; KURMA PURANA and BRAHMAVAIVARTA PURANA : the twig of this plant is good for brushing teeth; CHARAKA SAMHITA : used as

snuff to expel mucus from the nose ; *SUSHRUTA SAMHITA* : *Root-bark* : used in piles; *CHAKRADATTA* : used in haemorrhagia, deafness, acute opthalmia, etc.
HOMOEOPATHY : Astringent, diuretic, alterative; also useful in menorrhagia, diarrhoea, anasarca, acites and dropsy.

Modern use : *Plant* : pungent, purgative, diuretic, in dropsy, piles, boils, skin eruptions, colic, snake bite; *Infusion of root* : astringent; *Seed* : emetic, in hydrophobia.

Phytography : A stiff herb, common weed throughout India up to 1000 m; stem erect, grows up to 1 m, simple or branched; leaves simple, petiolate, very variable, generally thick, young tomentose on both surfaces, mature usually glabrate on upper side with closely appressed hairs on the lower surface, apex rounded or suddenly short acuminate; spikes rigid, lengthening during fruiting as much as 60 cm; flowers numerous, greenishwhite, 0.4–0.6 cm long, sharply reflexed against the stout pubescent inflorescence axis, bracts and bracteoles shorter than flowers, acuminate, spinescent, persistent; achene with oblong or ovate utricle, enclosed in the hardened perianth; seeds oblong, brown.

Phenology : *Flowering* and *Fruiting* : peak from March to June.

Distribution : Throughout India up to an elevation of 1000 m; Bangladesh and Pakistan.

Ecology and cultivation : Common in dry shady localities; wild.

Chemical contents : *Plant-ash* : rich in potash; *Root* : ecdysterone, inokosterone, insect moulting hormones; *Stem-extract* : sterol, saponins, ecdysterone; *Leaf* : ecdysterone; *Unripe fruits* : saponin C & D; *Seed* : saponin A & B.

Achyranthes bidentata Blume (Amaranthaceae)
(2n = 24)

Syn : *Achyranthes wightiana* and *A. lanceolata* Wall., A. aspera Wall.

Vernacular names : Asm : Apamarga, Bankhat; **Lus :** Vangvathus; **Tam :** Nagaranji, Singhappu naaiyuruvi.

Traditional use : *TRIBES OF CHOTANAGPUR* : *Root* : hemicarnia and scorpion sting; *Leaf-paste* : in cholera, on swelling; it is reported that sometimes it causes blister in mouth.

Modern use : *Plant* : diuretic, astringent.

Phytography : Herb; leaves variable, elliptic-ovate to oblanceolate, glabrescent to pubescent; petiole upto 2.5 cm; spike 9–20 x 0.7 cm; peduncle slender, puberulous; flowers 4 mm across, pink.

Phenology : *Flowering* : peak during November–March; *Fruiting* : almost throughout the year.

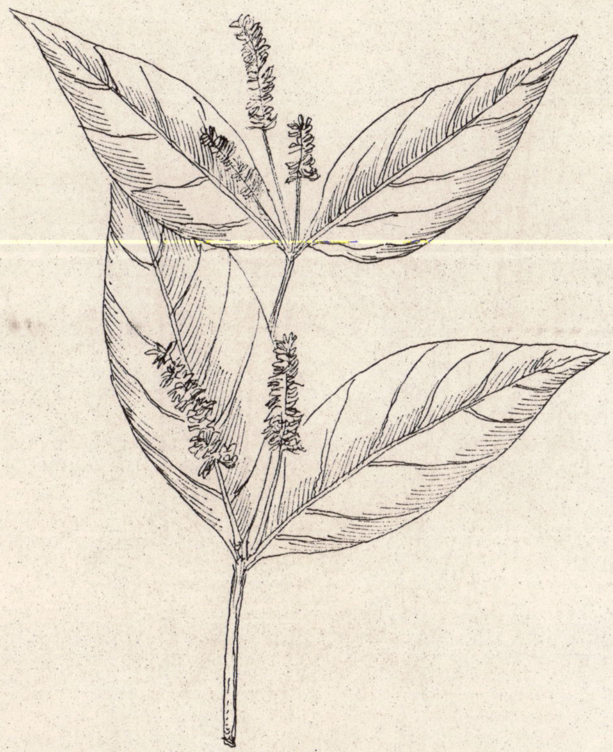

Distribution : Temperate and subtropical Himalaya, 1300–2000 m, Bihar, Karnataka, Tamil Nadu and the Nilgiri Hills.

Ecology and cultivation : Grows in shady places, forest borders, wastelands; wild.

Chemical contents : *EtOH* (50%) *extract of plant* : spasmol; *Root* : ecdysterone, inokosterone, rubrosterone; alkaloid contents of root and stem vary at different growth stages; *Seeds* : sapogenins which contain β-amyrine.

Aconitum chasmanthum Stapf. ex Holmes (Ranunculaceae)

(2n = 16)

Syn : *Aconitum dissectum* Wattnon Don, *A. napellus* L. var. spicatum Duthie p.p.
English name : Indian napellus.
Vernacular names : Hin : Balnag; **Kas :** Banbalnag, Mohand; **Pus :** Mohri.
Traditional use : PEOPLE OF KASHMIR : *Flower* : in rheumatism.
Phytography : The Indian aconite root is paired and tuberous, the daughter tuber conic to conic-cylindric from a broad base, 2.5–2.7 cm, rarely up to 5 cm in length and up to 18 mm in thickness with rootlet stubs mostly clustered at the basal end, dark brown externally with a cartilaginous fracture; its cambium is continuous and sinuous in cross section.
Distribution : Subalpine and alpine zones of the western Himalaya, from Chitral and Hazara to Kashmir between 3200 and 4000 m.
Ecology and cultivation : Plant of cold climate; wild.
Chemical contents : *Root* : chasmaconitine, chasmanthinine, chasmanine.
Adulterants : Roots poisonous, substitute for *A. napellus* L. and *A. ferox* Wall. ex. Ser.
Remark : Sometimes, it is mistaken as *A. napellus*, but *A. chasmanthum* is smaller.

Aconitum deinorrhizum Stapf. (Ranunculaceae)

(2n = 32)

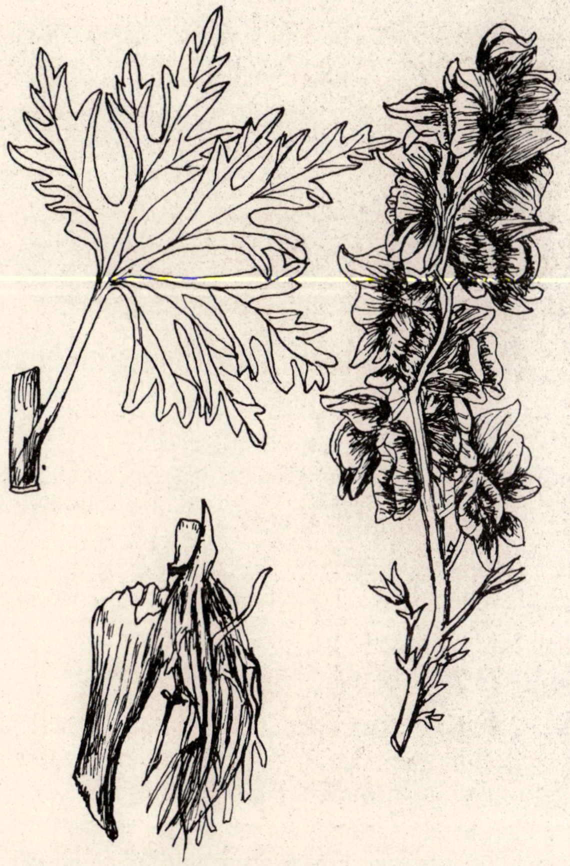

Vernacular names : Kas and **Pun :** Dudhiabish, Safed bikh.

Trade name : Indian aconite.

Traditional use : *Root* : poisonous; *Roots* and *Leaves* : in rheumatism, rheumatic fever and acute headache.

Phytography : It occurs as a paired biennial tuberous root; the daughter tuber being elongated conical, up to 6.5 cm long and up to 18 mm thick at the crown; its cambium is discontinuous being broken into circular, horse-shoe-shaped or flattened strands arranged into ring; the surface is dark brown and coarsely wrinkled; the mother tubers are similar but have longer filiform fibre.

Phenology : *Flowering* and *Fruiting* : June to October.

Distribution : Native to alpine Himalaya or Bashar; available throughout the central Himalaya; Nepal; between 3000 and 4500 m.

Ecology and cultivation : Not known in cultivation; *endemic* to the Himalaya.

Chemical contents : *Root* : pseudoaconitine—a respiratory depressant which is much toxic than aconite.

Remarks : Apart from *A. ferox* and *A. belfourii*, it appears to be the principal constituent of aconite of commerce; *vulnerable* due to excessive collection of roots for medicinal purposes.

Aconitum ferox Wall. (Ranunculaceae)

(2n = 20)

Syn : *Aconitum virosum* Don., *A. napellus* var. *rigidum* Hook. f. & T.

English names : Wolf's bane, Indian aconite.

Sanskrit names : Vatsanabha, Visa.

Vernacular names : Hin : Bish, Mahoor; **Guj** and **Mar :** Vachang; **Kas :** Mohra; **Tam :** Vasnumbi; **Tel :** Vasnabhi.

Trade name : Bish.

Traditional use : AYURVEDA : *Root* : used in the *mrityunjaya rasa* (used to treat the fever supposed to be caused by deranged *vayu*, i.e., wind, *sannipatika jvara* , i.e., remittent fever, *hingulesware-rasa*, *anandabhairav agnitundi vati*, etc.
Vatsanabha has been used in medicine from a very remote period. It is regarded as healing and stimulant. It is used in a great variety of affections, but is specially recommended in fever, cephalagia, affections of throat, dyspepsia and rheumatism.
HOMOEOPATHY : remedy for clotting of blood in heart or in lungs, pneumonia, lymptisis, pleurisy, eye trouble, earache, toothache and urinary trouble.

Modern use : Extremely poisonous; used in leprosy, fever, cholera, nasal catarrah, tonsillitis, sore throat, gastric disorders, debility, etc., also used as a sedative and diaphoretic; applied in the form of paste in cases of neuralgia and rheumatism.

Phytography : Perennial erect herb growing up to 2 m in height; roots look like the navel of children; leaves alternate, simple, rounded or oval, may be palmately 5-lobed; flowers borne on branched racemes, bracts and bracteoles present, large helmet-type, helmet vaulted with short sharp beak, pale dirty blue in colour, zygomorphic, floral parts arranged spirally on an elongated receptacle; follicles erect, usually densely villose—sometimes glabrous.

Phenology : *Flowering* and *Fruiting* : July–November.

Distributin : Alpine Himalaya including Nepal; *endemic.*

Ecology and cultivation : Temperate to alpine regions of the Himalaya in the altitude of 3300–5000 m.

Chemical contents : *Root* : pseudoaconitine (a toxic alkaloid), indactonitine, chasmaconitine, bikhaconitine.

Adulterants : Indian aconite root is known as 'bikh' or 'bish', the name which is applied to aconite from more than one species, and different authors have ascribed it to different species.

Remarks : *Vulnerable* due to excessive collection for medicinal uses. Collection in wild state should be banned and measures for cultivation should be initiated.

Aconitum heterophyllum Wall. (Ranunculaceae)

(2n = 16)

Syn : *Aconitum cordatum* Royle, *A. atees* Royle

Sanskrit names : Ataicha, Ativisa.

Vernacular names : Ben : Ataicha; **Hin :** Atis; **Tam :** Ati vadayam.

Trade name : Atis.

Traditional use : : *MATSYA PURANA* : the decoction of this plant can be used as universal antidote of poison; *AGNI PURANA* : refers this plant as medicinal and recommends the following uses : (i) to treat gastroenteritis, the patient should drink water containing the decoction of *A. heterophyllum* Wall., *Aegle marmelos* Corr., *Cyperus rotundus* L. and *vatsaka*; (ii) to cure children suffering from fever *A. heterophylum* should be used; *AYURVEDA* : *Root* of white, yellow, red and black varieties used; white is the best and is bitter tonic, hot, stomachic, digestive, alleviates dysentery and bilious complaints, good in periodic and intermittent fevers, useful in bites from snakes, scorpions, etc. Root is one of the constituents of medicine for ziardiasis, hysteria, piles, acute inflammations, throat diseases, loss of memory and antifertility agents.

UNANI : White and dark varieties—tonic, useful in dysentery, piles, bilious complaints, plethoric conditions; removes gases from the stomach.

HOMOEOPATHY : *Root* : to relieve pain and fever.

Modern use : *Root* : cures acute dysentery and chronic enteritis; used in splenic fever and gastric troubles; good astringent and tonic, but its efficacy as antiperiodic is disputed; considered as antifertility agent; used externally for rheumatism.

Phytography : Erect herb growing up to 1 m in height; stem simple or branched from the base, glabrous below and puberulous above; leaves simple, 5–10 cm in length, ovate or orbicular, cordate, more or less 5-lobed, toothed, upper leaves entire; flowers in panicled racemes, bracteate—bracts sharply toothed, flowers more than 2.5 cm in length, bright blue or greenish blue with purple veins, helmet half as high as long and shortly beaked; follicles downy; seeds smooth.

Phenology : *Flowering* and *Fruiting* : July–November.

Distribution : Common alpine and subalpine Himalaya from Indus to Kumaon, 2000–5000 m.

Ecology and cultivation : Prefers higher altitude; not known in cultivation.

Chemical contents : *Root* : atisine (0.4%) and heteratisine, hetisine, heterophyllisine, heterophylline, atidine, heterophyllidine, F-dihydroatisine, isoatisine, hetidine and hetisinone.

The root should break with a short starchy fracture, should taste bitter without producing any acridity and tingling sensation of the tongue.

Remarks : *Vulnerable* due to excessive collections for medicinal purpose; collection should be banned.

Aconitum napellus L. (Ranunculaceae)

English names : Monk's hood, Aconite root.

Sanskrit name : Visha.

Vernacular names : Ben : Kath bish; **Hin :** Mitha zahar; **Guj :** Shingadio, Vachang; **Kan :** Vashanabhi; **Mar :** Bacheng, Vashanavi.

Trade names : Aconite, Kath bish.

Traditional use : *MATSYA PURANA* : a king should eat, once in a month, following plant materials after boiling them in milk and then mixing them with *ghee*—roots of *A. napellus,*

Saccharum officinarum, and *Cynodon dactylon*; the person, who follows the above diction, can regain life even after being hit by a weapon.

UNANI : an ingredient for the medicine called Beesh' (mitha zahar).

HOMOEOPATHY : an excellent remedy for cerebro-spinal nervous system, for muscle, rheumatic pain, liver, heart and congestion of bronchial tract, fever and restlessness.

Modern use : Preparations of aconite and its principal alkaloid aconitine, when applied to skin, produce tingling followed by numbness. They are used in certain forms of neuralgia and rheumatism. Administered internally, aconite produces a steady fall of temperature, moistening of skin, increases in the amount of urine and lowering of sensibility; it is applied in cases of fever and pain, usually in the form of small doses of the tincture frequently repeated.

Phytography : Erect herb growing up to 1 m, often decumbent in small states, glabrous or slightly pubescent; leaves palmatipartite, ultimate divisions linear; racemes simple; rarely compound; flowers small, not more than 2.5 cm in length, irregular, bright or dull greenish blue, helmet three times as long as high, tapering to a slender beak; follicles few- to many-seeded; seeds albuminous.

Phenology : *Flowering* and *Fruiting* : July–December.

Distribution : It is a *rare* and poisonous species found in northern Himalaya; chiefly found in temperate and arctic Europe, Asia and America.

Ecology and cultivation : Prefers higher elevation; not in cultivation.

Chemical contents : Aconitine (0.4 %) is the chief alkaloid, large doses fatal to human. The alkaloid first stimulates and then paralyses the sensory nerves. Aconite root contains three closely allied alkaloids : aconitine, pieroconitine and aconine; there are also amounts of other alkaloids like misoconitine, hypercontine, neopelline, nepalline and nealine; vitamin C; plants from mountainous regions richest in aconitine, toxicity due to compounds other than aconitine.

Adulterants : The root of *Aconitum deinorhizum* Stapf., *A. chasmanthum* Stapf., Russian or Soviet aconite (*A. tianshanicum*), Atis root (*A. heterophyllum* Walliet) are used as substitute for *A. napellus.*

Remarks : *Vulnerable* and *threatened* due to ruthless exploitation for medicinal purposes; conservation measures should be enforced.

Aconitum palmatum D. Don (Ranunculaceae)

(2n = 48)

Syn : *Aconitum bisna* Buch. Ham.

Vernacular names : Hin : Bikhama; **Mar :** Wakhma.

Traditional use : *Non-poisonous roots* : antiperiodic, tonic, used in diarrhoea and rheumatism.

Phytography : Perennial erect herb; stem 60–150 cm high; leaves alternate, long petiolate, deeply palmatisect, 10–15 cm in diameter; panicles few-flowered; flowers long pedicelled, greenish blue or bluish white; follicles oblong, 2.5–3.7 cm long, glabrous; seeds black, obscurely winged with plaited testa.

Phenology : *Flowering* : June–August; *Fruiting* : September–October.

Distribution : Temperate Himalaya from Garhwal to Sikkim, between 3500 and 5500 m; Nepal and Tibet.

Ecology and cultivation : Alpine plant; wild.

Chemical contents : *Rhizome* : palmatisine.

Remark : *Threatened* and *vulnerable*; should be conserved.

Acorus calamus L. (Araceae)

(2n = 18, 24, 36, 48)

Syn : *Acorus griffithii* Schott., *A. belangeii* Schott., *A. casia* Bertol.

English name : The sweet flag.

Sanskrit name : Vacha.

Vernacular names : **Asm**, **Ben** and **Hin :** Boch; **Guj :** Godavaj, Vekhand; **Kan :** Baje, Baje gida; **Kon :** Waikhand; **Mal :** Vayambu; **Mar :** Vekhand; **Ori :** Bacha; **Pun :** Bari, Boj, Warch; **Tam :** Vasamboo; **Tel :** Vasa.

Trade name : Boch.

Traditional use : SANTAL : use the plant in the following ways : (i) they mix and grind black pepper, cloves, root of *Carissa carandus* L. along with little of the rootstock of *A. calamus* L., then stir the same in pure mustard oil—the emulsion, thus prepared is anointed daily over the whole body of the patient suffering form epilepsy with foaming and groaning, as soon as the fit comes on; a few drops of this emulsion should be poured into the nose of the patient; (ii) for the treatment of indigestion, they take pills made by grinding 100 black peppers, little amount of ginger and the root of *A. calamus* together; (iii) also use in the treatment of asthma, bronchitis, cold and cough, dry cough, epilepsy, haemoptysis, indigestion, phthisis; BIRHOR : *Rhizome* in alopecia, *Root* as massage, in fever, hysteria, pain in neck, teething trouble of children, malaria and cancer.
AGNI PURANA : this plant is of great medicinal value; it recommends the following uses : (i) for treatment of epilepsy, this plant should be boiled with *Costus speciosus*, *shankhapushpi*, along with the juice of *Bacopa monnieri* ; the substance thus obtained should be administered to the patient; (ii) drinking the decoction of this plant, *Piper peepuloides*, *Staphyles emodi* Wall., and *Cyperus parviflorus* Heyne and *pippalimula* is good for the patient of rheumatic arthritis; (iii) the powder or decoction of this plant helps curing chronic enlargement of spleen; (iv) decoction of the plant is beneficial for the patient of dropsy; AYURVEDA : *Rhizome* : bitter, healing, emetic, laxative, diuretic, carminative; improves voice and appetite; good for oral diseases, abdominal pain, epilepsy, bronchitis, hysteria, loss of memory, rat bite and worms in ear.
SIDDHA SYSTEM : fresh root for bronchial asthma.
UNANI : an ingredient of the medicine called 'Waje-Turki'; useful in flatulent colic, chronic dyspepsia, catarrhal, in burn wounds, carminative, anthelmintic and as bitter tonic.

Modern use : *Rhizome* : aromatic, bitter, carminative, emetic, stimulant, stomachic, useful in dyspepsia, colic, remittent fevers, nerve tonic, in bronchitis, dysentery, epilepsy and other mental ailments, glandular and abdominal tumours and in snake bite.

Phytography : Perennial, erect, aromatic herb, common on river banks and marshes, ascending to 3000 m; rhizome cylindrical or slightly compressed, about 2.5 cm in diameter, much-branched, externally light brown or pinkish brown but white and spongy within; leaves distichous, large, 1–2 m in length, base equitant, margin waved; spadix sessile, cylindric, densely flowered, not completely enclosed by spathe, spathe 15–75 cm in length, narrow, leaf-like; flowers small, bisexual; berries few-seeded; seeds oblong, albuminous.

Phenology : *Flowering* and *Fruiting* : July–August; fruiting very rare.

Distribution : Throughout India; ascending the Himalaya up to 2000 m; Sri Lanka, Pakistan and Bangladesh.

Ecology and cultivation : Probably introduced; found from the coast to 1200 m; often near village wells and along watercourses; confined to marshy areas; gregarious herb from a stout horizontal rhizome; wild and cultivated.

Chemical contents : *Dry rhizome* :1.5–3.5% of a yellow aromatic volatile oil—calamus oil; the oil contains β-asarone, small quantities of sesquiterpenes and sesquiterpenes alcohols; *Rhizome* : also contains choline (0.26%), flavone, acoradin, 2,4,5-tri-MeO-benzaldehyde, 2,5-di-MeO-benzoquinone, galangin, calameone, acolamone, isoacolamone, epoxyisoacoragermacrone; *Aerial parts* : lutcolin-6,8-c-diglucoside; chemical constituents vary in ecotypes and polyploides.

Adulterants : The powdered drug has been adulterated with siliceous earth, ground marsh mallow root and cereal flowers.

Remark : Rhizomes are valued for indigenous medicine.

Acorus gramineus Soland ex Ait. (Araceae)

(2n = 36)

Syn : *Acorus calamus* var. *terrestris* Engler, *A. terrestris* Spreng.

Traditional use : For colic pain, renal calculus and tonic.

Modern use : Similar to *A. calamus*; *Root* : stimulant, tonic, antiseptic, insecticide and antispasmodic.

Phytography : Much more variable in size than *A. calamus*, with shorter spathe and more slender spadix.

Distribution : The Japanese species is occasionally found in Sikkim Himalaya up to 2000 m, Khasia Hills (Meghalaya) up to 1600 m.

Chemical contents : Essential oil composition very variable. *Rhizome* : two chemotypes — (i) β-asarone (84–92%) type from Japan and Thailand, (ii) cis-Me-isoeugenol (52–55%) and shyobunone-like sesquiterpenes (12-24%) type from Hong-kong; *Leaf top oils* : very variable, may be classified at least into five chemotypes.

Adulterant : Rhizome is a substitute for *Acorus calamus* L.

Acrocephalus indicus O. Kuntze (Lamiaceae)

Vernacular name : Sin : Ustukudus.

Traditional use : *Plant* : expectorant.

Phytography : Annual herb with whorls of small flowers in dense heads with imbricating bracts; fruit of 4 smooth dry nutlets; basal scar small.

Phenology : *Flowering* and *Fruiting* : in cold season.

Distribution : All over India and Pakistan, ascending up to 2000 m in the Himalaya and other mountainous regions; tropical and subtropical Asia.

Ecology and cultivation : Grows in damp or shady places in forests; wild.

Acronychia pedunculata (L.) Miq. (Rutaceae)

(2n = 36)

Syn : *Acronychia laurifolia* Blume, *Cyminosma pedunculata* DC., *Clausena simplicifolia* Dalz., *Jambolifera pedunculata* Vahl.

Vernacular names : Asm : Loajan; **Mal** and **Tam** : Muttnari; **Nep** : Paowlay.

Modern use : *Root* and *Bark :* in external application for sores and ulcers, *Stem-bark* : tonic, used in scabies.

Phytography : Small glabrous tree or shrub, with somewhat silky or puberulous young shoots and inflorescence; leaves opposite or alternate, palmately compound, very variable, leaflets entire, obtuse or obtusely acuminate, multicostate; flowers small or rather large, fragrant, yellow, in long peduncled axillary and terminal corymbs, bracts and bracteoles small; fruit a 4-celled drupe or 4-valved loculicidal capsule, very variable in size; seeds black, albuminous, often exerted and pendulous from a slender funicle.

Phenology : *Flowering* : August–September.

Distribution : Dehra Dun, Maharashtra, Karnataka, Orissa, Meghalaya and the hilly forests of Sikkim, Assam, NE regions of India, up to 2000 m; Bangladesh (Chittagong) and Sri Lanka.

Chemical contents : *Bark* : bauerenol, K-oxalate, acrovestone; *Wood* : β-sitosterol, acronylin—a new phenolic compound.
The alkaloids—kokusginine and evolitrine are present in the leaves of the plants found in Sri Lanka but absent in Indian plants.

Remark : Stem-bark is poisonous to fish.

Actaea spicata L. (Ranunculaceae)

(2n = 16)

Syn : *Actaea acuminata* Wall.

English names : Cohosh, Heri christophir, Grape wort.

Vernacular name : Kan : Vishaphala.

Traditional use : *Root* is used for treatment of snake bite and asthma by tribes; FOLK OF MAHARASHTRA : Astringent, diarrhoea, dysentery; FOLK OF KONKAN : pulp with figs in asthma, sherbet in bilious dyspepsia.

HOMOEOPATHY: tincture is used for subacute rheumatic pain of the small joints, especially useful when sour stomach is present; 3x potency is used in tiredness and colic pain.

Modern use : *Rhizome* : for nerve disorders; *Rhizome-decoction* : used for ovarian neuralgia and uterine tenderness; also as a substitute for digitalis in the treatment of fatty and irritable heart; and as emetic and purgative, in rheumatic fever; *Fruit* : bitter, cooling, antipyretic, useful in biliousness, dysentery and excessive perspiration.

Phytography : Erect perennial pubescent herb; stem 60–84 cm from woody horizontal rootstock, covered at the base with leafless sheaths; leaves alternate, pinnately compound, nearly 30 cm in length; flowers white, small, in crowded racemes; fruits elliptic or subglobose berry, black, many-seeded; seeds smooth, depressed, with leathery seedcoat.

Phenology : *Flowering* and *Fruiting* : Autumn.

Distribution : Temperate Himalaya including Bhutan and Hazara.

Ecology and cultivation : Prefers cold climate.

Chemical contents : It contains trans-aconitic acid and cytostatic, toxic principle oil of barberry; *Seed* : tannin.

Adulterant : The plant is a substitute of *Digitalis purpurea* L.

Actiniopteris australis (L.f.) Link (Actiniopteridaceae)

Syn : *Actiniopteris dichotoma* Kuhn., *A. radiata* Link

English name : Peacock's tail.

Sanskrit name : Mayurashikha.

Vernacular names : Ben : Mayurpankhi; **Guj :** Dikthali; **Hin :** Marpath; **Mar :** Bhuitad.

Trade name : Mayurshikha.

Traditional use : TRIBES OF KHEDBRAHMA (Gujarat) : *Whole plant* : in ulcers and tuberculosis.
AYURVEDA : *Plant* : styptic, anthelmintic and antipyretic; *powdered* (along with seeds of *Ocimum americanum*): in antifertility; *Rhizome* (dipped in water) : tonic in tuberculosis.

Phytography : Tiny tough terrestrial fern; rhizome short, creeping, covered with stipe bases; stipes greenish or hay-coloured, 5–15 cm long; fronds dichotomously dissected into fine segments forming a small fan-like appearance, sterile fronds shorter than fertile ones; each segment linear, fertile ones bear dark brown sori on lower surface.

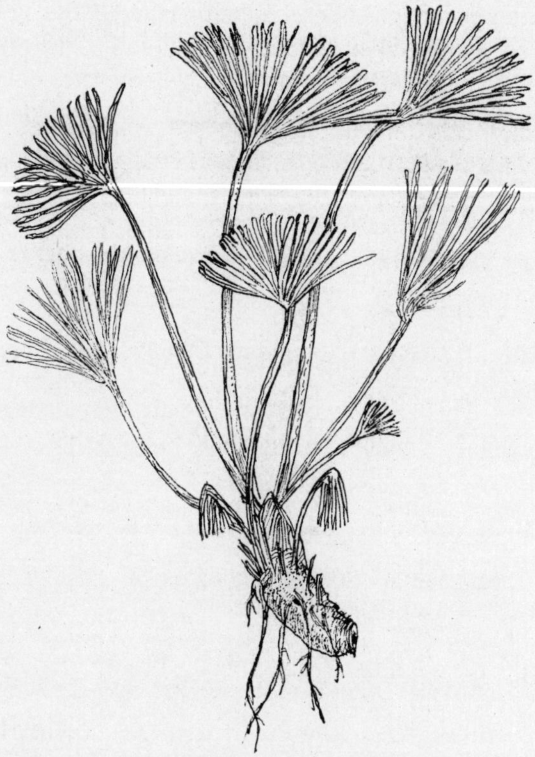

Phenology : Sori appear in late Autumn or Winter, mature in late Winter or Spring.

Distribution : Throughout India, especially in the Deccan peninsula below 1300 m; common in the Nilgiris (Tamil Nadu) up to 600 m and the Kumaon hills.

Ecology and cultivation : Prefers moist crevices of dry rocks; also grows along the slopes of black or reddish rocks and gravels; wild.

Chemical contents : *Mature sporophyte* : rutin; *frond* : hentriacontane, hentriacontol, β-sitosterol, its palmitate and β-sitosterol-D(+)-glucoside.

Actinodaphne hookeri Meissn. (Lauraceae)

Syn : *Actinodaphne angustifolia* Nees

Vernacular names : Asm : Petarichawa; **Kan :** Hoggodigimara; **Mal :** Malavirinya; **Mar :** Pisha, Pisa, Pichli; **Ori :** Jharchampa, Tudambo; **Tam :** Tali, Thali.

Trade name : Pisa.

Modern use : *Infusion of leaf* : in urinary disorders and diabetes; *Seed-oil* : application to sprains.

Phytography : Moderate-sized tree, branches almost woolly, buds large, silky; leaves whorled, rarely scattered or opposite, petioles tomentose, young leaves densely softly rusty-tomentose or villous, green when dry, smooth and often polished above, old leaves often rusty villous; flowers small, silky, dioecious, bracteate, males very shortly pedicelled, clustered, female flowers clustered, subsessile or pedicelled; fruits oblong, seated on the very small base of the perianth.

Phenology : *Flowering* : November.

Distribution : Evergreen forests of the Western Ghats, in the Konkan and North Kannada, common in Matheran and Mahabaleswar; also found on the east coast in the forests of Orissa.

Ecology and cultivation : Evergreen trees found in the forests near sea.

Chemical contents : *Bark* : an alkaloid—actinodaphnine; *Leaf* : spasmol, β-sitosterol, quercetin 3-rhamnoside, hydrocarbons; *Fruit-kernel* : monoflaurin, lauric acid, n-hexacosanol, β-sitosterol and its glucoside; *Seed-shell* : 25% of a reddish brown oil consisting largely of oleates mixed with resinates and some trilaurine; *Seed-kernel* : lauric acid.

Adansonia digitata L. (Bombacaceae)

(2n = 96)

English names : African calabash, Monkey-bread tree, Baobab, Sour gourd.

Sanskrit name : Gorakshi.

Vernacular names : Ben : Amli, Goraksha, Goraksha chakuli; **Hin :** Gorakamli; **Kan :** Brahmambika, Maggimavu; **Mar** and **Guj :** G orakh chinch, Gorak amla; **Tam** : Annaipuli, Papparapuli; **Tel :** Brahmamlica, Seemechinta.

Trade name : Baobab.

Traditional use : AYURVEDA : *Fruit* : sweet, bitter, cooling, antipyretic; useful in biliousness, dysentery, excessive perspiration, vomiting .

Modern use : *Leaves* : diaphoretic, polyphylactic against fever.

Phytography : Medium-sized deciduous tree with short thick trunk and spreading branches, girth may be 10 m or more, with hard, woody and brown outer bark and greyish, fibrous inner bark; leaves digitate, deciduous; flowers solitary, axillary, pendulous, long-stalked, pentamerous; fruit oblong, woody, indehiscent; seeds reniform with thick testa.

Phenology : *Flowering* : May–June; *Fruiting* : Winter.

Distribution : Native of tropical Africa; occasionally cultivated in many parts of India; Bangladesh and Pakistan.

Ecology and cultivation : Introduced; mainly in the gardens and roadsides.

Chemical contents : *Stem-bark* : adansonin, β-sitosterol; the ash of the outer shell of the fruit is rich in potash. About 50% of the *Fruit-pulp* is soluble in water, yields a mucilaginous liquid and an insoluble residue of cellular tissue—reducing sugars, acids, proteins comprise half this amount and the remainder is chiefly composed of pectic substances; *Seed* : devoid of starch, while the *Soft kernel* yields about 12% of a fatty oil, β-sitosterol, stigmasterol, cholesterol, etc; *Ash of seeds* : rich in potash (31%) and phosphates (34.2%).

Adenanthera pavonina L. (Mimosaceae)

English names : Coral pea, Red tree, Red bead tree, Coral wood, Red wood, Barbados pride.

Sanskrit name : Kuchandana.

Vernacular names : Ben : Ranjan, Rakta kambal; **Hin :** Bari ghumachi; **Kan :** Manjetti; **Mal :** Manjadi; **Mar :** Thorligunj; **Tam :** Anikundumani; **Tel :** Bandiguruvenda.

Trade names : Rakta kambal, Bari ghumachi.

Traditional use : ORAON : *Seed-paste* : on headache; *Wood-paste* : in pox; *Powder of seed* (mixed with honey) : used in colic; SOME OTHER TRIBES : pain; *Wood-powder* (mixed with water) : beneficial in headache when applied on the forehead; useful in orchitis; *Seed-decoction* : in cholera, general paralysis and pulmonary affection.

Modern use : *Root-powder* : emetic; *Decoction of leaves and bark* : remedy for chronic rheumatism, gout, astringent, in atonic diarrhoea and dysentery; *Powder of seeds* : anti-inflammatory, for pulmonary affection and ophthalmia.

Phytography : Deciduous tree, height varies from 20–67 m, girth 2–3 m, heartwood red; leaves alternate, bipinnate, pinnae short-stalked, 10–20 cm long, leaflets alternate, short-stalked, oblong; racemes axillary, simple and panicled at the end of branches, pedicels as long as the flowers; flowers minute, not exceeding 0.41 cm in length, hermaphrodite, usually pentamerous; pods 15–20 cm long, strap-shaped, twisting while opening; seeds small, hard, shining scarlet.

Phenology : *Flowering :* January–March; *Fruiting :* March onwards.

Distribution : Andamans, Karnataka, Maharashtra and the sub-Himalayan tracts; often planted, especially in South India.

Ecology and cultivation : Introduced.

Chemical contents : *Plant* : robinetin, chalcone, butein, ampelopsin, dihydrorobinetin; *Ether extract of wood* : 2,4-di-Off-benzoic acid as major compound; *Bark*: stigmasterol, glucoside, a saponin, oleonic acid, echinocystic acid; *Alcoholic extract of bark* : saponins which on hydrolysis and methylation yield Me-odeanolate and Me-echynocystate; *Leaf* : octacosanol; *Seeds* : dulcitol, stigmasterol.

Adulterant : Red heartwood is a substitute for red sandal wood (*Pterocarpus santalinus* L. f.)

Adenia palmata Engl. (Passifloraceae)

Syn : *Modecca palmata* Lamk.

Vernacular names : Kan : Undal; **Mal :** Mutakku; **Tel :** Gurivenda, Modikka.

Traditional use : *Root* and *Fruit* : poisonous; *Juice of leaf* and *root* : used externally for skin diseases.

Modern use : *Root* : used as an ingredient of tonic.

Phytography : Small, perennial, climbing herb, glabrous, branches sulcate; leaves simple, usually 3- to 5-lobed, sometimes entire, membranous, glabrous, cordate or ovate, acuminate, petiole 5 cm long with a gland on either side at the apex; flowers unisexual; fruits globose, capsular, big—like small apples; seeds large, pitted, ribbed at the edges.

Distribution : Maharashtra, Karnataka and in some parts of Andhra Pradesh.

Adenostemma lavenia O. Kuntze (Asteraceae)

(2n = 20)

Syn : *Adenostemma viscosum* Forst. f., *Verbesina lavenia* L.

Traditional use : *Leaf* : antiseptic; *Fresh juice of leaf* : stimulant and sternutatory; *Crushed leaves* : applied to cuts, wounds, bites of poisonous insects and caterpillars.

Modern use : *EtOH* (50%) *extract of plant* : diuretic.

Phytography : Erect, rarely decumbent, annual glabrous or glandular-pubescent, slender or robust herb; leaves opposite or the upper alternate; sessile or subsessile, linear to broadly ovate, serrated; heads few or many, in lax or dense corymbose panicles, involucre of bracts glabrous or glandular or scabrous, receptacle flat; florets white with a pinkish tinge, tubular with campanulate limb; achenes obtuse, 5-ribbed, glandular; pappus hairs short, rigid, often clavate, inserted in a short ring.

Phenology : *Flowering* : peak during November–March.

Distribution : Throughout India.

Ecology and cultivation : Hills above 1000 m, on the shola floor, by stream banks, in shade; also as a coloniser in the disturbed ground; not in cultivation.

Chemical contents : *Plants* : *ent*-11-hydroxylated kuranic acids.

Adhatoda vasica Nees (Acanthaceae)

Syn : *Justicia adhatoda* L., *Adhatoda zeylanica* Medic.

English names : Malabar nut, Vasaka.

Sanskrit names : Arusak, Vasika.

Vernacular names : Asm : Banhaka; **Ben :** Vasak, Bakash; **Guj :** Adulso; **Hin :** Arus, Amsha, Adalsa, Basak; **Kan :** Addalasa, Adusoge, Kurchigida; **Kon :** Odduso; **Mal :** Adalodakam; **Mar :** Adulsa; **Ori :** Basanga; **Pun :** Bansa subj, Basuti; **Tam :** Adatoday, Adatodaij; **Tel :** Atarushammu.

Trade name : Vasaka.

Traditional use : *ATHARVA VEDA* (*VAIDYAKAKALPA*) : prevents normal diseases; *AGNI PURANA* : for bile disorder, leprosy, tuberculosis; *AYURVEDA* : *Bark-decoction* : in worms, bleeding, respiratory troubles; *Leaf* : in asthma, bad body odour, bile insufficiency, scabies, piles, infection of small pox, bactericidal and antitumour.

Modern use : *Root* : expectorant, antiseptic, useful in mild bronchitis, pulmonary and catarrhal affections; *Root-extract* : bactericide; *Leaf* : glandular tumour, senile pruritus, antifatigue; *Alcoholic extract of leaf* : hypotensive, bronchodilator, respiratory stimulant, hypoglycemic, antispasmodic, antiviral against Ranikhet diseases; *Essential oil of leaf* : bronchodilator, anti-insect, juvenile hormone mimicking activity; *Shoot* : in liver enlargement.

Phytography : Gregarious shrub, 1.3–3.1 m high, sometimes arborescent, growing up to 7 m high; leaves simple, petiolate, 20 cm by 7.5 cm, minutely pubescent; spikes terminal, 2.5–7.5 cm long, often several together; flowers bracteate, bracteolate, white, 1.25 cm broad; capsules 1.8 cm long, pubescent, 4-seeded; seeds glabrous.

Phenology : *Flowering* : peak during November–March (plains) and July–October (hills).

Distribution : Throughout the plains of India and in the sub-Himalayan tracts ascending up to 1300 m.

Ecology and cultivation : Plains from the coast; lower than 1000 m; often planted along the fences.

Chemical contents : *Plant* : betaine, vascinone, vascinine, β-sitosterol, tri-triacontane; *Leaf* : viscine, a crystalline acid, small amount of essential oil; *Inflorescence* : vasicinone, δ-vasicinone, l-vasicinone; *Flower* : tri-triacontane, kaempferol and its glycerides, quercetin, β-sitosterol and its glucosides; *Seed* : fatty oil consisting of arachidic, behenic, lignoceric, cerotic, oleic, linoleic acids and sitosterol.

Remark : It is preferable to retain *Adhatoda* as distinct from the herbaceous *Justicia* species, despite the current tendency to refer this to the latter.

Adiantum capillus-veneris L. (Adiantaceae)

English names : Black maidenhair fern, Venus hair.

Vernacular names : Ben : Hangsapadi; **Hin :** Hansraj, Mubaraka; **Kas :** Dumtulli.

Trade name : Hansraj.

Traditional use : *TRIBES OF NORTH-EAST INDIA* : *Pinnae* : in cough and throat diseases; *RAJASTHANI TRIBES* : in bronchitis, fever and menstrual complaints.
AYURVEDA : *Plant* : demulcent, expectorant, diuretic, emmenagogue, febrifuge and hair tonic; *Frond* (pounded with honey) : useful in catarrhal affections.
UNANI : An ingredient of a medicine named 'Parsiaoshan'.

Modern use : Boiled with wine, given in case of hard tumours of the spleen, liver and other viscera.

Phytography : Delicate fern; stipes suberect, 10–23 cm long, blackish, glabrous, shining; fronds bipinnate having short terminal pinnae and many laterals on each side; segments cuneate, 1.5–2.5 cm wide; sori borne at the rounded sinuses of crenations.

Distribution : Introduced from America; common in Tamil Nadu (the Nilgiri Hills), Maharashtra, Punjab, western Himalaya ascending up to 2500 m, Manipur, Bihar; Pakistan.

Ecology and cultivation : Inhabits damp places, grows on walls and among rocks; frequent in greenhouses; variable in nature and cultivation.

Chemical contents : Adiantone, astragaline, 3α-4α-epoxyfilicane, 21-hydroxydiantone, isoquercitrin, kaempferol, luteol, nictiflorin, quercetol, querciturone, quinic acid, rutin, shikimic acid and an essential oil.

Adiantum incisum Forssk. (Adiantaceae)

Syn : *Adiantum caudatum* Forssk.

Sanskrit names : Mayurshikha, Nilkanthashikha.

Vernacular names : Ben : Mayurshikaha; **Hin :** Marshikha; **Pun :** Adhsarita kajari; **Tam :** Myle kondai.

Trade name : Marshikha.

Traditional use : *TRIBES OF RAJASTHAN* : (i) *Plant* : in cough, diabetes; (ii) *Leaf* : in skin diseases; *RURAL FOLK OF KASHMIR* : *Whole plant* : asthma, bronchitis, burning sensation inside body; *TRIBES OF NORTH-EAST INDIA* : *Leaf*: in fever and skin diseases. *BHAVAPRAKASHA* : sweet, helps digestion, reduces obesity, pacifies wind and phlegm, beneficial in diarrhoea, urinary troubles and in child diseases.

Modern use : *Fern* : antibiotic; *EtOH* (50%) *extract of plant* : hypoglycaemic and spasmogenic.

Phytography : Leafy fern, stipes dark chestnut-brown, wiry, tufted, spreading, 5–10 cm long; fronds elongated, rooting at the apex, pinnate, pinnae 1.5–2.0 cm long, lobed; sori borne on edges of lobes.

Phenology : Sori appear in late Autumn or Winter, mature in late Winter or Spring.

Distribution : Tamil Nadu, Maharashtra, Rajasthan, Punjab and West Bengal—in the plains and lower slopes.

Ecology and cultivation : Grows in shady and moist places; wild.

Chemical contents : *Plant* : adiantone, isoadiantone, β-sitosterol, fernene, hentriacontane, 16-hentriacontanone.

Adiantum pedatum L. (Adiantaceae)

English names : American maidenhair fern, Canadian maidenhair.

Modern use : *Rhizome* : stimulant, expectorant and demulcent.

Phytography : Rootstock slender, often elongate; fronds forked into strongly divergent branches, stipes purplish, polished, blades rounded to reniform, several pinnae pointing forward, horizontal; sori oblong to linear on ends of segment lobes.

Phenology : Sori appear in Autumn, mature in Spring.

Distribution : Maharashtra as a feral, from Kashmir to Sikkim.

Ecology and cultivation : Grows in shady and moist places; wild.

Chemical contents : *Fronds* : adiantone, adipedatol, fernene, isofernene, 7-fernane, filicine, filicenal; contain a bitter principle and volatile oil.

Remarks : In France and North America, used as a pectoral in chronic catarrah, demulcent, expectorant, tonic, astringent and emmenagogue.

Adiantum philipense L. (Adiantaceae)

Syn : *Adiantum lunulatum* Burm.

English name : Walking maidenhair fern.

Sanskrit name : Hamsavati.

Vernacular names : Ben and **Hin :** Kalijhant; **Bhil :** Kajaria ; **Guj :** Hansapadi; **Kan :** Nayalad; **Mar :** Hansraj, Mubarak; **San :** Dodhar, Dodhali.

Trade name : Kalijhant.

Traditional use : TRIBES OF RAJASTHAN : (i) *Plant* : in asthma, cough, fever; (ii) *Leaf* : balding, erysipelas, leprosy; BHIL : *Plant* : pimples; SIKKIMESE : *Leaf* : bronchitis; KHASIA and GARO : (i) *Plant* : bone fracture, (ii) *Leaf* : sprain; SANTAL : (i) *Plant* : in atrophy, cachexy, emaciation, watery diarrhoea, (ii) *Root* : in muscular pain, rabies, bite of a rabid jackal or dog.
AYURVEDA : *Plant* : antidysenteric, febrifugal, subdues burning sensation of body, useful as soothing agent (as lotion) in erysipelas and along with *Asparagus racemosus* in gonorrhoea; *Root* : in fever.

Modern use : Used in blood diseases and epileptic fits; *Rhizome* : prescribed for strangury and fever due to elephantiasis.

Phytography : Leafy fern; stipes dark chestnut-brown, wiry, 6–15 cm long, glabrous, shining, tufted; fronds often elongated, rooting at apex, pinnate, pinnae subdimidate; sori borne in a continuous line along the edge.

Phenology : Sori from Rainy Season to Winter.

Distribution : Found in the plains and lower slopes of hills in the western side of peninsula, ascending up to 1200 m.

Ecology and cultivation : Grows in moist places and under shade near swamps; wild.

Chemical contents : *Plant* : higher carotenoids and chlorophyll-degradation products.

Adiantum venustum G. Don (Adiantaceae)

English names : Black maidenhair, Venus hair.

Sanskrit name : Hansapadi.

Vernacular names : Ben and **Hin :** Hansraj; **Kas :** Sun raj; **Mar :** Mubaraka; **Tam :** Mayirsikki.

Trade name : Hansapadi.

Traditional use : *Fronds* : tonic, resolvent, expectorant, diuretic, emmenagogue, astringent, emetic, febrifugal and in scorpion sting.

Modern use : The drug *Hansraj* is used in Punjab as an anodyne in bronchitis and also against tumours, biliousness, inflammatory diseases of the chest and ophthalmia.

Phytography : Little fern; stipes glossy, ebony-coloured; fronds 3- or 4-pinnate, pinnules obovate-cuneate, finely dentate-serrate, 2- to 3-lobed, lobes notched—each notch has a large sorus underneath.

Phenology : Sori in Winter and Spring.

Distribution : Kashmir, Simla and in North-East Himalaya, between 1000 and 3000 m.

Ecology and cultivation : Prefers moist and shady places; grows luxuriously as undervegetation in shaded forest beds; wild.

Chemical contents : *Petroleum ether-extract of plant* : adiantone, 21-hydroxiadiantone, a carotenoid—possibly α-carotene monoepoxide, 3-filicene, kaempferol, leucopelargonidin and quercetin glucosides.

Adina cordifolia (Roxb.) Hook. f. ex Brandis (Naucleaceae)

Syn : *Nauclea cordifolia* Roxb.

English names : Yellow teak, Saffron teak.

Sanskrit name : Dhara kadamba.

Vernacular names : Ben : Dakom, Kelikadam, Petpuria; **Hin :** Haldu, Hardu; **Kan :** Arsintega, Yettega; **Mal :** Katampa; **Mar :** Heddi, Hedu; **Ori :** Holonda; **San :** Gunri-dare; **Tam :** Manjakadamba; **Tel :** Kamba, Pasupukadamba.

Trade name : Haldu.

Traditional use : SANTAL WOMEN : take the mucilage with paste of 21 black peppers in the days of menstruation for securing conception; LODHA : (i) use it in connection with physiology of reproduction; (ii) *Decoction of stem-bark* : in fever and to check yellow colour of urine; (iii) *Plant* : in biliary colic, carbuncle, cholera, and night blindness.
BRAHMA PURANA mentions it as a forest flora.

Modern use : *Bark* : antiseptic, antipyretic, antiinflammatory, digestive, febrifuge, tonic.

Phytography : Tall (up to 20 m) tree with broad crown and a straight fluted bole; leaves broadly ovate or orbicular cordate, 1.2–2.4 cm long, suborbicular stipules enclosing the terminal bud; heads 20–25 mm in diameter, yellow, fruiting heads dry, 1.0–1.25 cm.

Phenology : *Flowering* : June–July; *Fruiting* : February–May.

Distribution : Eastern Himalaya, Assam, Orissa, Maharashtra, Karnataka, Andhra Pradesh; Pakistan and Bangladesh.

Ecology and cultivation : Seedlings raised in seed boxes in April–May, may be planted during the following season.

Chemical contents : *Plant* : umbelliferone, 10-deoxyardifoline, and 10-deoxycordifoline, yellow colouring matter, napthaquinone adinin; *Defatted heartwood* : 2 flavanones; *Oleoresin from incision of trunk* : essential oil.

Remarks : Plant is used to treat rinderpest in cattle.
It is one of the best Indian timbers, suitable for flooring and panelling railway carriages. Used for making agricultural implements, canoes, cigar boxes, combs and toys. Rural people use the wood as housebuilding material, and leaves as fodder for livestock.
Santals worship the plant as abode of God during the *Karam-parab* in November–December; they do not fell living tree.

Adonis aestivalis L. (Ranunculaceae)

(2n = 32)

Syn : *Adonis inglishii* Royle

English names : Pheasant's eye, Flos adonis, Autumn adonis.

Modern use : *Plant* : used as cardiac stimulant and diuretic in Europe, poisonous to horses; *Flowers* : considered laxative, diuretic, lithotriptic.

Phytography : Annual erect herb, growing up to 60 cm in height, simple or branched, glabrous; leaves decompoundly pinnate, segments hair-like; flowers borne at the apices of branches, variable in size, scarlet with dark purple eye; achenes in ovoid or oblong head.

Distribution : West Himalaya, Kumaon to Kashmir; Pakistan (Hazara and Peshwar).

Ecology and cultivation : A weed of cultivation; prefers high altitude.

Chemical contents : *Root* : sitosterol and ceryl alcohol; *Stem* and *Leaf* : 0.216% of a glucoside $C_{25}H_{46}O_{10}$; *Flower* : 0.29% alkaloid. The herb also contains glucosides adonin, adonidin and adonilene.

Aegle marmelos Corr. (Rutaceae)

(2n = 18, 36)

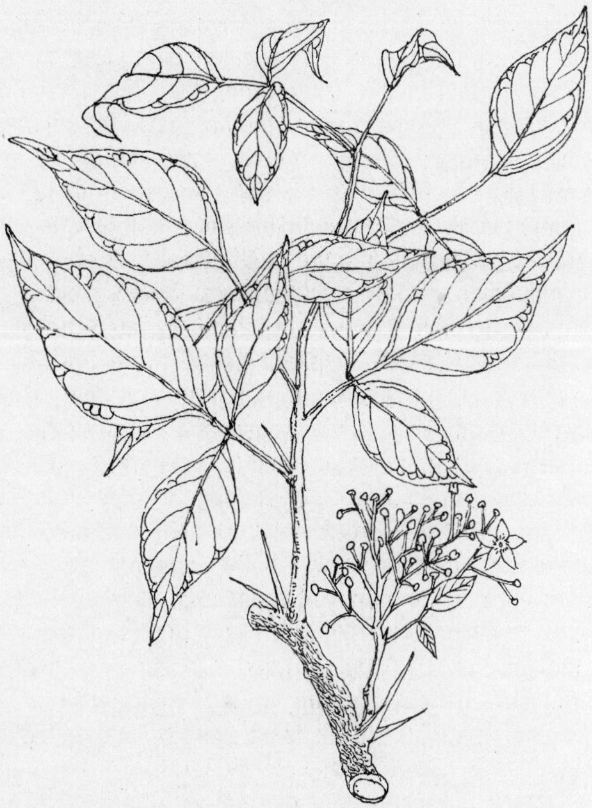

Syn : *Feronia pellucida* Roth, *Crataeva religiosa* Ainslie

English names : Wood apple, Bael tree.

Sanskrit names : Bilwa, Shriphal.

Vernacular names : Asm and **Ben :** Bel; **Guj :** Biliva phal, Bil; **Hin :** Shriphal, Bel; **Kan :** Bilwa patre, Bil patre; **Kon :** Belfoll; **Mal :** Vilvam, Koovalam; **Mar :** Bel; **Ori :** Bela; **Pun :** Bil; **Tam :** Vilvam, Vilvamaram; **Tel :** Bilambu, Maredu; **Urd :** Bel.

Trade names: Bel, Shriphal.

Traditional use : SANTAL : (a) *leaves* of this plant along with (i) many other plant materials for the treatment of continuous fever; (ii) the molasses and fruits of *Gryllus monostrosus* for the treatment of diseases without premonitory symptoms; (iii) the fruits of *Nigella indica* and black pepper to treat a patient of convulsions; (iv) the leaves of *Melia azadirac hta* and the seeds of *Semecarpus anacardium* to treat nausea; (v) the fruits of *Randia*

dumetorum and *Nigella indica*, oilcake of *Bassia latifolia*, salt and *Carum ajowan* for treatment of constipation with meteorismus; (vi) the bark of the following plants — *Pterocarpus marsupium*, *Bauhinia vahlii*, *Spatholobus roxburghii*, *Terminalia tomentosa*, *Dalbergia latifolia* and fried rice for treatment of chronic stomachache and (vii) the bark of *Odina wodier*, salt and *Carum ajowan* to treat watery diarrhoea; (b) *various parts of fruit* for making the following medicines : (i) for curing the patient of cholera, a medicine prepared with the following, is administered : the old fruit of *A. marmelos*, bark of *Melia azadirachta*, the resin of *Shorea robusta* and pulverized kernel of a mango stone; (ii) a sherbet prepared with the ripe fruit of *A. marmelos* and sugar candy is administered to the patient of dysentery; (iii) to treat the complaints of a mother after child-birth, a plaster made of the following is applied : top of the fresh unripe fruit of *A. mermelos*, leaves of *Ocimum sanctum*, little ginger and 84 black peppers; (iv) a patient of puerperal fever is given a liquid drink made with the entire contents scraped out of a ripe fruit of *A. marmelos*, juice of old dried fruits of *Mangifera indica*, two ounces of a kind of red ants and a little salt; (v) a thick mixture, made with twin fruits of *A. marmelos*, melted hail and earth excavated by rats, is smeared on the breasts of a young mother to check the flow of milk when the breasts pain or suppurate; and (c) the bark of old trees of *A. marmelos* and *Melia azadirachta* are used to prepare an antidote for the poison of all kinds of snake bite; *ATHARVAVEDA* : the unripe fruit medicinal, but ripe fruit harmful; *CHARAKA SAMHITA* : the unripe fruit increases the digestive power, but the ripe fruit is difficult to digest; *KURMA PURANA* : (i) regular drinking of fruit juice ensures long span of life and golden colour of the skin; (ii) brushing the teeth with young branch of this plant is beneficial; *MATSYA PURANA* : the paste of this plant is a good steriliser for clothings, beds, furniture, coat of armour, ornaments, canopy and fans made of wools and furs; *BRAHMAVAIVARTA PURANA* : consumption of ripe fruits of this plant helps to eliminate the ailment caused by excessive bile, but drinking water immediately after eating the fruit aggravates the bile-disorder; *AGNI PURANA* : (i) this plant is of great medicinal value; (ii) inhaling the oil, obtained from this plant, increases the poetic acumen of the inhaler and enables him to live a long life; (iii) for treatment of gastroenteritis, the patient should drink (a) the water containing the decoction of *A. marmelos*, *Aconitum heterophyllum* and *Holerrhena antidysenterica*, (b) the fruit juice of *A. marmelos*, *Mangifera indica*, *Dolicos lablab*, *Woodfordia fruticosa*, *shunthi* along with buttermilk and molasses; and (iv) it can cure a patient suffering from fever and its associate pains and cough.
UNANI : Ingredient for 'Belgiri'.
HOMOEOPATHY : Destroys phlegm and a good remedy in fevers associated with catarrhal symptoms, also good for dropsy, bleeding piles, dysentery, diarrhoea, bowel complaints and is a good laxative.

Modern use : *Fruits* : chiefly used in (i) chronic diarrhoea and dysentery; (ii) as a good laxative, and the sherbet prepared from the pulp is good for bacillary dysentery; *Half-ripe fruit* : astringent, digestive and used in diarrhoea. It is also antidiuretic, anthelmintic, antipyretic, carminative, tonic.

Phytography : Deciduous glabrous tree, up to 8 m in height and 1 m in girth, bark shallowly furrowed and corky; leaves alternate, with straight sharp axillary thorns (buds), trifoliate, aromatic; flowers in axillary panicles, greenish white, fragrant; fruit hard-shelled, up to 5 cm in diameter, greyish or yellowish, scented, pulp orange, sweet and gummy; seeds numerous, oblong, compressed, testa mucous.

Phenology : *Flowering* : April–May; *Fruiting* : persistent, ripening by February.

Distribution : Found in the plains and submontane regions of India, wild in the sub-Himalayan tract, often planted.

Ecology and cultivation : Wild; often planted.

Chemical contents : *Bark of root* and *stem* : umbelliferone, other coumarins, β-sitosterol, aurapten, marmin, lupeol, two unidentified alkaloids and two unknown compounds—their proportion varies with the age of the bark; *Stem* : yields a good gum; *Heartwood* : a ferro-quinoline alkaloid, dictamine, marmasin and β-sitosterol; *Leaves* : γ-sitosterol, aegeline, aegelenine, lupeol; sitosterol, rutin, marmesinin, β-sitosterol, glucoside and essential oil; *Fruits* : psoralen, aegelinol, furanocoumarin, marmelide, β-sitosterol and essential oil; *Ripe fruits* : xanthotoxol, scoparone, scopoletin, umbelliforone, marmesin, skimmin and β-sitosterol glucoside.

Adulterants : *Limonia acidissima* Correa, *Garcinia mangostina* L. and *Punica granatum* L. are used as substitute of *Aegle marmelos* Correa.

Aerva lanata (L.). Juss. (Amaranthaceae)

(2n = 16)

Syn : *Achyranthes lanata* L., *Aerva floribunda* Wight

Sanskrit name : Astmabayda.

Vernacular names : **Ben :** Chaya; **Guj :** Gorakhaganjo; **Hin :** Astmabayda, Chaya, Gorakhboonti; **Kan :** Bilihindisoppu; **Mar :** Kapur madusa; **Ori :** Paunsia; **Pun :** Buikallan; **Tam :** Sirupulai; **Tel :** Pindikimda.

Trade name : Chaya.

Traditional use : *Root-powder* : to relieve cough in bronchitis.

Modern use : *Plant* : anthelmintic and diuretic; *Root* : demulcent, diuretic, and used in cough, diabetes, headache, lithiasis and sore throat.

Phytography : A small herbaceous plant, ash-coloured and little tomentose; leaves ovate, obtuse, shortly mucronate, pubescent on both sides; spikes solitary or in twos and threes, sessile, horizontal; flowers bisexual, 1.5 mm across.

Phenology : *Flowering* : in Rainy season with a peak in July- August.

Distribution : Native to India and Sri Lanka; grows throughout the plains of India, Pakistan and Bangladesh, ascending up to 1000 m.

Ecology and cultivation : Mostly in wasteland, even in poor soils and sun-exposed places; locally abundant in arable and fallow fields.

Chemical contents : *Plant* : β-Litosteryl palmitate α-amyrin, and β-sitosterol.

Aerva tomentosa Forssk. (Amaranthaceae)

Syn : *Aerva javanica* Juss. ex Schult, *A. persica* Merrill

English name : Javanese wool plant.

Vernacular names : Guj : Bur; **Hin :** Dholimundi; **Pun :** Boi kalan; **Raj :** Buida; **Tam :** Perumbulai; **Tel :** Magbira.

Traditional use : *Decoction of leaves* : to remove smellings; *Decoction of flowers* and *seeds* : applied externally against swelling; *Woolly seeds* (stuffed in pillows) : relieve headache and protective against rheumatism.

Modern use : *Plant* : diuretic and demulcent.

Phytography : Deciduous undershrub; stem 60–100 cm, terete, branches thick; leaves alternate, 2.5–10 cm, flat or with recurved margins, sessile or short petioled; spikes often in large panicles, erect or nodding; flowers 0.25 cm long, unisexual, males few, covered in long wool.

Phenology : *Flowering* : peak in December–February.

Distribution : Punjab, Rajasthan, Gujarat, Madhya Pradesh, Karnataka and Tamil Nadu; Pakistan (Baluchistan, Sind) Sri Lanka, Myanmar.

Ecology and cultivation : Grows in plains, even in poor and gravelly soil, scrub jungle, wayside, wastelands; optimum size by arable lands.

Chemical contents : *Root* : aervanone (a new flavanone), chrysin-7-O-galactoside; *Plant* : α-amyrin, β-sitosterol, fatty acids, kaempferol-3-galactoside, kaempferol-3-rhamnogalactoside.

Aesculus hippocastanum L. (Hippocastanaceae)

(2n = 40)

English name : Horse chestnut tree.

Vernacular name : Pun : Pu.

Traditional use : *HOMOEOPATHY*: very useful in liver and portal system, piles, severe pain in the hips and waist regions (in macrotinum), in constipation, dyspepsia, gastralgia and in menstruation.
UNANI : used in 'Baloot'.

Modern use : *Extract of leaf* : in whooping cough; *Fruit* and *Bark* : antiperiodic, contains flavonic substances for treatment of circulatory venous systems.

Phytography : Tree, stem shows peculiar secondary growth; leaves alternate, exstipulate, digitate, deciduous; panicles terminal, thyrsoid; flowers large, polygamo-dioecious, i.e., although apparently bisexual but functionally unisexual, irregular, zygomorphic; capsules 3-chambered; chambers 1-seeded, valves loculicidal, coriaceous; seeds subglobose, exalbuminous, testa coriaceous, cotyledons thick.

Distribution : Indigenous in North India and occasionally cultivated in other places as an ornamental plant, especially in west Himalaya, Kashmir, Punjab, Himachal Pradesh, Tehri Garhwal and Kumaon.

Ecology and cultivation : Temperate and subtemperate moist climate; introduced as ornamentals.

Chemical contents : *Seed* : escin—a saponin, escigenin, esculin, esculetin and flavones. Sugar portion of the saponin contains glucose, xylose, glucuronic acid and flavonal glycoside. Flavonoids are present in *Stem* and *Flower*, *Leaf* : quercitrine; *Flower* also contains rutin, isoquercitrin, kaempferol-3-glucoside, kaempferol-3-arabinoside and kaempferol-3-rhamnoglucoside.

Remark : Used as substitute for coffee.

Aesculus indica Colebr. ex Camb. (Hippocastanaceae)

(2n = 40)

English names : Himalayan chestnut, Indian horse chestnut.

Vernacular names : Hin : Bankhor, Kanor, Pagla, Pankhar; **Kas :** Hanudun, Hane; **Pun :** Bankhor.

Modern use : *Root* : for leucorrhoea; *Bark-paste* : applied to dislocated joints; *Oil from seeds* : applied externally in rheumatism.

Phytography : Tree, growing up to 23 m in height, girth up to 37 cm only, branches glabrous and terete; leaves opposite, deciduous, digitate, leaflets obovate or oblong, serrate, unequal—the terminal larger than the laterals, petioles 7.5–15.0 cm in length, thickened at the base but sulcate above; flowers occur in terminal, thyrsoid panicles, large, irregular, calyx tubular, 5-cleft with unequal, imbricated but short and rounded lobes; capsules ovoid or subpyriform, reddish-brown, rough but without spine, 2.5–5.0 cm long; seeds dark, with coriaceous testa, subglobose, exalbuminous.

Distribution : Western Himalaya, Kashmir, Kulu, Chamba in Himachal Pradesh, Tehri Garhwal and Kumaon in Uttar Pradesh; Pakistan (Peshwar, Hazara, Baluchistan), on the hill tops.

Ecology and cultivation : Grows in temperate and subtemperate moist climate; wild.

Chemical contents : *Stem* : astragalin and rutin ; *Leaf* : aescin, quercetin, β-sitosterol, n-hentriacontane, n-hentriacontanol and palmitone.

Remark : Fruit is given to horses in colic pain.

Aganosma calycina A. DC. (Apocynaceae)

(2n = 22)

Syn : *Aganosma wallichii* G. Don, *A. roxburghii* G. Don, *Echites calycina* Wall.

Sanskrit name : Malati.

Vernacular names : **Ori :** Maloti; **Tel :** Palamalle.

Trade name : Malati.

Traditional use : *Plant* : tonic, useful in diseases caused by disordered bile and blood.

Phytography : Stout evergreen climber, branches stout, pubescent; leaves opposite, simple, petiolate, coriaceous, 7.5–10.0 cm, elliptic, glabrous, acuminate, veins 7–10 pairs, arching; cymes terminal, many-flowered, tomentose; flowers cylindrical, base broadly dilated, 1.25 cm long.

Phenology : *Flowering* : Autumn; *Fruiting* : Winter.

Distribution : Sub-Himalayan regions; Myanmar.

Ecology and cultivation : Grows in moist climate.

Aganosma dichotoma (Roth) K. Schum. (Apocynaceae)

Syn : *Aganosma caryophyllata* G. Don

Sanskrit name : Malati.

Vernacular names : **Ben, Hin** and **Tel** : Malati; **Mal :** Cheruparal; **Ori** and **Guj :** Maloti.

Trade name : Malati.

Traditional use : TRIBAL : (i) *Plant decoction* : for urinary trouble, tonic in fever and as emmenagogue; (ii) *Leaf* : in biliousness; (iii) *Flower* : in diseases of eye.
AYURVEDA : an ingredient of *Vishaghna Taila* which has anodyne and sedative properties.

Modern use : Used as tonic in fever, in paralegia, sciatica, neuralgia, to accelerate menstrual flow in dysmenorrhoea cases.

Phytography : A large and beautiful evergreen woody latex bearing climber; leaves ovate, elliptic and acuminate; cymes lax, pubescent, corymbose; flowers white; follicles densely yellow, tomentose, very variable in size, tapering gradually at the tip.

Phenology : *Flowering* : August; *Fruiting* : February–April.

Distribution : Bihar, lower Ganga belt, Maharashtra, Karnataka, Orissa, Andhra Pradesh; Bangladesh (Sylhet and Chittagong).

Ecology and cultivation : Occurs in moist rocky ravine, rocky valleys; often cultivated.

Chemical contents : *Leaf* : quercetin, kaempferol, phenolic acid.

Aganosma marginata G. Don (Apocynaceae)

Syn : *Aganosma acuminata* G. Don, *A. macrocarpa* A. DC., *Echites marginata* Roxb.

Modern use : *Root* : decoction for urinary troubles and as tonic for fevers, in convulescence; good medicine for oligomenorrhoea and in emmenagogue.

Phytography : Stout scandent or subscandent shrub; leaves opposite, petiolate, oblong, acute or acuminate or caudate, 5–10 cm in length, 1.8–3.8 cm in width, coriaceous, hard, polished above, pubescent beneath, nerves very strong beneath; cymes mainly axillary, peduncles slender, 2.5–5.0 cm long, flowers white, bracteate, bracts deciduous, small, 0.6–0.8 cm long; follicles long, 15–45 cm; narrow, width 0.8 cm, cylindric, straight or flexuous; seeds linear-oblong, glabrous, not beaked, albumen scanty.

Phenology : *Flowering* : Rainy season; *Fruiting* : Winter.

Distribution : Assam; Bangladesh (Sylhet, Chittagong), Java, Philippines, Malacca, Singapur, Sumatra.

Ecology and cultivation : Plant of humid tropical climate; hill above 800 m; also along forest border.

Agave americana L. (Agavaceae)

(2n = 60, 120, 180)

English names : American aloe, Century plant.

Vernacular names : **Ben :** Bilati anarash, Kantala; **Guj :** Jangli kunvar; **Hin :** Jangli keora, Kantala; **Kan :** Bhuttale budu, Kattale, Kattale neru; **Kon :** Gaipat; **Mal :** Nathukaita, Panamkatrazua; **Mar :** Ghaypat; **Ori :** Barabarasia, Muruga; **Pun :** Wilayeti kantala; **Tam :** Anakuttilai, Bontharakasi.

Trade name : Kantala.

Traditional use : RURAL FOLK : *Root* : diuretic and antisyphilitic; *Slices of fleshy leaf* : good poultice, in bruises and rheumatism; *Pulp of leaf* : good remedy for gonorrhoea.
HOMOEOPATHY : mother tincture is used in the sore in the mouth, in scurvy, in afflictions in teeth, pains and swelling in legs, in constipation, gonorrhoea.

Modern use : *Leaf-juice* : for warts, cancerous ulcers and putrid tumours. *Various parts of plant* : for ascites, dropsy, anasarca, venereal sores, syphilis and dysentery, diuretic, laxative, emmenagogue, antiseptic, diaphoretic and antipyretic, alterative, in toothache.

Phytography : Perennial herb with rhizome; leaves simple, alternate, large, densely crowded in massive rosettes, thick, fleshy, coated with wax, apex spiny; flowers come after many (5 to 68, sometimes 100) years of existence; gigantic racemose or paniculate inflorescence with ultimate branches of few-flowered cymes; flowers large, showy, bisexual; capsules.

Phenology : *Flowering* : almost throughout the year.

Distribution : Native of Mexico and Central America, grows wild in arid areas of India, Pakistan, Bangladesh and South-East Asia.

Ecology and cultivation : Xerophyte; wild.

Chemical contents : *Leaf* : hecogenin—precursor of steroid drug (content of hecogenin in various species grown by Sisal Research Station, Bamra, Orissa, summarized), piscidic acid, agave-saponin C, steroid saponins, steroid glycosides; *Flower* : chlorogenin, kaempferol-3-glucoside and 3-rutinoside; *Seed* : steroid saponins including hecogenin.

Agave angustifolia Haw. (Agavaceae)

Syn : *Agave wightii* D. & P.

English name : Dwarf aloe.

Sanskrit name : Kantala.

Vernacular names : Hin : Khetki; **San :** Murobba; **Tam :** Malaikkattalai; **Tel :** Kittanara.

Trade name : Khetki.

Traditional use : *Roots* : diuretic, diaphoretic; *Juice of the leaves* : applied to bruises. SANTALS apply the same with 21 black peppers on swelling of throat.

Modern use : *Plant-extract* : blood pressure depressor; *Tissue-culture extract* : anti-microbial.

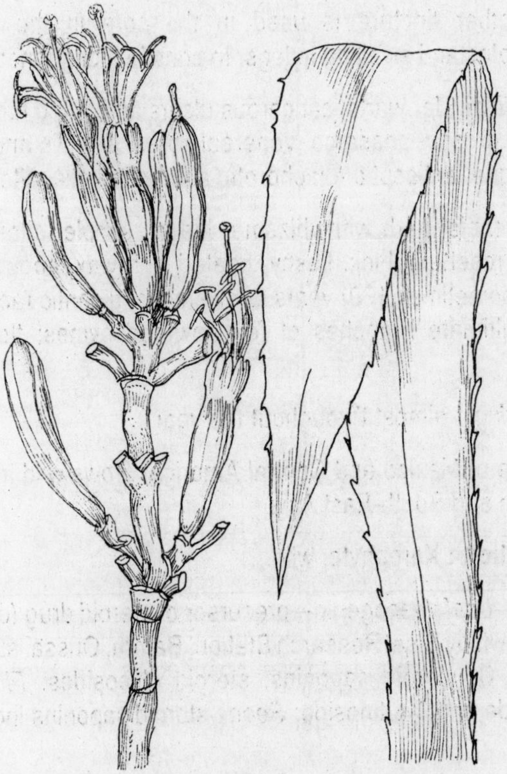

Phytography : Stem rhizome; leaves ensiform, linear lanceolate, many, forming a stiff even rosette, pale green, often tinged with ashy gray, broadly amplexicaul, marginal prickles weak.

Phenology : *Flowering* : Winter.

Distribution : Naturalised in sub-Himalayan and the outer Himalayan regions and many other parts of India; Pakistan and Bangladesh.

Ecology and cultivation : Prefers high altitude and semidry soil.

Chemical contents : *Leaf*: gitogenin, hecogenin and tigogenin; *Tissue -culture* : steroidal sapogenins.

Remarks : Usually planted for leaf-fibres.
Santals eat boiled flower as a vegetable.

Agave cantula Roxb. (Agavaceae)

(2n = 90)

English names : Bombay aloe, Cantala, Manila marguey.

Vernacular names : Ben : Bilati anarash; **Pun :** Kanatala, Kitki.

Trade name : Sisal hemp.

Traditional use : SANTAL : *various parts* for dropsy, dysentery, venereal diseases like syphilis, anasarca, ascites; *Root* : anthelmintic; *Leaf* : in burn, cut, wound and on boil.

Modern use : *Leaf* : Diuretic, antiseptic, blood purifier, laxative and analgesic, remedy for urinary infection, a general tonic; *Paste of leaf* : in relieving pain; *Decoction of leaf* : for treatment of scanty menstruation.

Phytography : Perennial herb with strong rhizome; leaves green, alternate, narrow, entire, thick, fleshy, rigid, ensiform, with a coating of wax, grow in dense rosette on a short woody axis, axis shows secondary growth; flowering rare, gigantic racemose inflorescence borne at the top of a naked scape; flowers large, showy, bisexual; capsules.

Phenology : *Flowering* : normally in Autumn.

Distribution : Mexican in origin, cultivated as hedge in Punjab; now naturalised in major parts of arid zones of India; Pakistan and Bangladesh.

Ecology and cultivation : Xerophytic plant; planted as a hedge.

Chemical contents : *Plant* : diosgenin; *Root* : steroidal constituents, β-sitosterol, β-sitosterol-B-D-guicoside, trigogenin, diosgenin, hecogenin, gitogenin, chlorogenin cantalasaponin-1; *Leaf* : cantanalin, a new glycoside of hecogenin; *Ethanalic extract of the inflorescence* : a new malluscicidal spirostanol glycoside; *Fruit* : two new steroidal glycosides, agaveside A and B, agaveside C.

Agave vera-cruz Mill. (Agavaceae)

English names : Blue elephant aloe, Railway aloe, Aloe.

Vernacular names : Hin : Kuwarbuti, Rakaspatta; **Tam :** Kattazha; **Tel :** Kattalai.

Trade name : Railway aloe.

Traditional use : *Plant* : purgative.

Phytography : A stout plant usually producing numerous shoots from the rhizome; leaves very deep green and glaucous, linear oblong, 1.25–2.0 cm long, attaining a width of 3 cm.

Phenology : *Flowering* : September–October.

Distribution : Native of Mexico, distributed throughout India; Pakistan and Bangladesh.

Ecology and cultivation : Prefers semidry regions.

Chemical contents : *Stem* : rich in and cheap source of polyfructosanes of the type of inulin and its isomers. *Juice of leaf* : hecogenin, dehydrohecogenin.

Remarks : Planted along railway embankments and roadsides; are suited for hedging, fencing and to check soil erosion.

Ageratum conyzoides L. (Asteraceae)

(2n = 20, 40)

Syn : *Ageratum cordifolium* Roxb.

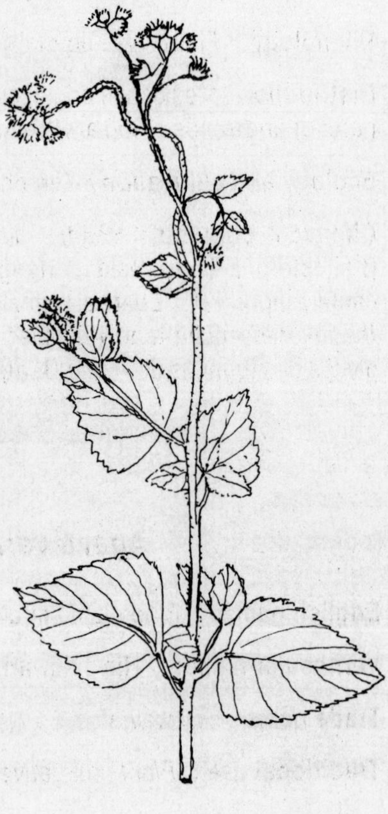

English names : Goat weed, White weed.

Vernacular names : **Ben** : Uchanti, Dochunty; **Guj** : Ajagandha; **Gar** : Gumdrya; **Kan** : Nayitulasi; **Mar** : Osari, Ghaneraosadi; **Ori** : Paksunga; **Tam** : Pumpillu.

Traditional use : *Plant* : boiled with oil and applied in rheumatism; *Root* : in kidney stone; *Leaf* : in cuts, leprosy, boils, burns, fever, piles, prolapsus anus and skin diseases, uterine disorders.

Modern use : *Leaf* : styptic, applied to cuts, sores and externally in ague; *Juice of leaf* : antilithic, used as a nervine tonic; *Decoction* or *infusion of leaf* : in diarrhoea, dysentery, colic with flatulence and other gastro-intestinal ailments; *Extract of flower* : useful in Asiatic form of Scwartz leukaemia and prolonged lifespan of mice.

Phytography : Erect hairy herb, growing up to 60 cm; leaves simple, petiolate, opposite, upper ones may be alternate, ovate, crenate; heads dense, terminal, corymbose; involucre companulate, bracts 2–3 seriate, linear, subequal; florets alike, tubular, bisexual, blue or white; achenes black, 5-angled, pappus scales 5-awned, often serrate below.

Phenology : *Flowering* : throughout the year, mainly December–May.

Distribution : American in origin, a common weed throughout India, ascending the Himalaya up to 165 m; also in Pakistan, Bangladesh and all warmer countries.

Ecology and cultivation : In hilly slopes, wastelands as well as arid zones, aggressive coloniser.

Chemical contents : *Herb* : friedelin, β-sitosterol, stigmasterol, hydrocarbons, α-spinasterol, coumarin, flavonoids, antigonadotropic hormones, precocenes 1 & 2, 5'-OMe-nobiletin, herb oil; *Leaf* : fresh herb from Dehra Dun yields essential oil, caryophyllene 7, γ-cadinine, several constituents; *Flower* : similar to leaf, γ-cadinene, caryophyllene, de-OMe-ageratochromene, ageratochromene.
Plants grown in North Vietnam contain some other chemicals.

Aglaia diepenhorstii Miq. (Meliaceae)

(2n = 162)

Syn : *Aglaia odoraissima* Blume

English name : Chinese rice flower.

Traditional use : AYURVEDA : *Root* : abortifacient, acrid, bitter, refrigerant, aphrodisiac, removes rheumatism (*vata*), biliousness, cures dysentery, leucoderma, skin diseases, leprosy, burning fever, burning of the body, blood impurities; *Bark* : emetic, cures dysentery, fever, skin diseases; *Leaf* : emetic, stop abdominal pain, influenza; *Leaf-decoction* or *infusion* : convulsion, menorrhagia; *Flower* : leprosy; *Fruit* : sweet and acrid, heavy to digest; astringent to bowels, remove phlegm (*kapha*), useful in uterine complaints, specially for women; *Seed* : sweet and acrid, cooling, tonic, removes biliousness and cough.

Modern use : *Leaf* : stimulant, antipyretic, used in convulsion; *Infusion of flower* : prescribed in eruptive fever and in venereal diseases.

Phytography : Tree reaching 15 m in height, young branches, petiolules and inflorescence covered with minute brown deciduous scales; leaves imparipinnate, leaflets opposite, thinly coriaceous, elliptic oblong, obovate; flowers globular, yellow; berry pyriform, velvety tomentose, buff-coloured, usually one-seeded; seeds ellipsoid.

Phenology : *Flowering* : November–December.

Distribution : Grows wild in Konkan, Karwar and other places in Maharashtra, Gujarat and Karnataka.

Ecology and cultivation : Prefers black cotton soil; wild.

Chemical contents : *Plant* : tetracycline, triterpenoids, aglaoil, aglaiondiol and two isomers of aglaitriol.

Aglaia elaegnoidea (A. Juss.) Benth. (Meliaceae)

Syn : *Aglaia roxburghiana* Miq., *Milnea roxburghiana* W. & A.

Sanskrit name : Priyangu.

Vernacular names : Ben and **Hin :** Priyangu; **Mal :** Punyava; **Tam :** Chokkala; **Tel :** Yerra aduga.

Trade name : Priyangu.

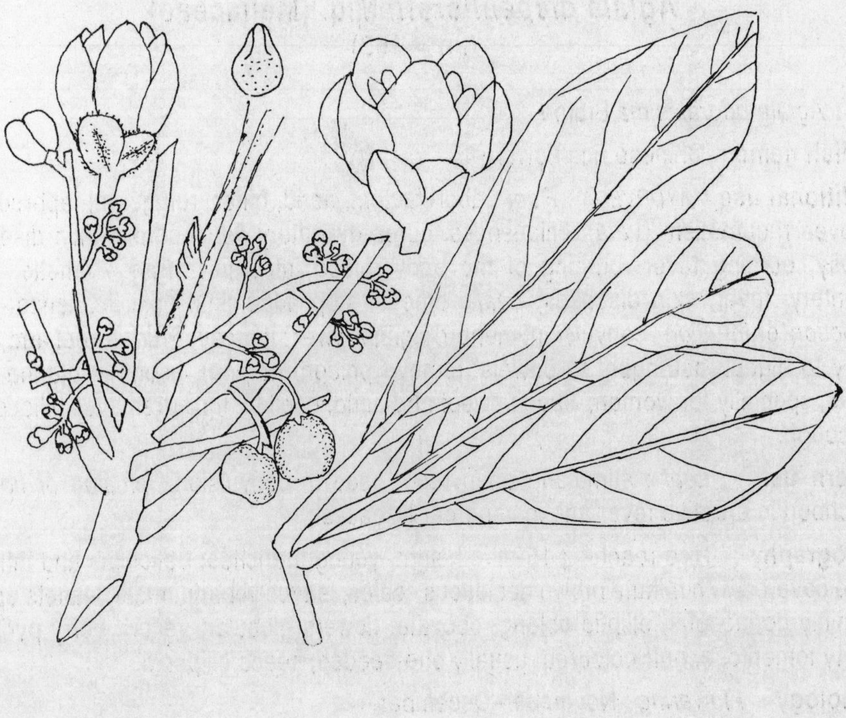

Traditional use : *MATSYA PURANA* : *Plant* : astringent; decomposition of plant along with some other decomposed plants and spleen of mongoose form antidote against poisons contacted by touch.

Modern use : *Fruit* : edible, cooling, astringent, used in inflammations; *Seed* : in painful micturition.

Phytography : Large tree, wood bright red; leaves shortly petiolate, pinnate, leaflets 5, rarely 7 or 3, elliptic, glabrescent, obtuse, sometimes acute; panicles rather supra-axillary, pyramidal, elongate; flowers shortly pedicellate, small, 0.2 cm in diameter, yellow; fruits subglobose, small, 0.25 cm in diameter, buff-coloured, orange when ripe; very minutely pilose, rusty, persistent.

Phenology : *Flowering* : November –February.

Distribution : Konkan and Medinipur southwards; Java, Sumatra and other Malaya Islands.

Ecology and cultivation : Hills (500–1000 m) by the forest border.

Chemical contents : *Leaf* : a bis-amide of roxburghilin and its two epimers.

Agrimonia eupatoria L. (Rosaceae)

(2n = 28)

Syn : *Agrimonia nepalensis* (Don) Nakai, *A. pilose* Ledeb.

English names : Agrimony, Cocklebur, Stickle wart.

Modern use : *Root* and *Leaf* : anthelmintic, astringent, bowel regulator, diuretic, expectorant and tonic.

Phytography : Slender perennial herb; rootstock woody, short or long; leaves stipulate, interruptedly pinnate, 7.5–21.0 cm, leaflets elliptic-ovate, obovate or lanceolate, obtuse or acute, serrated; racemes terminal and spike-like; flowers bracteate, pedicellate, small, yellow, calyx almost glabrous with erect spines; fruits pendulous, 1 or 2 achenes enclosed in the hardened spinous calyx.

Distribution : Kashmir to Kunawur, Sikkim, Meghalaya, ascending up to 2600 m; Pakistan (North-West Frontier Province).

Ecology and cultivation : Temperate and subtemperate humid climate.

Chemical contents : *Plant* : essential oil, flavones, hyperoside, luteolin, quercetin, also glycerides of the last two, apigenin 7-glucoside; *Rhizome* : vanillic acid, l-taxifolin, ellagic acid; *Root-sprout* : taenifuge agrimophol.

Remarks : Zulus use the finely powdered leaves for the treatment of tapeworm. Hottentots use infusion of roots as stomachic. Yosus and Sutos use the plant as a vermifuge.

Agropyron repens Beauv. (Poaceae or Gramineae)

Syn : *Agropyron acutum* Roem. e. Sch., *Triticum repens* L.

English names : Couch grass, Dog grass, Twitch grass.

Modern use : *Rhizome* : aperient, diuretic, demulcent, used internally as a demulcent, drink for irritable bladder and in cystitis; *Juice of root* : used for cirrhotic liver, tumours and cancerous growth.

Phytography : Grass, stem 30–60 cm, slender, erect or ascending, densely tufted at the base; leaves narrow, 0.2–0.3 cm in width, flat or convolute when dry; spikes slender, 7.5–15.0 cm long, erect, rachis slender, spikelets rather distant, 5- to 6-flowered, very pale, glumes rather spreading.

Distribution : Kashmir, Ladak, Zanskar, up to 4500 m.

Ecology and cultivation : Plant of cold climate.

Chemical contents : *Rhizome* : triticin, a carbohydrate allied to starch, K-salts, essential oil with antibiotic properties, possibly a saponin; *Ergot sclerotia grown on this grass* : ergothroneine, ergotannine, also yield an essential oil.

Ailanthus altissima (Mill.) Swingle (Simroubaceae)

Syn : *Ailanthus glandulosa* Desf.

English names : Ailanto, Tree of heaven.

Traditional use : *Bark-powder* : used in epilepsy; *Leaf* : produces dermatitis.

Modern use : *Root-bark* : recommended for cardiac troubles, epilepsy and asthma; *Stem-bark* : anthelmintic, antispasmodic, parasiticide and used in dysentery; *Leaf* : used in preparations of seborrhoea and in scabies; *Fruit* : in eye troubles and as an emmenagogue;

Phytography : Lofty tree; leaves alternate, unequally pinnate, exceeding 30 cm, leaflets numerous, pale beneath; panicles axillary or terminal, much-branched; flowers small, bracteate and bracteolate; samaras 3 cm long, 0.8 cm wide, wings large, membranous; seeds pendulous, 0.6 cm long, 0.25 cm wide.

Distribution : Native to China and Moluccas, probably introduced in northern India.

Chemical contents : *Plant* : 2,6-di-OMe-quinone, ailanthone, *Wood-alkaloid* : conthin-6-OMe-, its three oxides, 1-OMe-derivatives; *Root-bark* : four indole alkaloids, canthin-6-one and its 1-OMe-derivative, Me-4-OMe-B-carboline-1-carboxylate, 1-OH-canthin-6-one, shinjudilactone and some more alkaloids; *Dried stem-bark* : quassin, neoquassin; *Defatted bark* : ailanthone, amarolide and Ac-amarolide, shinjulactone B; *Leaf* : quercetin, caffeic acid, gallic acid, 2-hexenal, an antiviral phenolic protein, alkaloids; *Seed* : 2,6-di-OMe-benzoquinoailantholide.

Remarks : Dried root-powder is used in Africa as narcotic. It influences the nervous system, similar to that of *Nicotiana tabacum* L. leaf.

Ailanthus excelsa Roxb. (Simroubaceae)

(2n = 62, 86)

English name : Tree of heaven.

Sanskrit name : Mahanimba.

Vernacular names : Hin and **Mar :** Mahareikha; **Mal :** Mattipongilyam, **Tam :** Peruppi, Pee maram; **Tel :** Peddamanu.

Trade names : Mahanimba, Mahareikha.

Traditional use : *TRIBES OF WEST BENGAL AND BIHAR* : the powder from the resin of this plant mixed with curds as an astringent in diarrhoea and dysentery; *Bark* : abortifacient, useful in cold and cough, in fever, postnatal complaint; *Bark-decoction* : applied on wound; *Bark* and *leaf-paste* : applied on skin eruption; *Leaf-decoction* and *infusion* : rheumatism; in many parts of India, *decoction of barks and leaves* have been used as a tonic after debility and labour pain;
HOMOEOPATHY : for the treatment of diphtheria, follicular tonsillitis, stomach troubles, fever with restlessness, skin diseases, hysteria and hiccup.

Modern use : *Root-bark* : antitumour against P-388 lymphocytic leukaemia; *Leaf* and *Bark* : applied to skin eruptions, anthelmintic, antispasmodic, antipyretic, blood purifier, carminative and expectorant.

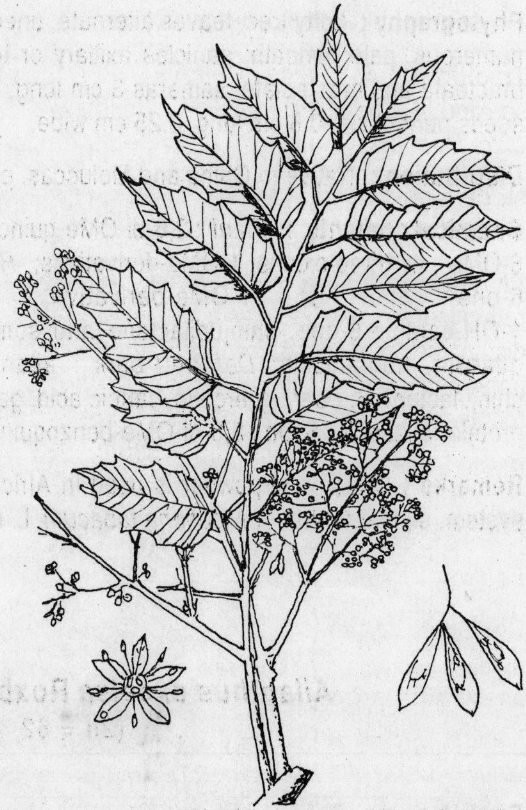

Phytography : Tree growing up to 27 m; leaves evil-smelling, alternate, petiolate, 30 cm or more, leaflets many, petiolule long, glandular, hairy, coarsely toothed; panicles axillary or terminal, large, much-branched, lax; flowers on long pedicels, small; samara large, red, twisted, blunt or pointed at both ends.

Phenology : *Flowering* : January to March; *Fruiting* : May.

Distribution : Orissa, Bihar, Madhya Pradesh, western peninsula and South India.

Ecology and cultivation : Prefers dry climate.

Chemical contents : *Root-bark* : quassinoids, ailanthinone, glaucarbunione, glaucarubol-15-isovalerate, 13,18-dehydroglaucarubol-15-isovalerate, canthin-6-one and its 1-OMe, 5-OMe- & 8-OH derivatives, triacontane, hexa-triacontane, excelsin; *Bark-extract* : excelsin, 13, 18-dehydroexcelsin, glaucarubol and quessinoids; *Leaf-extract* : protein-like stable inhibitor.

Remark : Stem-bark of this species is used as a substitute for *Holarrhena pubescens* (Buch.-Ham.) Wall. ex DC.

Ailanthus triphysa (Dennst.) Alston (Simaroubaceae)

Syn : *Ailanthus malabarica* DC.

English name : Malabar ailanto.

Sanskrit name : Pishachavriksha.

Vernacular names : Hin : Guggul dhup; **Kan :** Hal maddi; **Mal :** Mattipal; **Tam :** Perumaram; **Tel :** Peddaname.

Trade name : Guggul dhup.

Traditional use : SOME PEOPLE OF DIFFERENT PARTS OF INDIA : *Bark-powder*: after soaking in gingilly oil, as an antidote in snake bite; *VILLAGERS OF WEST BENGAL* : *Fresh extract of bark* : in migraine; *Fruits* : in different types of eye ointment.

Modern use : *Bark* : carminative, tonic and febrifuge; *Powder of bark* : used in bronchitis and asthma, also in dysentery; *Decoction of bark* : in typhoid and in constipation.

Phytography : Lofty tree, trunk tall, cylindrical, bark thick, rough, gray—often with bright red-coloured grains; leaves alternate, petiolate, pinnate, very large, leaflets on petiolule, distant, coriaceous, nearly glabrous, much lighter beneath, wavy, often thickened margin; panicles large, lax, much-branched; flowers large; samara broadly linear, 6.2 by 0.8 cm, reddish, rounded at both ends, not twisted.

Phenology : *Flowering* : Spring; *Fruiting* : Summer.

Distribution : Western Ghats, up to 1500 m; Sri Lanka.

Ecology and cultivation : Prefers black cotton soils and dry climate.

Chemical contents : *Root* : β-sitosterol, eight β-carboline alkaloids; *Stem-bark* : malanthin, eight β-carboline alkaloids; *Triterpenoids of trunk-exudate* : malabaricol, epoxy malabaricol, malabaricanediol.

Ajuga bracteosa Wall. ex Benth. (Lamiaceae)

Sanskrit name : Nilkanthi.

Vernacular names : Kas : Kauributti; **Kum :** Ratpath; **Pun :** Khurbanti.

Trade name : Nilkanthi.

Traditional use : TRIBES OF KUMAON : *Leaf-juice* : as blood purifier; *Leaf-powder* : on burns and boils; *Leaf-decoction* : in syphilis; *Plant* : in malarial fever; *Plant-extract* with that of *Centella asiatica* : in gonorrhoea and intermittent fever; *Root-bark* : anthelmintic and purgative.

Modern use : *Aromatic leaves* are considered as stimulant and diuretic, astringent, febrifuge, aperient, tonic; used in gout, rheumatism, palsy and amenorrhoea, tonic, useful in agues.

Phytography : Perennial herb, erect or ascending, rootstock 10–30 cm long, simple or branched; stem softly pubescent, villous or glabrate; leaves 2.5–10.0 cm long, lower petioled, upper sessile, oblanceolate or subspatulate, sinuate-toothed, obtuse;

verticillaster; flowers small, pale blue, bilabiate, upper lip erect and bifid, pubescent; nutlets 0.25 cm, ellipsoid, deeply roughly pitted.

Distribution : Kashmir to Nepal, up to 2300 m, Punjab plains (along the base of hills) to Oudh; Pakistan (Peshwar).

Chemical contents : *Plant* : ceryl alcohol, β-sitosterol, γ-sitosterol, cerotic acid, palmitic acid, a glucoside.

Remarks : In preliminary studies, ethanolic extract of the plant shows anticancer activity against sarcoma 180 in mice. It is also useful as a substitute for cinchona.

Alangium salviifolium (L.f.) Wang. (Alangiaceae)

Syn : *Alangium lamarckii* Thw., *A.latifolium* Miq.

English name : Sage-leaved alangium.

Sanskrit name : Ankola.

Vernacular names : Ben : Akarkanta; **Guj :** Onkola; **Hin :** Akola; **Kan :** Ankole; **Mal :** Irinjil; **Mar :** Ankol; **Tam :** Alangi; **Tel :** Ankolamu.

Trade names : Akola, Ankola.

Traditional use : RABHA (West Bengal) : *Leaf-decoction* : in skin diseases; *Leaf-paste* : in rheumatism; *Seed* : in haemorrhage and boils; SANTAL : *Root* : as abortifacient, *Dried root-powder* : in poisonous bites of animals, *Bark-extract* : in lowering blood pressure; *Flower* and *fruit-extract/infusion* : on eye ailments; TRIBES OF ANDHRA PRADESH : *Root* : in fever; *Leaf-decoction* : in reviving sense of taste.
AGNI PURANA : plant yields a sterilising effect against all poisons.
AYURVEDA : *Root* : acrid, bitter, slightly pungent, oily, heating, anthelmintic, alterative, cures erysipelas, biliousness, snake bite and fish poison; *Extract* : emetic, alexiformic, cures diseases of blood, hydrophobia, lumbago, dysentery and rat bite, biliousness and colic; *Root-bark* : in piles.
UNANI : *Stem* : good in vomiting and diarrhoea; *Fruit* : sweet, laxative, expectorant, carminative, alexiteric, useful in inflammation, diseases of the blood, burning of the body, spermatorrhoea, acute fever and lumbago.

Modern use : *An alkaloid from bark* : said to lower blood pressure, purgative, anthelmintic, useful in fever, skin diseases, emetic and diaphorbia; *Bark* : exhibits antitubercular activities; *Leaf* : as poultice in rheumatic pain; *Fruit* : acidic and astringent, laxative, tonic and refrigerant, used in haemorrhages; *Seed* : used in haemorrhage.

Phytography : Small deciduous tree or rambling shrub, armed or not, wood olive-brown, hard and close grained; leaves simple, alternate, petioled, 7.5–15.0 cm by 2.5–5.0 cm, oblong or elliptic, base inequal, somewhat rounded, acute or subobtuse, above nearly glabrous with pubescent veins, lower surface with scattered hairs or tuft of hairs or hollow glands in the axils or primary veins; flowers in small close fascicles, peduncles, pedicels and calyx tube usually woolly , silky white, ovary inferior; berries black, 1.5 by 0.9 cm, crowned by somewhat enlarged calyx-limb, pubescent or glabrous; seeds albuminous.

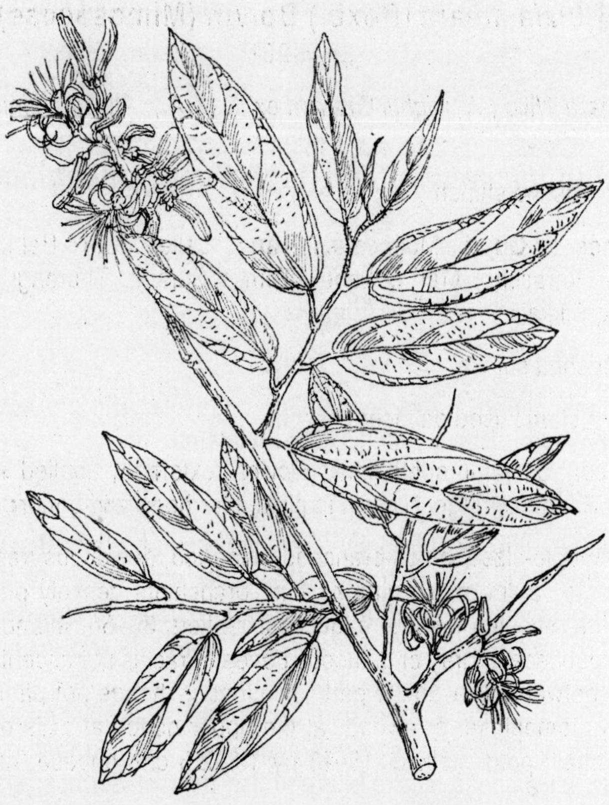

Phenology : *Flowering* : March–May; *Fruiting* : June–July.

Distribution : Drier parts of India, growing vigorously in the forests of South India.

Ecology and cultivation : Prefers dry climate; profusion in the vicinity of forests.

Chemical contents : *Plant* : venoterpine, di-salsoline, isocephaeline; *Root-bark* : emetine, cephaeline, psychotrine, marckidine, marckine, tubulosine and its Me-derivative, stigmasterol, β-sitosterol; *Stem-bark* : lamarchinine, cephaeline, de-Me-cephaeline, tubulosine, psychotrine and its de-Me-derivatives; *Leaf* : mild adrenolytic, pharmacological choline, ankorine, deoxytubulosine, alangi marckine, dehydroprotoemetine, steroids, stigmasta-5, 22, 25-trien-3 β-ol, myristic acid; *Seed* : emetine, cephaeline, N-Me-cephaeline, psychotrine, alangimide, alangimarine, alamarine, alangimaridine, deoxytubulosine, betulin, betulinaldehyde, lupeol, betulinic acid and its hydroxy lactone A, β-sitosterol.

Adulterant : Used as a substitute for Ipecac.

Albizia amara (Roxb.) Boivin (Mimosaceae)

(2n = 26)

Syn : *Acacia amara* Willd., *A. wightii* Graham ex W. & A., *Albizia affinis* Forn., *Mimosa amara* Roxb.

Sanskrit name : Krishna shirish.

Vernacular names : Guj : Moto sariso; **Hin :** Lallei; **Kan :** Bel khambi, Chigare; **Mal :** Oosulay, Varacchi; **Mar :** Lahei, **Tam :** Suranji, Thurangi, Usilai, Wunja; **Tel :** Nalla renga, Sikkai.

Trade name : Krishna shirish.

Traditional use : *Gum* : used on ulcers.

Modern use : *Leaf* : useful in ophthalmia; *Flower* : externally applied in inflammations, boils and ulcers; *Seed* : astringent, given in piles, diarrhoea and gonorrhoea.

Phytography : Middle-sized, much-branched, unarmed, deciduous tree with a crooked bole, height up to 3 m and a girth of 1 m; branchlets densely pubescent; leaves bipinnate, stipulate, stipules minute, caducous, leaflets 30–60, membranous, sessile, caducous, finely pubescent, gray or glaucous beneath, rachis pubescent, glands present on petiole and between the lower part of pinnae; heads not panicled, peduncles axillary, densely tomentose, cr owded at the upper nodes at axils of much reduced leaves; flowers small; pods stalked, 15–23 by 1.8–2.5 cm, opaque, gray-brown, 6- to 10-seeded.

Phenology : *Flowering* : March–May; *Fruiting* : Autumn.

Distribution : Drier regions of South India, occasionally in Central India, sometimes planted as roadside trees; also in Sri Lanka.

Ecology and cultivation : Prefers dry climate; dominant at foothills and up to an elevation of 1200 m.

Chemical contents : *EtOH extract of aerial parts* : spasmogenic; *Bark* : β-sitosterol; *Heartwood* : melacacidin and 3-Me-ether, melanoxetin and its 3-Me-ether; *Leaf* : flavonoid glycosides, 4-OMe-quercetin-3-rutinoside; *Seed* : sapogenin, echinocystic acid; *EtOH extract of seeds* : antibacterial.

Adulterant : *Leaf* : used as an adulterant of tea.

Remark : Tree is a source of gum.

Albizia chinensis (Osbeck) Merrill (Mimosaceae)

(2n = 26)

Syn : *Albizia stipulata* Boiv.

Vernacular names : Ben : Amluki, Chakua; **Hin :** Kanujera, Siran; **Kan :** Hotta bege, Kal bege; **Mal :** Potta vega; **Mar :** Laeli, Udala; **Tam :** Pili vegei, Katturinjil; **Tel :** Chinduga, Konda chiragu.

Trade name : Siran.

Modern use : *Plant* : fish poison; *Infusion of bark* : used as a lotion for cuts, scabies and skin diseases.

Phytography : Tall, unarmed deciduous tree with cylindrical bole of 10 m and girth of 2.6 m, crown broad-spreading and flat topped, branchlets downy, finely gray; leaves bipinnate, leaflets 40–80, small, sessile, membranous, sensitive, glaucous beneath, slightly recurved, acute, stipules persistent, membranous, downy, cordate, acute; heads in copious axillary simple or compound racemes with densely pubescent erect branches; pods subsessile, thin, 12–15 cm by 2.5 cm, pale brown, indehiscent, 8- to 10-seeded.

Phenology : *Flowering* : May–June; *Fruiting* : October, April.

Distribution : Sub-Himalayan tract up to 1300 m in Assam, West Bengal, South India, the Andaman and Nicobar Islands.

Ecology and cultivation : Prefers moist situation, grows in the damper forests near streams and ravines; extensively cultivated for shade in tea and coffee gardens; propagated by line and cuttings.

Chemical contents : *EtOH extract of aerial parts* : spasmogenic and diuretic; *Gum* : saponin.

Remark : Cultivated as a shade tree in coffee and tea plantations.

Albizia lebbek (L.) Willd. (Mimosaceae)

(2n = 26)

Syn : *Acacia lebbek* Willd., *Albizia latifolia* Bois.

English names : Fry wood tree, Silk flower tree, Siris tree, Walnut.

Sanskrit names : Kapitana, Sukhapushpa.

Vernacular names : Asm, Ben, Hin, Mar and **Pun** : Shirish; **Guj** : Harreri, Pilo sarshio; **Kan** : Bagemara, Shirish bage, Home bage; **Kon** : Xirxi; **Mal** : Vaga; **Mar** : Chichola, Motha siras; **Ori** : Siris, Tiria; **Tam** : Vaghei, Vagail; **Tel** : Dirasana.

Trade names : Kokko, Shirish.

Traditional use : NATIVES OF DIFFERENT PARTS IN INDIA : in the treatment of leprosy and syphilis; *Root-powder* : as anthelmintic; PEOPLE OF MAHARASHTRA : *Leaf* : as poultice for the treatment of rheumatic pain; TRIBES OF BIHAR : frequently use this drug for treatment of different kinds of skin affections; ; SANTAL AND MUNDA : for treatment of habitual constipation; PEOPLE OF BIHAR AND WEST BENGAL : for symptomatic treatment of snake bite.

AGNI PURANA : antidote of poison available in root, bark, leaf, flower and fruit; BRAHMA PURANA : twigs good as tooth brush; MATSYA PURANA : for proper health care, one should consume this plant fortnightly along with *Citrus medica, Ficus glomerata, Prosopis spicigera* and *ghee;* AYURVEDA : *Root* : in hemicrania; *Bark* : bitter, cooling, alexiteric, anthelmintic, cures diseases of blood, leucoderma, itching, piles, skin diseases, erysipelas, bronchitis, good in rat bite; *Leaf* : ophthalmia; *Flower* : in asthma, snake bite.

UNANI : *Root* : astringent, prescribed for ophthalmia; *Bark* : anthelmintic, relieves toothache, strengthens gums, used in leprosy, deafness, biles, scabies, syphilis, paralysis; *Leaf* : good in night blindness; *Flower* : aphrodisiac, emollient, maturant, smell useful in hemicrania; *Seed* : tonic to brain, used for gonorrhoea, tubercular gland; *Oil* : applied in leucoderma.

Modern use : *Root-bark* : in powdered form used to strengthen gum; *Bark* and *Seed* : astringent, given in piles and diarrhoea, tonic, restorative; *Leaf* : in night blindedness.

Phytography : Tall, erect, unarmed, deciduous tree, crown umbrella-shaped, bole usually 6.6–10.0 m, but in the Andamans, reaches up to 33 m, girth normally 2 m, while in the Andamans, it is up to 3 m; leaves bipinnate, leaflets 8–18, short-stalked, subcoriaceous, oblique, obtuse, rachis glabrous or downy, a large gland present near the base of petiole; heads round, many-flowered, short-peduncled, 3–4 together from crowded upper nodes; flowers small, fragrant, yellow or greenish yellow; pods strap- shaped, yellow-brown, firm, 15–30 cm by 2.0–3.7 cm; 6- to 10-seeded.

Phenology : *Flowering* : April–June; *Fruiting* : January.

Distribution : Throughout India, ascending to 2000 m, in the Andamans; Pakistan and Bangladesh.

Ecology and cultivation : It prefers moist situations, grows on a variety of soils. It can be grown by line sowings on well-loosened soil; transplantation of nursery plants and propagation by cuttings yields satisfactory results.

Chemical contents : *Stem* : gum; *Bark* : 7–11% condensed tannin, δ-catechin, δ-leucocyanidin, friedelan-3-ore, γ-sitosterol; *Heartwood*: malanoxetin, δ-pinitol, okanin, leucopelargonidin; *Wood* : saponin, lebbekanin E, *Leaf*: echinocystic acid, β-sitosterol flavonoids, vicenin II, *Flower* : 4 saponin glycerides, lebbekanins D, F, G & H; *Pod* : saponin, lebbekanin C, albigenic acid; *Seed* : saponins lebbekanin A & B, echinocystic acid, oleanolic acid.

Adulterant : Tree gum is adulterant of gum arabic.

Albizia odoratissima Benth. (Mimosaceae)

(2n = 26)

Syn : *Acacia odoratissima* Willd., *Mimosa odoratissima* L., *Albizia leblecoides* Benth.

English name : Fragrant albizia.

Vernacular names : Asm : Jati korvi; **Ben** : Kakur shirisha, Koroi tetura; **Guj** : Kalo sarasio; **Hin :** Kala shirish, Siran, Bhandir, Bersa, Bas, Bassein, Bansa; **Kan** : Bellawar, Bilvara, Godhunahi, Pullibaghi; **Mal** : Puli vega, Karu vegii; **Mar** : Borhi chinchua, Chikunda; **Pun** : Karambru, Lasrin, Polach; **Tam** : Kal thuringi, Karuvagei, Karraghe, Ponnai murankai, Sela vanjai, Solamanim; **Tel** : Chinduga, Shinduga chinduga, Telsu, Yerju, Chinta, Karu vage.

Trade names : Black shirish, Kala shirish.

Modern use : *Bark* : applied externally on leprosy and inveterate ulcers; *Leaf* : boiled with *ghee* and used as remedy for cough.

Phytography : Tall, erect, unarmed tree, bole 8–10 m; leaves large, dark green, drooping, bipinnate, leaflets 16–40, sessile, very oblique, subcoriaceous, glaucous beneath, strongly veined, midrib parallel with the upper edge at a short distance from it, obtuse, gland at the base of petiole; heads small, copiously panicled, few-flowered; flowers 0.4 cm in length, gray-silky; pods firm, 15–20 cm by 2.5–3.7 cm, opaque or glossy; 8–12 seeds.

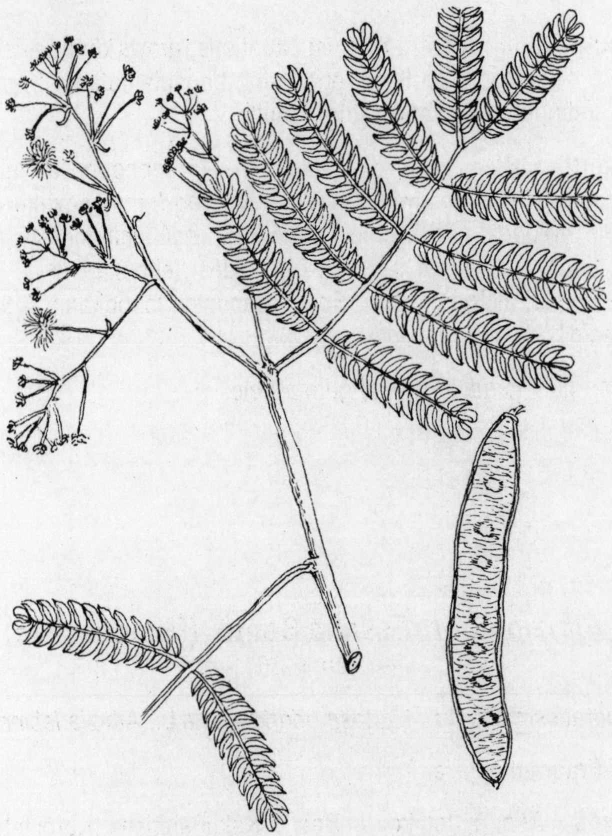

Phenology : *Flowering* : May–June; *Fruiting* : November–February.

Distribution : Throughout India, common on hill slopes, ascending up to 1650 m in the sub-Himalayan tract.

Chemical contents : *Stem* : dark brown insoluble gum in form of rounded tears; *Heartwood* : d-O-penta-Me-dihydro-melanoxetin; *Seed* : two trioterpenic acids, machacrinic acid, adoratissimia.

Remark : A host for lac insect.

Albizia procera (Roxb.) Benth. (Mimosaceae)

(2n = 26)

Syn : *Acacia procera* Willd., *Mimosa procera* Roxb.

English names : Stone sirissa, Chite siris.

Vernacular names : Ben : Koroi, Medilova; **Hin :** Safed shirish, Gurar, Karra, Karo, Karanji, Gurbai, Gurkur, Kalsis, Kardu, Garso; **Kan :** Bellate, Chikul; **Mal :** Karun thagara, Kothunaga; **Mar :** Kinhai; **Tam :** Konda vagei; **Tel :** Tella chinduga.

Trade name : Safed shirish.

Traditional use : *TRIBAL* : *Bark-paste* : on body pain and backache.
MATSYA PURANA : paste of the plant can remove all poison from human body.

Modern use : *Leaf* : insecticide, made into poultice for application to ulcers.

Phytography : Large deciduous tree, 20–22 m high, 2 m or more in girth; leaf bipinnate, rachis glabrescent, large gland at the base of petiole, leaflets 12–24, short-stalked, subcoriaceous, gray beneath, glabrous, obtuse; heads copiously panicled; flowers small, white, scented; pods brown, 10–20 cm by 1.25–2.5 cm, glabrous, 8- to 12-seeded.

Phenology : *Flowering* : August–September; *Fruiting* : December–May.

Distribution : Common in sub-Himalayan tract from Jamuna eastwards to Bengal, western peninsula, Andamans; occasionally planted as an ornamental/avenue tree; also found in Bangladesh.

Ecology and cultivation : Grows on alluvial soil along streams, prefers moist climate; cultivated in gardens and along roadside.

Chemical contents : *Root* : saponin; *Bark* : β-sitosterol, biochanin A, formonoetin, genistein, diadzein; *Heartwood* : pterocarpan and 4 isoflavones; *Leaf* : α-spinasterol, hentriacontane, hexacosanol; *Seed* : saponin and sapogenin.

Remark : A host for lac insects.

Aleurites moluccana (L.) Willd. (Euphorbiaceae)

(2n = 22)

Syn : *Jatropha moluccana* L.

English names : Candle nut tree, Belgium walnut tree, Varnish tree.

Sanskrit name : Akshota.

Vernacular names : Ben and **Hin :** Jangli akhrot; **Mar :** Akrod, Jaiphal; **Tam :** Nattakkarott, Kotakrote; **Tel :** Natakrotu.

Trade name : Jangli akhrot.

Traditional use : TRIBAL USE : *Root-bark paste* : on wounds; *Bark-decoction* : in unconsciousness; *Bark* and *Nut* : in asthma, ulcers, sores, swollen belly; *Bark* : antitumour, its decoction to treat unconsciousness; *Nuts* : laxative, stomachic; *Fruit-powder* : as stomachic; *Kernel* and *Oil* : laxative.

Modern use : *Oil from seed* : purgative, substitute for castor oil.

Phytography : Evergreen tree, growing up to 20 m in height; stem and young leaves stellately puberulous or tomentose; leaves alternate, long petioled, simple, polymorphous, crowded at the end of branches, 10–30 cm long; cymes tomentose; flowers white, 1.25 cm long; fruits ovoid, 5.0–6.5 cm in diameter, olive-green, fleshy, smooth, containing 1 or 2 large hard seeds.

Phenology : *Flowering* : December–February on a dense, rusty inflorescence; *Fruiting* : massive, February onwards.

Distribution : Native of Malaya; found wild in Wynaad and Assam.

Ecology and cultivation : Planted in higher elevations also.

Chemical contents : *Skin of the fruit* : essential oil; *Seed* : fixed oil; *Kernel* and *Oil* : toxalbumin.

Remarks : In Hawaii, flowers, nuts and bark are used in general debility, to treat asthma, sores, ulcers, swollen womb and as laxative.

Alhagi pseudalhagi (Bieb.) Desv. (Fabaceae)

Syn : *Alhagi camelorum* Fisch., *A. maurorum* Desv., *A. napaulensiam* DC., *Hedysarum pseudalhagi* Bieb.

English names : Camel thorn, Persian manna plant.

Sanskrit name : Durlava.

Vernacular names : Guj : Jawaso; **Hin :** Jawain, Turanjbin, Bharbharra; **Kan :** Billiduruva; **Pun :** Tamiya; **Tel :** Tellaginiya.

Trade names : Jawain, Turanjbin.

Traditional use : TRIBES OF RAJASTHAN : *Leaf* : asthma, fever; *Flower* : as a blood coagulant.
UNANI : Constituent of 'Jwansa', 'Turajabeen'.

Modern use : *Plant* : laxative, diuretic, expectorant, antiemetic; *Infusion* : diaphoretic, antibillicus; *Oil from leaf* : used in rheumatism; *Flower* : used in piles.

Phytography : Low shrub, armed with copious hard pungent prickles; leaves simple, drooping, stipules spiny, oblong, coriaceous, glabrous, obtuse; flowers axillary, pedicels 0.6–0.9 cm, reddish; pods linear.

Phenology : *Flowering* : March.

Distribution : Plains of Punjab, Rajasthan, Gujarat, Maharashtra, upper Ganga, ascending up to 1000 m on the Kishengunge; Pakistan (Sind, Baluchistan).

Ecology and cultivation : Prefers hot climate.

Chemical contents : *Plant :* seven 24-alkylsterols; *Root* and *Stem* : B-Ph-Et-amine, hordemine, salsolidine, tetrahydroisoquinoline, which show sympathomimetic activity.

Remarks : It is a fodder for camel, hence the name. Freshly cut plants are used in Gujarat for making tattis in the hot season.

Allamanda cathartica L. (Apocynaceae)

English names : Allamanda, Campanilla, Golden trumpet.

Vernacular names : **Ben :** Harkakra; **Kon :** Kanangam; **Mar :** Jaharisontakka; **Mal :** Kolaambi; **Mun :** Arba; **Tel :** Allemandathiga.

Traditional use : *Root* : is active *in vivo* in mice and *in vitro* against human carcinoma of nasopharynx; *Alcoholic* and *aqueous extract of root* : hypotensive; *Bark* : hydragogue in ascites; *Leaf* : possess valuable athartic properties; *Extract of leaf* (with sodium bicarbonate solution) shows high inhibition of ehrlich ascites tumour cells, antifungus.

Phytography : Woody climbing shrub; leaves simple, opposite or whorled; racemes lax; flowers large, showy, golden yellow; drupe subglobose, spiny, 4 cm across; seeds minutely winged, without a tuft of hairs.

Phenology : *Flowering* : mainly in Autumn.

Distribution : Native of Central America and Brazil; cultivated in India for showy flowers, found wild in Karnataka.

Ecology and cultivation : Plains to sea coast; planted.

Chemical contents : *Root* : allamandin, allamandicin, allamdin, antileukaemic iridoid lactone; *Root-bark* and *Stem* : plumericin, isoplumericin; *Stem* and *Leaf* : ursolic acid, β-amyrin, β-sitosterol, 2-triterpine lactones; *Flower* : kaempferol, quercetin, hesperitin.

Allium ampeloporasum L. var. *porrum* (L.) Gay. (Alliaceae)

(2n = 32)

Syn : *Allium porrum* L.

English name : Winter leek.

Vernacular names : Ben : Paru; **Hin :** Vilaiti lasan, Gandina, Kirath; **Mar :** Khorat.

Trade name : Vilaiti lasan.

Traditional use : RURAL FOLKS : *Bulb* : as stimulant and applied externally to promote suppuration of boils.

Modern use : *Bulb* : stimulant, expectorant, used to hasten suppuration of boils.

Phytography : Tall, hardy, biennial; bulb white, narrowly ovoid; leaf linear, flat-keeled, shorter than tall terete scape; head globose, densely many-flowered; flower pedicellate, white, campanulate.

Phenology : *Flowering* and *Fruiting* : Winter.

Distribution : Native of Europe and West Asia, cultivated in India as a garden plant.

Ecology and cultivation : Winter season crop; in India thrives well in high altitudes but not in moist localities.

Chemical contents : *Bulb* : 0.4% alliin, traces of alanine, arginine, aspartic acid, asparagine, histidine, leucine, methionine, phenyl-alanine, proline, serine, threonine, trytophan and valine.

Allium ascalonicum L. (Alliaceae)

(2n = 16)

Syn : *Allium sulvia* Ham.

English name : Shallot.

Vernacular names : Ben : Gundhun; **Hin :** Gandana.

Trade name : Gandana.

Traditional use : UNANI : Ingredient of 'Gandana'.

Modern use : *Alcoholic extract of bark* : anticoagulant, fibrinolytic, hypocholesteremic; *Bulb* : aphrodisiac, in earache.

Phytography : Foetid scapigerous herb, bulbs coated; leaves basal; scape stout fistular; head globose, densely many-flowered; flowers white, pedicels very short; clusters of greenish white or red cloves occur at the base of hollow cylindrical leaves.

Phenology : *Flowering* and *Fruiting* : Winter.

Distribution : Native of Palestine, extensively cultivated in India, mainly in the South.

Ecology and cultivation : Propagated by bulbs and seeds.

Chemical contents : *Aqueous extract of bulb* : fructose, glucosamine, lysine, uronic acid; *Brown skin of bulb* : quercetol and its glucosides; *Fresh leaf* : flavanoids.

Allium cepa L. (Alliaceae)

(2n = 16, 32)

Syn : *Allium cumaria* Herb.

English name : Onion.

Sanskrit names : Palandu, Sukandaka.

Vernacular names : Asm : Ponoru; **Ben :** Pianj, Pyanj; **Guj :** Dungari; **Hin :** Piyanj; **Kan :** Nirulli, Irulli, Ullagaddi; **Kon :** Piao, Kandeapat; **Mal :** Chuvannaulli, Irulli, Ulli; **Mar :** Kanda; **Ori :** Piaja; **Pun :** Piyaz, Ganda; **Tam :** Vengayam; **Tel :** Nirulli, Ullagaddi, Ulligaddalu.

Trade names : Onion, Piyanj.

Traditional use : *SANTAL* : *bulb-paste* with that of black pepper in malarial fever; *BHOXA* (a tribe of Garhwal) : *Fresh onion-paste* : on blister, boils, eye complaints; *TRIBES OF SAGAR DISTRICT* (Madhya Pradesh) : *Bulb* as diuretic, expectorant, rubifacient, skin diseases, sedative, in insect bites, wounds; *Bulb-juice* : applied in bleeding from nose, ear complaints, in vomiting and on piles; *RURAL FOLK OF INDIA* : *fresh onion* in faintness, convulsions, epileptic fits, applied on insect sting and circular alopecia; *Seed-paste* (with that of root of *Punica granatum*, *Cajanus cajan*, *Piper rubrum* and honey) is given for abortion.

UPAVARHANA SAMHITA : *Stem* delicious, energiser, stimulant; *CHARAKA SAMHITA* and *SUSHRUTA SAMHITA* : destroyer of wind, stimulant, helps in digestion but increases bile and phlegm; *AYURVEDA* : *Tuber* is aphrodisiac, pungent, taste-improver, tonic useful in biliousness, bleeding piles, body pain, tumours, vomiting; *Seed* : fattening, useful in caries of teeth and urinary discharges.

UNANI : *Bulb* is appetiser, pungent, stomachic, tonic, used in asthma, spleen disorders, vomiting, scabies, piles and night blindness.

HOMOEOPATHY : For different ailments of eye, conjuctivitis and for relaxation of colic pains.

Modern use : *Bulb* : antimalarial, antirheumatic, antidiabetic, antibacterial, antiatherogenic, antihypertensive, diuretic, expectorant, stimulant, useful in flatulence and dysentery, hypocholesteremic, hypolipaemic, increases clotting time and fibrinolytic activity in man; *Onion without peel* : remedy for circular alopecia; *Onion-extract* : reduces blood sugar.

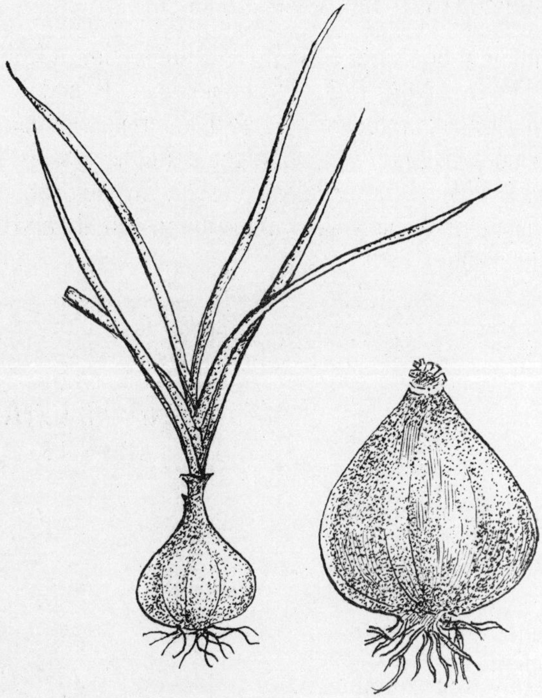

Phytography : Foetid scapigerous biennial bulbous herb; lower part of the plant swollen at maturity to form a large bulb, bulbs vary in size, colour (white, yellow, red, brown), shape (round, flat, conical), firmness, keeping quality, period of maturity and strength of flavour; leaves subdistichous, green, linear, fleshy, hollow cylindrical; scape inflated; umbels bear flowers and bulbils; flowers pedicellate, white, trimerous; capsules 3-chambered containing small black seeds.

Phenology : *Flowering* : December–January.

Distribution : Extensively cultivated in India, Bangladesh and Pakistan; originated probably in West Asia; cultivated in New and Old World.

Ecology and cultivation : It can be grown from seeds, bulbs or bulbils—bulbs are preferred. A cool season crop, requires high temperature and long photoperiod, dry climate, grows best in a light manured soil, sandy loam with pH 5.8–6.5 is the most suitable.

Thorough ploughing before planting and fortnightly irrigation required. Finely powdered fish manure, sulphate of ammonia potash, nitrate of soda and superphosphates are among the fertilisers. The crop is harvested in about 90 days. Extensively cultivated in India as a field crop and a garden crop. Being an irrigated crop of short duration, the onion can be grown as a cash crop in the midst of crops like sugarcane and turmeric and can be rotated with rice, coriander and garlic.

Chemical contents : *Onion* : 1.2% proteins, 11.6% carbohydrates, lesser than 1% fat, 0.18% Ca, 0.5% P.O., Fe, vitamins A, B and C; *Bulb* : caffeic acid, ferulic acid, polyphenols, protocatechuic acid, kaempferol, quercetin and its derivatives; *Onion skin* : kaempferol; *Dried skin* : quercetin, sterols, cholesterol, β-sitosterol, stigmasterol ; *Bulb* and *Fresh herb* : on steam distillation yield an essential oil, called onion oil, chief constituent of onion oil—various mono-, di-, tri- and tetra-sulphides, thiols and thiophene derivatives.

Allium sativum L. (Alliaceae)

(2n = 16)

English name : Garlic.

Sanskrit names : Arishta, Lashuna, Mahausada, Rason.

Vernacular names : **Asm :** Naharu; **Ben :** Rasun; **Guj :** Lasan, Shunam; **Hin :** Lashun; **Kan :** Belluli; **Kon :** Lossun; **Mal :** Velluli, Velluthulli; **Mar :** Lasun; **Ori :** Rasun; **Tam :** Vellaipundu; **Tel :** Tellagadda.

Trade names : Rasun, Lashun.

Traditional use : *UPAVARHANA SAMHITA* : good for health; *AGNI PURANA* : a mixture of this plant, ginger and *Moringa oleifera* is helpful for eye and also as eardrop.

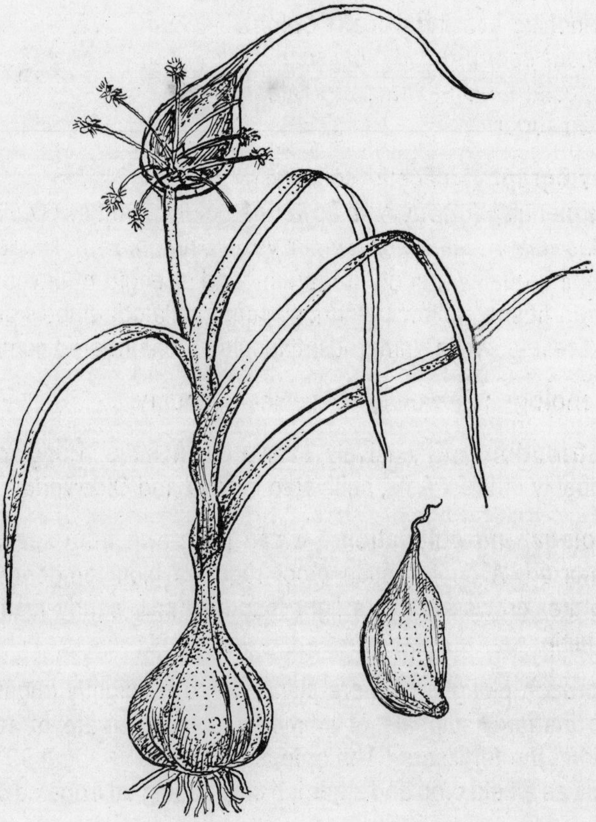

MATSYA PURANA : of great medicinal value, should be stored in forts.
UNANI : an ingredient of 'Seer (Lahsan)'.
HOMOEOPATHY : Mother tincture, a good remedy for cough and cold, pain in the chest, swelling of glands, *psoas*, *iliacus* and erosion of old diseases at the adimanic stage.

Modern use : *Bulb* : antidiabetic, antiinflammatory, anticancer, effective in rheumatism, catarrhal conditions; *Raw garlic* : decreases glucose, cholesterol, phospholipids, triglycerides, useful in dyspepsia, cryptococcal meningitis, rickettsia; applied externally as resolvent to indolent tumours, internally given with common salt in nervous diseases, headache; *Liniment* : beneficial in infantile convulsion and other spasmodic affections; *Juice* : applied to bruises and sprain, relieves earache, allays pain in otorrhoea; preparation have been given to pulmonary phthisis, bronchitis, gangrene of the lung and whooping cough; *Garlic juice* : good for treatment of laryngeal tuberculosis, lupus and duodenal ulcer; *Inhalation of fresh garlic juice* : useful in pulmonary tuberculosis, dyspepsia, flatulence and colic; in external application, the *juice* is used as a rubifacient; in skin diseases, as an eyedrop and in earache.

Phytography : Hardy perennial bulbous scapigerous herb; stem flat, dry, lower portion of the plant forms bulb which consists of several smaller bulbs called *cloves*, surrounded by a thin white or pinkish sheath; leaves flat, narrow green; heads bear small white flowers and bulbils.

Phenology : *Flowering* and *Fruiting* : Winter.

Distribution : Cultivated throughout India since ancient days; also cultivated in Pakistan, Bangladesh and almost in all tropical countries.

Ecology and cultivation : It favours a richer soil and higher elevation (1000–1300 m), well-drained, moderately clayey loam is best suited for cultivation; requires a cool, moist period during growth and a relatively dry period during maturing of the crop; takes 4–5 months to mature, harvested during February to April.

Yields good results, if treated with farmyard manure and top-dressing of ammonium sulphate mixed with superphosphate.

It grows as a late irrigated crop; in South India, it is rotated with ragi.

Chemical contents : *Garlic* : protein 6.3%, fat 0.1%, carbohydrate 29%, Ca 0.03%, Fe 1.3 mg/100 g, vitamin C 13 mg/100 g, and also Cu-peptides, 2 mercapto-L-cysteins, anthocyanins, glycosides of kaempferol and quercetin, polysaccharides, allinase, sterols, hydrocarbons, alliin (nonvolatile sulpher amino acid), sativin I & II, scordinines A & B; *Essential oil* : the bulbs, on distillation, yield 0.06–0.1% of an essential oil containing allyl-propyl-disulphide, diallyl disulphide and two sulphur containing compounds.

Allium schoenoprasum L. (Alliaceae)

Syn : *Allium sibricum* L.

English names : Chives, Cives.

Vernacular name : Hin : Piazi.

Trade name : Piazi.

Traditional use : Like *Allium cepa*.

Modern use : *EtOH* (50%) *extract of bulb* : used as cardiac depressor.

Phytography : Hardy perennial growing in dense clumps; bulbs narrow, at the base of leaves, scales membranous; leaves grass-like, light green or purple tinted, 10–25 cm long, terete or grooved above, smooth or scaberulous, sheaths elongate; scapes 15–35 cm long; heads subglobose, densely-flowered; flowers pink or pale purple, campanulate, pedicellate; capsules small, globose; chambers 2-seeded.

Phenology : *Flowering* and *Fruiting* : Winter.

Distribution : Kashmir to Kumaon between 2650 and 3650 m; grown as garden crop.

Ecology and cultivation : Like *Allium cepa*.

Chemical contents : Volatile oil contains several compounds.

Allium tuberosum Rottl. ex Spring. (Alliaceae)
(2n = 32)

English names : Chinese chive, Fragrant onion, Indian leek.

Vernacular names : Hin : Bangagandina; **Mar :** Bungaghunduna.

Traditional use : TRIBES OF NORTH-EAST INDIA : *Seed* : in spermatorrhoea.

Phytography : Bulb 1–3 together, attached to a horizontal rhizome, brownish; scape 30–45 cm tall, leafy at the base; leaves 2–3, flat, 6 mm wide, somewhat shorter than the scape; spathe one-bracted, shorter than the many-fid umbel, pedicels equal, two to three times the perianth, white or pink stellate flowers.

Phenology : *Flowering* and *Fruiting* : Cold season.

Distribution : Western Himalaya, Meghalaya as ferals; cultivated in West Bengal.

Chemical contents : *Leaf* : volatile oil, flavonoids, glycerides.

Remark : Khasis and Jaintias eat the bulb as a vegetable.

Allophylus serratus (Roxb.) Kurz (Sapindaceae)

Syn : *Allophylus cobbe* Blume (in part).

Sanskrit name : Triputah.

Vernacular names : Ben : Rakhalphul; **Kan :** Togaratti, Sidisali, Eruala; **Mal :** Mukhanumperu; **Mar :** Tipani; **Ori :** Khondokali; **Tam :** Amalai; **Tel :** Eravelu.

Trade name : Rakhalphul.

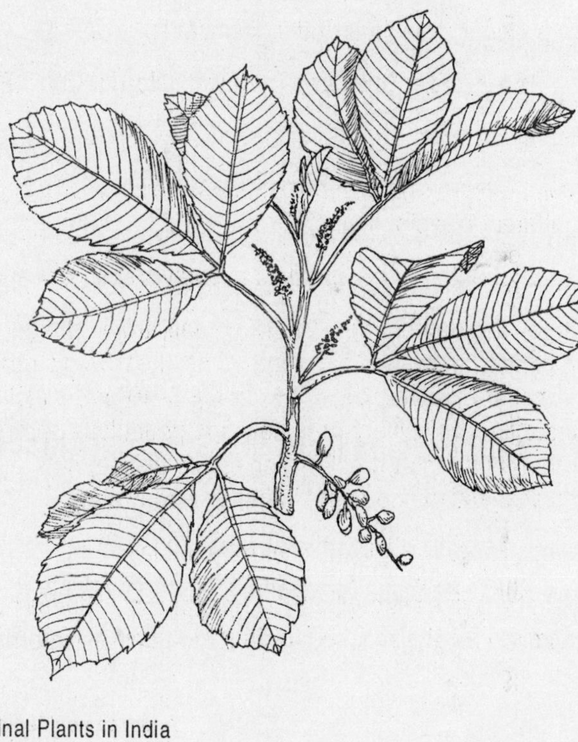

Traditional use : *Bark* and *Leaf* : for elphantiasis; *Crushed* and *pulverised leaf* : inhaled as a sternulatory.

Modern use : *Root* : anti-inflammatory, astringent, bitter, carminative and constipating, digestive, used to check diarrhoea, sweet, vulnerary, useful in bone fractures, dislocations, inflammations, ulcers, wounds, dyspepsia and anorexia; *Fruit* : sweet, cooling and nourishing tonic.

Phytography : A shrub or a small tree with whitish bark; leaves alternate, trifoliate, crowded at the end of the branchlets; pubescent; flowers small, white, in fascicles; fruits smooth, globose, red when ripe.

Phenology : *Flowering* : June–July; *Fruiting* : September.

Distribution : Throughout India, especially in Assam and the peninsular India; Bangladesh (Sylhet).

Ecology and cultivation : Generally on sandy soil.

Alnus nepalensis D. Don (Betulaceae)

(2n = 56)

Syn : *Betula leptostachya* Wall., *Clethropsis nepalensis* Spach.

English names : Indian alder, Nepalese alder, Himalayan alder.

Vernacular names : Hin : Udis, Utis; **Lep :** Kowal; **Nep :** Udis; **Pun :** Kohi.

Trade names : Indian alder, Udis.

Traditional use : *HOMOEOPATHY* : good for digestive system, glands, imperfect secretion of gastric juice and different types of injury.

Modern use : *EtOH* (50%) *extract of stem-bark* : hypoglycaemic and spasmogenic.

Phytography : Large deciduous tree, 22–33 m tall, trunk straight, cylindrical with 2.0–2.2 m girth; branchlets glabrous, sapwood white, heartwood reddish, turns light brownish gray on exposure; leaves 7.5 –17.5 cm by 5–10 cm, subglaucous beneath; flowers unisexual, male flowers solitary or in panicled pendulous spikes, bracteate and bracteolate; female flowers in short erect or pendulous spikes; nuts 1.25–2.5 cm long, nuts with persistent bracts and membranous wings in woody cones.

Phenology : *Flowering* : February–April; *Fruiting* : Apri–May.

Distribution : Moist shady ravines throughout the Himalaya at altitudes ranging from 1000 to 3000 m, from Kashmir to Meghalaya; southwards to Kachin Hills in Myanmar.

Ecology and cultivation : Prefers moist, shady ravines in high altitudes.

Chemical contents : *Bark* : 7% tannin.

Alnus nitida Endl. (Betulaceae)

Syn : *Clethropsis nitida* Spach.

English name : West Himalayan alder.

Vernacular names : Hin : Kunis, Utis; **Kas :** Chamb; **Pun :** Sharol.

Trade name : Utis.

Modern use : *EtOH* (50%) *extract of bark* : has antiviral properties.

Phytography : Large (27–33 m high, 2–4 m girth) deciduous tree, branchlets pubescent, wood reddish brown to light brownish gray or roseate gray; leaves alternate, petiolate, elliptic, acuminate, 10–20 cm long and at times 8 cm wide; flowers unisexual, male flowers in terminal erect spikes, female spikes shorter; nuts ovoid, 0.8–3.7 cm long, margin thick, narrow.

Phenology : Like *Alnus nepalensis.*

Distribution : Gregariously along banks of rivers and moist ravines from Kashmir to Kunawar, in the Himalaya between 1000 and 3000 m, sometimes descending into the plains.

Chemical contents : *Bark* : 3% tannin.

Alocasia indica (Roxb.) Schott. (Araceae)

(2n = 28)

Syn : *Arum indicum* Roxb., *Colocasia indica* Kunth.

English name : Giant taro.

Sanskrit name : Manaka.

Vernacular names : Asm and **Ben :** Mankachu; **Guj :** Alavu; **Hin :** Mankanda; **Kan :** Manaka; **Kon :** Allu, Auum; **Mar :** Alu; **Ori :** Manasaru; **Pun :** Arvi; **Tam :** Merukantalangu; **Tel :** Charakanda.

Trade names : Mankachu, Manakanda.

Traditional use : This plant has been mentioned in UPAVARHANA SAMHITA; CHARAKA SAMHITA : useful to treat digestive disorders; SUSHRUTA SAMHITA : medicinal plant; AYURVEDA : digestive, diuretic, good for piles; *Extract of petiole* : removes puss from ear; *Ash of rhizome* : cures buccal ulcer.

Modern use : *Flour* obtained is a light nutritious food, suitable for invalid; *Leaf-juice* : astringent; *Rhizome* : said to act as a mild laxative and diuretic, considered useful in anasarca and piles.

Phytography : Tall (up to 2.7 m) aroid with succulent (10–20 cm) underground rootstock; leaves alternate, petiolate, large (up to 1 m), deeply sagittately cordate, tip deflexed, midrib thick; spadix large (20–54 cm), peduncle short, spathe pale yellow-green.

Phenology : *Flowering* : April–May.

Distribution : Native of India, cultivated in Assam and West Bengal as a food crop; grown in Maharashtra and South India.

Ecology and cultivation : Moist soil, especially in rainy season, occasionally cultivated near residential houses; some are ornamentals.

Chemicai contents : *Rootstock* : yields a pure white starch.

Alocasia macrorrhiza Schott. (Araceae)

Syn : *Alocasia odora* C. Koch., *Arum macrorrhizum* L.

English name : Giant taro.

Sanskrit name : Hastikarni.

Vernacular name : Asm : Baro mankaahu.

Trade name : Boro mankachu.

Traditional use : TRIBAL : *Root-paste* : in skin irritation; *Leaf-paste* : in tumour.

Modern use : *Stem-juice* : to treat stings of scorpion and giant nettle; *Root* : skin irritant; *Extract of tuber* : diuretic; *Leaf* : rubifacient, used against tumour; *Petiole* : toothache, cough; *Seed-extract* : antibacterial, antifungal.

Phytography : Large (2.0–5.3 m in height and 0.3 m in diameter) aroid; leaves large (15–45 cm by 5–10 cm), broadly ovate-sagittate, veins very stout, petioles 0.6–2.1 cm, stout; spathe 15–25 cm, hooded and cuspidate at the top; berries large like cherry.

Phenology : *Flowering* : Winter; *Fruiting* : Summer.

Distribution : Native of Sri Lanka; wild and cultivated in tropical and subtropical India.

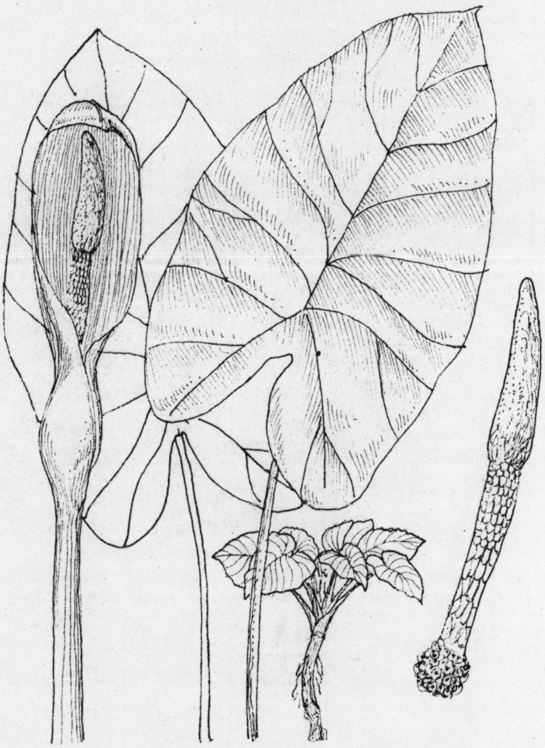

Ecology and cultivation : Frequently met with as feral near ponds, ditches and heaps of the rubbish near moist places.

Chemical contents : *Plant* : a flavonoid and saponin; *All parts of plant except tubers* : a cyanogenic principle, a mixture of triglochinin and isotriglochinin, β-glucosidases; *Rhizome* : fructose, glucose, a phytosterol-like compound, Ca-oxalate; *Leaf, Stalk, Tuber* and *Root* : high content of soluble oxalates; *Petiole* : HCN (up to 0.018%).

Aloe barbadensis Mill. (Liliaceae)
(2n = 14)

Syn : *Aloe indica* Royle, *A. littoralis* Koening., *A. vera* Tourn. ex Linn.

English names : Barbados aloe, Curacas aloe, Indian aloe, Jafarabad aloe.

Sanskrit name : Ghritakumari.

Vernacular names : Asm : Chalkunwari; **Ben :** Ghritakumari; **Guj :** Kumarpathu, Kunvar; **Hin :** Ghee kunvar; **Kan :** Lolesara; **Kon :** Kantikkor, Katkunvor; **Mal :** Kattarvazha kumari; **Mar :** Korphad; **Ori :** Gheokunri; **Pun :** Ghikur, Kawargandal; **Tam :** Alagai, Chirukuttali, Kuttilai; **Tel :** Chinnakata banda, Kala banda, Kittanara.

Trade names : Ghritakumari, Ghee kunvar.

Traditional use : TRIBAL : *Leaf-pulp* : in liver troubles, jaundice, fever, gonorrhoea, spleen disorder, rheumatism, piles, dysmenorrhoea, sterility in women; *Leaf-mucilage* : mild laxative, to cure hardening of breast tissues, in insect stings.

AYURVEDA : alternative, bitter, cooling, purgative, sweet, tonic, anthelmintic, useful in eye diseases, tumours, enlargement of spleen, liver troubles, vomiting, skin diseases, biliousness, asthma, leprosy, jaundice, strangury, ulcer; *Flowers* : anthelmintic.
UNANI : Gheekawar is useful in inflammation of spleen, lumbago, muscular pain, ophthalmia, digestive, purgative; *Leaves* good for piles and biliousness.

Modern use : *Aloe* : in menstrual diseases, stomach pain, tonic after pregnancy, uterine disorders, high fever; *Pulp* : menstrual suppressions, nervous imbalance; *Aloe compound* : in treatment of women sterility; *Mucilage* : painful inflammation; *Root* : colic pain; *Aloe mixture with other plant extracts* : for treating obstruction of lymphatic system..

Phytography : A coarse-looking plant with a short (30–60 cm high) stem; leaves succulent, green, large (37 cm long, 10 cm broad, 2 cm thick), densely crowded; flowers in racemes, bright yellow, tubular, stamens frequently projected beyond the perianth tube.

Phenology : *Flowering* : September–December; *Fruiting* : scarce.

Distribution : A native of North Africa, Canary Islands and Spain; naturalised in India; many varieties are found in a semi-wild state in all parts of India; also cultivated in pots and gardens.

Ecology and cultivation : Xerophyte; propagated by suckers.

Chemical contents : *Plant* : aloin, aloe-emodin and resins.

Adulterant : *Aloe candelabrum* Berger is used as substitute for *Aloe barbadensis* Miller.

Alpinia calcarata Rosc. (Zingiberaceae)

(2n = 48)

Syn : *Alpinia bracteata* Rosc., *Renealmia calcarata* Haw.

Vernacular names : Mal : Kattchenu; **Ori :** Toroni; **Tam :** Amkolinji.

Traditional use : *TRIBAL* : *Rhizome-decoction* : against inflammation.

Modern use : *Herb* : accredited with antitubercular properties, good in rheumatism, stomach disorders, bronchial catarrh, tonic and stimulant; *Seed* : used for colic, diarrhoea, vomiting and as a sternutatory; *Rhizome-decoction* : gives anti-inflammaion action on human body.

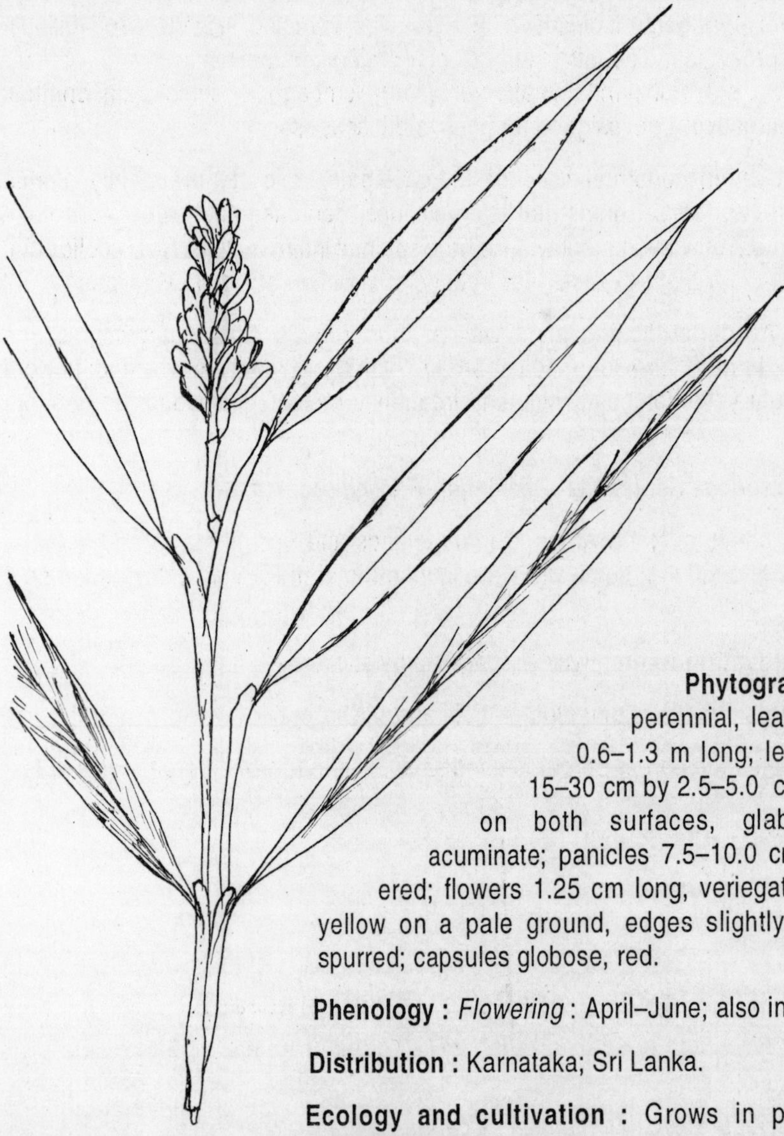

Phytography : Rootstock perennial, leafy stem slender, 0.6–1.3 m long; leaves lanceolate, 15–30 cm by 2.5–5.0 cm, green, glossy on both surfaces, glabrous beneath, acuminate; panicles 7.5–10.0 cm, densely flowered; flowers 1.25 cm long, veriegated with red and yellow on a pale ground, edges slightly incurved, base spurred; capsules globose, red.

Phenology : *Flowering* : April–June; also in Winter.

Distribution : Karnataka; Sri Lanka.

Ecology and cultivation : Grows in plains and hills; cultivated.

Chemical contents : *Rhizome* : yields an essential oil.

Adulterant : Substitute for Ginger.

Alpinia galanga (L.) Willd. (Zingiberaceae)

(2n = 48)

Syn : *Alpinia viridiflora* Griff., *Maranta galanga* L., *Amomum galanga* (L.) Lour.

English name : Greater galangal.

Sanskrit name : Kulanjan.

Vernacular names : Ben and **Hin :** Kulanjan; **Kan :** Dumbarasme; **Mar :** Kosht kulinjan; **Tam :** Pera rattai; **Tel :** Pedda dumparashtram.

Trade name : Kulanjan.

Traditional use : *Rhizome* : used in rheumatism.
UNANI : rhizome used against impotency and kidney troubles, bronchitis and dyspepsia.

Modern use : *Rhizome* : yields an essential oil which possesses well-marked insecticidal properties; also bitter, acrid, thermogenic aromatic, revulsive, carminative, disinfectant, aphrodisiac, expectorant, bronchodialator, febrifuge, diuretic; used in medicines for rheumatism, stomach disorders, bronchial catarrh, as tonic and stimulant.

Phytography : Plant 2.0–2.1 m high, bears perennial tuberous, slightly aromatic, deep orange-brown, pungent, bitter rootstock; leaves 30–60 cm by 10-15 cm, green, glossy on both sides; panicles copiously branched, densely flowered, 15–30 cm long, rachis densely pubescent, branches short; flowers 0.8–1.25 cm, greenish white; fruits orange red, 1.25 cm long, cherry-sized.

Phenology : *Flowering* : April–May.

Distribution : Native to Indonesia and Malaysia, India, mainly in the eastern Himalaya and south-west India.

Ecology and cultivation : Cultivated in North India for its rhizome which is used as spice for flavouring food, curries and medicines.

Chemical contents : *Rootstock* : α-pinene, β-pinene, limonene, cineol, terpinen, α-terpineol and some other monoterps; *Seeds* : antiulcer agents, 1'-OAc-chavivol-OAc and 1'-OAc-eugenol-OAc, caryophyllene oxide, caryophyllenol I & II, pentadecane, 7- heptadecane, fatty acid Me esters.

Adulterant : Rhizome of this plant is a substitute for ginger.

Alpinia malaccensis Rosc. (Zingiberaceae)

(2n = 48)

Syn : *Costus malaccensis* Koenig., *Maranta malaccensis* Burm.

Vernacular name : Tam : Saliyeridumpa.

Traditional use : *Rhizome* : for sores; *Fruit* : emetic.

Modern use : *Plant* : enters into composition of Malayan *ipoha*; *Rhizome* : used for the treatment of sores; *Fruits* : with a little salt, used as emetic.

Phytography : Tall (up to 2 m) plant, rootstock perennial; leaves 66 cm–1 m, oblong-lanceolate, pubescent beneath; racemes erect, 15–30 cm long, rachis very stout, densely pubescent, pedicels short; flowers 2.5 cm, oblong, white, variegated with red and yellow; capsules globose, yellow, 2.5 cm in diameter.

Phenology : *Flowering* : April–May.

Distribution : Eastern Himalaya, ascending up to 1500 m, Assam, Meghalaya, Eastern and Western Ghats.

Ecology and cultivation : Usually in damp forests.

Chemical content : *Rhizome* and *Leaf* : volatile oil which contains 74–80% methyl cinnamate.

Alpinia officinarum Hance (Zingiberaceae)

English names : Chinese ginger, China root, East Indian root, Galangal.

Vernacular names : Hin : Kulanjan; **Mar :** Sugandha bacha.

Trade name : Sugandha bacha.

Modern use : *Rhizome* : stomachic, stimulant and carminative; also used in flatulence, dyspepsia.

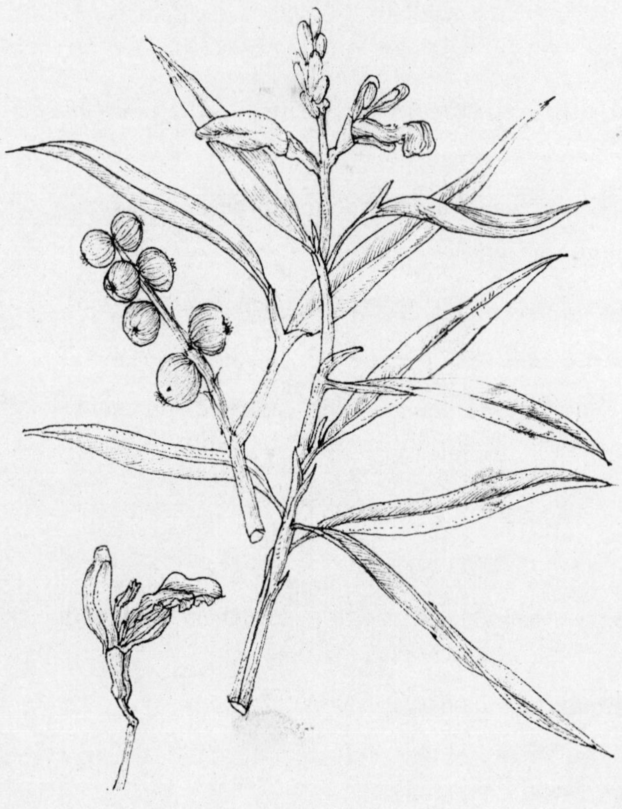

Phytography : A reed-like plant attaining about a metre in height, consists of branched rhizome, frequently cylindrical, sometimes tapering, enlarged or branched, dull, reddish-brown, longitudinally striated or shrivelled.

Phenology : *Flowering* : Cold season.

Distribution : Native to China and cultivated in North India.

Ecology and cultivation : The rhizome is dug up in Autumn, washed, trimmed, cut into pieces and dried; preserved for sowing in future.

Chemical contents : Essential oil, galangin, cineol, methyl cinnamate and a pungent oily body galangal.

Remark : The rhizome is exported in bales.

Alpinia speciosa K. Schum. (Zingiberaceae)

(2n = 48)

Syn : *Alpinia nutans* Rosc., *Zerumbet speciosum* Jacq.

English name : Light galangal.

Vernacular names : Ben : Punagchampa; **Tam :** Sittarattai.

Modern use : Same as *A. galanga.*

Phytography : Plant grows up to 3.3 m, rootstock perennial; leaves 30–60 cm by 7–15 cm, oblong-lanceolate, finely pubescent beneath, panicles 15–30 cm, rachis very hairy; flowers bracteate, bracteolate, 2.5 cm, white tipped with pink; capsules red, globose.

Phenology : *Flowering* : Rainy season.

Distribution : Native to East India; also cultivated in gardens for its sweet-scented flowers.

Ecology and cultivation : Plains and hills in cultivation.

Chemical contents : *Rootstock* : ethereal oil; *Leaf* : 0.05% ethereal oil, 30% δ-camphor, 17% δ-camphene, cineol, cinnamic ester.

Alstonia scholaris R. Br. (Apocynaceae)

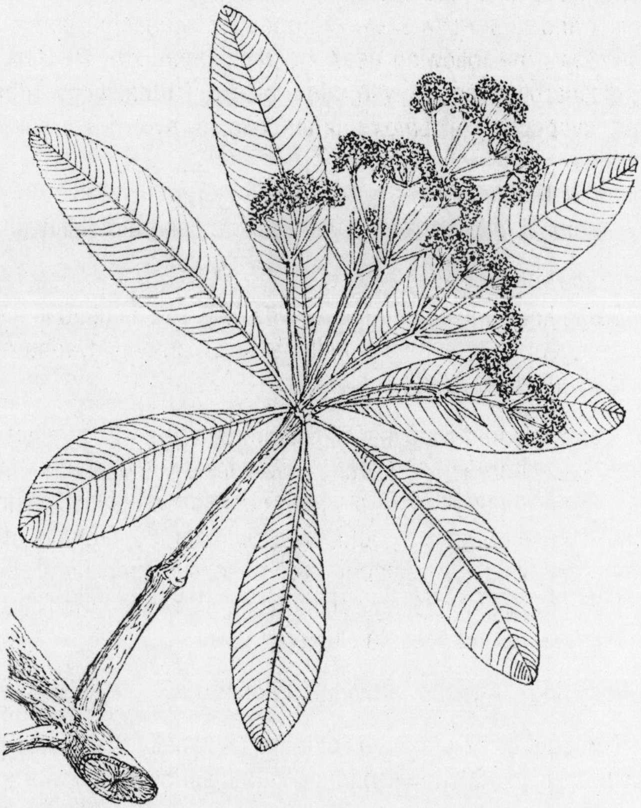

Syn : *Echites scholaris* L., *E. pala* Ham.

English names : Devil's tree, Dita bark.

Sanskrit names : Saptaparni, Saptaparna, Sarada, Vishalalvaka, Vishamachhda, Ayugmaparna, Gandhiparna, Payasya, Jivani, Kshalrya, Madagandha, Grahashi, Grahanashana.

Vernacular names : Asm : Chatiar; **Ben :** Chhatim; **Hin :** Chatian, Satni chatian; **Kan :** Saptaparna, Maddale, Kodale, Elele kale, Janthalla, Hale; **Kon :** Santnarukh; **Mal :** Ezhilampala, Mukkampala, Pala; **Mar :** Satvin; **Ori :** Chhatiana, Chhanchania, Silgandha; **Pun :** Satona; **Sin :** Rukattana; **Tam :** Elilaipillai, Mukumpalei, Pala, Wedrase; **Tel :** Eda kula, Pala garuda.

Trade names : Chatiyan, Shaitan wood, Saptaparni.

Traditional use : MUNDAS OF CHOTANAGPUR : *Bark* : in colic pain; SOME PARTS OF INDIA : *Plant* : used in the treatment of leprosy; *Twig* : hung in the room of the newly confined woman to lessen the activities of evil spirit on the new born.
ATHARVA VEDA : preventive and curative of diseases caused by change of season. CHARAKA SAMHITA and SUSHRUTA SAMHITA : good for headache, sores, and some other diseases; AYURVEDA : the following uses are recommended : (i) *Bark* : dermal sores, ragging fever, discharge of sperm with urine, hiccup, insufficiency in breast milk, gout, cold congestion, dyspepsia; (ii) *Latex* : caries, pimple, pyorrhoea; (iii) *Flower* : asthma, respiratory troubles.
UNANI : Ingredient of 'Kashim'.
HOMOEOPATHY : Malarial fever, anaemia, indigestion, general debility and other stomach ailments.

Modern use : *Bark* : known in commerce as *Dita bark* and is used in medicine as bitter, febrifuge and astringent, in treatment of malarial fever, chronic dysentery, diarrhoea and in snake bite; *Milky juice* : applied to ulcers.

Phytography : Large (20 m high, 3 m girth), evergreen tree with straight, often fluted and buttressed base, branches whorled, bark yellow inside and exudes milky bitter latex; leaves simple, whorled—usually 7 in a whorl, coriaceous, whitish beneath, obovate or elliptic or oblong, obtuse rounded or obtusely acuminate, 30–60 pairs of horizontal veins joining an intramarginal one; cymes peduncled or sessile, umbellately branched; flowers aromatic, 0.8–1.25 cm in diameter, greenish white, pubescent; follicles 30–60 cm long and 0.3 cm in diameter, pendulous, in clusters.

Phenology : *Flowering* : Autumn; *Fruiting* : Winter.

Distribution : Throughout moist regions of India, especially in West Coast forests, in the Himalaya it ascends up to 1000 m; also found in Bangladesh and Pakistan. Planted in the gardens.

Ecology and cultivation : Also grown as an ornamental.

Chemical contents : *Root* and *Root-bark* : echitamine chloride, α-amyrin, lupeol-OAc, stigmasterol, β-sitosterol, campesterol, alkamicine—its Nb-oxide and Nb-methiodide, ψ-akummicine, Nb-di-Me-echitamine, tubotaiwine; *Stem-bark* : hydrochloride of echitamine, echitamidine, a glyceride of venotarpine, sterols, two isomeric lactones; *Latex* : caoutchouc and resins; *Leaf* : picrinine, nareline, akuammidine, picralinal, akuammigine, betulin, ursolic acid, β-sitosterol, flavonoids, phenolic acids, scholarine; *Flower* : picrinine, strictamine, tetrahydroalstonine, n-hexacosane, lupeol, β-amyrin, palmitic acid, ursolic acid.

Adulterants : *Alstonia macrophylla* Wall. and *A. venenata* R. Br. are used as substitute for *A. scholaris.*

Alstonia venenata R. Br. (Apocynaceae)

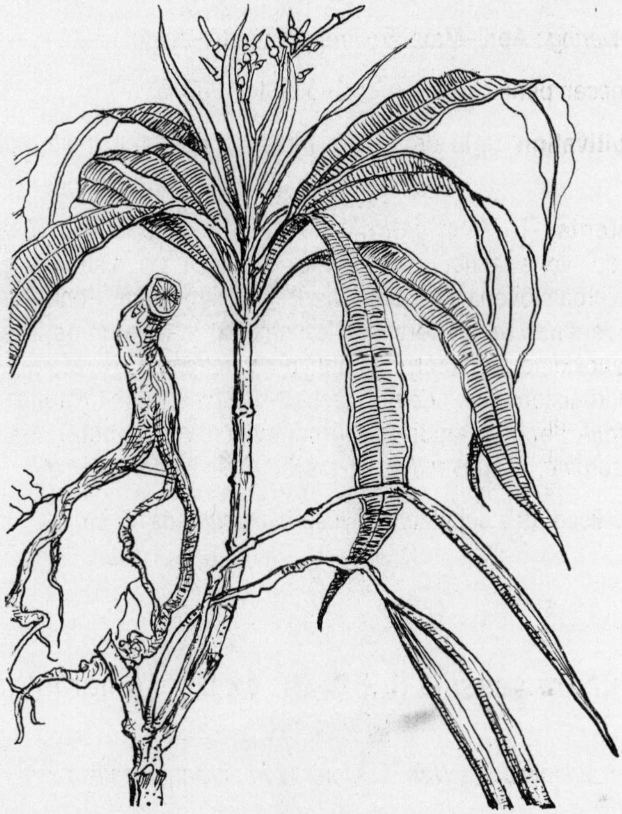

Syn : *Alstonia venenatus* Brown., *Blaberopus venenatus* A. DC., *Echites venenata* Roxb.

Sanskrit names : Vishaghni, Rajaadana, Anandana.

Vernacular names : Kan : Addasarpa; **Mun :** Mangor jita; **Ori :** Ranajita; **Tam :** Pazhamunnipala.

Trade name : Vishaghni.

Traditional use : Useful in skin diseases, erysipelas, leprosy, cobra bite and other venomous bites, epilepsy, fatigue, fumes and otalgia.

Modern use : *Ripe fruit* : in syphilis, insanity, epilepsy, as tonic, antiperiodic and anthelmintic; *Root* : astringent, bitter, thermogenic, depurative, antitoxic, febrifuge and anodyne.

Phytography : Shrub, up to 2.6 m; leaves whorled, simple, 10–20 cm by 1.2–3.8 cm, lanceolate, finely acuminate, petiole short, slender; cymes subumbellate; flowers often racemose on the branches, 2.5 cm long, slender, top barrel-shaped; follicles 10–12 cm by 0.8 cm, sword-shaped, coriaceous, long beaked.

Phenology : *Flowering* : April–May; *Fruiting* : February–August.

Distribution : Deccan penninsula, ascending up to 1500 m.

Ecology and cultivation : In deciduous forests throughout India, hills, normally in cultivation.

Chemical contents : *Root-bark* : 5, 22-dioxikopsane, 16-epivenenatine, 16-epialstovenine, venenatine, alstovenine; *Root* : alstovenine, venenatine, reserpine, 3-dehydroalstovenine; *Stem-bark* : alstovenine and its C-3 epimer, venenatine, isovenenatine, reserpine, kopsinine, venalstonine, venanstorudine, echitovenine, venoxidine, veneserpine, dehydroalstovenine, Δ^3-alstovenine, tri-OMe-gallamide, anhydroalstonatine; *Leaf* : echitovenaldine, echitoserpidine, 19-*epi*-(+)-echitoveniline; *fruit* : echitovenidine, (+)-minovincinine, venoterpine, ursolic acid, β-amyrin, *echitoserpine, echitoveniline,* β-amyrin-OAc, lupeol ester of a β-hydroxi acid.

Adulterant : It is used as a substitute of *Alstonia scholaris* R. Br.

Alternanthera sessilis (L.) R. Br. ex DC. (Amaranthaceae)

(2n = 34)

Syn : *Alternanthera denticulata* Wall. A. Don, *Gomphrena sessilis* L.

Sanskrit names : Matsyaksi, Putturah.

Vernacular names : Ben : Chanchi; **Hin :** Gudri sag; **Kan :** Honagone soppu; **Mal :** Ponnanganni keeray, Kozhuppa; **Mar :** Kanchari; **Ori :** Madranga; **Sin :** Mukunu wenna; **Tam :** Ponnanganni, Ponnanganni keeray; **Tel :** Pannaganti, Ponnaganta kura.

Trade names : Chanchi, Gudri sag.

Traditional use : Sweet, astringent, acrid, cooling, constipating, depurative and digestive.

Modern use : *Plant* : galactagogue, cholagogue, febrifuge; *Stem* and *Leaf* : useful in eye trouble.

Phytography : Prostrate herb, branches 7.5–45 cm long, glabrous, nodes often villous; leaves linear or oblong or lanceolate or elliptic, 2.5–7.5 cm long, longer in watery places, rather fleshy; heads axillary, sessile, 0.6–1.2 cm long, white, glistening; flowers minute with corky wings.

Phenology : *Flowering* : throughout the year.

Distribution : Throughout hotter and damp places in India as a weed, ascending up to 1300 m in the Himalaya; in Bangladesh, Pakistan and all other tropical and subtropical countries.

Ecology and cultivation : Throughout Indian subcontinent in moist places up to 2100 m; sometimes cultivated.

Chemical contents : *Young shoots* : 5% protein, 16.7 mg/100 g hydrocarbons, ester, sterols, palmitates of sterols, α-spinasterol, β-sitosterol, enecycloartanol, campesterol, 5-α-stigmasta-7-eneol.

Althaea ludwigii L. (Malvaceae)

Syn : *Malva malwensis* Edgew.

Modern use : *Plant* : Aperient.

Phytography : Annual herb, up to 30 cm in height, slender, branching; leaves long petioled, orbicular, 5–7 lobed, lobes wedge-shaped; flowers shortly peduncled, few or tufted, whitish, 1.25–2.0 cm in diameter, bracteate, bracteolate; fruit doubly keeled, not winged, wrinkled at the sides, glabrous; seeds smooth, brown.

Phenology : *Flowering* : March–July.

Distribution : Northwestern India, Punjab, South India; Pakistan (Sind, Baluchistan); temperate regions of the world.

Ecology and cultivation : Xerophyte.

Althaea officianlis L. (Malvaceae)

(2n = 40 to 44)

English names : Marsh mallow, White mallow.

Sanskrit name : Silihaka.

Vernacular names : Hin : Gulkhera; **Mar :** Gulkhiro; **Tam :** Simaitutti.

Trade name : Gulkhera.

Traditional use : Emollient, used in making absorbent pills and pastilles; *Leaves* in the preparation of a soothing ointment; *Seed* (dissolved in vinegar) : used to remove toothache.
UNANI : ingredient of 'Khatmi'.

Modern use : *Root* (aqueous extract) : anti-inflammatory; *Gummy matter of root*: used in hoarseness, erosion of stomach and intestine, in relaxing the passage in nephritic and calculus complaints; *Root*: demulcent and emollient; *Infusion of flowers* : recommended in bronchial catarrh and bronchitis, considered to be very useful herbal cough remedy.

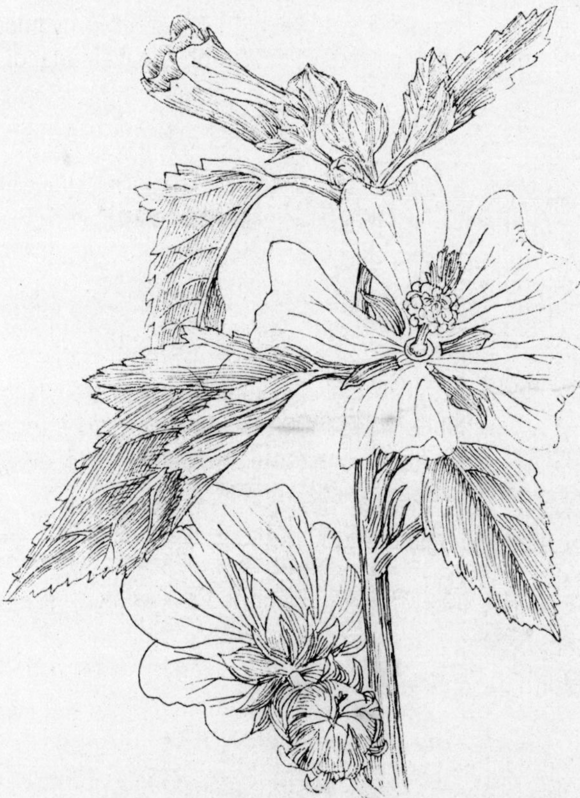

Phytography : Perennial herb, stem erect up to 1 m in height; leaves simple, ovoid, unequally toothed, stipules linear-subulate; flowers peduncled in axillary clusters, rosy, 2.5 – 5.0 cm in diameter, bracteate, bracteolate.

Phenology : *Flowering*: March–July.

Distribution : Native to eastern Europe; found in Punjab and Kashmir.

Chemical contents : *Root*: 35% mucilage, 37% starch; *Plant-mucilage* : polysaccharide compounds; *Seed-oil* : malvalic acid, fatty oil, butyric acid, phytosterin.

Adulterant : The roots of other Malavaceous plants, especially the roots of the Hollyhock (*Althaea rosa* Cav.).

Althaea rosea Cav. (Malvaceae)

(2n = 26)

Syn : *Althaea flexuosa.*

English names : Hock herb, Hollyhock, Round dock.

Vernacular name : Hin : Gul-e-khara.

Trade name : Gul-e-khara.

Traditional use : *Mucilage of fresh petioles, stems and roots* (shaken in water) : applied in bruished and wounded parts; in Tamil Nadu, used in dysentery; *Flower* : for treatment of rheumatism.

Modern use : *Root* : astringent, demulcent, in dysentery; *Flower* : cooling, diuretic, use for rheumatism; *Seed* : demulcent, diuretic, febrifuge, used in chest complaints.

Phytography : Tall, cultivated annual; stem hairy; leaves alternate, stipulate, petiolate, simple, trilobed; large racemes or spikes; flowers sessile or shortly pedicellate, large, coloured; schizocarpic carcerulus.

Phenology : *Flowering* : Winter.

Distribution : Originated in China; it is cultivated in gardens of India, Bangladesh and Pakistan.

Ecology and cultivation : Prefers semidry soil.

Chemical contents : *Flower* : cyanidin; *Seed* : 11.9% of a dry oil.

Altingia excelsa Noronha (Altingiaceae)

Syn : *Sedgioickia cerasifolia* Griff., *Liquidamber altingia* Blume

Vernacular names : Asm : Jutuli, Duang; **Hin** and **Guj :** Silaras; **Mal :** Rasamala; **Mar :** Shilaras; **Tam :** Neriyurishippal; **Tel :** Shilarasamu.

Trade names : Silaras, Rasamala.

Modern use : *EtOAc extract of roots* : prevent gastric ulcer; *Resin* : carminative, expectorant, stomachic, antiscorbutic, antipyretic, used in leucoderma, scabies, vesicular calculi, lumbago, menorrhagia, hydrocile, renal and pulmonary troubles; *Stem* : source of resin which is used in medicines as expectorant, carminative and stomachic.

Phytography : Tree, reaching even 33 m in height; leaves alternate, petiolate, 7.5–10.0 cm, oblong or ovate, glabrous on both surfaces, tufts of hair may be present in the axils of veins underneath, stipules deciduous or persistent; flowers unisexual, small, reduced, male racemose, wrapped by large bract; female heads of 10–20 flowers, fruit globose, harsh.

Distribution : Assam; Bhutan.

Chemical contents : *Plant*: 0.17% essential oil, resin, benzaldehyde, cinnamic aldehyde.

Alysicarpus longifolius W. & A. (Fabaceae)

Syn : *Hedysarum longifolium* Rottl., *H. bracteatum* Herb., *Alysicarpus vaginalis* Wall., *A. spicatus* Royle

Vernacular names : Guj : Dhodasmervo; **Mar :** Jangli gailia, Motha dampta.

Modern use : *Root*: substitute for liquorice.

Phytography : Herb, stem stout, ascending, subglabrous, reaching 1.6 m; leaves trifoliate, leaflets oblong or lanceolate; 7.5–15.0 cm long, ciliated on the veins below; racemes 15–30 cm long, rachis hairy, flowers adpressed to the rachis, pedicels 0.2 cm, papilionaceous; pods moniliform, 0.2 cm wide.

Distribution : Scattered throughout the plains of India.

Chemical contents : *Leaves* : myricyl alcohol, β-sitosterol, pinitol, rutin and flavone.

Alysicarpus vaginalis DC. (Fabaceae)

(2n = 16, 20)

Syn : *Alysicarpus rubibarna* Wall., *A. diversifolius* Wall., *A. harnieri* Schweinf., *Hedysarum vaginale* L.

English name : Alyce clover.

Vernacular names : Guj : Gaadar; **Mar :** Dhakta dhampta; **Mun :** Birbut; **Tam :** Namappoondu.

Modern use : *Plant* : skeletal fracture; *Decoction of roots* : for cough; *Leaf-extract* : anticancer.

Phytography : Herb, stem robust, ascending up to 1 m; trifoliate, glabrous, 1.25-2.5 cm, oblong, obtuse, petiole 0.3–0.6 cm; racemes 5.0–7.5 cm long, 6- to 12-flowered; flowers small, papilionaceous; pods 1.25–1.8 by 0.2 cm, indistinctly moniliform, a little thickened at the end, faintly pubescent, rugose.

Distribution : Himalaya—ascending up to 1300 m..

Amaranthus blitum L. (Amaranthaceae)

(2n = 16)

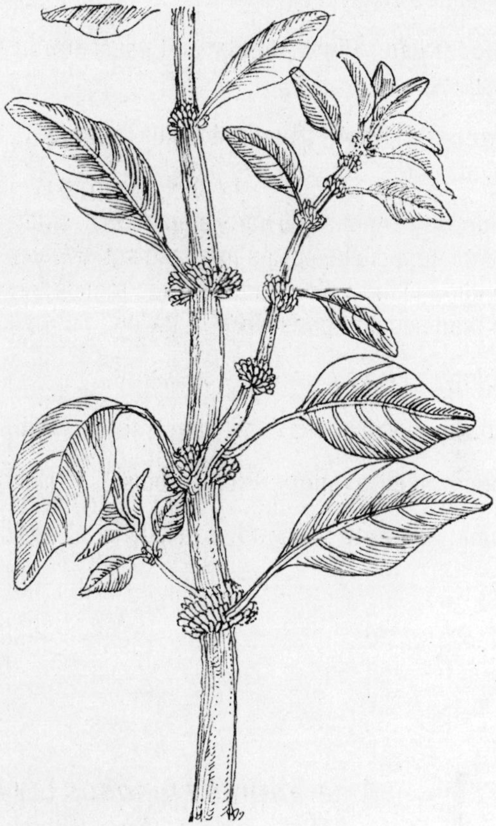

Syn : *Amaranthus oleraceus* L., *Euxolus oleracea* Moq.

English name : Amaranth.

Sanskrit name : Alpa marisha.

Vernacular names : Ben & Hin : Sada natya, Van natya; **Guj :** Undi bhaja; **Mar :** Ranta durja, Tandulja.

Trade name : Sada natya.

Modern use : *Plant* : cooling, emollient, astringent.

Phytography : Annual herb, stem procumbent, glabrous; leaves log-petioled, small, bilobed; spikes axillary, in small clusters; flowers reduced, utricle indehiscent.

Phenology : *Flowering* : Rainy season.

Distribution : Weed in wastelands; common in Punjab and other places.

Ecology and cultivation : Cultivated in India and elsewhere in the warmer regions.

Amaranthus caudatus L. (Amaranthaceae)

(2n = 32)

Syn : *Amaranthus cruentus* Willd., *A. paniculatus* L.

English names: Tassel flower, Lovelies bleeding.

Sanskrit name : Rajagiri.

Vernacular names : Ben : Chuko, Natya, Anantmul, Kaholabhoji, **Hin :** Ramdana, Chaulai, Chaumarsa; **Kas :** Bustanafroz; **Tam :** Pungikirai, Siru kirai; **Tel :** Keikera.

Trade names : Natya, Ramdana.

Traditional use : *Plant* : dropsy and anascara; *Leaf* : abortifacient, relieving pulmonary congestions.

Modern use : *Plant* : diuretic, in piles, strangury; also given in scrofula and applied to scrofulous sores; for purifying blood.

Phytography : Annual tall herb, stem robust, striate; leaves long petioled, elliptic or ovate or lanceolate, obtuse; spikes in dense soft thyrses, central one and laterals usually very long and pendulous; flowers unisexual, reduced, red or green or white, bracteate, bracts longer than sepals; seeds white or yellow.

Phenology : *Flowering* : Rainy season.

Distribution : Cultivated in hilly tracts throughout Indian subcontinent up to 3000 m.

Ecology and cultivation : Prefers moist wastelands.

Chemical contents : *Leaves* : oxalic acid; *Seeds* : carbohydrate, oil.

Amaranthus spinosus L. (Amaranthaceae)

(2n = 32, 34)

English names : Prickly amaranth, Spiny amaranth, Spiny pigweed, Foxtail.

Sanskrit name : Tanduliya(ka).

Vernacular names : Asm : Kata khutura; **Ben :** Kanta notey; **Guj :** Kantamidant, Tanjaljo; **Hin :** Kanta miris, Chulai; **Kan :** Mullu keere, Harive soppu; **Kon :** Dovi-bhaji; **Lod :** Janum-leper ara; **Mal :** Mullancheera; **Mar :** Kate math; **Ori :** Kantaneutia, Kantamarisha; **Pun :** Chulai; **San :** Lal-bhaji ara; **Tam :** Mullukkerai, **Tel :** Mulla tokakura, Mundla thotakura.

Trade names : Kanta miris; Kanta notey.

Traditional use : LODHA : (i) *Plant-decoction* (fresh) : to reduce the menstrual flow; (ii) *Fresh young leaf* (boiled) : to children as laxative; SANTAL : *Root-paste* : on boils; TRIBES OF JHARGRAM (West Bengal) : as an antidote to snake bite, as diuretic, in the treatment of eczema and whitlow.

MATSYA PURANA : of great medicinal value; AGNI PURANA : poison-killer, drinking of *root-extract* with the same of *Convolvulus trupethum* helps one to overcome the effects of poisons of insects and snakes.

Modern use : *Whole plant poultice* : applied to buboes and abscesses to hasten suppuration; *EtOH* (50%) *extract of plant* : spasmogenic; *Decoction of herb* : mouth-wash for toothache, poultice mixed with native soap as whitlow remedy; *Root* : in gonorrhoea, colics; *Leaf* : as enema, in stomach troubles, piles, leprosy, vomiting; *Leaf-infusion* : diuretic; *Pollen-extract* : allergic asthma, allergic rhinitis; it is a very potent drug for the treatment of leucorrhoea and menorrhagia; *Decoction of leaves* and *roots* : given to the children as laxative.

Phytography : Perennial, erect, strong herb, up to 60 cm in height; stem green, hard, glabrous, branched, armed with prickles; leaves simple, petioled, axils with 5 spines, 3.7–10.0 cm by 2–5 cm, ovate or oblong, obtuse; spikes axillary, usually densely flowered; flowers reduced, 0.1 cm long, unisexual, monoecious, short, bracteate, bracts setaceous; seeds very small, black, shining.

Phenology : *Flowering* and *Fruiting* : throughout the year.

Distribution : Throughout Indian subcontinent in wastelands, fields, gardens.

Ecology and cultivation : Prefers wastelands; wild or semiwild.

Chemical contents : *Root* : β-spinasterol, octacosanoate, saponins; *Stem* and *Leaf* : hentriacontane, spinasterol.

Remarks : Lodhas smoke root bark as hallucinogen. Eating root-paste can cause insanity. Some tribes of Medinipur tie a piece (2 om) of root in black thread on left arm of a pregnant to cure piles.

Amaranthus tricolor L. (Amaranthaceae)

(2n = 34)

Syn : *Amaranthus gangeticus* L., *A. oleraceous* Roxb., *A. mangostanus* L.

English names : Tampala, Chinese spinach.

Vernacular names : Ben : Dengo, Lal shak; **Hin :** Lal sag; **Tam :** Serikkeerai; **Tel :** Chirakura.

Trade name : Lal sag.

Traditional use : *Plant* : used in the treatment of snake bite, consumption, cough and bronchitis; *Fresh extract of root* : may be used in acute diarrhoea along with rice water; *Powder of root* : may be used in the treatment of paranechia as local application.

Modern use : *Plant* : poultice in ulcerated throat, mouth, toothache and as a cooling agent, astringent, in diarrhoea, dysentery, haemmorrhage.

Phytography : Erect, stout herb, up to 1 m; stem and leaf green or pink or liver-coloured or bright red; leaves long petioled, 5.0–12.5 cm, form very variable, tip obtuse and often notched; spikes squarrose; bracts awned; flowers reduced, unisexual, monoecious; sepals long awned; seeds lenticular, pitch-black, shiny, 2 cm in diameter.

Phenology : *Flowering* and *Fruiting* : throughout the year.

Distribution : Cultivated throughout India.

Ecology and cultivation : Occurs in plains, in low-lying moist grounds, river banks and village wastelands; cultivated as a vegetable.

Chemical content : *Plant* : fatty oils.

Amaranthus viridis L. (Amaranthaceae)

(2n = 34)

Syn : *Amaranthus polystachyus* Ham., *A. gracilis* Desf., *A. fasciatus* Roxb., *Euxolus caudatus* (Jacq.) Moq.

English names : Wild amaranth, Green pigweed.

Sanskrit names : Tanduliya, Vishaghna.

Vernacular names : Ben : Bon notey; **Guj :** Dhinmado; **Hin :** Karund; **Kan :** Chilikirai, Soppy, Daglisppu; **Lod :** Notya-ara; **Tam :** Kuppaikiral; **Tel :** Mullatotakura.

Trade names : Bon notey, Karund.

Traditional use : LODHA : *Root-paste* : on boils for suppuration.
AYURVEDA : *Root* : an antifertility agent.

Modern use : *Herb* : in snake bite; *Leaves* : emollient, in scorpion sting.

Phytography : Herb, erect, rarely decumbent below, up to 60 cm in height; leaf simple, petiolate, green, 2.5–5.0 cm long, width variable, tip rounded or notched, base truncate or cuneate; flowers in slender axillary or terminal panicled spike-like racemes, small, unisexual, monoecious, bracteate and bracteolate, greenish; seeds minute, pitch-black, shining.

Phenology : *Flowering* : peak season December–April.

Distribution : Very common in wastelands, fields and gardens; practically cosmopolitan in the tropical and subtropical regions of the world.

Ecology and cultivation : Wild or semiwild, plains from the coast to 1500 m, abundant in arable land as a weed of cultivation.

Chemical contents : *Plant* : shows vitamin B_{12}-like action, saponin; *Petroleum ether extract of plant* : juvenomimetic activity.

Amberboa divaricata Kuntze (Asteraceae)

Syn : *Volutarella divaricata* Benth. et Hook. f.

Vernacular names : Hin : Badaward; **Mar :** Sakayi.

Trade name : Badaward.

Traditional use : *Plant* : tonic, aperient, deobstruent, febrifuge, slight mucilaginous, used in coughs and fevers.

Phytography : Annual straggling stiff weed, dichotomously branched, branches 30–60 cm long, angled, smooth or scabrid; leaves alternate, sessile, oblong or obovate, entire or toothed or pinnatifid, size very variable—from 2.5 x 0.6 cm to 15 x 7.5 cm; heads hard, heterogamous, pale purple, 1.29 cm in diameter, involucre of bracts glabrate, reddish, spinous; achenes 0.5 cm long, narrow, acutely angled, grooved, pappus spiny of many small unequal scaberulous hairs, silvery brown, 3 or 4 innermost long and flattened.

Phenology : *Flowering* : Winter.

Distribution : Bihar, upper Ganga plains, Punjab (ascending up to 1000 m), Gujarat, Maharashtra, Madhya Pradesh and South India; Pakistan (plains of Punjab and Baluchistan).

Ammannia auriculata Willd. (Lythraceae)

Syn : *Ammannia senegalensis* Lamk.

English name : Blistering ammania.

Vernacular names : Ben and **Hin :** Dadmari; **Guj :** Jata agio; **Mal :** Kullaravanchi; **Mar :** Bharjambhul; **Pun :** Dadmari, Fauglimehndi; **Tam :** Kallurivi; **Tel :** Agnivednapaku.

Trade name : Dadmari.

Modern use : *Plant* : hypotensive and as blistering agent, antityphoid and antituberculosis properties, applied on ringworm and skin diseases.

Phytography : Annual, glabrous, erect herb, varies from 15–60 cm in height; branches quadrangular; leaves simple, opposite, sessile, subauriculate at base, elongate-oblong, 2.5–5.0 cm long; cymes compound; flowers campanulate with 4 or 8 green lines, bracteate, bracteolate; capsules red, globose, 0.2 cm in diameter.

Phenology : *Flowering* : Rainy season.

Distribution : Plains of Punjab, Rajasthan, Kurram valley, Chitrol, Kashmir, upper Ganga plains, ascending up to 1600 m; Tibet, Bangladesh.

Ecology and cultivation : Wild in semiarid places.

Chemical content : *Plant* : rich in vitamin C (88.9 mg/100g).

Ammannia baccifera L. (Lythraceae)

(2n = 26)

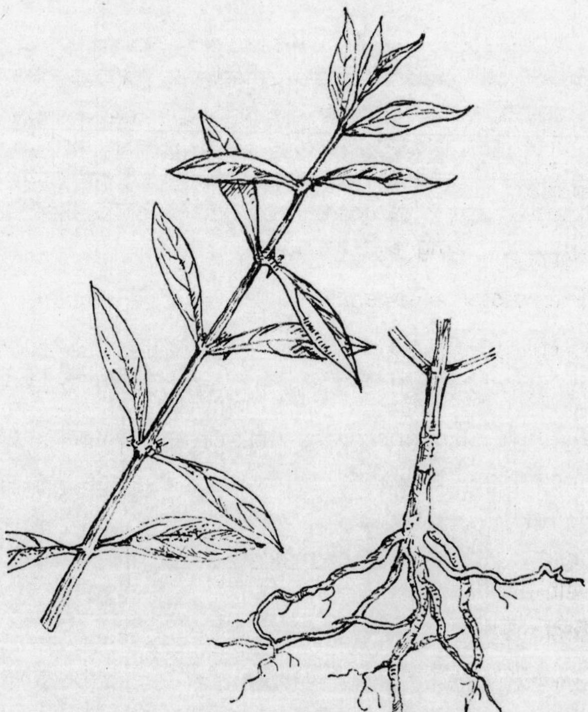

Syn : *Ammannia vesicatoria* Roxb., *A. indica* Lamk.

Sanskrit name : Agnigarbha.

Vernacular names : Ben : Dadmari; **Hin :** Jangli mehndi; **Mal :** Kallur vanchi; **Mar :** Bharjambal; **Pun :** Dederbuli; **Tam :** Kalluriva; **Tel :** Agnivandrapaku.

Trade names : Jangli mehndi; Agnigarbha.

Traditional use : *Leaves* : used for ringworm and other parasitic skin affections.

Modern use : *Plant-extract* : antibacterial; *Leaves* : acrid, used to raise blisters, in rheumatism, pains, fevers and as rubefacient in skin diseases.

Phytography : Annual, glabrous erect herb, normally 20 cm tall, may be even 60 cm; lower branches and leaves opposite; upper leaves alternate, 2.5–6.7 cm long, narrow, obtuse; cymes axillary, very short; flowers hemispherical; capsules globose, depressed; seeds black, subhemispheric.

Phenology : *Flowering* : Rainy season.

Distribution : Very common in moist places throughout tropical India; Bangladesh and Pakistan.

Ecology and cultivation : Abundant in moist places, mainly in the cultivated fields, in the vicinity of *bunds*, *nallahs,* etc.

Chemical content : *Leaves* : lawsone.

Ammi majus L. (Apiaceae)

English names : Greater ammi, Bishop's weed.

Traditional use : *Fruits* : tonic, stomachic, carminative and diuretic; *Seeds* : in vitilago and also in the formulation of sun-tan lotions.

Modern use : *Dried fruit-powder* or *plant-extract* : in the treatment of vitilago or leucoderma; the drug relaxes smooth muscle and lowers the toxicity of the water; *Decoction* : used for the passage of kidney calculi; it is also a source of khellin; prescribed for angina pectoris.

Phytography : Herb, up to 1.5 m in height; branchlets glabrous; leaves 2- or 3-pinnate, up to 7 cm long, upper leaves gradually become deeply segmented, leaflets elliptic, chartaceous, base subacute, margin serrulate, apex acuminate; umbels compound; flowers 5–20 per ray, white; fruit oblong, 2 mm, laterally compressed.

Phenology : *Flowering* : beginning of March; *Fruiting* : end of April.

Distribution : Indigenous to Egypt and widely distributed in Europe, Mediterranean regions, Abyssinia and West Africa. Experimental cultivation is being done in Jammu, Delhi and Himachal Pradesh.

Ecology and cultivation : Thrives well in loamy soil. Water should not be allowed to stand in the field as it is very sensitive to water.

Chemical contents : Chief active constituent is bitter principle khellin and furocoumarin ammoidin; *Leaves* : xanthotoxin, isopimpinellin, bergapten, imperatorin, umbelliferone; *Fruits* (also referred to as *seeds*) : ammoidin (xanthotoxin), ammidin (imperatorin), majudin (bergapten), four furocoumarins, viz., marmesin, isopimpinellin, xanthosin and bergapten; ammajin from Romanian fruits.

Adulterant : Admixed/substituted for *visnaga.*

Remark : Growing well in Jammu.

Amomum aromaticum Roxb. (Zingiberaceae)

(2n = 48, 52)

Syn : *Geocallis fasciculata* Horan.

English names : Assam cardamom, Bengal cardamom.

Sanskrit names : Ela, Upakunchika.

Vernacular names : Ben : Elach, Bara elach; **Guj :** Elachi; **Hin :** Elachi, Bari elaichi; **Mal :** Elam, **Mar :** Elayachi; **Ori :** Alaicha; **Pun :** Khusboodar ilaichi; **Tel :** Pedda elakulu.

Trade names : Cardamom, Elachi.

Traditional use : RURAL FOLKS : *Seeds* : in vomiting tendency caused by worms or acidity, congestion of liver, certain digestive disorders and in combination with melon seeds as diuretic; *Decoction of seeds* : as a gargle in affections of teeth and gum; *Powdered seeds* : in gonorrhoea.
MATSYA PURANA : antidote of poison from plant; SUSHRUTA SAMHITA : *Seeds* in urinary troubles.

Modern use : *Seeds* : used as fragrant adjuncts to other stimulants, purgative and bitters; *Decoction of seed* : used as a gargle in affections of teeth and gum; *Seed-oil* : aromatic and used as stimulant, stomachic and applied to eyelids to allay inflammation.

Phytography : Herb, 1.0–1.3 m high; leaves simple, alternate, 15–30 cm by 5–10 cm, oblong-lanceolate, glabrous; spikes small, pendant; flowers tubular, 2.5 cm long, white tinged with brown, lip pale yellow; capsules narrowly obovoid, 3.7 cm long, 3-valved, aromatic, many-seeded; seeds black.

Phenology : *Flowering* : May; *Fruiting* : September.

Distribution : Indigenous to West Bengal and Assam, and distributed to surrounding areas.

Ecology and cultivation : Cultivated in the wetter areas of West Bengal and Assam, at the foothills of eastern Himalaya.

Chemical contents : *Seeds* : 1–1.2% oil containing a large quantity of cineol.

Adulterants : Cardamom capsules are chiefly adulterated with immature fruits, mouldy or insect-infested fruits and cardamom 'splits' (partially open fruits); the principal adulterants of cardamom seeds are pebbles (small); seeds of *Amomum* species and ground pericarps of cardamom.

Amomum constatum Benth. (Zingiberaceae)

Syn : *Alpinia costata* Roxb., *Cardamomum medium* Schult., *Elettaria costata* Horan., *Elettaria media* Link ex Horan., *Hornstedtia costata* (Roxb.) K. Schum.

Modern use : *Seeds* : employed for stomachic ailments, general debility, asthma and other pulmonary affections.

Phytography : Herb, 1.6–2.0 m high, rootstock 1.25 cm in diameter; leaf simple, alternate, 0.6–1 m by 7.5–10.0 cm, oblong-lanceolate, pubescent beneath; spike shortly pedunculate, oblong, 5.0–7.5 cm long, bract pink, 3.7 cm long; flower cylindric, 5 cm, red-yellow; capsule 1.25 cm, ovoid, strongly ribbed; seed many, aromatic, brownish, obovoid.

Phenology : *Flowering* and *Fruiting* : Almost throughout the year.

Distribution : Eastern tropical Himalaya; Bangladesh (Sylhet), Sri Lanka.

Ecology and cultivation : Occasional small plantations; grows on hills.

Amomum subulatum Roxb. (Zingiberaceae)

English names : Large or greater cardamom, Nepal cardamom.

Sanskrit name : Brihata ela.

Vernacular names : Asm : Danger elasi; **Ben :** Elach; **Guj :** Mothi elachi; **Hin :** Bara elaichi; **Kan :** Dodda yelakki; **Mal :** Kattu elam; **Mar :** Mothi elachi; **Ori :** Bada alaicha; **Pun :** Wadi illachi; **Tam :** Periya elakkai; **Tel :** Pedda elakulu.

Trade name : Bara elaicni.

Traditional use : *UNANI* : Constituent of 'Heel-kalan', exhilarant for vital spirit, beneficial in fainting, insanity, nausea and gas.

Modern use : *EtOH* (50%) extract of rhizome and roots : hypoglycaemic; *Essential oil from seeds* : active against growth of keratinophilic fungi; *Seeds* : used as fragrant adjuncts to other stimulants, bitter and purgative; *Seed-oil* : aromatic and used as stimulant, stomachic and applied to eyelids to allay inflammation.

Phytography : Herb, 1–1.3 m high; leaf simple, spiral, 30–60 cm by 7.5–10.0 cm, oblong-lanceolate, green, glabrous; spikes globose, very dense, short peduncled; bracts red; flowers 2.5 cm, yellowish white; capsules 2.5 cm, globose, red or brown.

Phenology : *Flowering* and *Fruiting* : Almost round the year.

Distribution : Grows in swampy places in Assam, Sikkim, Tamil Nadu and West Bengal.

Ecology and cultivation : Prefers higher elevations.

Chemical contents : *Fruits* : volatile oil containing a terpenoid; *Seeds* : chalcone cardamonin, flavones, alpinetin, petunidin-3,5-diglycoside, leucocyanidinglucoside, subulin and essential oil.

Adulterant : Seeds of this species are used as substitute for *Elettaria cardamomum* Maton.

Amoora cucullata Roxb. (Meliaceae)

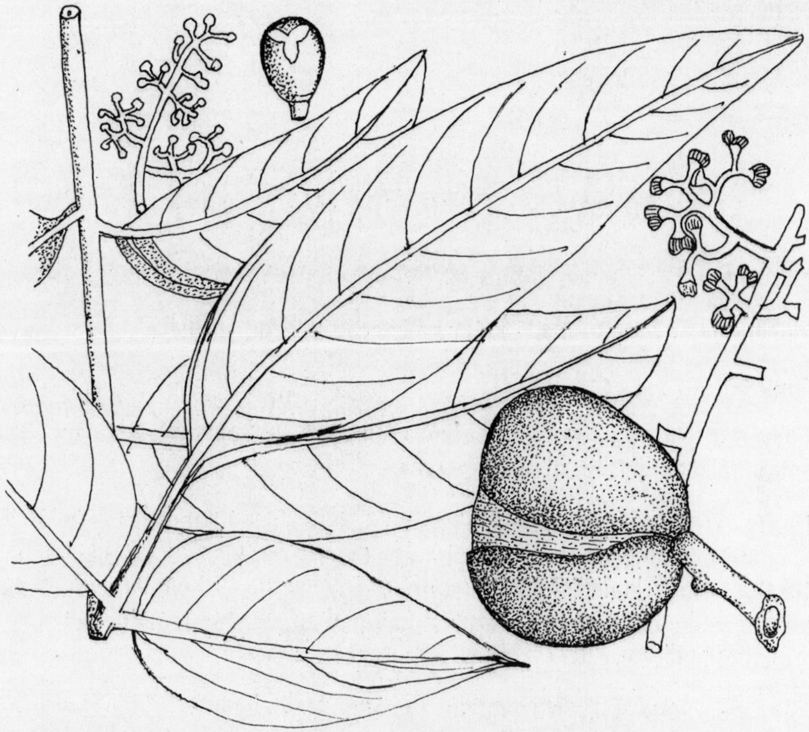

Syn : *Amoora auriculata* Miq., *Andersonia cucullata* Roxb., *Sphaerosaeme rohituka* Wall.

Vernacular names : Ben : Amur, Latmi.

Trade name : Amur.

Modern use : *Bruised leaves* : applied to remove inflammation.

Phytography : Tree with slow growth, cinereous bark, branches smooth; leaves alternate, imparipinnate, 15–40 cm long, leaflets 7.5–18.7 cm by 3.3–6.2 cm, opposite or subopposite, obliquely ovate-oblong, glabrous, shining; male panicles drooping, male flowers yellow, 0.3 cm; female racemes few-flowered; capsules subglobose, trilobed, 6.2 cm in diameter; seeds three-quarters covered with a fleshy bright, orange-coloured aril.

Phenology : *Flowering* and *Fruiting* : October to January.

Distribution : Sundarbans, Andamans; Nepal, Myanmar, Bangladesh, Pakistan, Sri Lanka.

Ecology and cultivation : Occasionally found in mangrove forests and tidal areas of the Sundarbans along the sides of the creeks and riverbanks; also found throughout the mangrove forests of India.

Amoora wallichii King (Meliaceae)

Syn : *Amoora spectabilis* Hiem.. non Miq., *Aglaia spectabilis* Hort.

Vernacular names : Asm : Amari; **Ben :** Lalee, Pitraj, **Hin :** Lalehini.

Trade name : Amari.

Modern use : *EtOH* (50%) *extract of stem* : antivirus, anthelmintic, anticancer against human epidermoid carcinoma of nasopharynx.

Phytography : Tree; leaves alternate, unipinnate, 37.5 cm long, petioles angular, leaflets 11–13, opposite or subopposite, oblong, 10–20 cm by 3.7–5.0 cm, acutely subacuminate, glabrescent, shining, glaucescent beneath, petiolules 0.6–0.8 cm long, male panicles 45 cm long; pedunculate with alternate unequal branches, female panicles short; capsules obvoid-pyriform, 3.0–4.5 cm by 2.5–3.0 cm.

Phenology : *Flowering* : July–November; *Fruiting* : May–June.

Distribution : Scattered in evergreen forests of Sikkim, West Bengal, Assam, Manipur and the Andamans.

Ecology and cultivation : Grows in humid tropical forests; not cultivated.

Chemical content : *Heartwood* : β-sitosterol.

Amorphophallus campanulatus (Roxb.) Bl. ex Decne (Araceae)

(2n = 26, 28, 36)

Syn : *Amorphophallus virosus* N.E. Br., *Arum campanulatum* Roxb., *Candarum roxburghii* Schott.

English names : Elephant-foot yam, Teliga potato.

Sanskrit names : Arshaghna, Balukanda.

Vernacular names : Asm and **Ben** : Ol, Ol kachu; **Guj** : Suran; **Hin** : Ol, Zamirkand, Suran; **Kan** : Suvarna (or Churna) Gadde; **Kon** : Shotri, Suran, Luttiezhadd; **Mal** : Chaena; **Mar** : Suran; **Ori** : Ola, Olua; **Pun** : Zamin kanda; **Tam** : Shaenai, Karnai kilangu; **Tel** : Kānda, Manchi kandha, Theeya kanda.

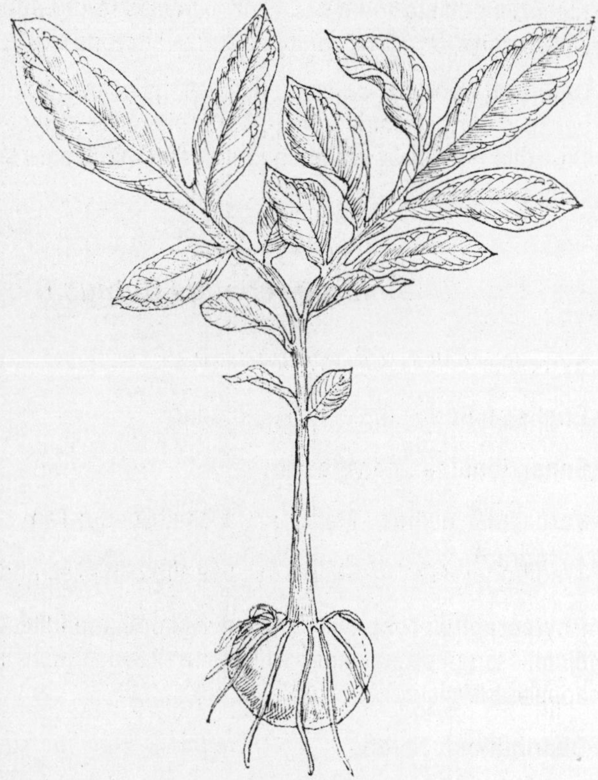

Trade names : Yam, Ol.

Traditional use : *RURAL FOLKS* : use in earache, swelling of throat, breathing difficulty, pimples, constipation, cholera, haemorrhage, septicemia, pain, kala-azar, enlarged spleen, intercostal neuralgia, pleurosis, pneumonia, puerperal fever, diarrhoea; *SANTAL* : (i) *Powdered corm* : to treat earache, throat troubles, as an ingredient of medicines for cholera and constipation; (ii) *Plant-juice* : in sores. *AGNI PURANA* : stomachic, tonic; *AYURVEDA* : *Corms* for piles and dysentery.

Modern use : *Corm* : against abdominal tumours.

Phytography : Robust herbaceous plant with a dull brown hemispherical and depressed corm (dia 20–25 cm) weighing few kg, and an erect solitary stem usually 96 cm –2.5 m in height bearing at the top a large solitary mottled leaf; a large terminal spadix with foul odour sprouts after 4 years; each corm produces 5–10 cormels at the end of each growing season.

Phenology : *Flowering* : April–June; *Fruiting* : November.

Distribution : Indian in origin and found throughout the plains of Indian subcontinent.

Ecology and cultivation : It prefers sandy loam soil, and requires well-distributed rainfall or irrigation, warm and humid conditions for vegetative growth, while for corm development, it requires cool and dry conditions. The cormels are planted during April–June. They reproduce vegetatively. New corms may be harvested after 5–7 months.

Corms can be stored for a few months in dry and well-ventilated place as they cannot sprout immediately. This is often cultivated mixed with ginger or banana. Groundnut and some such crops are often used as cover crops in the fields of this yam.

Chemical contents : *Corm* : glucose, galactose, rhamnose, xylose, triacontane, lupeol, betulinic acid, stigmasterol, β-sitosterol and its palmitate, vitamins A, B_1, with Fe and Ca; Ca-oxalate crystals are abundant in the wild plants but less abundant or absent in the cultivated ones.

Amorphophallus dubius Blume (Araceae)

Syn : *Draconitum polyphyllum* Denst.

English name : Purple-stalked dragon.

Sanskrit name : Kanana kanda.

Vernacular names : Mal : Kattuchena, Shena; **Tam :** Kattukarmei.

Traditional use : RURAL FOLKS : use *herb* in piles and apoplexy.

Phytography : Tuberous herb; leaves large, tripartite, petiole long, rough, green with pale blotches; spadix subsessile, spathe 7.5–12.5 cm in diameter, 7.5 cm long, green, appendage globosely ovoid.

Distribution : Kerala.

Ecology and cultivation : Wild; not known in cultivation.

Amorphophallus sylvaticus (Roxb.) Kunth (Araceae)

(2n = 26)

Syn : *Synantherias sylvaticus* Schott., *Amorphophallus zeylanicus* Blume, *Arum sylvaticum* Roxb., *Brachyspatha sylvatica* Schott., *B. zeylanica* Schott.

Sanskrit name : Vajra kanda.

Traditional use : RURAL FOLKS : use against tumours.

Modern use : *Seeds* : cure toothache and gland enlargement.

Phytography : Tuberous herb, tubers subglobose, bulbilliferous; leaves 1 or 2, petiole 30–60 cm in length, pinnatifid, leaflets 5–15 cm, long-acuminate; spathe 5–10 cm long, streaked with green and pale pink.

Distribution : Western Deccan peninsula, Tamil Nadu and Coorg.

Ampelocissus araneosa (M. Laws) Planch (Vitaceae)

(2n = 80)

Syn : *Vitis araneosus* Dalz. & Gibbs ex M. Laws.

Vernacular names : Hin : Kauraj, Kuraji; **Mar :** Bandervel, Chamarmuli, Ghorvel.

Trade name : Chamar musli.

Traditional use : RURAL FOLKS : use as astringent and cooling.

Phytography : Slender climber, whole plant covered with deciduous hairs; stem flattened; leaves alternate, petiolate, membranous, compound, leaflets shortly petiolulate, serrated, glabrous above but felted below; cymes small, compact, umbellate on long woolly peduncles; flowers dark brown or red, small; fruits globose, black, 3- to 4-seeded; seeds elliptic, 0.8 by 0.4 cm, margins puckered.

Phytography : *Flowering* : July–September; *Fruiting* : Winter.

Distribution : Gujarat, Maharashtra, South India, up to an elevation of 1500 m.

Ecology and cultivation : In jungles and scrub jungles; wild.

Ampelocissus arnottiana Planch (Vitaceae)

Syn : *Vitis indica* W. & A.

Vernacular names : Ben : Amaluka; **Hin :** Jangli angur; **Kan :** Palkandi; **Mal :** Semparavalli; **Mar :** Randraksha; **Tam** and **Tel** : Sambaravalli.

Trade name : Jangli angur.

Traditional use : *Juice of root* : with kernel of coconut employed as a depurative and apierient; used as an alterative in form of decoction; considered as a purifier of blood and diuretic rendering the secretions healthy.

Modern use : Used in eye troubles and in ulcers.

Phytography : Climber; stem slender, woolly, tendrils usually leaf-opposed; leaves alternate, cordate-ovate, glabrous and shining above, denticulate-serrate; spikes cylindric; flowers greenish purple; fruits globose, 2- to 4-seeded; seeds 0.8 by 0.6 cm, elliptic, slightly curved on the back, otherwise flattish.

Phenology : *Flowering* : July–August; *Fruiting* : December–January.

Distribution : Gujarat, Maharashtra, Karnataka and Kerala, up to an elevation of 350 m.

Ecology and cultivation : Wild in jungles, wastelands, rocks and trees.

Ampelocissus barbata (Wall.) Planch (Vitaceae)

Syn : *Vitis barbata* Wall., *V. latifolia* Hb. Ham., *V. lanata* Hb. Roxb.

Vernacular names : Lep : Mikrum-rik; **Nep :** Jarila laha.

Traditional use : LEPCHA : rub crushed *stem* and *tubers* against swellings and body pains.

Phytography : Climber; branches stoutish, hollow, dark brown or black, covered with numerous long spreading glandular hairs; leaves alternate, simple, petiole flattened, 10–20 cm long, bears a long, forked, slender tendril above its middle, leafblade slightly lobed, 20–30 cm long, cordate-ovate, puberulous or densely tomentose beneath, petiole and leafblade covered with spreading glandular hairs; cymes large, regularly paniculate; flowers sessile, small, 4-cleft or 5-merous; fruits black with short pedicels.

Phenology : *Flowering* : June; *Fruiting* : July–August.

Distribution : In the hills of Assam and Meghalaya up to 1200 m; Bangladesh (Sylhet).

Ecology and cultivation : Wild in wastelands, walls, trees, crevices, etc.

Ampelocissus latifolia (Roxb.) Planch (Vitaceae)

(2n = 40, 60)

Syn : *Vitis latifolia* Roxb., *V. glabrata* Heyne, *V. indica* Wall., *V. zeylanica* Russell

Vernacular names : Ben : Govila; **Hin :** Pani bel; **Mal :** Karantavalli; **Mar :** Golinda; **Ori :** Dibroi; **Tam :** Kat tukkodimundirigai; **Tel :** Bedasativva.

Trade name : Pani bel.

Traditional use : RURAL FOLK : *Root* : bone-fracture; *Root-decoction* : chronic dysentery; *Stem-bark decoction* : stomach pain; *Plant* : muscular pain, sores, pneumonia, snake bite and bone fracture.

Modern use : *Plant* : resistant to pest, *Juice of tender leaves* : used in dental troubles, detergent for indolent ulcers.

Phytography : Climber, tendrils leaf-opposed, stem weak, hollow, glabrous; leaves alternate, petiolate, simple, 15–20 cm in length and maximum width almost equal; flowers in compact thyrsoid cymes, very small, peduncle bears a long wiry tendril below the cyme; fruits black, 2-seeded.

Phenology : *Flowering* : June–July; *Fruiting* : August–October.

Distribution : Kumaon to Moradabad, Assam, Konkan to Coromandal; Bangladesh (Sylhet).

Ecology and cultivation : Grows in hilly terrain; wild.

Ampelocissus tomentosa (Heyne) Planch (Vitaceae)

(2n = 40)

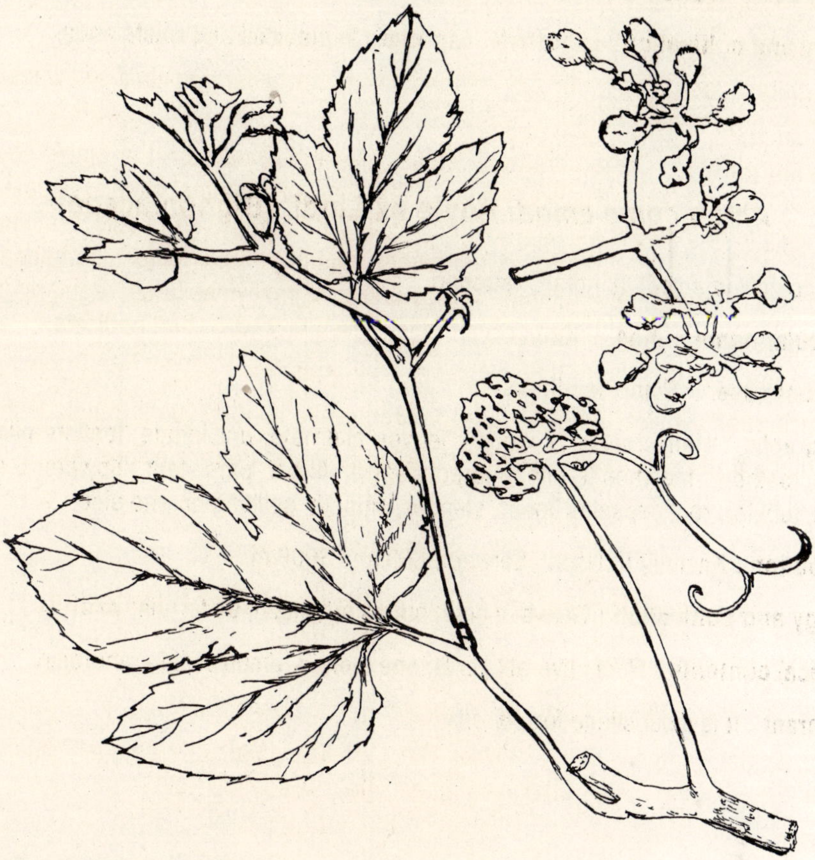

Syn : *Vitis tomentosa* Heyne

Vernacular names : Ori : Katobhangonoi; **San :** Ghoralidi; **Tam :** Sirunaralai; **Tel :** Atukulabaddu.

Traditional use : *HILL TRIBES OF WESTERN PENINSULA* : *Plant* in bone-fracture, swelling of head and limbs, chest pain, anasarca, dropsy, spleen complaints, dysentery, fistula, sores and carbuncle, pneumonia, pleurisy and neuralgia.

Modern use : *Root* and *Bark* : used for haemorrhoids.

Phytography : Climber; whole plant covered with thick, woolly, persistent tomentum; leaves alternate, simple, 7.5–22.5 cm long, palmately 3–5 angled or lobed; cymes compact, densely woolly peduncle 7.5–15.0 cm long, stout, woody, bears a stiff woody forked tendril below the inflorescence; flowers sessile, scarlet, dark-brown when dry; fruits 0.8 cm, black, round.

Phenology : *Flowering* : July–August; *Fruiting* : December–January.

Distribution : Western peninsula of India.

Ecology and cultivation : Lithophyte, especially in crevices and wastelands.

Amphicome emodi Royle ex Lindl. (Bignoniaceae)

Syn : *Incarvillea emodi* (Lindl.) Chatterjee

Vernacular name : Kas : Kaur.

Traditional use : *Plant* : febrifuge.

Phytography : Herb, rootstock woody; leaves alternate, unipinnate, leaflets elliptic or oblong, toothed; racemes terminal; bracts small, linear, persistent; flowers 5 cm by 1.5 cm, tubular, red; capsules linear, slender, smooth, splitting on one side.

Distribution : Kashmir to Nepal, between 660 and 3000 m.

Ecology and cultivation : Grows in cold, humid mountainous terrain; wild.

Chemical contents : *Root* : five alkaloids, one steroid, amphicoside, androsin.

Adulterant : It is a substitute for *chiretta.*

Amygdalus communis L. (Rosaceae)

(2n = 16)

Syn : *Prunus amygdalus* Batsch.

English name : Almond.

Vernacular names : Ben, Hin, Pun, Mar and **Tel :** Badam; **Kan :** Badami; **Mal :** Vatam kolta; **Tam :** Vadam kottai.

Trade names : Almond, Badam.

Traditional use : *Seed* : demulcent, stimulant, nervine, tonic; *Seed-oil* : various medicinal purposes.

Phytography : Tree; leaves alternate, simple—conduplicate in bud, glandular-serrulate, petiole often 2-glandular, stipules fimbriate, leaves appear after the flowers; flowers subsessile, white or red, solitary or fascicled corymbose or racemed; drupes 1-seeded; seeds pendulous, albumen scanty.

Distribution : Cultivated in Punjab, Kashmir; Pakistan (Baluchistan).

Ecology and cultivation : Native of eastern Mediterranean region; well-adapted to subtropical, nearly frost-free climate.

Chemical contents : *Seed* : almond oil; *Fruit* : HCN-glucoside.

Amygdalus persica L. (Rosaceae)

(2n = 16)

Syn : *Prunus persica* Batsch., *Amygdalus collinus* Wall., *Persica vulgaris* Mill., *P. saligna* Royle

English name : Peach

Vernacular name : Hin : Aru.

Trade name : Peach.

Modern use : *Flower* : diuretic, purgative; *Fruit* : stomachic, demulcent and antiscorbic; considered to be useful as an ascaricide.

Phytography : Branched, unarmed glabrous tree; leaves alternate, simple, serrate, petiole glandular, appearing after or with the flowers; flowers solitary or in small cymose clusters, pink, perigynous; drupes fleshy, downy, 1-seeded; seeds hard, furrowed, non-endospermic, covered with shallow wrinkles and minute holes.

Distribution : Native of China; grown in Jammu and Kashmir and Uttar Pradesh, up to 3300 m and in the Nilgiris.

Ecology and cultivation : Short-lived tree, grows in cool climate but susceptible to frost and low temperatures; propagated by budding; pruning is practised to shorten the plant and for its spreading.

Chemical contents : *Stem* : gum; *Fruit* : protein (1.6%), fat (0.2%), minerals (0.6%), carbohydrates (7.6%), vitamin B (20 mg/100 g), vitamin C (1 mg/100 g), nicotinic acid (0.2 mg/100 g), riboflavin (1 mg/100 g); *Seed* : fixed and volatile oil.

Amyris commiphora Roxb. (Burseraceae)

Syn : *Amyris agallocha* Roxb., *Commiphora agallocha* Engl., *C. roxburghii* (Arn.) Engl.

Sanskrit name : Aguru.

Vernacular names : Ben : Gugala; **Mar :** Gugal; **Tam :** Kungula; **Tel :** Aguru.

Trade name : Gugal.

Traditional use : *Gum-resin* : astringent, antiseptic, expectorant, aphrodisiac, enriches blood, demulcent, aperient, carminative, in snake bite and scorpion sting.

Phytography : Tree, growing up to 2 m in height; branches slightly ascending; leaves alternate, 1- to 3-foliate, leaflets sessile, thin; flowers small, polygamous, nearly sessile, a few in each fasicle, brownish red; drupes red, containing 2-celled stones.

Phenology : *Flowering* : Summer.

Distribution : Possibly African in origin, found in Assam; Bangladesh (Sylhet).

Ecology and cultivation : Prefers semidry, dry and xerophytic habitat.

Chemical contents : *Stem* : resinous fluid—yellowish white at first, soon hardens to a reddish brown mass; commercial product contains 1.45% essential oil besides gums and resins.

Remark : Resin was used in olden days as incense and perfume.

Anacardium occidentale L. (Anacardiaceae)

(2n = 24, 40, 42)

English name : Cashew nut.

Vernacular names : Asm : Kaju badam; **Ben :** Hijli badam, Kaju badam; **Guj, Hin, Mar** and **Pun :** Kaju; **Kan :** Gerupappu; Godambi; **Kon :** Kaz; **Mal :** Kashumaru, Parankimara, Andiparuppau; **Ori :** Lanka badam; **Tam :** Mundiri, Munthirikai; **Tel :** Jidimamidi, Muntha mamidi.

Trade names : Cashew nut, Kaju.

Traditional use : *Cashew nut shell-oil* : mild purgative, used in folk medicine for treatment of hookworm, cracks on soles of feet, warts, corns, leporus sores.

Modern use : *Cashew nut* : used in mental derangement, sexual debility, nervous prostration following seminal emission, morning sickness in pregnancy, palpitation of heart, rheumatic percarditis, loss of memory as a sequel to small pox; *Kernel* : good for week patients suffering from incessant and chronic vomiting; *Kernel-oil* : antidote for irritant poisons; *Liquor made from fruit* : diuretic. *HOMOEOPATHY* : for boils, warts, wounds and different types of cracks in legs; used sometimes in case of leprosy.

Phytography : Small tree with short, thick, crooked trunk; leaves simple, alternate, petiolate, entire, hard, 10–20 cm by 7.5–12.5 cm; panicles terminal, bracteate, pubescent; branches long, naked to the tips; flowers yellow with pink stripes, 0.8 cm in diameter; cashew apple red or yellow—is the swollen, enlarged pedicel which bears the nut.

Phenology : *Flowering* : March–April; *Fruiting* : April–May.

Distribution : Native to tropical America, naturalised in the hotter sea-shores of India; Pakistan, Bangladesh and Malaysia.

Ecology and cultivation : Introduced; grows in plains, especially towards the coast, very occasionally ascending up to 1200 m; being extensively planted by clearing scrub jungles in the plains.

Chemical contents : *Bark* : exudes gum; *Flower* : polyphenols; *Cashew shell* : yields gum, oil and liquid (CNSL); *Liquid-free nutshell* : syringic and gallic acids, galocatechin; *Defatted nutshell* : naringenin, prunin-6"-O-p-coumarate; *Cashew apple liquor* : vitamin C, vitamin E; *Reddish brown testa* : D-catechin, gallic acid, caffeic acid, quinic acid, polyphenols, bioflavone.

Anacyclus pyrethrum DC. (Asteraceae)

Syn : *Anacyclus officinarum* Hayne

English name : Spanish pellitory.

Sanskrit name : Akarakarava.

Vernacular names : Ben, Hin and **Mar** : Akarakara; **Mal** : Akkira karam.

Traditional use : *Plant* : in sexual disorders in male; *Root* : along with roots of *Withania somnifera* and *Vitis vinifera* in epilepsy, as gargle in sore throat; constituent of indigenous drug, *Amber mezhugu*, used in Tamil Nadu for treatment of many types of rheumatism.
UNANI : used in 'Aaqaqarha'.

Modern use : Pellitorine, an active principle of roots, intensely pungent; produces sialogogue effect.

Phytography : Annual herb; stem 5–10 cm; leaves 1- to 2-pinnatifid; heads small, subhemispheric; florets small; achenes oblong, flattened.

Distribution : Indigenous to North Africa, then to South America; scatteredly spread in different parts of India.

Ecology and cultivation : Widespread weed.

Chemical contents : *Root* : pellitorine, anacyclin, enetriyne alcohol, hydrocarolin, inulin, (+)sesamin, tetradeca-*trans*-2, *trans*-4-dienamides, a new series of tryamine amides.

Anagallis arvensis L. (Primulaceae)

(2n = 22, 40)

Syn : *Anagallis latifolia* L.

English names : Blue pimpernel, Scarlet pimpernel.

Vernacular names : Hin : Jonkmari; **Guj :** Anagallideae, Morgellina; **Pun :** Dhabbar.

Trade name : Jonkmari.

Traditional use : *Herb* : used in gout and dropsy; also used as a home remedy as diaphoretic, diuretic, resolvent, and in gall stones, liver cirrhosis and lung ailments.
HOMOEOPATHY : in hydrophobia, irritations in the urinary tract, cutaneous diseases and prickles in the body.

Modern use : *Herb* (MeOH extract) : inhibits herpes and poliomyelitis virus development; *Herb- and leaf-extract* : antifungal; *Herb* (50% EtOH extract) : antiviral, diuretic, and hypotensive.

Phytography : Annual herb, erect or procumbent, glabrous; stem 15–45 cm, branching from the base, branches gland-dotted; leaves opposite, rarely whorled, 2–4 cm, ovate or lanceolate, gland-dotted, acute; flowers axillary, solitary, bracteate, peduncles 2–5 cm long, slender, regular, blue or red; capsules pea-sized; seeds many, minute, endospermic.

Phenology : *Flowering* and *Fruiting* : January–March.

Distribution : West Bengal, Nepal westwards, ascending up to 1500 m, Central India, Nilgiri Hills; Sri Lanka.

Ecology and cultivation : Common as a winter weed in cultivated fields, in fallow lands, and hills above 800 m.

Chemical contents : *Plant* : triterpenes anagalligenone B, anagalligenin B, bitter cucurbitacin glycosides, arvenin I & II, cucurbitacin B, D, E and probably cucurbitacin I, L and R; *Root* : saponins of triterpene series; *Aerial parts* : n-hexacosane, rutin, β-sitosterol, stigmasterol, β-amyrin, lacceric acid, anagallingen, glucose, arabinose, xylose; *Flower* : stigmasterol, β-sitosterol, α-spinasterol 3-glucoside, kaempferol, quercetin, rutin; *Seed* : oil, higher alkanols, stigmasterol, sitosterol.

Anamirta cocculus W. & A. (Menispermaceae)

(2n = 16, 24, 26)

Syn : *Anamirta paniculata* Coleb., *A. flavescens* Miers., *Cocculus lacunosus* DC., *C. suberosus* DC., *C. populifolius* DC., *Menispermum cocculus* L.

English names : Levant berry, Fish berry, Crow killer.

Sanskrit name : Kakmari.

Vernacular names : **Ben, Hin** and **Guj** : Kakmari; **Mal :** Garaphala; **Mar :** Kakaphala; **Tel :** Kakmari, Kadilige.

Trade names : Indian berry, Kakmari.

Traditional use : *Berry* : used externally in indigenous medicine; also useful in some cases of ringworm.

Modern use : *Plant* : in skeletal fractures; *EtOH* (50%) *extract of aerial parts* : anticancer and diuretic; *Seed* : in night sweats of phthisis.

Phytography : Climbing shrub with corky bark; leaves simple, petiolate, petioles 5–15 cm long, lamina 10–20 cm long, ovate or ovate-cordate, glabrous above, hairy in the vein-axils beneath; panicles 30–45 cm, pendulous; flowers unisexual, 0.6 cm in diameter; kidney-shaped drupes on trifid gynophore, 1.6 cm, black, turn red on ripening.

Phenology : *Flowering* : April–May; *Fruiting* : June.

Distribution : Meghalaya, Orissa, Cuddapah, Malabar, Karnataka; Bangladesh, Sri Lanka and Pakistan.

Chemical contents : *Root* and *Stem* : berberine, palmatine, magnoflorine, columbamine; *Seed* : pierotoxin, cocculin or anamirtin, a fixed oil; *Perisperm* : menispermine, para-menispermine.

Remarks : Berries are fish poison; ointment prepared from drupes is used as insecticide.

Ananas comosus (L.) Merr. (Bromeliaceae)

(2n = 50, 100)

Syn : *Ananas sativus* Schult. f., *Bromelia comosa* L.

English name : Pineapple.

Sanskrit name : Ananasa.

Vernacular names : **Ben** : Anaras; **Hin** and **Mar :** Ananas; **Tam :** Anassappalam; **Tel :** Anasapandu.

Trade names : Ananas, Anaras

Traditional use : *Juice of leaf* : anthelmintic and helps in digestion, antifertility and in ringworms.

Modern use : *Leaf* : highly anthelmintic; *Unripe fruit* : abortifacient and purgative; *Juice of fruit* : antiscorbutic, diuretic, digestive, diaphoretic, promotes vitality.

Phytography : Shrubby or aborcus; leaves spirally arranged with the base, mostly with clustered radical rosettes, often with white glistening scales and margin very frequently prickly; flowers bisexual, regular, perianth separable into trimorous calyx and corolla; fruit baccate, sometimes combined below into a syncarp.

Phenology : *Flowering* : January–March; *Fruiting* : mainly in the Rainy season and in Autumn.

Distribution : Native to South America, now cultivated mostly in Assam, Tripura, West Bengal, coastal Andhra Pradesh, Tamil Nadu, Kerala, Karnataka, Orissa; Bangladesh, Pakistan, Sri Lanka and many other tropical and subtropical countries.

Ecology and cultivation : Tropical and subtropical; seminaturalised in some parts of warmer and moist areas; some ornamental-leaved pineapples are grown as garden plants.

Chemical contents : *Stem* : ananasic acid, bromelain, glyceryl esters of caffeic and p-coumaric acids; *Leaf* : β-sitosterol, campestanol, campesterol, ergosterol peroxides, stigmasterol, 5-stigmastene-3β, 7α-diol; *Fruit* : bromelin, bromolin A & B.

Remarks : Pineapple waste is a source of oxalic acid; and canned fruit is commercially more important than in its fresh form.

Anaphalis neelgerriana DC. (Asteraceae)

Syn : *Gnaphalium neelgerrianum* DC., *G. nilagiricum* Sch.

Sanskrit names : Raktaskandana, Varnapatra, Vrana pata.

Vernacular name : Tam : Katplaster.

Trade name : Varnapatra.

Traditional use : *Leaf* : bruished and applied to wounds and cuts as a plaster.

Phytography : Perennial, herbacious, wholly clothed with cottony wool, branches many, crowded on a stout woody stock, some very short, densely leafy, flowerless, others 10–25 cm long and flower-bearing; leaves narrow linear, slightly spreading and reflexed on the flowering branches; capitulums 0.4–0.6 cm in diameter, sessile, densely crowded or solitary or in corymbose clusters, bracts 0.4 cm long, variously shaped, white, opaque.

Phenology : *Flowering* : Winter.

Distribution : Nilgiri mountains, between 2000 and 3000 m.

Ecology and cultivation : Wild.

Andropogon contortus L. (Poaceae)

Syn : *Andropogon allionii* DC., *A. bellardi* Bub., *A. messanensis* Biv., *H. contortus* Beauv. ex Roem. & Sch.

English name : Spear grass.

Vernacular names : Ben : Kher; **Guj :** Dabhjuliyun; **Hin :** Shurval; **Mar :** Gantegawta, Kantigawta; **Ori :** Dauria; **Pun :** Suryala; **Tam :** Oosipullu; **Tel :** Eddigaddi.

Traditional use : *Plant* : in toothache, fever, atrophy, emaciation or cachexy, muscular pain, haematuria, dysentery and scorpion sting; *Oil* (distilled from awns) : in asthma.

Modern use : *Root* : stimulant and diuretic.

Phytography : Stem 0.3–2.0 m, erect or decumbent below, densely tufted, branched, compressed towards the base; leaves 15–45 cm, narrow, glabrous or hairy specially near the base, rigid, margins and upper surfaces scabrid, ligules truncate, ciliate, sheaths compressed, keeled; spikes 3.7–7.5 cm, lower spikelets contain male florets, male spikelets have coherent joints, upper spikelets sessile and contain female florets; joints very short and curved, callus long rufous-bearded with brown hairs.

Phenology : *Flowering* : September–December; *Fruiting* : October–January.

Distribution : Throughout India, ascending the Himalaya up to 1600 m; Sri Lanka, Myanmar.

Ecology and cultivation : Common in slopes, foothills up to 1200 m.

Chemical contents : *Plant* : myo-inositol, galactinol and raffinose.

Remark : Cellulose content of the grass is sufficiently high to warrant its utilisation for paper manufacture.

Andrographis echioides Nees (Acanthaceae)

Syn : *Justicia echioides* L.

Vernacular names : Guj : Kalu kariyatu; **Hin :** Birbhubat; **Mar :** Banchimani; **Mal** and **Tam :** Gopuramthangi; **Mun :** Gusumpuru; **Orn :** Nadnaur.

Traditional use : *Plant* : febrifuge, bitter tonic, astringent, anodyne, used in dysentery, cholera and diabetes.

Modern use : *EtOH* (50%) *extract of plant* : diuretic, in sluggishness of liver, jaundice.

Phytography : Annual herb, 20–60 cm in height, closely pubescent with patent white hairs; leaves sessile, oblong or subelliptic, 75 cm by 2.5 cm, obtuse; racemes numerous, axillary, as long as the leaves, pedicels very short; flowers pubescent, pink or white, lower lip dark purple, 0.8–1.25 cm; capsules 0.8 cm by 0.4 cm, acute at both ends; seeds glabrous.

Phenology : *Flowering* and *Fruiting* : July–January.

Distribution : Drier districts of tropical India from Punjab to Chotanagpur; Sri Lanka.

Ecology and cultivation : Plant of drier climate; wild.

Chemical contents : *Plant* : echioidin, echioidinin.

Andrographis paniculata (Burm. f.) Wall. ex Nees (Acanthaceae)

Syn : *Justicia paniculata* Burm. f.

English names : The great king of bitters, the creat.

Sanskrit names : Bhunimba, Kirata.

Vernacular names : Ben : Kalmegh; **Guj :** Kariyatu; **Hin :** Kirayat; **Kan :** Nelabaru; **Mar :** Olikiryata; **Tam** and **Tel :** Nelavemu.

Trade names : Kalmegh, Kirayat.

Traditional use : *Plant* : febrifuge, alterative, anthelmintic, anodyne, useful in debility, diabetes, consumption, influenza, bronchitis, itches and piles; in Bengal, household medicine known as 'Kalmegh', made from leaves, is given to the children suffering from stomach complaints.
HOEMOEPATHY : used for treatment of different ailments of head, mind, eyes, nose, mouth, tongue, throat, abdomen, stool, urine, fever and other modalities.

Modern use : *Drug constitute stem, leaf* and *inflorescence* : as a tonic and in the treatment of fevers, worms, dysentery and also beneficial to liver and digestive ailments; it is reported that it has some antityphoid and antibiotic activity; *Decoction* : used for sluggishness of liver and in jaundice.

Phytography : An erect herb with square stem, glabrous below, glandular hairy above; leaves linear, lanceolate, glabrous and distinctly pedicelled; flowers white or pale purple; capsules compressed transversely; seeds bony.

Phenology : *Flowering* and *Fruiting* : September–May.

Distribution : Throughout India in the plains and hills; Bangladesh, Pakistan, all South-East Asian and SAARC countries.

Ecology and cultivation : Common in stony lines in forests and in wastelands. Cultivated as an ornamental.

Chemical contents : *Plant* : kalmeghin, bitter principle andrographolide; bitterness is due to nonbasic principle.

Adulterants : It is used as adulterants for *Chirata*, and is a substitute for quinine.

Remark : Whole plant is bitter.

Aneilema conspicuum Kunth (Commelinaceae)

Syn : *Commelina conspicua* Blume

Traditional use : *Plant* : Emmenagogue.

Phytography : Herb, stem 30–60 cm, suberect or creeping and rooting below; leaves simple, alternate, petioled, lanceolate; panicles short, subsessile, pyramidal, branches horizontal or decurved, many-flowered; flowers bracteate and bracteolate, small, hermaphrodite, trimerous; capsules globose, glabrous, chambers 1-seeded; seeds rugose and hemispheric.

Distribution : Assam, Patkoye mountains; Penang, Malaya Islands.

Aneilema lineolatum Kunth (Commelinaceae)

Syn : *Aneilum elatum* Kunth, *Commelina lineolate* Blume

Traditional use : *Plant* : said to be abortifacient.

Phytography : Herb; stem stout, erect; leaves simple, alternate, large, narrowly (1.3–4.0 cm by 2.5–3.0 cm) oblong or ensiform, acuminate, margins white; panicles stout, glabrous effuse; flowers small, blue, bracteate, bracteolate, bracts marcescent; capsules ellipsoid or subglobosely trigonous, 0.6 cm in diameter; chambers 3- to 4-seeded; seeds smooth or reticulate or glandular puberulous.

Distribution : Sikkim, Khasia Hills at 660–1300 m altitude, Western Ghats from Maharashtra to Karnataka; Bhutan, Myanmar and Malaya Islands.

Ecology and cultivation : Grows in hills up to 1200 m; wild.

Aneilema scapiflorum Wight (Commelinaceae)

Syn : *Aneilema tuberosum* Ham.., *A. serotinum* Dor., *Commelina scapiflora* Roxb., *Murdannia tuberosa* Royle

Vernacular names : Ben : Kureli; **Hin :** Siyahmusli; **Guj** and **Mar :** Sismulia.

Traditional use : *Root* : astringent, tonic, in snake bite; dried and powdered, mixed with sugar as aphrodisiac and with Tulsi leaves in spermatorrhoea.

Modern use : *Root-bark* : used in colic, piles, infantile convulsions, asthma and incontinence of urine.

Phytography : Herb; roots elongated, tuberous; stem short; leaves simple, alternate, 10–25 cm long, narrow, ensiform, finely acuminate; scape erect with narrow panicle, panicles elongate, bracts large, sheathing, upper bracts small, amplexicaul; flowers small; capsules 0.6 cm, ellipsoid, trigonous, mucronate; seeds in a triangular column, angled, straw-coloured with white, minutely reticulate and glandular epidermis.

Phenology : *Flowering* : March to May.

Distribution : Temperate and tropical Himalaya, upper Ganga plains, Karnataka; Shan Hills (Myanmar) and Sri Lanka.

Ecology and cultivation : Frequently found in grass tract annually burnt and also in sparse forest; wild.

Anemone obtusiloba D. Don (Ranunculaceae)

Syn : *Anemone govaniana* Wall., *A. discolor* Royle, *A. micrantha* Klotzsch.

Vernacular names : Jau : Ageli; **Kum :** Kakriya, Ratamjota; **Pun :** Padar, Rattanjog.

Traditional use : *Seed-oil* : for rheumatism, as emmenagogue, antiinflammatory, analgesic and tonic.
UNANI : antidote to snake bite when taken with wine and good for sores in the mouth.

Modern use : *Rootstock* : for concussions; *Root mixed with milk* : used externally as a blistering agent, good in spleen and kidney diseases.

Phytography : Perennial herb; rootstock clothed with old leaf-sheaths, woody, fibrous; leaves many, stalked, suborbicular, tripartite, segments broad, cuneate, variously cut and lobed, rarely shortly petiolate—usually sessile; scapes 15–30 cm, 1- to 3-flowered; flowers white, purplish or golden, pedicellate, pedicels long, slender; achenes strigose, rarely glabrous.

Distribution : Kashmir to Sikkim at altitudes from 3000 to 5000 m.

Echology and cultivation : Plant of cold climate; wild.

Chemical contents : New triterpenoid saponin, obtusilobinin and obtusilobin, obtusilobicinin.

Anemone narcissiflora L. (Ranunculaceae)

Syn : *Anemone crinita* Juz.

Traditional use : *Plant* : said to be poisonous.

Modern use : Protoamnonin—an antibiotic obtained from the plant.

Phytography : Perennial herb; rootstock woody, stout, stem 30 cm; leaves petiolate, petioles long, lamina 2.5–5.0 cm in diameter, palmately 5-partite, segments deeply cut into linear lobes; scapes erect; flowers umbelled; achenes nearly orbicular with an incurved beak.

Distribution : Kashmir; West Asia, Siberia, Central Europe, northwestern America.

Echology and cultivation : Plant of cold climate; wild.

Chemical contents : *Herb* : betulinic acid, oleanolic acid and its rhamnoside, β-amyrin, β-sitosterol, 3 triterpenoid saponins, narcissiflorine, narcissiflorinine, narcissifloridine.

Anethum graveolens L. (Apiaceae)

(2n = 22)

Syn : *Peucedanum graveolens* Benth.

English name : Dill.

Vernacular names : Guj : Sawa; **Hin :** Soya; **Mar :** Sowa, Balan shep; **Ori :** Juani; **Pun :** Soye; **Tel :** Seemasopu, Sadhakuppapi,

Trade names : Dill, Soya.

Traditional use : Hill people of India give the *tea made from boiled seed* to newborn child suffering from loose bowels.

Modern use : *Dill herb oil* : a good carminative; a mixture of Dill herb oil, alcohohol and water is a common household remedy for flatulence in children.

Phytography : Annual or biennial herb; stem erect, slightly grooved, grows up to 1.2 m in height, contains oil; leaves alternate, light green, decompound, contains oil; flowers small, yellow, grow in compound umbels; fruits brown, oval, dry, grooved, consist of 2 compressed, indehiscent parts, 3–4 mm long, 2–3 mm broad, 1 mm thick, winged; odour and taste characteristically aromatic.

Distribution : Native of Abyssinia (Ethiopia), Egypt, Iran, southern USSR and adjoining European countries; cultivated in Germany, Hungary, Holland, England and USA. In India, it is cultivated on a small-scale in Jammu and Kashmir.

Ecology and cultivation : Plant of cold climate, grows well in any good garden soil. Fruits are sown in the field in spring. Frequent weeding of the field is essential. The crop is susceptible to weather hazards.

Chemical contents : *Root* : lipid and essential oil; *Aerial parts* : an essential oil, the composition of which varies in different ecotypes; sopoletin and a new coumarin, vitamin C, carvone, d.a.-phellendrene; *Leaf* : flavonoids, glycerides; *Fruits* (*so called seeds*) : essential oil, 3 phenolic acids, 6 flavonoids including vicenin, a new xanthone glyceride.

Remark : It is closely related to the Indian Dill, i.e., *Anethum sowa.*

Anethum sowa Kurz (Apiaceae)

Syn : *Peucedanum graveolens* L.

English name : Dill.

Sanskrit name : Shatapushpi.

Vernacular names : Ben and **Hin :** Sowa; **Guj :** Surva; **Kan :** Sabsige; **Tam :** Sata kuppi.

Trade name : Sowa.

Traditional use : *Essential oil* : given to children in flatulence.
UNANI : Ingredient of 'Shibt (soya)'.

Modern use : *Dried ripe fruits* : used as carminative and stomachic.

Phytography : A graceful annual, about 1 m high with striated darker and lighter green stems covered with a whitish bloom and decompound leaves cut into filiform segments; flowers yellow, in umbels often enlarging into fruits.

Phenology : *Flowering* and *Fruiting* : Cold season.

Distribution : Cultivated all over India; Bangladesh and Pakistan.

Chemical contents : *Stem, Leaf* and *Fruit* : Essential oil, carrone content is low.

Angelica archangelica L. (Apiaceae)

Syn : *Archangelica officinalis* Hoffm.

English names : Angelica, Archangel, Holy ghost.

Trade name : Angelica.

Traditional use : *Root*, *Rootstock* and *Fruit*: stimulant, expectorant and diaphoretic.

Modern use : *Herb* : used in medicines for digestion, bringing instant relief from colic, flatulence and heart burn; *Infusion of leaf* and *stem* : provides fairly good gargle for the treatment of sore tonsils and throats.

Phytography : A perennial herb with smooth, purple-green, hollow stems; leaves twice pinnate; flowers green-white; fruits cremocarps oval, externally greenish yellow or pale yellowish brown.

Distribution : Native of Syria, and northern Siberia; now cultivated in Kashmir.

Ecology and cultivation : Plant of cold climate.

Chemical contents : Musk-like odour of *root-oil* is generally attributed to pen-tadecanolide which is one of the most widely used fixatives in first class perfume; *Dry rootstock* : 0.35% essential oil; *Fruit* : 1.2–1.3% essential oil; volatile oil containing a terpene with a pepper odour, etc.

Angelica glauca Edgw. (Apiaceae)

English name : Angelica.

Vernacular names : Hin : Chora; **Mar :** Chora, Chura.

Traditional use : *Herb* : in dyspepsia, constipation, ulcer of palate, infantile atrophy, dysentery, menorrhagia and rinderpest; *Root-powder* : in stomach trouble of children, vomiting, and as cardiac stimulant; *Dal* or curry flavoured with this : gives vitality and strength to women after delivery.

Modern use : *Herb* : as stimulant and also in dyspepsia and constipation.

Phytography : Glabrous herb, 1–4 m high; leaves uni-, bi- or tripinnate, pinnae large, toothed; umbels compound with many rays; many bracts and bracteoles; florets white or lurid purple; fruits 1.25 cm by 0.6 cm, subquadrate; seeds small.

Distribution : From Kashmir to Himachal Pradesh at altitude from 2700 to 3000 m.

Ecology and cultivation : Plant of sub-Alpine climate; wild.

Chemical contents : *Dried roots* (of plants in Jammu and Kashmir) : an essential oil (0.4%); it contains lactones, sesquiterpenes, δ-α-cadinene, umbelliprenin and a terpene alcohol; *Fruit* : an essential oil.

Remark : Roots are used as spice and condiment.

Anisochilus carnosus (L.) Wall. (Lamiaceae)

English name : Thick-leaved lavender.

Sanskrit names : Ajapada, Ulapalabhenda.

Vernacular names : Hin : Tanjiri-ka-pat, Sitaki; **Mal :** Patukurkka; **Mar :** Kapurli; **Tam** and **Tel :** Karpuravalli.

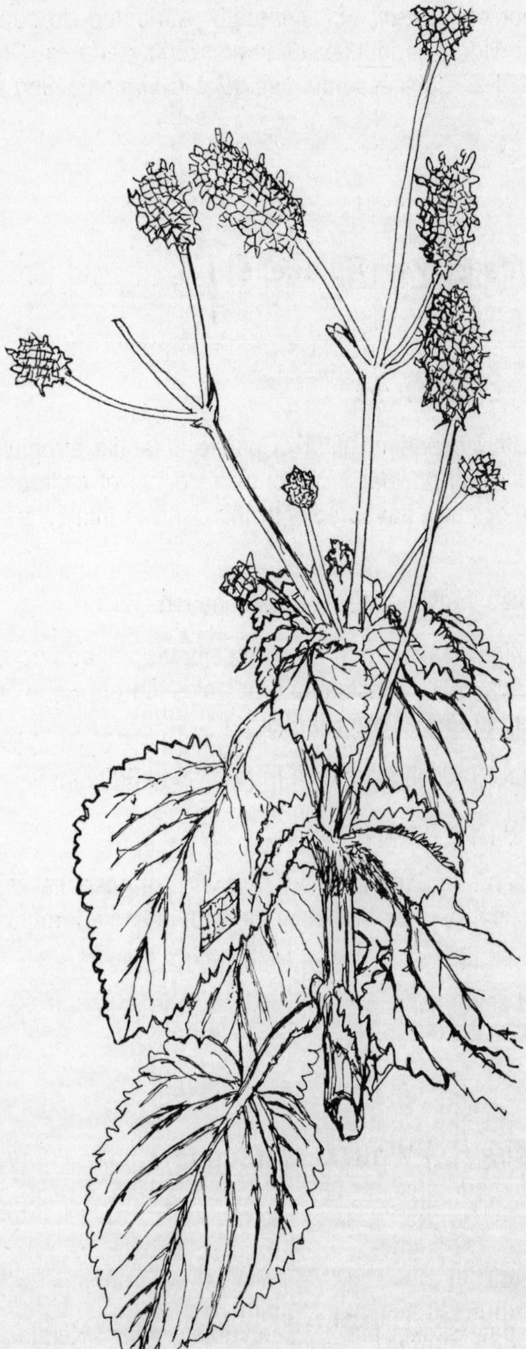

Traditional use : *Leaf* : fresh juice in urticaria associated with liver disorder; mixed with sugar candy for coughs and cold; mixed with sugar candy and gingelly oil as cooling liniment for the head.

Modern use : *Plant* : diaphoretic, expectorant, stimulant and used for cold and cough; *Alcoholic extract of plant* : antibacteria; *Essential oil* : antitubercular.

Phytography : Erect herb; stem 66 cm–1.0 m, variable indumentum; petiole long, leaf simple, ovate; panicled spike with long peduncles, whole inflorescence covered with small red glands; flowers pale purple; fruit 0.25 cm long.

Phenology : *Flowering* : October–November.

Distribution : Introduced from East Indies; occurring in West Himalaya, Central and South India.

Ecology and cultivation : Lithophyte—specially on limestone rocks; wild.

Chemical contents : *Plant* : essential oil, diasmetin and diosmin; *Leaf* : luteolin and apigenin.

Anisomelis indica (L.) O. Kuntze (Lamiaceae)

Syn : *Anisomelis ovata* R. Br., *Nepeta indica* L.

Vernacular names : Hin : Kalabhangra; **Mar :** Gopali.

Traditional use : *Plant* : tonic; *Oil* : in uterine affections.

Modern use : *Plant* : used as carminative astringent; *EtOH* (50%) *extract of herb* : hypothermic.

Phytography : A small, smelling, softly pubescent herb, 90–180 cm high with ovate coarsely crenate acuminate leaves, 25–75 mm long and purplish flower on axillary dense-flowered whorls.

Phenology : *Flowering* : September–November; *Fruiting* : December–January.

Distribution : Throughout India, up to 2000 m; Bangladesh and Pakistan.

Ecology and cultivation : Occurs in wastelands and scrub jungles; wild.

Chemical contents : *Root* : stigmasterol, β-sitosterol, paraffins and fatty acids; *Leaf* : diterpenes, ovatodiolide and its derivatives.

Remarks : Tribals use plant-decoction in animal dysentery; crushed leaf as ointment on the inflammation of neck of bullock; leaf-extract as anti-inflammatory and anti-arthritic in albino rats.

Anisomeles malabarica R. Br. ex Sins (Lamiaceae)

Syn : *Anisomeles intermedia* Wight, *Nepeta malabarica* L.

English name : Malabar catmint.

Traditional use : *Leaf-infusion* : used in catarrah, intermittent fever and bowel disorder; *Essential oil* : for rheumatism.

Modern use : *EtOH* (50%) *extract of aerial parts* : diuretic, anticancer, spasmogenic and spermicidal.

Phytography : Erect, 1.3–6.0 m tall, branched herb; branches sometimes very stout, densely clothed with snow-white appressed trichomes; leaves opposite, short-petioled, 5–15 cm, basically oblong, very thick; flowers in verticillaster or lax-flowered branched cymes, bracts filiform, teeth very slender; calyx woolly, petals purple; nutlets pale.

Distribution : Deccan peninsula.

Ecology and cultivation : Grows in hot, dry places; wild.

Chemical contents : *Plant* : β-sitosterol, 2 diterpeniod lactones, ovatodiolide, anisomelic acid, anisomelolide, malabaric acid and derivatives, anisomelol; *Aerial parts* : an isoflavonoid, anisomelin identical with cirsillineol.

Remark : Deserves commercial cultivation.

Annona muricata L. (Annonaceae)

(2n = 14, 16)

English names : Prickly custard apple, Soursop, Guanabana.

Vernacular names : Hin : Mamphal; **Mal :** Mullanjakka, Vilatinunna; **Mar :** Mamphal; **Tam :** Mullu chitta, Pulipala.

Trade name : Mamphal.

Traditional use : *Green leaf* : purgative; *Leaf* and *Flower* : used in kidney troubles; *Fruit* : antiscorbutic; *Seed* : emetic, astringent.

Modern use : *Leaf* : sedative.

Phytography : Erect, small, evergreen tree; stem branched; leaf alternate, simple, entire, oil duct present; flowers solitary, trimerous, regular, velvety; fruit an aggregate of berries, united to the receptacle, more or less heart-shaped, the dark green surface rough; seeds large, endospermic.

Phenology : *Flowering* : May–June; *Fruiting* : Winter.

Distribution : Native of West Indies; cultivated to a less extent in Assam, Meghalaya and the adjacent areas as well as in South India; Sri Lanka.

Ecology and cultivation : Wild in the jungles and on the scrub hills.

Chemical contents : *Bark* : two crystalline alkaloids—muricine and muricinine, both belonging to aporphine group; *Leaf* : traces of essential oil and alkaline substance.

Remarks : Fruits edible. Seeds used as insecticide and fish poison.

Annona reticulata L. (Annonaceae)

(2n = 14)

English name : Bullock's heart.

Vernacular names : Asm : Atlas; **Ben :** Nona; **Guj, Hin, Mar** and **Pun :** Ramphal; **Kan :** Rama phala; **Kon :** Anon; **Mal :** Atha; **Ori :** Neua, Nena, Badhiala; **Tam :** Rama sita; **Tel :** Rama sita, Rama pandu.

Trade name : Ramphala.

Traditional use : *INDIAN RURAL FOLK* : *Bark-powder* : in dysentery and as vermifuge; *Fruit-pulp* and *seed-paste* : to kill lice.

Modern use : *Fruit* : anthelmintic, antidysenteric.

Phytography : Small tree, reaches up to 8 m in height; leaves simple, 12.5–20.0 cm long, 3.7–5.0 cm wide, oblong, upper surface smooth, lower surface roughish, petiole 1.2 cm long; flowers 2–3 together on lateral peduncles, trimerous; fruits are fleshy syncarps—formed by the fusion of numerous ripened ovaries and receptacles, subglobose, exterior roughish, with pentagonal areoles.

Phenology : *Flowering* : June–July; *Fruiting* : December onwards.

Distribution : Native of tropical America and West Indies, naturalised in West Bengal and some other parts of India; occasionally cultivated.

Ecology and cultivation : Grows up to 1000 m.

Chemical contents : *Root-bark* : liriodenine, oxoushinsunine, anonaine, michelalbine, reticuline; *Seed* : a resinous mixture—highly toxic to aphides.

Annona squamosa L. (Annonaceae)

(2n = 14, 28)

English names : Custard apple, Sugar apple, Sweet sop.

Sanskrit name : Gandagatra.

Vernacular names : Asm : Atlas; **Ben** and **Ori :** Ata; **Guj :** Sitaphal; **Hin** and **Pun :** Sarifa, Sitaphal; **Kan :** Seethaphala; **Kon :** Ater; **Mal :** Sitapalam, Seema atha; **Tam :** Attamaram; **Tel :** Sitaphalamu, Sitapandu.

Trade names : Sitaphal, Sarifa.

Traditional use : *Leaf*, *Fruit* and *Seed* : irritation to conjunctive and ousterus abortifacient.

Modern use : *Root* : purgative.

Phytography : Shrub or small tree with oblong-lanceolate leaves, nearly glabrous and slightly scented; flowers drooping, yellowish green; fruits tubercled.

Phenology : *Flowering* : March–May; *Fruiting* : July–September.

Distribution : Native of South America and the West Indies; now cultivated throughout India; Banagladesh and Pakistan; sometimes may be seen as feral.

Ecology and cultivation : Mesophyte; deciduous; leaf fall in February; new foliage–March onwards.

Chemical contents : Amorphous alkaloid, toxic resin and oil; hydrocyanic acid in leaves, bark and root.

Remarks : Seeds insecticide, used to remove lice from head and as fish poison.

Anogeissus acuminata (Roxb.) Wall. ex Guill & Perr. (Combretaceae)

Syn : *Anogeissus hirta* Wall., *Conocarpus acuminata* Roxb.

English name : Button tree.

Vernacular names : Ben : Chakwa; **Hin** and **Ori :** Pansi(a); **Kon :** Dehra, Gora hasel; **Raj :** Dhauria; **Tam :** Nunnera; **Tel :** Panchman, Pansi.

Trade name : Yon.

Traditional use : *Bark-powder* with fruit of *Terminalia chebula* and common salt : appetiser.

Modern use : *Seed-extract* : antibacterial and antifungal.

Phytography : Deciduous tree, sometimes attaining 20 m in height; trunk may be erect, armed with spines when young; bark 1.25 cm thick, rough, dark grey; leaves alternate, simple, shortly petiolate, leaf blade 5.0–6.25 cm by 1.25–2.5 cm, elliptic or oblong; heads globose, peduncles solitary, rarely clustered; flowers small, ovary inferior; fruits small, winged and beaked.

Phenology : *Flowering* : March–April; *Fruiting* : April–May.

Distribution : Almost throughout India, ascending up to 1000 m; Bangladesh(Chittagong); large supplies are available in the forests of Orissa.

Ecology and cultivation : Forest plant; often found along the river banks on alluvial ground; a characteristic species of freshwater swamp forests in the plains; may thrive in dry regions.

Anogeissus latifolia (Roxb.) Wall. ex Bedd. (Combretaceae)

Syn : *Conocarpus latifolia* DC.

English names : Axle wood, Button tree, Dindiga tree, Ghatti tree.

Sanskrit name : Dhava.

Vernacular names : Ben : Dhoaya; **Hin :** Bakla, Bakli, Bankli, Dhaura, Dhauri; **Kan :** Dinduga, Dindler; **Mal :** Mala kangiram, Marukinchiram; **Mar :** Dhavada; **Ori :** Dhobi, Dohu; **San :** Hesel; **Tam :** Vellay naga; **Tel :** Chiriman, Chirimanu, Sheriman, Yella maddi.

Trade names : Axle wood, Gatty gum, Ghatti gum.

Traditional use : *Plant* : in toothache, spleen enlargement, dysuria, diarrhoea, cold, cough, cholera, excessive perspiration, piles, urinary disorders, in scorpion sting and snake bite; *Bark* : bitter, astringent; *Gum* (called Ghatti or Indian gum) : as tonic after delivery.
UNANI : A constituent of the drug 'Gul-e-Dhawa'.

Modern use : *Gum* : gives stable oil-in-water emulsion and patented for incorporating oil-sol, vitamin preparations; has an excellent emulsifying properties.

Phytography : Usually small tree, but may be 21 m high and a girth of 2 m; leafless during most of the hot season; trunk erect, branchlets drooping; bark 0.6 cm thick, whitish grey, smooth, having shallow irregular depressions; leaves alternate or apparently opposite, petiole 1.25 cm or smaller, leafblade may be 12.5 cm long, usually much smaller, broad elliptic, deep red leaves are shed in February and the tree remains leafless till April–May; flowers in dense, globose heads on axillary peduncles; peduncles 1 or more from the same axil, often branched; flowers small, bracteate, ovary inferior; fruits small, sometimes 0.9 cm long, beaked, coriaceous, rusty-pubescent when young, compressed, 2-winged, 1-seeded.

Phenology : *Flowering* : September–January; *Fruiting* : December–March.

Distribution : Common in dry deciduous forests of India, Pakistan and Sri Lanka; ascending to 1000 m; but not found in the trans-Ganga plains.

Ecology and cultivation : Wild; may be cultivated by sowing seeds on well-irrigated raised beds; seeds produced after years of drought are fertile—others usually do not respond to cultivation; seedlings should be transplanted without injury. The tree coppices well and pollarding increases leaf-crop.

Chemical contents : *Bark* : quinic acid, shikimic acid, steroids, 3, 3', 4-tri-OMe-flavellagic acid and two glycerides; *Heartwood* : ellagic acid and its derivatives, gallic acid, quinic acid, shikimic acid, quercetin, myricetin, steroids; *Plant-mucilage* : aldobiouronic acid, aldotriouronic acid; *Gum* : calcium salt of ghattic acid; *Leaf* : quinic acid, shikimic acid.

Adulterant : It is used as substitute for gum arabic.

Remarks : Plant is used in treatment of rinderpest in cattle. *Tasar* silkworm is fed on it. Gum is extensively used in cloth-printing, used in the food industry as an emulsifier. Wood is used for making axles of carts, cart-wheels and agricultural implements. Santals use decoction of leaves for preparing black dye.

Anogeissus pendula Edgw. (Combretaceae)

Syn : *Anogeissus myrtifolia* Wall.

Vernacular names: Guj : Dhau, Dhaunkara; **Hin** and **Raj :** Dhao, Dhaunkara, Kala dhaura; **Mar :** Kardhahi.

Modern use : *EtOH* (50%) *extract of aerial parts* : diuretic and cardiovascular stimulant; *Seed-extract* : haemagglutinating against human A, B and O red cells.

Phytography : Gregarious bush or small tree; leaves alternate, simple, petiolate, 1.8–2.5 cm long, elliptic or obovate, acute or obtuse; flowers in heads, peduncles solitary, simple, sometimes with an elliptic leaf-like bract; flowers small, ovary inferior; fruit subquadrate, glabrous, beaked.

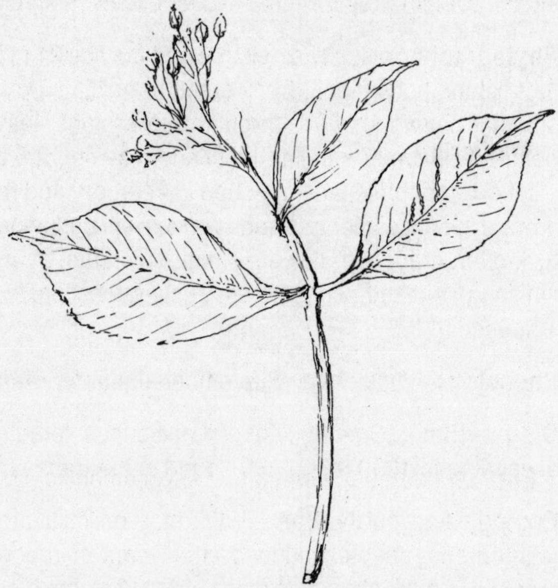

Distribution : Common in Rajasthan, found in Gujarat, western Uttar Pradesh; Pakistan (NW Frontier Province).

Ecology and cultivation : Xerophyte; wild.

Adulterant : May substitute 'Ghatti', or 'Indian gum'.

Anthemis cotula L. (Asteraceae)

(2n = 18)

Syn : *Maruta cotula* DC.

English names : Dog fennel, Mey weed, Wild chamomile.

Traditional use : *Plant* : in inflammation, acrid, blisters of the skin; *Leaf* : unguent for old sores, causes contact dermatitis, insecticide; *Leaf* and *Flower* : antiseptic, tonic, emmenagogue, emetic and useful in headache.

Phytography : Perennial, erect, glabrous or hairy, foetid herb, 20–45 cm high; stem corymbosely branched; leaves alternate, pinnatisect, segments very narrow, gland dotted; heads 2.5–3.7 cm in diameter, terminal, peduncles slender, involucre of bracts multiseriate, oblong, margins scarious; florets small and heterogamous; achenes oblong, glabrous, 4- to 5-angled, 8- to 10-ribbed, tubercled, pappus absent or very short, paleaceous or membranous.

Distribution : Native of Europe and the Mediterranean regions; found in arid zones and desertic parts of Rajasthan; Pakistan (Baluchistan and Sind).

Ecology and cultivation : Xerophyte; grows in arid and rocky regions; wild.

Chemical contents : *Plant* : alkaloids; *Leaf* : flavanoids; *Leaf* and *Flower* : quercetin, 5-glyceride; *Dried flower-head* and *flowering-plant* : volatile oil, mucilage, an antispasmodic substance.

Adulterant : Used as adulterant of *Anthemis nobilis* L. (Roman chamomile).

Anthemis nobilis L. (Asteraceae)

(2n = 18)

English names : Roman or True chamomile, Double chamomile.

Trade name : Chamomile.

Traditional use : *Dried flower-head* : stimulant, carminative, and nervine tonic, an infusion known as 'chamomile tea' is domestic remedy for indigestion; it helps offset some reactions to alcoholic drinking; *Chamomile in composition* : liquid bactericide, deodorant, softer for skin hygiene; *Flower* : in sprains, bruise and rheumatism.
UNANI : Chamomile is an ingredient of 'Majnoon-e-flasfa'.
HOMOEOPATHY : excellent remedy for cough and cold, pain in throat and for eye troubles.

Modern use : *Essential oil* : improves appetite, but violent poison in large doses; prolonged use cause mental disturbance and madness.

Phytography : Perennial aromatic herb; dried flower head or capitula is hemispherical, white or nearly white in colour, becoming yellowish or buff coloured later on.

Distribution : Native of Europe and Mediterranean regions; found chiefly in the vicinity of arid and rocky zones of western Rajasthan; Pakistan (Baluchistan, Sind).

Ecology and cultivation : Xerophytic; grows in arid and rocky zones; wild; suitable for cultivation in stiff loam.

Chemical contents : *All aerial parts* : essential oil; *Leaf* : hydroxyisonobilin; *Flower* : essential oil, sesquiterpene lactone, nobilin, 3-epinobilin, 3-dehydronobilin, 1,10-epoxynobiline, glucosides, n-nonacosane, n-heptacosane, lignoceryl alcohol, β-amyrin and ψ-taraxasterol. Chemical contents of flowers of different countries and various ages differ considerably.

Adulterants : *Anthemis cotula* is an adulterant of *A. nobilis* L., *Single chamomiles, Matricaria chamomilla* L. and *Chrysanthemum parthenium* Bern are used as its adulterants.

Anthericus cerefolium Hoffm. (Apiaceae)

Syn : *Scandix cerefolium* L., *Cerefolium cerefolium* Britt.

English name : Salad chervil.

Trade name : Atrila (Indian bazar).

Traditional use : Diuretic and stomachic

Phytography : More or less hairy, 5-leaved, annual with erect branching, stem, ± 30–60 cm high, leaves decompound—ultimate leaflets 7–8 mm long, ovate to oblong and deeply cut; flowers white, minute in compound umbel; fruits black, smooth.

Phenology : *Fruiting* : Summer.

Distribution : Native of Caucasus, South Russia and extending up to West Asia, Quattar and Pakistan (Peshwar); also found in Rajasthan.

Ecology and cultivation : Suited to higher elevations in the tropics; wild.

Anthocephalus chinensis (Lamk.) A. Rich ex Walp. (Rubiaceae)

Syn : *Anthocephalus cadamba* (Roxb.) Miq., *Anthocephalus indicus* A. Rich. *Nauclea cadamba* Roxb.

English name : Kadam.

Sanskrit names : Kadamba, Neepa.

Vernacular names : Asm : Raghu, Kadam; **Ben :** Kadam, Neepa; **Guj :** Kadamba; **Hin :** Kadam, Kadamba; **Kan :** Bale, Kadamba mara, Kadavela; **Kon :** Kodomb; **Mal :** Kadambu, Katampu; **Mar :** Kadamb, Kadamba; **Ori :** Kadamba; **Pun :** Kadam; **San :** Kadam dare; **Tam :** Vellaikadamba; **Tel :** Kadambamu, Kadambe, Kallambamu.

Trade names : Kadam, Kadamba.

Traditional use : SANTALS : (i) *Bark* : paste is given to the patients of continuous fever to drink, juice squeezed out from the paste obtained by grinding this bark along with barks of mango tree and sal tree is given to the patient of cholera with a little shell lime; body of patient suffering from anaemia is plastered with the paste obtained by grinding this bark with the same of *Celastrus paniculatus*; (ii) *Decoction of stem-bark* : given to the patient of dyspepsia; (iii) *A piece of twig* : as a constituent of a number of herbal compounds

used for treatment of wounds and ulcers; (iv) *Leaf*: to cover sores and wounds; (v) *Juice of leaf with common salt*: in stomach pain.
SHUKLA YAJURVEDA: pollens strengthen body and mind; CHARAKA SAMHITA: prescribes the following uses : (i) *Bark* : strengthens body and removes fever; *Bark-powder* along with equal quantity of alum and opium : useful in eye troubles; drinking *extract of bark* stops vomiting; *extract* is also useful as analgesic, in rectifying defects of semen, to cure urinary troubles; (ii) *Leaf* : cures pimples and wounds, analgesic; *Leaf-juice* : stops burning sensation of palms and feet and scaling off of skin; *Extract of leaf*: for gurgle; (iii) *Fruit*: good to quench thirst during high fever; SUSHRUTA SAMHITA: *Leaf*: analgesic, useful in diseases caused by phlegm, rheumatism and in inflammation of glands; BRAHMAVAIVARTA PURANA : brushing teeth with twig of this plant beneficial for teeth; AYURVEDA : *Root* and *Flower* : abortifacient; *Fruit* : cooling, analgesic and decoction in stomatitis.

Modern use : *EtOH* (50%) *extract of stem-bark* : anthelmintic and hypoglycaemic.

Phytography : Large deciduous tree with a straight stem about 20 m high and up to 2 m girth with a beautiful contour of drooping branches; leaves simple, opposite, stipulate, petiolate, coriaceous, 12.5–23.0 cm long, shining above, pubescent beneath; flowers in globose heads, orange coloured, scented at night; fruits confluent into a fleshy globose mass of many few-seeded coriaceous pyrenes.

Phenology : *Flowering* : May–July; *Flowering* : August–October.

Distribution : Karnataka, Tamil Nadu, Andhra Pradesh, Orissa, Bihar, West Bengal, Assam; sub-Himalayan tract from Nepal eastwards to Myanmar and Bangladesh.

Ecology and cultivation : Common in moist deciduous forest; thrives well in well-drained alluvial soil; grown as roadside tree and also in gardens; grows rapidly in the first 6–8 years, attains maximum size by 20 years.

Chemical contents : *Root*: quinovic acid; *Stem-bark*: cadambagenic acid, quinovic acid, β-sitosterol, saponins A, B, C, D; *Leaf* : cadambine, 3d-dehydrocadambine, 3β-dihydrocadambine, isodihydrocadambine, hentriacontanol, β-sitosterol, cadamine.

Remarks : Timber used in making stools, benches, boards, etc.; fruits edible.

Anthoxanthum odoratum L. (Poaceae)

(2n = 10)

English name : Sweet vernal grass.

Traditional use : Specially provocative of hay fever and hay asthma; a medicinal tincture from it sniffed well into the nose and throat for immediate attack.

Phytography : Perennial, sweet-scented grass; stem 15–45 cm, tufted, erect or ascending; leaves flat and hairy; panicles 2.5–12.5 cm, pubescent, spikelets 0.6–0.8 cm, fascicled, green, 2-flowered, narrow, persistent on short pedicles; grains terete, free within the shining glume.

Distribution : A native of Central and North Asia, found specially in the hilly regions as cultivated and as feral.

Chemical contents : *Grass* : yields an aromatic essential oil, named 'Fluove oil' with esters and coumarins as chief constituents—aroma is due to coumarins; coumarin content varies with season and parts of the grass; *Leaf* and *Node* : also contain flavone.

Remark : Aromatic hay.

Antiaris innoxia Bl. (Moraceae)

Syn : *Antiaris toxicaria* (Pers.) Lesh., *A. saccidora* Dalz.

English names : Upas tree, Sack tree.

Sanskrit name : Valkala.

Vernacular names : Hin : Jasund; **Kan :** Ajjanapati; **Mar :** Chandla, Karvat; **Mal** and **Tam :** Nettavil.

Trade name : Upas tree.

Traditional use : *Wood-powder boiled with Canjee* : in diarrhoea; *Seed* : febrifuge, in dysentery.

Modern use : *EtOH* (50%) *extract of aerial parts* : diuretic, stimulates intestine and uterine contraction; in mild doses, the *juice* is cardiac and circulatory stimulant.

Phytography : Majestic tree, attaining 82 m in height; branchlets glabrous; leaves alternate, stipulate, petiole very short, leaf blade oblong or elliptic, 10–20 cm long, glossy; flowers monoecious, males crowded on the surface of an axillary, fleshy, shortly peduncled receptacle, females solitary in a pyriform involucre of bracts; fruits fleshy, pericarp confluent with involucre of bracts, purple or scarlet or crimson, pyriform, velvety, tipped with a few bracts, intensely bitter, 1-seeded; seed hard, exalbuminous.

Distribution : Evergreen forests of Maharashtra, Karnataka and Kerala, specially up to 700 m. It also occurs in Sri Lanka, Myanmar and Malaysia.

Ecology and cultivation : Plant of subequatorial forest; wild.

Chemical contents : *Cardenolides from latex* : antioside, α-antiarin, β-antiarin, desglucocheirotoxin, convallatoxin, convallatoxol; *Seed* : 34 cardenolides including cymarin, cymarol, strophanthidin, periplogenin, peripalloside, strophalloside, etc.

Remark : Latex is used as arrow and dart poison by the Adivasis.

Antidesma alexiteria L. (Stilagianaceae)

Syn : *Antidesma zeylanicum* Lam.

Vernacular name : Tam : Nolaidali.

Traditional use : *Leaf* : antidote to snake bite.

Phytography : Much-branched slender shrub or small tree; leaves alternate, simple, stipulate, very variable in form and size, coriaceous, pubescent, shining on both surfaces; spikes simple or panicled, slender, pubescent, bracts short, broad; male flowers sessile, female flowers very shortly pedicelled; drupes 0.6 cm in diameter, gibbously orbicular.

Phenology : *Flowering* : March–July; *Fruiting* : November–December.

Distribution : Orissa, Andhra Pradesh, Tamil Nadu, Kerala, Karnataka; Sri Lanka.

Ecology and cultivation : Occurs in valleys in hilly areas and in the sides of streams.

Antidesma bunius (L.) Spreng. (Stilaginaceae)

Syn : *Antidesma ciliatum* Presl., *A. sylvestre* Lamk., *Stilago bunius* L.

English names : Chinese laurel, Salamander tree.

Vernacular names : Aru : Sonkongasing; **Asm :** Bor heloch, Pani heloch; **Gar :** Bol-aborak; **Kha :** Dieng-soh-silli; **Lep :** Kantjer, Kunchur-kung; **Mal :** Nulittali; **Mar :** Amati; **Nep :** Himalcheri; **Tam :** Nolaiali; **Tel :** Anepu.

Traditional use : *Leaf* : antidote to snake poison; *Boiled young leaf* : in syphilitic cachexia; *Bark* : poisonous.

Modern use : *Leaf* : diaphoretic.

Phytography : Small evergreen tree; shoots tomentose; leaves alternate, petiolate, petioles 0.4–0.6 cm, stipulate, stipules deciduous, simple, 7.5–15.0 cm by 3.7–6.2 cm, very variable in form; spikes simple or panicled, stout, pubescent, often 10.0–12.5 cm long, slightly rusty-tomentose, bracts minute; male flowers sessile or subsessile, female flowers shortly pedicelled; drupes small, compressed.

Phenology : *Flowering* : April–May; *Fruiting* : August.

Distribution : Native to Malaysia; found throughout the hotter parts of India, mainly in Bihar, Sikkim, Assam on the high mountains only; Nepal, Bangladesh, Pakistan and Sri Lanka.

Ecology and cultivation : Found in evergreen forests; wild.

Chemical contents : *Bark* : an alkaloid; *Stem* : friedelin, dammaro-20, 24-dien-3-ol; *Leaf* : friedelin.

Antidesma diandrum (Roxb.) Heyne ex Roth (Stilaginaceae)

Syn : *Antidesma acidum* Retz., *A. lanceolarium* wall., *Stilago diandra* Roxb.

Vernacular names : Ben : Archal, Mutta; **Hin :** Matta ara, Dhakki, Aamli khatua, Katma; **Kan :** Karrihulipa; **Mal :** Areepazham; **Mar :** Ghondurili; **Ori :** Kundui; **Tam :** Ariporian; **Tel :** Pellagumudi.

Traditional use : HILL TRIBES : in whitlow, sores, dropsy, anasarca, muscular pain, carbuncle, gravel, dysentery, pneumonia, pleurisy, adenitis axillaris, intercostal neuralgia and bite of rabid jackal.

Modern use : *EtOH* (50%) *extract of aerial parts* : spasmolytic.

Phytography : Bushy small tree; shoots glabrous or rusty-pubescent; leaves alternate, simple, stipulate, stipules slender, shortly (0.26–0.30 cm) petioled, size and shape very variable, 2.5 by 1.6 to 12.5 by 5 cm in size, thin, yellowish green when dry; racemes very slender, usually branched, glabrous, bracts short—lower often the longest, glabrous; flowers minute, sessile or subsessile; drupes 0.3–0.4 cm long, ovoid.

Phenology : *Flowering* : May–June; *Fruiting* : November–January.

Distribution : Tropical Himalaya from Garhwal to Sikkim, up to 1200 m, West Bengal, common throughout peninsular India; Sri Lanka.

Ecology and cultivation : Tropical alpine plant; wild, chiefly in valleys and near streams.

Chemical contents : *Trunk-extract* : sitosterol; *Leaf* : vitamin B complex.

Antidesma ghesaembilla Gaertn. (Stilaginaceae)

Syn : *Antidesma paniculatum* Roxb.

Engish name : Black currant.

Vernacular names : Asm : Heloch; **Ben :** Judijamb; **Hin** and **Pun :** Umtao; **Kan :** Pullampurasi, Gida; **Lep :** Chipli; **Mik :** Theng-merok-arong; **Ori :** Nuniare, Jamula; **Tel :** Lona.

Traditional use : *Bark* : astringent and tonic; *Wood* : emmenagogue.

Modern use : *EtOH* (50%) *extract of aerial parts* : diuretic and hypotensive.

Phytography : Tree, about 14 m high; branches, leaves and spikes often rusty-tomentose; leaves alternate, simple, petiolate, petioles 0.4–1.25 cm long; leaf blade 5.0–7.5 cm long, orbicular or obovate or oblong, obtuse; spikes rather slender, panicled, bracts minute; flowers sessile, male flowers clustered; berries dark purple with pleasant subacid taste, subglobose, 0.6 cm long.

Phenology : *Flowering :* Summer; *Fruiting* : Winter.

Distribution : Tropical Himalaya, Simla to Assam, West Bengal and South India; Bhutan and Bangladesh.

Ecology and cultivation : Grows in semiarid and drier zones on hill sides and scrub jungles; wild.

Antidesma menasu Miq. ex Tul. (Stilaginaceae)

Syn : *Antidesma pubescens* H.f. & T.

Vernacular names : Kan : Kadivalasoppu; **Mal :** Putharaval; **Tam :** Kalakalattha.

Modern use : *Extract of aerial parts* : anti-inflammatory and possibly diuretic.

Phytography : Small tree, growing up to 7 m; bark of branches grey, lenticellate, branchlets tomentose or pubescent; leaves simple, alternate, petioles short, 0.3–0.6 cm, glabrous or puberulous, stipules lanceolate; leaf blade usually 20.0 by 7.5 cm, shape and apex variable, shining beneath; racemes or spikes usually long, pubescent, solitary or sparingly branched, bracts very small; flowers sessile or shortly pedicelled, small, monoecious; berries 0.4–0.6 cm long, ovoid or ellipsoid, acute.

Distribution : Found in wet, deciduous and semievergreen forests of the Deccan peninsula from Konkan southwards, ascending up to 1800 m.

Ecology and cultivation : Wild.

Chemical contents : *Aerial parts* : antidesmanol, n-tritriacontane, friedelin, canophyllal, cyanophyllol, 16 α-oH-3- and 3β-oH-16 ketoisomultiflorene.

Aphanamixis polystachya (Wall.) Parkar (Meliaceae)

Syn : *Amoora ruhituka* W. & A.

Sanskrit name : Rohituka.

Vernacular names : Ben : Pittaraj, Royna, Tiktaraj; **Hin :** Harinhara; **Kan :** Mullumunthala; **Mal :** Chemmarom; **Mar :** Rahada; **Tam :** Malapuluvan; **Tel :** Chawamanu.

Trade names : Amoora, Rohituka.

Traditional use : TRIBES OF ORISSA : *Plant-extract* : in eye diseases; RURAL FOLK : *Seed* : as laxative and anthelmintic.
CHARAKA SAMHITA : *Paste of bark* : useful in diseases of liver, spleen, abdominal complaints and diseases caused by impurities in blood; *Paste of flower along with honey* : diseases caused by imbalance of phlegm and bile; AYURVEDA : *Decoction of stem-bark* : in diseases of liver and spleen and in obesity; *Decoction of root-bark* : in piles, menorrhagia, intestinal worms; *Oil of seed* : as liniment in rheumatism..
HOMOEOPATHY : for control of mind, for treatment of head, eyes, face, stomach, abdomen, fever, etc.

Modern use : *EtOH* (50%) *extract of stem* : anticancer; *Bark-extract* : appears to be immunosuppressive; *Seed-extract* : antibacterial and antifungal.

Phytography : Large evergreen tree having a straight cylindrical trunk, about 16 m high and a girth of about 2 m, a heavy crown of leaves; leaves usually unequally pinnate, 33 cm–1 m, leaflets 9–15 cm, elliptic or ovate, 7.5–22.5 cm by 3.7–10.0 cm, shortly petiolulate; spikes simple in female and panicled in male; flowers bracteate, subsessile, white, male 0.6 cm long and female 0.3 cm long; capsules smooth, pale yellow or reddish, subglobose, about 2.5 cm in diameter, rather soft and fleshy, 3-chambered and 3-valved, grey to yellowish, dehiscing by 3 valves, exposing the red seeds.

Phenology : *Flowering* : August–September; *Fruiting* : May–June.

Distribution : Sub-Himalayan tract from Gonda (Uttar Pradesh) to Sikkim, West Bengal, Assam, Andamans, western peninsula from Konkan to Travançore; Bangladesh (Sylhet).

Ecology and cultivation : Majestic shola tree; occasional.

Chemical contents : *Bark* : tetranortriterpene, aphanamixin, aphanamixinin; *Leaf* : diterpene alcohol, aphanamixol and β-sitosterol; *Fruit-shell* : triterpene, aphanamixin; *Seed* : rohitukin, limonoid polystachin, a saponin, a new glyseride, a reddish brown semidrying oil and a fixed oil.

Apium graveolens L. var. *dulce* (Mill.) DC. (Apiaceae)

(2n = 22)

Syn : *Apium dulce* Mill.

English name : Garden celery.

Sanskrit names : Ajamoda, Brahmakoshi, Brahmamusha, Hastikavari, Hayagandha, Mayura, Markati, Vastamoda.

Vernacular names : **Ben** : Chanu, Chilury, Randhuni; **Guj** : Bodiajamuda; **Hin** : Ajmud, Boriajmud, Karafs; **Mar** : Ajmoda; **Pun** : Bhulijhata.

Trade name : Celery, Salari.

Traditional use : *Root* : alterative, diuretic, in anasarca and colic; *decoction* used in rheumatism and in rheumatoid arthritis.
UNANI : Ingredient of 'Karafs'.

Modern use : *Seed* : used as stimulant, carminative, nervine sedative, tonic, in cordial, emmenagogue, antiseptic, in asthma, bronchitis, liver and spleen diseases. In India, the celery oil is utilised as a tonic and nerve stimulant.

Phytography : Plants are biennial; stem 0.3–2.4 m, erect, branching; leaves pinnate with large, deeply lobed segments, upper ones tripartite, others once or twice bifid, coarsely toothed at the apex; the inflorescence is a compound umbel of small white flowers; umbel rays 5–10; fruit 1.6–2.0 mm, ridged, narrow, vitta broad, taste aromatic, warm and pungent.

Phenology : *Flowering* : December to February.

Distribution : Native of Europe, found at the foot of North-West Himalaya and in the hilly regions of Uttar Pradesh, Himachal Pradesh, Punjab and South India.

Ecology and cultivation : Grows best in cool weather; prefers sandy loam soil, well-supplied with organic matter; requires considerable moisture in the soil; adequate weed control is necessary.

Chemical contents : *Stem* and *Leaf* : essential oil, but inferior to that from seed; *Fruit* : a number of furo coumarins, rutaretin, apiumetin, bergapten, isoimpinellin, isoimperatorin, coumarins; dehydrofluro coumarin glucoside; *Seed* : fatty acids, stigma sterol, pheromone steroid, an essential oil—called celery seed oil; *Seed-husk* : graveobioside A & B.

Remarks : Extensively used as a condiment and to flavour food products, sauces, soups, pickles, pastries, perfumes, soaps and liquors.
Two varieties of celery known : *A. Graveolens* var. *dulce* (Mill.) Pers. and var. *rapaceum* (Mill.) Gaudich. In India, the first variety is mainly cultivated.

Aquilaria malaccensis Lamk. (Thymelaeaceae)

Syn : *Aquilaria agallocha* Roxb.

English names : Agar wood, Aloe wood, Calambac, Eagle wood, Mallaca eagle wood.

Sanskrit name : Aguru.

Vernacular names : Asm : Sasi; **Ben :** Aguru; **Hin :** Agar.

Trade names : Indian eagle wood, Agar.

Traditional use : TRIBAL : leaves boiled in oil is used for removing fish-spines from throat; *Agar* : a pathological product formed as a result of a fungal disease in the wood of the tree is considered medicinal. Agar is used as liniment in various skin diseases.
AYURVEDA : it is considered as a contraceptive.
UNANI : Constituent of the medicine 'Ood Hindi' (Agar).

Modern use : *Agar* : stimulant, cordial, tonic and carminative and enters into several compound preparations. It is largely used in preparing fumigators and pastilles.

Phytography : Large evergreen tree, up to 23 m in height and 2.3 m in girth; stem moderately straight and often fluted; leaves simple, alternate, petiolate, 5.0–7.5 cm, thinly coriaceous, shining; umbels many-flowered, sessile or shortly peduncled; flowers white, perianth persistent in fruit; capsules 0.8 cm long and 0.6 cm broad, oblanceolate, compressed, coriaceous or hard and woody.

Distribution : The hill forests of Arunachal, Meghalaya (particularly Khasi and Garo hills), Nagaland, Assam (Cachar), parts of North Bengal; Bangladesh (Sylhet) and Bhutan.

Ecology and cultivation : Plant of damp hilly tropical forests; wild.

Chemical contents : *Wood* : essential oil containing agarospirol, a chromone—agarotetrol, two sesqui terpenes—gmelofuran and agarol; *Heartwod* : 1,7-oxoaporphine, liriodenine; *Stem-bark* : two cytotoxic compounds.

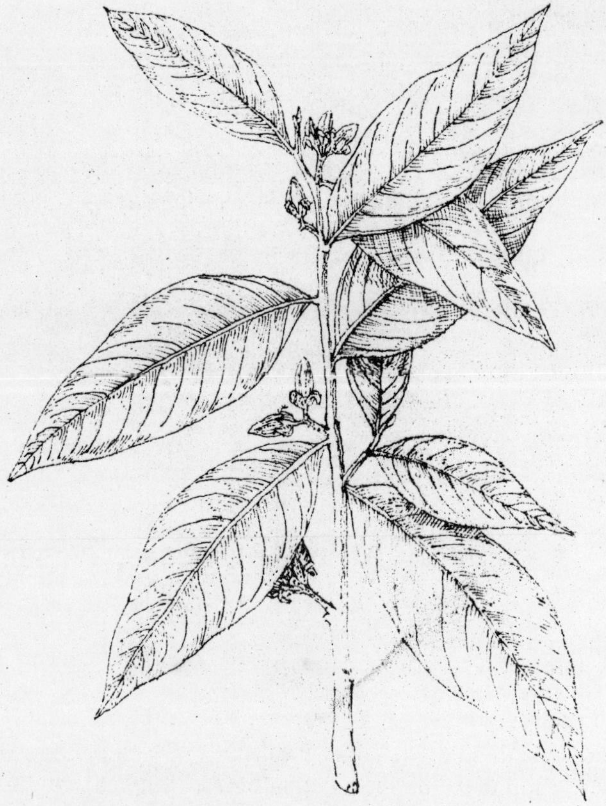

Remarks : In India, *Aguru,* i.e., the *Agar attar*, has been highly prized as the otto of roses, and is being used in religious rituals and perfumes since hoary past.
Tribals use wood as insect repellent.
The species has become *endangered* owing to indiscriminate felling of the trees.

Aquilegia vulgaris L. (Ranunculaceae)

(2n = 14)

Syn : *Aquilegia kaelini* Baker

English name : Common columbine.

Traditional use : *Herb* : diuretic, diaphoretic, antiscorbutic and tranquilising; *Root* : used for removal of kidney stone; *Seed* : used as an oxytocic and for jaundice.
HOMOEOPATHY : used in hysteria, somnolency, dysmenorrhoea and chronic skin troubles as well as in gastric troubles.

Modern use : *EtOH* (50%) *extract of aerial parts* : hypotensive.

Phytography : Plant stout, finely pubescent or the stem sparsely short hairy; leaves many fid, basal leaves biternate, ultimately leaflets broad-cuneate, the lobes shallow; flowers nodding, blue or purple or white; fruit (follicle) densely pubescent.

Phenology : *Flowering* : Summer.

Distribution : Temperate and subalpine Himalaya.

Ecology and cultivation : Plant of cold climate; wild.

Chemical contents : *Plant* : di-OH-di-MeO-aporphine; *Seed-oil* : a trienoic acid, ranenculeic acid.

Arachis hypogea L. (Fabaceae)

(2n = 20, 40)

English names : Ground nut, Monkey nut, Peanut.

Sanskrit name : Buchanaka.

Vernacular names : **Asm** and **Ben** : China badam; **Guj** : Bhuichana, Bhui shing, Maffali; **Hin** and **Pun** : Mungphali; **Kan** : Kadalachay, Kallekai, Nelagadale, Shenga; **Kon** : Mushbibiknamzhadd; **Mal** : Nela kadalak, Nilakkadalai; **Mar** : Bhui mug; **Ori** : China badam; **Tam** : Verakadalai, Nilakkadalai; **Tel** : Veru senaga.

Trade names : Ground nut, Mungphali, China badam.

Traditional use : *Shell* : for hypertension; *Unripe fruit* : lactalogue; *Fruit* and *Oil* : astringent to bowels; *Oil* : aperient, emollient, used as a substitute of olive oil.

Modern use : *Peanut-skin* : hoemostatic tincture, haemostatic agent; *Peanut-meal* (extract) : increases thromboplastic and prothrombin activity and higher antihaemophilic globulin in blood; *EtOH* (50%) *extract of defatted peanuts* : has excitatory effects on smooth muscles of experimental animals but inhibitory effect on isolated frog-heart; *Peanut-oil* : for haemophilia, as purgative, in nephropathy and dislocated joints, it is a substitute for olive oil in preparation of liniments, ointments, plasters and soap, a base for several medicines, suitable for making medicine for the injection of bismuth salicylate and mercury; used for preparing concentrated solutions of vitamins A and D.

Phytography : Herbaceous annual; taproot system, root laterals contain nodules, number of nodules may be between 800–4000, size of the nodule 0.4 cm; stem cylindrical, hairy; flowers papilonaceous, orange-yellow; fruits pod, containing 1–3 seeds with white cotyledons rich in oil and proteins.

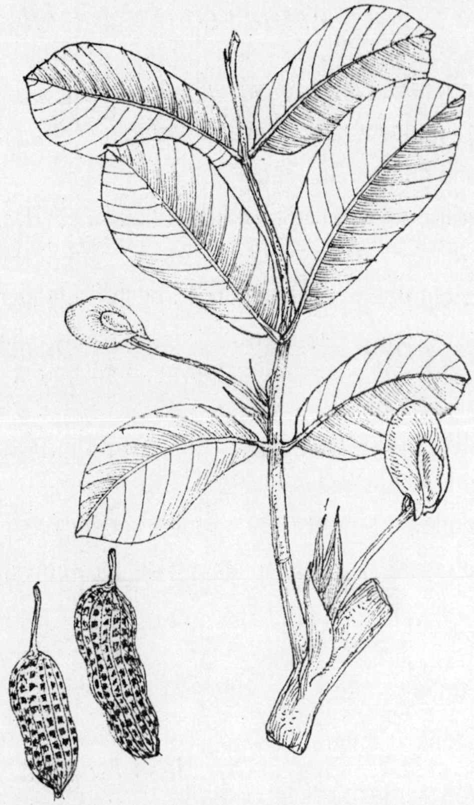

Phenology : *Sowing* : June–July; *Harvesting* : October–November.

Distribution : Native of Brazil; widely cultivated in India; Pakistan and Bangladesh.

Ecology and cultivation : Tropical plant, thrives even under subtropical regions between 45° N and 30° S latitudes and up to 1050 m, thrives best on sandy loam soils; it is a rainfed *kharif* crop, grows best under irrigation during summer; requires a fine loose seedbed.

Chemical contents : *Shell* : luleolin, β-sitosterol, daucosterol, eriodictyol, luteolin, 5,7-di-OH-chromene and a saponin; *Red skin* : arachidoside; *Peanut-cotyledon* : arachins A & B, α-arachin, milk substances, proteins, vitamins A and B (thiamin), some members of B_2 group (Riboflavin), vitamin E, lecithin, nicotinic acid and oil; *Peanut-oil* : fatty acid compounds, glycerides, carotenoids, tocopherols, triterpenes, sterols, phophatides, elaidic acid, erucic acid.

Adulterants : *Peanut-shell*, in the finely ground condition, has been found as adulterant in ground spices and drugs. *Peanut-skin* : a rich source of thiamine.

Remarks : Ground nut is important as human food and cattle feed. Ground nut milk and curd are prepared from the kernels in addition to the ground nut oil. The oil may be used in soap making, as an illuminant, lubricant and as insecticide; useful in controlling insect pests on plants. The oil-cake is a good manure for many crops and fruit trees. Artificial fibre, called *ardil,* is made from ground nut protein. Ground nut shell is used for manufacture of carbon, acetic acid, acetone, alcohol; finely ground peanut shell is used for cleaning carbon from aircraft engines, fuel, plastics and bedding.

Aralia pseudo-ginseng Benth. (Araliaceae)

Syn : *Panax pseudo-ginseng.*

Vernacular name : Mar : Tapamari.

Trade name : Ginseng.

Traditional use : Aphrodisiac, stimulant in dyspepsia and vomitting, expectorant, antipyretic.

Modern use : Ginseng is used by the laity as a stimulant and aromatic bitter.

Phytography : Perennial herb with fleshy, often forked roots; stem bears a single whorl of 3-palmetely compound leaves; a solitary stalked umbel of greenish white flowers and bright red fruits.

Distribution : Meghalaya, Arunachal Pradesh and Sikkim , between 2000 and 3000 m; also in Nepal and Pakistan.

Chemical content : Glucoside (panaquilon).

Adulterant : *Panax repens* is used as a substitute.

Arctium lappa L. (Asteraceae)

(2n = 32, 36)

Syn : *Lappa major* Gaertn., *Lappa minor* Schkuhr., *Lappa tomentosa* Lamk.

English names : Great burdock, Begger's buttons.

Traditional use : *Root* : diuretic, diaphoretic and alterative, used in gout and skin diseases.
HOMOEOPATHY : very useful for small boils, eczema on head, shoulder and mouth and for angina; very good as a blood purifier.

Modern use : *Plant-extract* : causes sharp, long-lasting reduction of blood sugar and increases carbohydrate tolerance; *Alcoholic extract of plant* : for treating psoriasis, seborrhic eczema and alopecia areata; *Root* : inhibits tumour growth; *EtOH* (50%) *extract of root* : cardiac stimulant, diuretic and spasmolytic.

Phytography : Tall (60–130 cm), coarse, erect, biennial, branched herb; leaves simple, alternate, petiolate, 30 cm, ovate-cordate, margin sinuate, lower surface cottony, petioles solid and deeply furrowed; heads terminal, stoutly peduncled, solitary or fascicled, sub-corymbose, 1.8–3.7 m in diameter, involucre of bracts slender, angled, spreading, all florets similar, purple with white styles; achenes angled and ribbed, odour indistinct, taste mucilaginous, sweetish, then bitter.

Distribution : Native of North Europe; found extensively in western Himalaya from Kashmir to Simla at 2000 to 3000 m; western Tibet (at about 4000 m altitude) and Pakistan (North-West Frontier Province and Baluchistan).

Ecology and cultivation : Plant of cold, dry regions; found along roadsides and waste-lands; wild.

Chemical contents : *Root* : 45% insulin, S-containing acetylenic acid, acetic acid, γ-guanidine-n-butyric acid, essential oil and several volatile compounds; *Leaf* : Germa-kranolides, bitter sesquiterpenoid lactone, arctiopicrin, eremophilene, fukinone, petasitolone, fukinanolide, β-dudesmol, taraxasterol and its palmitate esters, dehydrofukinone or 9(10)-fukinone, arcitol or 8 α-hydroxy-eudesmol; *Fruit* : arctin, arctigenin, mateiresinol, lappaol C, D & E and a new sesquilignan derivative; *Seed* : sesquilignans, lappaol A & B, and a glucoside.

Remark : Deserves commercial cultivation.

Ardisia solanacea Roxb. (Myrsinaceae)

Syn : *Ardisia humilis* Vahl

Vernacular names : Ben : Banjam; **Mar :** Mayarawa; **Mal :** Molakka; **San :** Garagulanchi; **Tam :** Manipudbam; **Tel :** Kondamayuri.

Traditional use : *SANTAL* : apply paste of stem-bark on swellings; *Root* : febrifuge, in diarrhoea and in rheumatism; *Root* along with flower of *Bauhinia purpurea* mixed in *cow-ghee* : given to asthma patients.

Modern use : *Root* : in fever.

Phytography : Evergreen small tree attaining about 8 m; leaves large (15 by 5 cm), bright green, coriaceous or fleshy at the ends of the branchlets; flowers about 0.4 cm, pink or rose-coloured, waxy; berries small, depressed, globose, black on ripening.

Phenology : *Flowering* : April–May; *Fruiting* : October–January.

Distribution : Throughout India except desert regions, occurring up to 1700 m; Bangladesh and Pakistan.

Ecology and cultivation : Grows along streams, frequent in forests; wild.

Chemical contents : *Root* and *Stem* : baucrenol; *Leaf :* baucrenol, β-amyrin.

Remark : Ripe fruits are eaten by the Santals.

Areca catechu L. (Arecaceae)

(2n = 32)

English names : Areca palm, Areca nut, Betel nut, Pinang palm.

Sanskrit names : Gubak, Phalam, Poag, Pooga, Poogi.

Vernacular names : Asm : Tambul; **Ben :** Supari, Gua; **Guj :** Supaari; **Hin :** Kasaili, Supari; **Kan :** Adike, Bette; **Kon :** Maddi; **Mal :** Adakka, Pugam, Pakka; **Mar :** Supari **Ori :** Gua; **Tam :** Kamubu, Pakku; **Tel :** Poka, Vakka.

Trade names : Areca nut, Betel nut, Supari. There are over 150 trade types.

Traditional use : *SANTAL* : (i) a patient of small pox is given to eat the areca nut when the pustules subside; (ii) a mixture for biliary colic is prepared with areca nut as a constituent; (iii) an ointment for chancre and syphilis is made by pestling areca nut with the root of *Gymnema hirsutus,* leaf of *Piper betel* and then cooking the same in mustard oil or butter; *TRIBALS* also use this plant in rhagadas, venereal sores, syphilis, dysentery, cholera, small pox and for fractured bones.
References to this plant are found in the *BHAGVATA* along with the plants of Musa paradisica and *Borassus flabellifer.* *CHARAKA SAMHITA* : *Fruit* : useful in the diseases caused by bile; *SUSHRUTA SAMHITA* : *Fruit* : beneficial in the diseases caused by phlegm; but overuse of this may distort voice of a man; *CHACRADATTA* : *Paste of unripe fruit* : may

be used as liniment; *Extract of unripe fruit* : useful in small pox; HARITA SAMHITA : sesame oil in which extract of unripe fruit has been boiled should be used; AYURVEDA : various preparations of *unripe* and *ripe nuts* are useful in toothache, pyorrhea, gum diseases, in treatment of worms, while *extract of young leaf* mixed with mustard oil is useful as liniment in rheumatism; BRAHMAVAIVARTA PURANA : brushing the teeth with twig of this plant is beneficial; AGNI PURANA : (i) immortality can be attained by consuming decoction of this plant along with the powder of root, bark, leaf and fruit of margosa and juice of *Wedelia calendulacea*; (ii) alkaloids of this plant are beneficial medicine.
UNANI : Ingradient of 'Futal (Chalia)'.

Modern use : *Nut* : chewing facilitates salivation, it being a good source of fluoride prevents tooth decay, but constant use might cause oral carcinoma; shows antimicrobial activities; *Aqueous extract of nut* : exhibits vascoconstriction and adrenalin potentiation in rats; *Extract of leaf* and *fruit* : spasmogenic.

Phytography : Tall, slender, unbranched palm with a crown of leaves; stem annulate; leaves pinnate with a conspicuous sheet; flowers in spadix, male many at the upper portion, female much longer and a few at the base; fruits are single-seeded berries with flesh and fibrous pericarp and a stony seed, 3.8–5 cm long, smooth, orange or scarlet when ripe.

Phenology : *Flowering* : August–January; *Fruiting* : about a year later.

Distribution : Cultivated in the coastal regions of India, Bangladesh, Pakistan, Sri Lanka, Myanmar and other tropical and subtropical countries.

Ecology and cultivation : This palm requires a moist tropical climate with heavy (500 cm/year) rainfall.provided with good drainage. It can be grown in drier areas (rainfall 50 cm/year), if properly irrigated. It is a shade-loving plant, especially in the earlier stages and is very sensitive to drought. It grows on a variety of soils, but saline or alkaline soil, light and sandy soil are not suitable for it, but slightly saline sandy soil is good.

It is generally cultivated as a mixed crop with coconut and plantain or along with *Erythrina indica*. The betel nuts are sown in October/November with a distance of 10–15 cm between two nuts. Transplanting is normally done after two years, occasionally after 3 or 4 years. Transplantation is done in July in the highlands and from February to April in lowlands. The second transplantation takes place when the first have come into bearing. In a fully planted grove, a distance of about 2 m each way is kept between the betel nut tree.

Chemical contents : *Nut* : alkaloids—arecoline, arecaidine, guvacine and isoguvacine.

Adulterants : Fruits of *Areca triandra* Roxb. and *Areca nagensis* Griff. are substitutes for *Areca catechu* L.

Remark : Stem and leaves are used in various ways.

Arenga obtusifolia Mart. (Arecaceae)

Vernacular name : Malaya : Langkap.

Traditional use : *Plant* : used as fish poison.

Phytography : a cluster-forming pleonanthic palm; stem erect; 8 m long, about 30 cm in diameter near base, exposed portion of the stem dull grey, distinctly ringed with disatinct leaf scar mark, leaves pinnate, slightly ascending; bracts boat-shaped, leathery, flowering branches simple, fastigated deep green in colour, male in pairs, female subterbinate; freuit subglobose.

Distribution : introduced in Inida—cultivated in the Indian Botanic Gurden, Howrah; Indonesia, Java.

Ecology and cultivation : Can be grown in pot.

Remarks : Fruit is anticoagulant (Chinese Materia Medica); juice from mesocarp of fruit is highly irritating to the skin.

Arenga pinnata (Wurmb.) Merr. (Arecaceae)

(2n = 26 ,32)

Syn : *Arenga saccharifera* Lab., *Gomutus saccharifera* Spreng.

English names : Gomuti palm, Sugar palm of Malacca.

Vernacular name : Tam : Kichilippanai.

Traditioanal use : *Root* : in bronchitis, stomachic; *Juice of outer fleshy covering of fruit* : corrosive; *Fresh nut* : intoxicating; *Juice of tender fruit* : laxative; *Burnt nut* : used in dentifrices.

Phytography : A single-stemmed monocarpic palm, ± 10 m long; stem massive, completely covered with black coarse leaf-sheath fibres, bristles and portions of persistent leafbases; leaves pinnate, massive, ± 10 m long, upper leaves ascending, deep green, rounded, leaflets narrowly oblong, margins distinctly toothed; inflorescence terminal and axillary; male flower oblong, female flower subglobose; fruit acute, trigonous, obovoid; seeds narrowly ovoid, longitudinally flattened.

Distribution : Assam, Andaman Islands; Malaysia, Indonesia and Philippines.

Ecology and cultivation : Cultivated in Indian Botanic Garden, Howrah, since the early part of the 19th century.

Remark : Outer fleshy covering of fruit is used as fish poison.

Argemone mexicana L. (Papaveraceae)

(2n = 28)

English names : Mexican prickly poppy, Yellow thistle.

Sanskrit name : Bramhadandi.

Vernacular names : Ben : Shialkanta; **Hin :** Sialkanta, Kateli, Kandiari, Bharband; **Kan :** Datturi; **Mal** : Pannummattam; **Tam** : Brahmadandu, Kudiyoetti; **Tel :** Brahmadandu.

Trade name : Shialkanta.

Traditional use : SANTAL : *Oil* : use the for burning; *Juice* (which exudes when the plant is injured) : in scabies, ophthalmia and as diuretic; *Powder of the seed* : on catarrhal affections of throat, pulmonary mucous membrane, in asthma, headache and stomach complaints; RURAL PEOPLE OF KARNATAKA : *Juice of the plant* : in ulcers and itch.

Modern use : *Root* : useful in chronic skin diseases; *Juice of the plant* : for jaundice and cutaneous affections, anthelmintic; *Seed* : laxative, emetic, expectorant, demulcent, antidote to snake poison.

Phytography : A prickly herb; ± 60–90 cm with yellow latex; leaves sinuate pinnatifid, green and white, half amplexicaul; flowers ± 50 mm in diameter, yellow, sepals cuspidate; capsules 20–38 mm long.

Phenology : *Flowering* and *Fruiting* : More or less throughout the year, but peak season from February to June.

Distribution : Naturalised from America, throughout India; Pakistan, Bangladesh, Sri Lanka, Nepal and all the adjacent countries up to an elevation of about 3000 m.

Ecology and cultivation : Grows in wastelands and roadsides; wild.

Chemical contents : Contains several alkaloids, principally protopine and allocryptomine, berberidine, sanguinarine and ceryl alcohol; C_6H_6 is most toxic to mice.

Adulterants : Seeds of argemone advertently or inadvertently mixed during harvesting of mustard and usually mustard oil contains a small percentage of argemone oil also, which causes epidemic dropsy.

Argyreia nervosa (Burm. f.) Boj. (Convolvulaceae)

(2n = 30)

Syn : *Argyreia speciosa* Sweet, *Convolvulus nervosus* Burm. f., *Lettsomia nervosa* Roxb.

English names : Elephant creeper, Woolly morning glory.

Sanskrit name : Samudrapalaka.

Vernacular names : Ben : Bichtarak; **Hin :** Ghabel, Samandar sokh, Samandar kapat; **Mar :** Guguli; **Mal** and **Tel :** Samudrapala; **Tam :** Samutrappachai.

Traditional use : *Plant* : in stomach complaints, sores on foot, small pox, syphilis, diarrhoea, dysentery; *Leaf* : antiphlogistic, used as emollient, poultices for wounds, externally for skin diseases, gleet gonorrhoea, strangury and chronic ulcers.

Modern use : *Root* : as tonic, alterative, in rheumatism and disease of the nervous system; *Seed* : show significant hypotensive and spasmolytic activity.

Phytography : Climbing shrub.

Phenology : *Flowering* : Rainy and Cold seasons; *Fruiting* : January to April.

Distribution : Throughout India (except in dry western regions) up to an elevation of 400 m.

Chemical contents : *Leaf* : 1-triacontanol, β-sitosterol, epifriedelinol; *Seed* : ergoline, ergine, isoergine, isolysergic acid amide, pennidavine, caffeic acid, Et-caffeate, lysergic acid amide,

Remark : Grown as an ornamental.

Arisaema speciosum (Wall.) Mart. (Araceae)

Syn : *Arum speciosum* Wall.

English names : Cobra lily, Snake lily, Snake plant.

Vernacular names : Hin : Sanpbuti; **Pun :** Kiralu, Samp-ki-khumb, Kiri-ki-kuri; **Sik :** Gurba.

Traditional use : SIKKIMESE : *Root* : antidote to snake poison, in wounds.

Phytography : Tuberous herb, rootstock oblique or shortly creeping and rooting, about 12.5 cm in diameter; leaf solitary, petiole very stout, green, smooth, often marked with brown or purple, lamina pinnate, 40–46 cm, leaflets petiolulate, petiolules 1.25–5.0 cm long, leaflets edged with red or purple, acuminate; spadix dioecious, dark purple with pink or yellowish tail, base of appendage usually ovoid, tube of spathe 5–10 cm, striped with purple, limb of spathe 5–15 cm long, ovate-lanceolate, incurved, banded with white and purple.

Distribution : Temperate Himalaya from Kumaon to Sikkim at 2000 m to 3000 m; Bhutan (3000–3300 m).

Ecology and cultivation : Grows in cold hilly terrain; wild.

Chemical contents : *Tuber* : Calcium oxalate crystals.

Remarks : Tuber is applied to sheep as a remedy for colic and to kill worm which infest cattle; also causes dermatitis.

Aristolochia bracteolata Lamk. (Aristolochiaceae)

Syn : *Aristolochia bracteata* Retz.

English names : Bracteated birthwort, Worm killer.

Sanskrit names : Bhringi, Dhumrapatra, Sulabha.

Vernacular names : Hin : Kiramar; **Mar :** Kidamari; **Ori :** Paniri; **Raj :** Aulosa, Jufa; **Tam** and **Mal :** Aduthinapalai; **Tel :** Adumuttadagida.

Traditional use : *Plant* : emmenagogue; *Juice of leaf* : vermifuge, applied to foul and neglected ulcers; *Bruised leaf* (mixed with castor oil) applied to eczema on children's legs; *Powdered root* : in fertility control; *Seed* : purgative.

Modern use : *Plant* used medicinally as purgative and anthelmintic; *Decoction of root* : used for expelling roundworms.

Phytography : Prostrate bitter herb; root perennial, slender; stem slender, decumbent, angled, striated; leaves petiolate, reniform or broadly cordate, glaucous beneath;

peduncle bracteate; flower solitary, tubular with globose base, dark purple, villous nearly 2.5 cm long; fruit pyriform, many-seeded; seeds triangular.

Phenology : *Flowering* : August–September.

Distribution : Uttar Pradesh, Madhya Pradesh, Maharashtra, Andhra Pradesh, Karnataka, Tamil Nadu, West Bengal; Pakistan (Sind).

Ecology and cultivation : Plains from the coast, common in wastelands, along with river banks, bunds of paddy fields, etc.; wild.

Chemical contents : *Root* : aristolochic acid and aristo-red are the two crystalline principles; they are also present in the seeds in minute quantities; *Seed* : yields α-non-drying fatty oil, unsaponified matter with β-sitosterol, mangoflorine and two more alkaloids.

Remark : Plant is poisonous to man and livestock.

Aristolochia indica L. (Aristolochiaceae)

(2n = 12)

English name : Indian birthwort.

Sanskrit name : Ishvari.

Vernacular names : Ben and **Hin** : Isharmul; **Mal** : Isvaramuli; **Mar** : Sapasan; **Tel** : Eswaramuli.

Trade name : Iswarmul.

Traditional use : *Root* : tonic, stimulant, emetic, emmenagogue, in fever, in powder form is given with honey for leucoderma; *Root-decoction* : in impotency; *Crushed root* : applied on itching; *Juice of leaf* : in snake bite, used for cough; *Seed* : inflammations, biliousness and dry cough.
UNANI : a constituent of 'Majnoon-e-Flasfa'.

Modern use : *Plant* : used as abortifacient; *EtOH* (50%) *extract* : diuretic and anti-inflammatory; *Dried stem* and *root* : used as drug, which should be used in minimal doses; the drug promotes digestion and controls menstruation; in higher doses, it may prove lethal, it is used as a stimulant, tonic and for fevers; in moderate doses, it is used as a gastric stimulant and in dyspepsia; *Root* : considered as a stimulant, tonic and emmenagogue and also used in intermittent fever and in bowl troubles of children; shows antifertility activity in experimental animals.

Phytography : Twining herb, semiwoody, having more or less swollen nodes; leaves cordate or ovate, exstipulate; flowers irregular, often offensively smelling, perianth globose with a purple dilated and trumpet-shaped mouth with a strap-shaped brown purple appendage or lip behind; fruit a subglobose capsule.

Phenology : *Flowering* : June to October; *Fruiting* : November to March.

Distribution : Found throughout the subcontinent, mainly in the plains and lower hilly regions from Nepal to Bangladesh.

Ecology and cultivation : Found in open scrub jungles; wild.

Chemical contents : *Root* : a crystalline substance—probably a glucoside, a micro-crystalline principle glucosidic in nature named isoaristolochic acid, allantoin, 0.05% carbonyl compounds and a small amount of an oil, with the odour of isovanillin, ishwarone, ishwarane, aristolochene.

Aristolochia tagala Cham. (Aristolochiaceae)

Syn : *Aristolochia roxburghiana* Klotzsch.

Vernacular names : Asm : Belikol, Panipipuli; **Meg :** Krah-kshiang; **Mik :** Chohu; **Tel :** Nallayiswari.

Traditional use : WAR JAINTIA : *Whole plant* : in bone-fracture; KHASI and JAINTIA : (i) *Root* : in toothache; (ii) *Leaf* : in stomachache, fits, rheumatism, as antidote to snake bite; BODO : (i) *Root* : in dyspepsia, colic, snake bite; (ii) *Leaf* : in flatulence, fever and malaria.

Phytography : Robust twinner with large lanceolate, ovate-lanceolate leaves, finely acuminate at the apex; flowers in very short racemes, purple; capsules subglobose.

Phenology : *Flowering* : April–May; *Fruiting* : May.

Distribution : Sikkim, North Bengal, Bihar.

Ecology and cultivation : Grows in open jungle; wild.

Chemical contents : *Root* : aristolochic acid.

Adulterant : Root is adulterant of *Aristolochia indica* L. roots.

Arnica montana L. (Asteraceae)

Syn : *Doronicum arnica* Desf., *D. montanum* Lam.

English names : Arnica, Celtis nard, Leopard's bane.

Trade name : Arnica.

Traditional use : Sedative, stimulant and resolvent;.
HOMOEOPATHY : for the treatment of fall and injury, epileptic conditions, loss of memory, paralysis, abortion and unconsciousness.

Modern use : *Tincture of arnica*, when injected, causes drop in blood pressure followed by marked rise; internally it is stimulant and irritant; but is now seldom used.

Phytography : Perennial herb; stem erect, 30–60 cm high, simple or slightly branched, hairy; basal leaves forming a rosette, oblong, 50–125 mm long, sessile, entire, short-hairy, upper leaves few, smaller; heads deep yellow, solitary or few, 50–75 mm across, involucre hairy and glandular; pappus yellowish.

Distribution : Northern hemisphere, especially central Europe. It is found in the lower elevations, especially in the meadows. Efforts are on to grow it in temperate Himalaya.

Ecology and cultivation : High altitudinal plant.

Chemical contents : *Rhizome* and *Flower* : essential oil, bitter principle; *Flower-head* : 0.1% choline, arnidendiol, alloarnidendiol and a petroleum ether solution fraction of arnicin; *Flower* : gave 0.1% and *Root* 0.5–1.5% essential oil which are not identical; *Flower* : contains a flavone, an ardenaline-like precursor substance, a cardiotonic substance, choline, betaine; essential oil is irritating to skin; 2 new substances—carnaubyl alcohol, stearine, besides paraffins, faradiol and annidendiol.

Adulterants : Flowers of *Calendula officinalis* L., *Anthemis tinctoria* L., *Inula britannica* L., *Taraxacum* sp., *Scorzonera* sp. and *Daronicum pardalianches* L. and roots of *Geum urbanum* L., and *Hiercium* sp. are used as adulterants.

Artabotrys odoratissimus R. Br. (Annonaceae)

(2n = 16, 18)

Syn : *Artabotrys hexapetalus* (L.f.) Bhandari, *A. uncinatus* (Lam.) Merr.

English name : Climbing ylang ylang.

Sanskrit name : Harachampaka.

Vernacular names : Asm : Kothali champa; **Ben :** Kantali champa; **Hin :** Kantalichampa, Harichampa; **Kan :** Manoranjini, Kandala sampige; **Kon :** Hirvo champo; **Mal :** Manoranjini; **Mar :** Hirva chapha; **Ori :** Chini champa; **Pun :** Champa; **Tel :** Monoranjana, Manoranjitham.

Trade name : Kantali champa.

Traditional use : *Decoction of leaf*: in cholera.

Modern use : *EtOH extract of leaf*: weak estrogenic and in higher concentration (150 mg/kg) antioestrogenic; *Benzene extract of leaf*: strong antioestrogenic in rats; *Volatile fraction of the extract* : antifungal; *Flower-extract* : strong antifungal; *EtOH extract of fruit* : initially cardiac depressor, followed by cardiac stimulation, hypotensive, spasmogenic.

Phytography : Large climbing ornamental shrub; branches long; leaves oblong or lanceolate, 5–20 cm by 5 cm, glabrous, flowers solitary or paired, green flowers turning yellow, on hooked or circinate ultimately woody peduncles, highly fragrant; fruit berry.

Phenology : *Flowering* : April to September.

Distribution : Indigenous to India and Sri Lanka, mostly found in western peninsula; cultivated in gardens throughout India.

Chemical contents : *Root*: yingzhaousu A & B.

Remark : Fruits are eaten by tribal and rural folk.

Artabotrys suaveolens Blume (Annonaceae)

Vernacular name : Mal : Koluvaravalli.

Traditional use : RURAL PEOPLE OF ASSAM: *Bark-decoction*: as emmenagogue and also after parturition.

Phytography : A large woody climber.

Distribution : Assam; Bangladesh and eastwards to the Philippines.

Ecology and cultivation : Plant of damp tropical climate; wild.

Chemical contents : *Root* and *Stem-bark* : two crystalline alkaloids—artabotrine and squaveoline; a third alkaloid has also been isolated which is known as artabotrinine; *Seed*: essential oil.

Remark : In the East Indies, a decoction of leaf is said to be beneficial in cholera.

Artemisia absinthium L. (Asteraceae)

English names : Absinthi, Madder wood, Worm wood.

Sanskrit names: Indhana, Amar.

Vernacular names : Ben and **Guj :** Mastaru; **Hin :** Vilayati afsanthin; **Kan :** Urivalu, Urittige; **Kas :** Tethwen; **Mal :** Nilampala; **Mar :** Serpana; **Tam :** Machipattri; **Tel :** Tartiha, Moshipatri.

Trade name : Afsanteen.

Traditional use : Used in chronic fever, sweating and inflammation of liver; *Warm wood oil* : used in rheumatism; *Poultice of leaf* : applied for fomentation to rheumatic or gouty joints; used in the diseases of digestive system, nocturnal pollution, anaemia; *Infusion of leaf* : used for dispersing the yellow bile of jaundice; *Leaf* : antiseptic and detergent in skin diseases.
UNANI : Ingredient of 'Absanteen (Qaisoom)'.

Modern use : *Flowering-tops* : as vermifuge, stimulant and tonic.

Phytography : Perennial herb; stem erect, 30–100 cm, ribbed, pubescent, almost silky hairy, very aromatic; leaves petiolate, 2.5–5.0 cm, ovate or obovate, unequally 2- to 3-pinnatifidly cut, hoary on both surfaces; heads numerous but rarely crowded, 0.6–0.8 cm in diameter, peduncled, hemispheric; flowers yellow; achenes elliptic-oblong, 0.2 cm long.

Distribution : Kashmir and Kumaon at 1600–2400 m elevations; Europe, Asia and Northern Africa.

Chemical contents : *Plant* : artimisetin, essential oil, absinthe, fatty oil azulene and santonin.

Adulterant : *Artimisia dracunculus* L. (Tarragen) is used as its adulterant.

Remarks : Flavouring agent for beer; source of the famous poisonous French intoxicant —absinthe. Artemisias are among the causes of hay fever.

Artemisia annua L. (Asteraceae)

English name : Sweet worm wood.

Modern use : *Plant*: antipyretic, may be antimalarial.

Phytography : Sweet-scented glabrous annual erect herb, 0.7 cm–1.75 m high, branches deeply grooved; leaves 0.25–0.60 mm long, 7 cm broad, 1–2 pinnately divided into lanceolate, oblong, deeply cut segments; heads yellow, subglobose, involucre of bracts glabrous, inner orbicular, green, with scarious margin and shining.

Distribution : Native of America; occurs in Punjab—ascending up to 2000 m; Pakistan (Peshwar to Waziristan).

Ecology and cultivation : Xerophyte; wild—reported to be recently cultivated in Jammu and Kashmir.

Chemical contents : *Plant* : artemisinin, artemisininic acid, arteannuin, arteannin B, benzyl isovalerate, coumarins, flavonoids, β-famesene, lactone, and essential oil composed principally of artemisia ketones, cineole, α-camphor, pinene, etc.

Remarks : In Indochina, the herb is considered as a good stomachic, diuretic, and prescribed in jaundice and skin diseases.

Artemisia caruifolia Roxb. (Asteraceae)

Traditional use : *RURAL PEOPLE OF ASSAM* : *Plant* : in fever, chronic diarrhoea, phthisis, perulent scabies and intestinal troubles.

Phytography : Annual herb, 60–120 cm in height; stem stout, glabrous, branched, upper parts densely leafy; leaves alternate, sessile, deep green, multifid, segments very narrow and weak; heads green, hemispheric, 0.3–0.6 cm in diameter, involucre of bracts glabrous with narrow scarious margins, many-flowered.

Distribution : Assam; Bangladesh, Myanmar, Nepal.

Ecology and cultivation : Grows in damp, moist climate; wild.

Artemisia dracunculus L. (Asteraceae)

(2n = 18)

English names : Dragon mugwart, Estragon, Tarragen.

Traditional use : *Aromatic leaves* : as an aperient, stomachic, stimulant and febrifuge.

Phytography : A perennial aromatic herb, green, glabrous, 50–60 cm high with erect, branched stem; basal leaves 3-parted at tips, linear, lanceolate, entire or small-toothed; heads whitish green; flowers of two kinds—pistillate and bisexual.

Distribution : Western Himalaya.

Ecology and cultivation : Plant of cold climate; wild.

Chemical contents : *Plant* : an essential oil called 'Oil of Tarragon', coumarine, methyl chavicol, estragole, phellandrene, ocimine, rutin, artemitin, hyperoside; *Root* : artemidiol, acetylenes and dehydrofalcarnione derivatives; *Leaf* : sopoletin, herniarin.

Adulterants : It is used like *Artemisia absinthium*.

Remarks : Oil of Tarragon is used for flavouring vinegar, sauces, salad dressings, etc.

Artemisia maritima L. (Asteraceae)

(2n = 18)

Syn : *Artemisia brevifolia* Wall., *A. fragrans* Willd.

English names : Worm seed, Santonica.

Vernacular names : Gar : Purcha; **Hin :** Kirmala; **Jam** and **Kum** : Seski; **Kas :** Moorni; **Mar :** Kirmaniova.

Trade name : Kirmala.

Traditional use : *Herb* : stomachic and laxative; *Decoction* : febrifuge, in argue, intermittent and remittent fevers;
UNANI : ingredient of 'Dirmana Turki'.

Modern use : *Dried immature leaves* and *flower-heads* constitute the drug. It is collected when the flower heads have not fully matured, since the proportion of santonin is said to be maximum just before the flowers begin to open. The drug is mainly used for expelling threadworms from the stomach. It is also used in fevers, dropsy and as a stimulant.

Phytography : Rootstock woody, branched, stem 15–45 cm, woody or wiry, erect or ascending, much branched from the base; leaves simple, alternate, petiole slender, lamina often white, ovate, 2-pinnatisect, segments small, spreading; heads 3- to 8-flowered, often reddish, mostly erect or suberect in spicate fascicles, outer bracts tomentose while inner scarious, glabrous.

Distribution : Western Himalaya from Kashmir to Kumaon and western Tibet between 2000 and 5000 m.

Ecology and cultivation : Plant of cold climate; wild.

Chemical contents : *Plant*: Santonin, β-santonin, pseudosantonin, essential oil containig carboxylic with four 7 C atoms, camphene, cineole β-thujone and (-)camphor, together with p-cymene and possibly β-Thuja-Ketonic acids.

Adulterants : Ground mustard cake, from which the volatile oil had been distilled, has been reported in admixture with powdered santonica to the extent of 40–50 per cent. The unexpanded flower heads of *Artimisia pauciflora* Weber and *A. maritima* var. *stechmanniana* Besser may be mixed as substitute.

Remarks : There are two varieties : *neercha* and *seski*, the latter yielding more oil than the former.

Artemisia nilagirica (Clarke) Pamp. (Asteraceae)

(2n = 18)

Syn : *Artemisia vulgaris* Auct. non L., *A. vulgaris* var. *nilagirica* C.B. Clarke

English name : Indian wormwood fleabane.

Vernacular names : **Hin :** Nagduana, Dona, Gathivana, Majtari; **Kan :** Davana, Manjipatri; **Mal :** Appa, Damanal; **Mar :** Banjirce, Gathena, Surband; **Ori :** Doyone; **Tam :** Machipatri, Mashi battiri, Tirunama; **Tel :** Davanamu, Maschipatri.

Trade name : Nagdona.

Traditional use : *Expressed juice of herb* : applied on head/forehead to prevent convulsions, *Smoke of the plant* : helps cure burn; *Weak decoction of plant* : given to children suffering from measles and used as fomentation in skin diseases and ulcers; *Leaf* : haemostatic and also used to reduce/remove burning sensation in conjunctivitis; *Leaf* and *flowering-top* : in asthma.

Modern use : *Herb* : anthelmintic, antispasmodic and stomachic; *Root* : tonic and antispasmodic; *Infusion of leaf* and *flowering-top* : given in nervous spasmodic affections and in asthma.

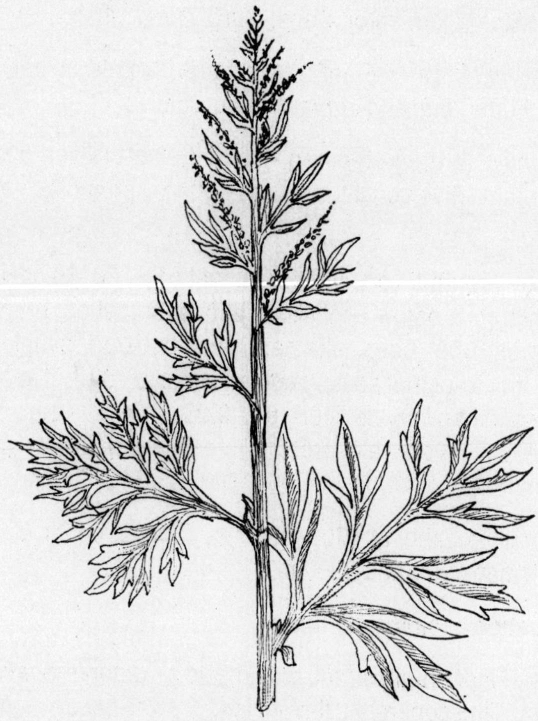

Phytography : Tall perennial herb; leaves alternate, pinnatisect, lobes deeply incised, smooth above and white tomentose beneath; heads in greenish panicled racemes; flowers whitish yellow.

Phenology : *Flowering* : Rainy season; *Fruiting* : Autumn.

Distribution : Kashmir, Sikkim, Assam, Meghalaya, West Bengal (Darjeeling), Manipur, Maharashtra, Karnataka, all mountainous tracts of India; Nepal and Bhutan.

Chemical contents : *Plant* : yields an essential oil; it contains p-cymène, l-linalool, p-thujone azulene, thujyl alcohol and an unidentified hydrocarbon. Yield of the chemicals vary with the altitude.

Artemisia scoparia Waldst. & Kit. (Asteraceae)

Syn : *Artemisia tricophylla* Wall.

Vernacular names : Mar : Churisaroj, Danti; **Pun** : Dona, Jhan.

Traditional use : RURAL FOLK : *Plant* : as a cure for pain in ear, and the smoke is considered good for burns; *Infusion of plant* : as purgative.

Modern use : *Artemisol* (a drug prepared from the plant) : used against unrolithicasis; *p-OH-acetophenone* (an active constituent of the plant) : promotes secretion of biles and bile salts.

Phytography : Faintly scented, very slender annual herb, 30–60 cm high; rootstock may be perennial; stem glabrous below and villous above, branched, branchlets often almost capillary; leaves almost from base, alternate, long-petioled, simple, 2.5–7.5 cm long, broadly ovate, uni- to tri-pinnatisect, segments linear, distant, upper leaves filiform; heads sessile or on short capillary pedicels, 0.2–0.25 cm in diameter, solitary or fascicled; florets female of bisexual, involucre of bracts glistening with narrow golden disk; achenes very minute, 0.04 cm long.

Distribution : Kashmir to Lahul at 1700–2250 m, Punjab, upper Ganga plain; Pakistan (Sind) and Tibet between 2200 and 3500 m.

Ecology and cultivation : Alpine plant; wild.

Chemical contents : *Root* : β-sitosterol, oxalic acid, a lactone; *Leaves* : capillin, 1-Ph-2, 4-hexadiyne-1-ol, vanillin,scoparone, flavonoids; *Flower-heads* : capillarin, coumarine, scopanone; *Air-dried flowering tops from Jammu* : carvone, dillapiol, an essential oil with characteristic odour or eugenol and various other constituents.

Remark : Possibilities of therapeutic use of this plant as tranquiliser and sleep-inducing agent are high; researches are in progress.

Arthrocnemum indicum Moq. (Chenopodiaceae)

(2n = 18)

Syn : *Salicornia indica* Willd., *Arthrocnemum glaucum* Woodr. non Ung. Sternb.

Sanskrit names : Subhar, Suvar.

Vernacular names : Ben : Jadu palang; **Guj :** Bholado; **Hin :** Machola; **Mar :** Machura; **Tam :** Umari; **Tel :** Koyyapippali.

Traditional use : *Plant* : alexipharmic; *Ashes of plant* : in scorpion sting.

Phytography : Fleshy, leafless herb, 2.5–7.5 cm high; stem prostrate, woody, more than 1 m long, branches short, diffuse, ascending, nodes 0.4–0.8 cm long and 0.6–0.8 cm in diameter; spikes 1.25–2.5 cm long, cylindric, broader than nodes; flowers minute, bisexual, 2–3 together in the axils of scales, bracts conceal flowers; fruits very spongy with thin margins, laterally compressed; seeds free, erect, orbicular, testa membranous.

Phenology : *Flowering* and *Fruiting* : July to November.

Distribution : A common mangrove of the Sundarbans, Goa and Maharashtra coast; Bangladesh and Sri Lanka.

Ecology and cultivation : Halophyte; wild, on the sides if the rivers, creeks.

Chemical contents : *Plant* : Stigmastane type 3C, 29 sterols including fucosterol.

Remark : Coastal people of Maharashtra eat the plant as a vegetable.

Artocarpus heterophyllus Lam. (Moraceae)

(2n = 56)

Syn : *Artocarpus integrifolia* L.f., *A. integra* (Thunb.) Merrill

English name : Jack tree.

Sanskrit name : Panasa.

Vernacular names : Ben and **Hin :** Kathal; **Kan :** Halasu; **Mal :** Chakka; **Mar :** Phanas; **Tam :** Pilapalam; **Tel :** Panasa.

Trade name : Kathal.

Traditional use : SANTAL : (i) *Red excresence of richly fruit-bearing tree* : for the treatment of barrenness in women; (ii) *Paste of young shoot* or *flower* : in treatment of post-delivery complications; (iii) *Leaf* : as plaster on the sores on the skin, thigh and some other parts of the body, as powder on the body of a patient suffering from small pox, as ingredient of an ointment to treat carbuncles and chronic progredient sores; (iv) *Young leaf* and *shoot* with *gur* : as lactagogue in women; RURAL FOLK : (i) *Root* : in skin diseases, asthma and diarrhoea; (ii) *Gum* : to remove fatigue, weakness; (iii) *Leaf-decoction* : in ringworm, eczema and other skin diseases; (iv) *Inflorescence-axis* : in muscular rheumatism; (v) *Juice of unripe fruit* : in weakness of heart; (vi) *Decoction of unripe fruit* : in redema; (vii) *Seed-decoction* : in spiting blood.
RAMAYANA : ripe fruit nutritious; CHARAKA SAMHITA : ripe fruit sweet, slightly astringent, useful in pile disorder, increases semen, purifies blood, enhances skin colour, but slightly

increases phlegm; *SUSHRUTA SAMHITA* : ripe fruit tasty, slightly astringent, cooling, but slightly difficult to digest; *AYURVEDA* : (i) *Root* should be eaten in diarrhoea; (ii) *Gum of stem* : useful in glandular swelling and in suppuration of abscesses; (iii) *Leaf* : beneficial in skin diseases; *VAYU PURANA* : regular consumption of fruit juice increases longevity; *KURMA PURANA* : consumption of fruit juice regularly enhances life-span and improves skin colour.

Modern use : *Latex* : Bacteriolytic; mixed with vinegar, promotes healing of abscesses; *Leaf* : estrogenic.

Phytography : Large tropical fruit tree with a dense crown reaching a height of 18 m or more; the straight cylindrical stem is covered with rough bark which exudes milky latex.

Phenology : *Flowering* and *Fruiting* : in Summer and Rainy seasons.

Distribution : Indigenous to India, probably in the Western Ghats, where it grows wild; grown extensively throughout the warmer parts of the country; Bangladesh, Sri Lanka and Myanmar.

Ecology and cultivation : For successful cultivation, it requires a moist tropical climate and a deep rich soil.

Chemical contents : *Bark* : cycloheterophyllin and its iso-derivatives; *Heartwood of Indian plants* : flavonoids, artocarpesin, nor-artocarpetin, di-Me ethers of artocarpin and isoartocarpin, cycloartocapesin, oxyhydroartocarpesin and some phenolics; *Leaf* : cycloartenone, cycloartenol and β-sitosterol; *Seed* : 2 isoelctins, dupeptide, aurantiamide-OAc.

Remarks : Unripe fruits are eaten as vegetable. Timber is useful. The plant is of religious value to the Hindus.

Artocarpus hirsutus Lam. (Moraceae)

Syn : *Artocarpus pubescens* Willd.

English name : Wild jack.

Vernacular names : Mal : Annili; **Mar :** Ran phanas; **Tam :** Anjali

Traditional use : *RURAL PEOPLE OF SOUTH INDIA* : *Dry leaf* and *juice* : together with zedoary and camphor, used as application to buboes and swelled testicles.

Modern use : *EtOH* (50%) *extract of aerial parts* : diuretic.

Phytography : Lofty evergreen tree, branchlets strigose; leaves alternate, petiolate; petioles 1.8–2.5 cm long, stipules almost of same length, lanceolate, simple, lamina blackish green, coriaceous, ovate, entire, glabrous generally; heads pendulous, 10–15 cm long, cylindrical; fruits erect, ovoid, size of a lemon, spine straight, 0.6 cm long; seeds 1.25–1.80 cm long, ovoid.

Distribution : Gujarat, Maharashtra, Karnataka, Kerala.

Chemical content : *Seed* : fixed oil (16-17%).

Artocarpus lacucha Buch. Ham. (Moraceae)

Syn : *Artocarpus lakoocha* Roxb.

English names : Monkey jack, Wild jack.

Sanskrit name : Lakucha.

Vernacular names : Asm : Chama, Dewa; **Ben :** Delomadar, Dephal, Dehua; **Hin :** Barhal, Dahua, Lakucha; **Kan :** Vatehuli, Wents; **Mar :** Badhar, Wotomba; **Ori :** Jetua;

Pun : Deheo; **San :** Dau-dara, Dahu; **Tam :** Ilagusam; **Tel :** Kammaregu, Laku chamma, Nakka renu; **Urd :** Barhal.

Trade names : Dahua, Dephal.

Traditional use : SANTAL : (i) *Exudation of plant* : rubbed over spleen in complaints of spleen; (ii) *Bark-paste* : to drink in troubles of spleen of man, (iii) *Powder of seed* : as purgative; TRIBES OF PURULIA (West Bengal) : (i) *Seed* : in stomach disorder.

RAMAYANA : beautiful tree; fruits edible; CHARAKA SAMHITA : eating unripe fruit along with honey and milk harmful, consumption of ripe fruits with juice of *Phasealus radiatus,* molasses and *ghee* forbidden; SUSHRUTA SAMHITA : fruit sour, causes imbalance of wind, phlegm and bile and distension of stomach, harms semen; RAJNIGHANTU : fruit (unripe) bitter, astringent, energiser, reduces corpuscular obesity, helps in digestion, alleviates troubles caused by phlegm, causes burning sensation, helps stool formation; BHABAPRAKASHA : immature fruit leads to storing up of energy, just ripe fruit difficult to digest but sufficiently ripened fruit sweet-sour, alleviates disorders of wind and bile; AYURVEDA : (i) *Powder of bark* : externally applied to sores; (ii) *Paste of bark* : lotion like application on boils and pimples; (iii) *Juice of unripe fruit* (in small doses) : in constipation, obesity; (iv) *Juice of ripe fruit* : (in small doses) : in anorexia, wind in stomach, insufficiency of sperms; (v) *Seed-powder* : in diarrhoea.

Modern use : *Seed* : haemagglutinating.

Phytography : Deciduous tree, attaining 20 m in height, with a spreading dense crown and a clear straight bole; bark dark brown, exfoliating in small round peels, inside reddish; branches villousely tomentose; leaves alternate, stipules 1.25 cm, lanceolate, pubescent —but some portion glabrous, leafblade 10–30 cm by 6–20 cm, elliptic or ovate, thinly coriaceous, pinnatifid, lower surface pubescent or tomentose, upper surface glabrous;

inflorescence axillary, solitary, unisexual, male flowers subsessile, orange-yellow, pubescent, bracts stoutly stalked; female heads irregularly subglobose, 1.0–1.5 cm, bracts peltate, fruits syncarp, subglobose, shallowly lobed, 6–12 cm long, unripe green, yellow when ripe, kernel red, surface pubescent; seeds 10 by 6 mm, endocarp horny.

Phenology : *Flowering* : December–April; *Fruiting* : May and October, November.

Distribution : Sub-Himalayan tract up to 1500 m from Kumaon to Assam, Bihar, Orissa, Madhya Pradesh, Karnataka, Kerala, Tamil Nadu and the Andamans; Bangladesh, Myanmar.

Ecology and cultivation : Occurs in evergreen, semi-evergreen and moist deciduous forests, found frequently in rural areas and waste places.

Chemical contents : *Aerial parts* : latex containing water solubles, carutchouc and coagalum; *Root-bark* : a glycoflavonol, lupeol, β-sitosterol; *Stem-bark* : lupeol-OAc and cycloartenol, cycloartenone; *Stem* : 2, 4, 3', 5'-tetra-OH-stilbene; *Wood* : artocarpin, cycloartocarpin, norartocarpin, norcycloartocarpin, resorcinol and oxyreserveratrol.

Remarks : Santals use *bark-paste* in bone fracture of cattle and give *leaf* to cows as fodder for quick expulsion of foetus.
Santals eat inflorescence, flower and fruit; Tribes of Purulia, Sikkim, Khasi and Garo regions eat fruits; Mikirs use bark as masticatory.

Arundo donax L. (Poaceae)

Syn : *Arundo sativa* Lamk., *Donax aurandinaceus* P. Beauv., *Cynodon donax* (L.) Rajpail.

English names : Giant bamboo reed, Spanish bamboo reed, Spanish cane.

Vernacular names : Ben : Gahanal; **Hin :** Baranal, Narhal, Narsal; **Miri :** Nalani; **Pun :** Bansi.

Traditional use : *Decoction of rhizome* : emollient, diuretic, stimulates menstrual discharge, and diminishes the secretion of milk.

Modern use : *EtOH extract of rhizome* : hypotensive and antispasmodic; *Flower* : hypotensive, causes respiratory depression.

Phytography : Tall and stout, perennial, reed-like grass; stem fistular, 2–3.5 m high, drooping, creeping below; leaves 30–60 cm by 25.5–50.00 mm, broad, ensiform with amplexicaul base, drooping, ligules of long hairs; spikelets 0.8 cm, green or yellowish.

Phenology : *Flowering* : October.

Distribution : Lower Himalaya, from Kashmir to Meghalaya, ascending up to 1500 m, Punjab, Bihar, the Nilgiris range between 2000 and 2500 m; Nepal, Bangladesh (Sylhet).

Ecology and cultivation : Grows in ditches and streams; wild.

Chemical contents : (i) *Rhizome* : indole-3-alkylamine bases including bufotenidine and dehydrobufontenine; (ii) *Plant* : indole-3-alkyl-amines; (iii) *Leaf* : sterols and triterpenoids; (iv) *Flower :* gramine, its methohydroxide, eleagnine, N,N-di-Me-tryptamine methohydroxide, 3,3'-bis(indolyl-Me)-di-Me-ammonium hydroxide.

Asclepias curassavica L. (Asclepiadaceae)

(2n = 22)

English names : Curassavian shallow-wart, West Indian ipecacuanha, False ipecae, Blood flower, Milk weed, Silk weed.

Vernacular names : Hin : Kakatundi, Kaura dodi; **Mar :** Kurki; **Pun :** Kakatundi.

Trade name : *Root* : West Indian or Bastard Ipecacuanha.

Traditional use : *Root* : Cathartic; *Latex* : used to remove warts and corns; *Juice of leaf* : anthelmintic, antidysenteric, sudorific and used as anti-cancerous drug; *Powder of leaf* : treatment of sores and wounds.

Modern use : *Root* : emetic, purgative, used in piles and gonorrhoea; *Root-extract* : cardiovascular ; *Root-powder* (mixed with *Acorus* root) : cures chronic ulcer; *Latex* : bactericidal, used against warts and cancer; *Extract of root, stem, leaf* and *flower* : can destroy human cancer cells in culture; *Alcoholic extract of plant* (from Dehra Dun) : cardiotonic; *Leaf-juice* : anthelmintic, sudorific, useful in abdominal pain; haemorrhages, gonorrhoea, externally for piles.

Phytography : About 60 cm high with lanceolate or oblong-lanceolate, glabrous or pubescent leaves, umbels many-flowered; flowers orange or scarlet, moderate-sized with reflexed corolla and 5 erect spoon-shaped coronal processes adante to the stipitate column; follicles inflated.

Distribution : Native of West India; often found as ornamentals for showy flowers.

Ecology and cultivation : Commonly cultivated plant.

Chemical contents : *Root* : vincetoxin, isolated from root, resemble emetine and aconitine, asclepin (chief active principle)—a glucoside with cardiac activity; *Plant* : the

steroidal sapogenins uzarigenin crotoxigenin, coroglaucigenin, calatropin and curassavicine along with a new acylated geninaselepin and its four glycosides.

Adulterant : Root used as Trinidad ipecacuanha.

Remarks : Plant used as fish poison. In Jamaica, it is called 'blood flower' owing to the efficacy in dysentery (George Watt).

Asparagus adscendens Roxb. (Liliaceae)

Sanskrit name : Musali.

Vernacular names : Gar : Jhirna; **Guj :** Dholi musali, Ujli musli; **Hin** and **Mar :** Safed musli, Satavar.

Trade name : Safed musli.

Traditional use : *Root* : tonic, cooling, diaphoretic and used in general debility; *Powder of root* (boiled with milk and sugar) : a reputed drug for non-specific leucorrhoea.

Modern use : *Root* : used as demulcent, in diarrhoea and dysentery.

Phytography : An excessively branching, suberect tall plant with densely crowded cladophylls.

Distribution : Western Himalaya and Punjab, up to 4000 m; Afghanistan.

Chemical contents : *Plant* : Asparagin, 3-heptadecnone, 8-hexadecenoic acid, methyl pentacosanoate, tetratriacontane, tricontane, methyl palmitate, palmitic acid, stearic acid.

Asparagus filicinus Buch. Ham. ex D. Don (Liliaceae)

Traditional use : Astringent, tonic and diuretic.

Modern use : *EtOH* (50%) *extract of plant* : antiprotozoal, spasmogenic; *Plant-extract* : antifungal.

Phytography : Tall, erect plant with stout rootstock, aerial stem flexuous, fistular, much-branched, lower branches spreading, upper internodes short, smooth, unarmed; leaves minute scales, often spinescent, bearing in their axils tufts of green needle-like or flattened branchlets called cladodes; cladodes 0.25–0.6 cm long; flowers axillary, small or minute, solitary or in racemes, sessile or short-pedicelled, pedicels 0.6–1.2 cm long, jointed at middle; berries globose, 0.6–0.8 cm in diameter; seeds black.

Distribution : Temperate and tropical Himalaya—from Kashmir (1300–2300 m) eastwards to Assam, Meghalaya (1700–2000 m); Bhutan, Myanmar.

Ecology and cultivation : Plant of temperate and cold tropical regions; wild.

Asparagus gonocladus Baker (Liliaceae)

Vernacular names : Ben : Shatamuli; **Hin :** Shakakul; **Mar :** Shatavari, Shatavarimul; **Tam :** Kilavari; **Tel :** Philipichara.

Trade name : Shakakul.

Traditional use : *Root-extract* : given in gonorrhoea.

Modern use : *Root* : aphrodisiac, boiled with oil—applied to cutaneous diseases; *Leaf* : (along with other ingredients) : strong steriliser.

Phytography : Excessively branched undershrub; branches and branchlets curved, green, strongly angled; cladodes 1.6–2.5 cm long and 0.3 cm in diameter, green, ensiform, flat, falcate, narrowed to the acuminate tip; racemes 2.5–7.5 cm long, often fascicled, sometimes connate, lax-flowered; pedicels 0.12–0.20 cm long, jointed below the middle; flowers bracteate—bracts short, cymbiform, 0.2 cm in diameter, perianth segments spreading.

Distribution : Maharashtra, Karnataka and Tamil Nadu.

Chemical contents : *Aerial parts* : apigenin, kaempferol, rutin, 4,6,4'-tri-OH-aurone, a chalcone glyceride; *Flower* : malvin, asparagine.

Asparagus officinalis L. (Liliaceae)

English name : Garden asparagus.

Vernacular names : Ben : Hillua; **Hin :** Nagdown, Halyan, Seetmuli.

Trade name : Seetmuli.

Traditional use : *Plant* : Diuretic; used in cardiac dropsy and chronic gout; *UNANI* : Ingredient of the medicine called 'Halyan'.

Modern use : *Berries* (decoction) : contraceptive.

Phytography : Perennial plant with an erect branching stem, nearly 1.2–2.8 m in length; roots of many cord-like fleshy parts, glabrous, slightly glaucous; cladode thread-like; flowers 1–4, in the axil of cladodes, bell-shaped; berry hanging, red.

Distribution : Cultivated in India; indigenous to temperate Asia and Europe; introduced in various countries.

Ecology and cultivation : Thrives best in well-drained, deep, loose soil; requires liberal use of fertilisers. Seeds are soaked in warm water for a day before sowing and vigorous seedlings, when a year old, are transplanted.

Chemical contents : *Root* : β-sitosterol, sarsasapogenin, asparagosides A-1, hexasaccharides, and nine steroidal glycerides; *Stem* : furostanol saponins, asparasaponins I and II, two cysteine derivatives, rutin, quercetin, kaempferol; *Etiolated shoot* : diosgenin, yamogenin, asparagusic acid and its derivatives; *Leaf* : rutin; *Flower* : flavonoids, rutin, quercetin, hyperoside and their glycerides; *Fruit* : glycerides; *Seed* : polysaccharides.

Asparagus racemosus Willd. var. *javanicus* Baker (Liliaceae)

Syn : *Asperagus tetragonus.*

English name : Asparagus.

Sanskrit names : Shatavari, Shatamuli.

Vernacular names : Asm : Shatamul; **Ben :** Shatamuli; **Guj :** Satawari; **Hin :** Shakakul, Sada bori, Shatamuli, Satwar; **Mal :** Shatavali; **Mar :** Shatavari, Shatavarimul; **Ori :** Shatamuli; **Tam :** Chattavari, Cadumullam, Shimai shadavari; **Tel :** Challagadda.

Trade names : Shatamuli, Asparagus.

Traditional use : *SOME TRIBES OF INDIA* : *Root* : refrigerant, antiseptic, alterative and galactogogue; roots form a constituent of medicinal oils used for nervous and rheumatic complaints; powder boiled with milk and sugar has been used in different types of neurological disorders, urinary calculi; rub the root of the plant in different diseases of urinary organs; *Powder of leaf* : enhances breast milk, used for the treatment of colon syndrome.

AYURVEDA : useful in different types of cardiac abnormality; beneficial for intelligence as well as on memory modulatory; *Powder of root* (boiled with milk) : for treatment of night blindness;

UNANI : Ingredient of 'Satawar'.

Modern use : *Root* : used as demulcent, aphrodisiac, diuretic, antidysenteric, in impotency, azoospermea.

Phytography : A slender, scrambling/scandent suffruticose perennial, woody, prickly shoots with reflexed spines; cladode more or less acicular and triquetrous falcate 13–26 mm long; divaricate, finely acuminate; flowers white, sweet-scented, ± 6 mm on filiform articulate pedicels, 6–12 mm long in very short racemes; berries 2.5–6 mm in diameter, scarlet, red on ripening.

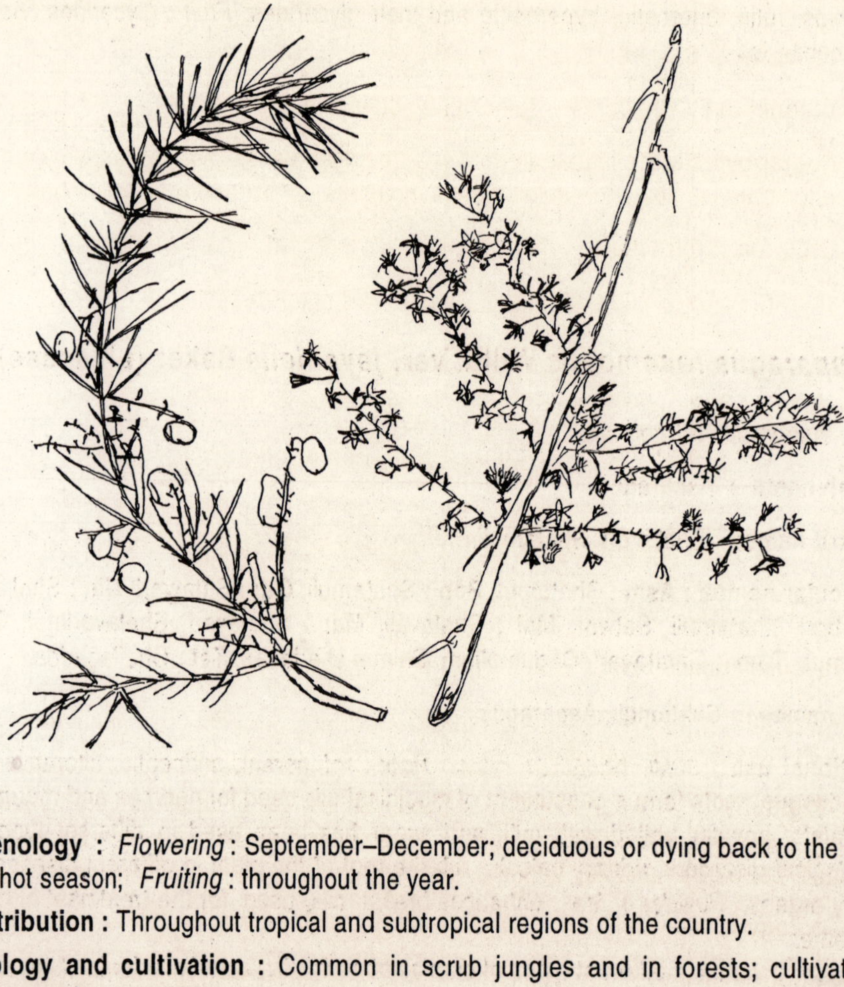

Phenology : *Flowering* : September–December; deciduous or dying back to the root in the hot season; *Fruiting* : throughout the year.

Distribution : Throughout tropical and subtropical regions of the country.

Ecology and cultivation : Common in scrub jungles and in forests; cultivated as ornamentals.

Chemical contents : *Leaf* : quercetin-3-glucoronide, sitosterol, stigmasterol, sarsasapogenin, sito-sterol D-glucoside, stigmasterol-B-D-glucoside, two sirostanolic and two furostandic saponins.

Remark : Root is used in veterinary medicine.

Asparagus sarmentosus L. (Liliaceae)

English name : Asparagus.

Sanskrit name : Satamuli

Vernacular name : Ben : Shatamuli; **Hin :** Shatawar; **Gar :** Shatawari; **Kan :** Majjige-gadde; **Mar :** Shatawarmul, Shatavali; **Tam :** Shimai shadavari; **Tel :** Challgadda, Pilli tegalu.

Trade name : Shatamuli.

Traditional use : *Fresh root* : demulcent, nourishing and aphrodisiac.

Phytography : Herb, climber; stem short, creeping; leaves linear to lanceolate or ovate, venation parallel; flowers regular, symmetrical, always six-androus; fruit a berry.

Distribution : North India.

Adulterant : Used as adulterant of *Asparagus racemossus.*

Asphodelus tenuifolius Car. (Liliaceae)

Syn : *Asphodelus clavatum* Roxb.

Vernacular names : Hin : Bakat, **Pun :** Piari, Bokat.

Traditional use : *Seed* : diuretic, applied externally to ulcers and inflamed parts.
UNANI : Ingredient of 'Gardana'.

Phytography : A weed-looking annual with slender erect semiterete leaves; often much fastigiately branched scapes, bearing white flowers on the elevate pedicels, jointed below the middle; capsule globose.

Phenology : *Flowering* and *Fruiting* : May–June.

Distribution : Indo-Ganga plains from Punjab to West Bengal; Bangladesh and Pakistan.

Ecology and cultivation : In cultivated fields; common in cornfields.

Chemical contents : *Plant* : lupeol, quercetin; *Tuber, Leaf* and *Seed* : anthraquinones; *Seed* : 1-0-17-Me-stearyl-myoinositol; *Seed-oil* : β-amyrin, β-sitosterol, campesterol, stigmasterol.

Asterocantha longifolia (L.) Nees (Acanthaceae)

Syn : *Hydrophila spinosa* T. Anders., *Hydrophila auriculata* Heine.

Sanskrit names : Ishugandha, Kokilaksha.

Vernacular names : Ben : Kuley khara; **Guj :** Ekharo; **Hin :** Gokul kanta, Talmakhana; **Kan :** Kabija, Kalavam; **Mal :** Vayalchulli; **Mar :** Kolasanda, Kotista; **Ori :** Kanta kalia; **Tam :** Nirmull(i); **Tel :** Nirugubbi, Nirguriveru.

Trade name : Kuley khara.

Traditional use : *ATHARVA VEDA* and *CHARAKA SAMHITA* : improves quality of semen; *SUSHRUTA SAMHITA* : *Root-powder* : improves sexual vigour, removes obstructions from seminal passage; *HAREETA SAMHITA* : *Juice of Root* : in insomnia; *BAGBHATTA* : *Plant-extract* : in purification of blood; *AYURVEDA* : *Root, Leaf* and *Seed* : diuretic and

useful in jaundice, dropsy, rheumatism, anasarca, urinogenital diseases, gonorrhoea, spermatorrhoea, gall stone; *Leaf*: to stop bleeding in varicose vein, herpes.
SIDDHA: the drug *Neermulli* is obtained from the plant.
UNANI: source of the drug 'Talmakhana'.

Modern use : *Essential oil from the plant*: antibacterial.

Phytography : Spinescent herb; fascicled, with many roots at the base of stem; stem 60–50 cm, rusty-green, hispid, quadrangular, pith holed, leaves simple, opposite, subsessile, lanceolate, 7.5–17.0 cm, entire, sparsely hispid with long white hairs, whorls of straight stout spines present at nodes; flowers sessile, mostly in axillary whorls—at times in terminal heads, bracteate and bracteolate, tubular, 2.5 cm, pale blue-purple, nearly glabrous; capsules 0.8 cm, 4- to 8-seeded; seeds like the eye of a cuckoo.

Phenology : *Flowering*: October–November.

Distribution : Throughout India; Bangladesh, Pakistan and Sri Lanka.

Ecology and cultivation : Grows in moist and marshy places, crop fields and ditches; may be propagated by cuttings and seeds.

Chemical contents : *Plant*: lupeol, stigmasterol, hydrocarbons, essential oil; *Flower*: apigenin glucoronide; *Seed*: sterol, oil, compounds of unsaponified matters.

Adulterant : Young plant looks like *Enhydra fluctuens*.

Astragalus homosus L. (Fabaceae)

(2n = 32, 34, 48)

Vernacular names : Hin : Purtuk; **Mar :** Akhlilelmalik; **Pun :** Akhilulmalik.

Traditional use : *Plant*: emollient, demulcent, useful in irritation of mucous membrane, laxative, lactagogue, used in nervous affections and catarrhal affections.
UNANI: Ingredient of 'Iklil-ul-malik'.

Phytography : Robust herb; stem 30–60 cm long; leaves alternate, petioles 10.5–15.0 cm long, pinnate, leaflets stalked, pale green, 1.25–1.80 cm long, thinly matted with silvery hairs on both sides; heads peduncled, dense, 6- to 20-flowered; flowers pale yellow, 0.6 cm long, covered with hairs; pods 1.25–2.5 cm long, cylindrical, glabrous.

Distribution : Plains of Punjab; Pakistan (Baluchistan and Sind).

Ecology and cultivation : Xerophytic; wild.

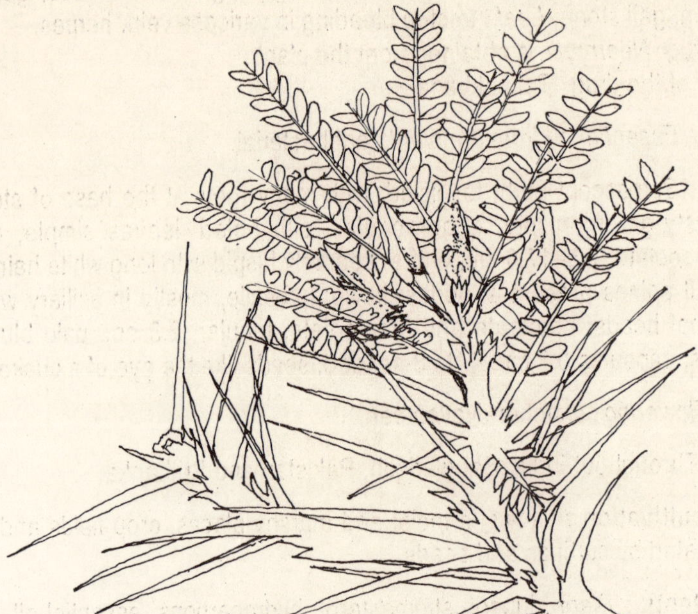

Chemical contents : *Plant* : a gum-like tragacanth, saponin.

Remark : In pharmacy, tragacanth is used as demulcent and emulsifier.

Astragalus multiceps Wall. (Fabaceae)

Vernacular names : Pun : Kandiara, Sarmul.

Traditional use : RURAL FOLK OF PUNJAB : *Seed* : demulcent, emollient in colic and leprosy.

Phytography : Undershrub, branches covered with hairs, nodes densely crowded; leaves alternate, petiolate, stipules lanceolate, pinnate, 3.7–7.5 cm long, leaflets 21–31, mostly crowded, thick, clothed with grey silky hairs; flowers 1 or 2 together in leaf-axils, 1.8 cm long; pods sessile, oblong, bilocular, clothed with fine, grey, silky hairs, 12- to 14-seeded.

Distribution : Western Himalaya, in temperate zone, between 3000 and 4000 m, Garhwal, Kumaon, Himachal Pradesh.

Astragalus tribuloides Delile (Fabaceae)

Vernacular name : Pun : Ogai.

Traditional use : *RURAL PEOPLE OF PUNJAB* : *Seed* : demulcent, emollient.

Phytography : Herb; stem trailing, 7.5–30 cm, densely clothed upwards with ascending white hairs; leaves alternate, stipules minute, petioles 2.5–6.0 cm long, pinnate, leaflets mostly close, densely covered on both surfaces with long white hairs; heads 6- to 10-flowered; flowers pale yellow; pods 0.9 cm long, densely pubescent, cluster often spreading from stem like the spoke of a wheel from axis.

Distribution : Plains of western and central Punjab.

Asystasia gangetica T. Anders. (Acanthaceae)

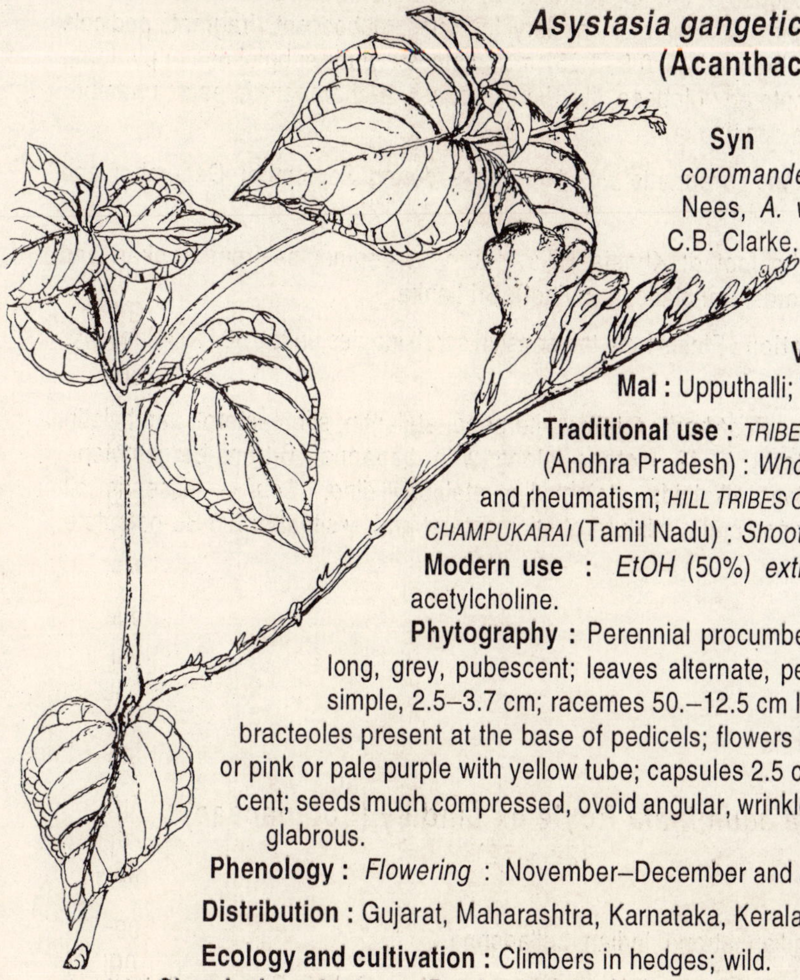

Syn : *Asystasia coromandeliana* Wight ex Nees, *A. violacea* Dals. non C.B. Clarke.

Sanskrit name : Lavana valli.

Vernacular names : **Mal :** Upputhalli; **Tam :** Meddykeerai.

Traditional use : *TRIBES OF EAST GODAVARI* (Andhra Pradesh) : *Whole plant* : in swelling and rheumatism; *HILL TRIBES OF DHOOMANOOR* and *CHAMPUKARAI* (Tamil Nadu) : *Shoot* : in wounds.

Modern use : *EtOH* (50%) *extract of plant* : anti-acetylcholine.

Phytography : Perennial procumbent herb, 30–60 cm long, grey, pubescent; leaves alternate, petioles 0.8 cm long, simple, 2.5–3.7 cm; racemes 50.–12.5 cm long, lax, bracts and bracteoles present at the base of pedicels; flowers 2.5–3.0 cm, yellow or pink or pale purple with yellow tube; capsules 2.5 cm, cylindric, pubescent; seeds much compressed, ovoid angular, wrinkled or subtubercular, glabrous.

Phenology : *Flowering* : November–December and again in Summer.

Distribution : Gujarat, Maharashtra, Karnataka, Kerala; Sri Lanka.

Ecology and cultivation : Climbers in hedges; wild.

Chemical contents : *Petals* : luteolin 7-glucoside, isosalipurposide and its glycerides.

Atalantia monophylla (Roxb.) DC. (Rutaceae)

Syn : *Limonia monophylla* Roxb. non L.

Traditional use : *Root*, *Leaf* and *Berry* : antispasmodic, antiseptic, antipyretic and stimulant.

Modern use : *Essential oil from leaf* : antitubercular, antifungal; *Oil from fruit* : used in chronic rheumatism and paralysis.

Phytography : Large glabrous shrub or tree; branches rigid, flexuous, tips often pubescent; leaves alternate, short petioled, unifoliate, 2.5–7.5 cm, ovate or oblong or elliptic, obtuse, notched or bilobed at the tip, coriaceous, bright green, stipular scales subulate, axillary bud often spinous; flowers in axillary fascicles, pubescent, fragrant, pedicels 0.6–1.2 cm, minutely bracteolate, 0.8–1.25 cm in diameter, buds subglobose or pyriform; berries large like nutmeg, globose like lime, usually 4-chambered, each chamber 1-seeded, orange on ripening.

Phenology : *Flowering* : fragrants with two peaks during September–December and February–March.

Distribution : At the foot of Khasia Hills, southern peninsula—from Konkan and Coromondal southwards; Bangladesh (Sylhet), Sri Lanka.

Ecology and cultivation : Plains from the coast in scrub jungles up to 750 (1000) m; wild, not very common.

Chemical contents : *Root-bark* : tetranortriterpene, atalantin, stigmasterol, xanthyletin, marmesin, γ-sitosterol, C-12 lactone along with capaene, β-trans-bergamotene, β-bisabolene, acridone alkaloids, atalaphyllin, atalaphyllidine; *Leaf* : essential oil, friedelin, epifriedelanol, and a mixture of stigmasterol and γ-sitosterol, n-Bu-palmitate; *Fruit* : severine.

Atropa acuminata Royle ex Lindley (Solanaceae)

(2n = 72)

English names : Indian atropa, Indian belladonna.

Vernacular names : Ben : Jebrui; **Hin :** Sag angur; **Mar :** Girbuti; **Pun :** Bantamaku.

Trade name : Indian belladonna.

Traditional use : It is used in manufacture of tinctures and plasters. *Leaf* and *Root* : narcotic, sedative, diuretic and mydriatic, used as anodyne; *Berry* : poisonous.
UNANI : Ingredient of 'Luffah'.

Modern use : The drug (Belladonna) is collected when the plant is in flowering. The drug decreases secretion of sweat, gastric and salivary glands; acts as strong antispasmodic in intestinal colic; used in asthma and whooping cough; *Root* : said to have some poisonous substances and are thus prescribed for external applications in neuralgia, inflammations and rheumatism; may serve as a source of atropine.

Phytography : An erect-branched perennial herb; leaf pale green, oblong-elliptic, tapering at both the apex and base of the lamina; flowers yellow and funnel-shaped; berry bilocular and contain numerous seeds.

Distribution : Kashmir, Himachal Pradesh and Chamba at an elevation of about 4500–6000 m; Pakistan (Baluchistan).

Ecology and cultivation : The best period for raising nurseries is from the last week of June to first week of July; bed should be 3 m x 1 m, should be shaded and watered regularly; should be planted and supervised by expert personnel.

Chemical contents : Total alkaloid contents in the Himalayan plant—0.81% in root and 0.5% in leaf. The chief constituent of the drug is the alkaloid hyoscyamine. Atropine is absent in the young leaves but is profusely present (5–40%) in the older leaves. Belladonna leaves contain a fluorescent substance (β- Methyl-aesculetin) and yield about 14% of ash. Alkaloids present in all parts of the plant body.

Adulterants : Leaves of *Phytolacca decandra* L., (the Pokeweed, family *Phytolaccaceae* offered under the name of Portuguese Belladonna), leaflets of *Ailanthus glandulosa* Desf. (the tree of heaven, family *Simaroubaceae*, sometimes described as Italian Belladonna, leaves of *Scopolia carniolica* Jacquin (family *Solanaceae*), leaves of *Solanum nigrum* and *Datura stramonium* are sometimes substituted for belladonna leaves.

Remarks : There is great potentiality of its export. This is being extensively cultivated in Jammu and Kashmir and Tamil Nadu. To save the plant from complete annihilation, the Forest Department of Jammu and Kashmir has enforced a legislation, popularly known as 'Kuth Act'.

Atropa belladonna L. (Solanaceae)

(2n = 50, 72)

English names : Belladonna, Deadly night shade, Sleeping night shade.

Vernacular names : Ben : Yebruj; **Hin :** Angurshefa, Sag angur; **Mar :** Girbuti; **Pun :** Suchi.

Trade name : In commerce, the root is called *Belladonna radix* and leaves *Belladonna folium*; commonly called *Belladonna.*

Traditional use : *Root* and *Leaf* : narcotic, diuretic, sedative and mydriatic; *Leaf* : anodyme in febrile conditions, night sweating, cough, whooping cough, spermatorrhoea; externally used in gout and rheumatic inflammation.

HOMOEOPATHY : excellent action on nervous system, pain in liver, colic pain, eyesight and puerperal convulsion.

UNANI : Ingredient of 'Luffah'.

Modern use : Belladonna acts as a local anaesthetic and anodyne, and is used externally to relieve pain. Internally, it is given to check the sweating in phthisis, as a sedative to respiratory nerves, to relieve spasmodic cough and in numerous other conditions. It has the property of dilating pupils of the eyes and preparations like atropine have been largely employed as eye drops and ointments at the time of testing eyes and other ophthalmic treatments. Also used as an antidote in certain types of poisoning cases.

Phytography : An erect glandular pubescent herb of about 2 m height; leaves stalked, ovate-lanceolate, entire, long-pointed, upper one with a much smaller leaf springing from the same point; flowers pale purple tinged with yellow, green, single or drooping, usually axillary stalks; berry globose, purple black, surrounded at the base by the enlarged spreading calyx.

Distribution : Native to Europe, now cultivated in Kashmir. The plant grows wild in central and southern Europe and is cultivated in many countries of eastern Europe, Bulgaria, Hungary, Germany, etc.

Ecology and cultivation : It is found in chalky and limestone areas. Roots are collected from the plants which have been grown to yield belladonna herb and they are removed by ploughing up the ground at the end of the season.

Chemical contents : *Root* : hyoseyamine, traces of scopolamine (hyocine); other contents are starch, tannin, etc.; total alkaloid content depends upon the stage of development of the plant—low during flowering and high when setting fruits; alkaloid content in the leaves is influenced by plant spacing; *Germinating seeds* and *seedlings* : usual alkaloids are absent but Cuscohygrine, a new normal alkaloid and bellardine are obtained

Adulterants : Scopalia rhizome, Japanese belladonna root, Indian belladonna root, phytolaccae root and leaves of *Xanthium strumarium* L. (Asteraceae), *X. macrocarpon* DC. (Asteraceae), *Carthamus helenoides* Desf. (Asteraceae), *Solanum nigrum* L. (Solanaceae), *Chenopodium hybridum* L. (Chenopodiaceae) and *Hyoseymus niger* L. (Solanaceae) are used as adulterants of *Atropa belladonna*.

Atylosia goensis Dalz. (Fabaceae)

(2n = 22)

Syn : *Atylosia barbata* Baker, *Cajanus goensis* Dalz., *Dunbaria barbata* Benth.

Sanskrit name : Mashaparni.

Vernacular name : Tam : Peruvidukol.

Traditional use : *Root* : in rheumatism, biliousness, fever, consumption, swelling, and as febrifuge.

Phytography : Woody climber; branches densely pubescent; leaves alternate, petiolate, stipules caducous, leaflets 7.5–10.0 cm broad, roundish, subcoriaceous, green above but paler and greyer beneath, both surfaces minutely pubescent, petiolule 1.25–2.50 cm, stipules minute; racemes long peduncled, 10–15 cm long, rachis densely hairy; flowers bracteate, 1.25 cm long; pods 3.5–5.0 cm long, straight, densely clothed with grey hairs, 5- to 6-seeded.

Distribution : Assam, Karnataka.

Avena sativa L. (Poaceae)

(2n = 42)

English name : Oat.

Vernacular name : Hin : Jai.

Trade name : Oat.

Traditional use : *Seed* : nerve tonic, stimulant, laxative and antiseptic.

Phytography : Tufted annual, 1.2 m with culms smooth or scabrous beneath; leaves flat, more or less scabrous, sheaths long and loose, ligule short and jagged; panicles terminal and spreading, branches hanging from one side; fruits narrow, grooved.

Phenology : *Flowering* : Autumn; *Fruiting* : Winter.

Distribution : Cultivated in India, Bangladesh and Pakistan up to an elevation of 400 m.

Ecology and cultivation : Prefers semidry land.

Chemical contents : *Grains* : oxalic acid, iodine, vitamin B, fat and proteins.

Averrhoa bilimbi L. (Averrhoaceae)

English names : Bilimbi, Cucumber tree.

Vernacular names : Ben and **Hin** : Bilimbi; **Guj** and **Mar :** Blimbu; **Tam :** Pulicha; **Tel :** Bilumbi.

Trade name : Bilumbi.

Traditional use : *Fruit* : astringent, stomachic, refrigerant; in the form of curry useful in piles and scurvy.

Phytography : A medium-sized tree; leaves alternate, petioled, paripinnate with 5–17 pairs of leaflets; flowers small; berries cucumber-shaped, greenish yellow, 5-lobed, about 50 mm long, sour and juicy, few flattened seeded.

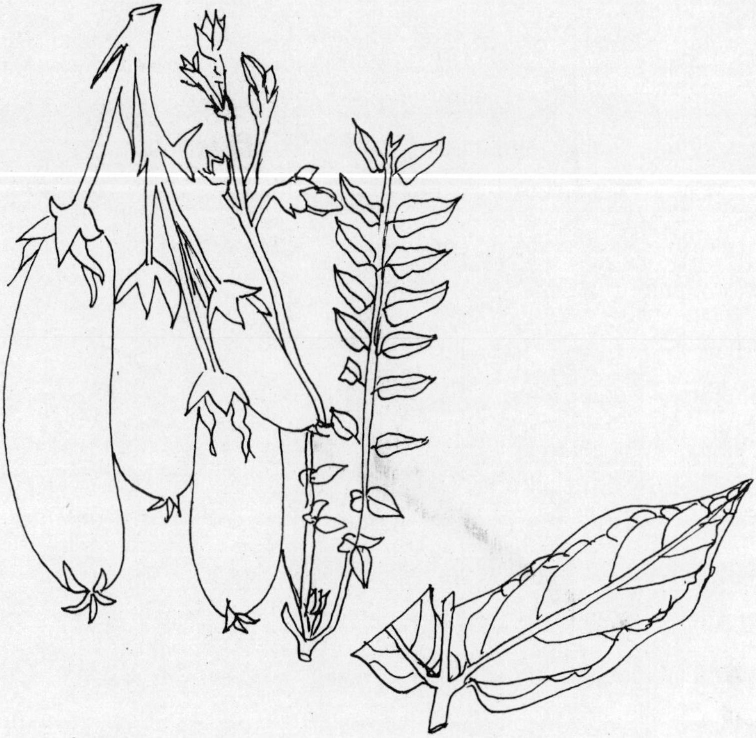

Distribution : Naturalised and wild in tropical parts of India; Bangladesh and Pakistan.

Ecology and cultivation : Native country not known; widely cultivated.

Chemical contents : *Fruit-juice* : acidic (pH 4.47).

Averrhoa carambola L. (Averrhoaceae)

(2n = 24)

English names : Carambola, Chinese gooseberry.

Sanskrit names : Karmar, Shiral.

Vernacular names : Asm : Kordoi tenga; **Ben :** Kamranga; **Guj :** Karmarakha; **Hin :** Kamrak, Kamrakh, Karmal; **Kan :** Kamarak, Kamaraxi; **Kon :** Kormongarukh; **Mal :** Irimpanpuli; **Mar :** Camaranga, Karamara, Kemarakha; **Ori :** Karmanga; **Pun :** Kamrakh; **Tam :** Kamarankai, Tamarattai; **Tel :** Tamarta.

Trade name : Kamrakh.

Traditional use : MIRIS (Mishing) : *Fruit* : in dysentery; TRIBES OF SAURASHTRA (Gujarat) : *Fruit* : in piles; RURAL FOLK OF ASSAM : *Fruit* : in jaundice.
RAMAYANA : beautiful forest plant; CHARAKA SAMHITA : *Fruit-juice* : appetiser, digestive, decreases wind and phlegm, keeps urinary tract clean, but excessive use of sour juice

may cause blood vomiting and harm to semen; *BHAVAPRAKASH* and *RAJNIGHANTU* : *Fruit* : acidic, laxative, cures the diseases caused by wind and phlegm; *AYURVEDA* : (i) *Immature fruit* : acidic, increases body temperature, causes pain in chest; (ii) *Ripe fruit* : sweet and sour, increases bile and body strength, useful in piles, stomachic; (iii) *Leaf-powder* : in nagging fever.

Phytography : Evergreen tree, 5–15 m high, densely branched, bark grey; leaves alternate, petiole stout, pubescent, lamina pinnate with a terminal leaflet; panicles axillary —usually found in young branches; flowers white and purple, pentamerous; berries green when young, golden yellow or amber when ripe, 7.5 cm long, 5-angled and 5-furrowed, fragrant, watery, translucent, sweetish sour.

Phenology : *Flowering* : June–September; *Fruiting* : September–November.

Distribution : Throughout the hotter part of Indian subcontinent as a garden plant.

Ecology and cultivation : Plains; planted tree.

Chemical contents : *Flower* : quercetin 3-glucoside, rutin; Fruit : amino acid compounds, proteins, vitamins A, B, C, oxalic acid, Ca, P, Fe.

Avicennia officinalis L. (Avicenniaceae)

Syn : *Avicennia tomentosa* Jacq., *A. resinifera* Forst.

English name : White mangrove.

Sanskrit name : Tuvara.

Vernacular names : Ben : Bina, Bain; **Guj :** Tavariyan; **Hin :** Bina, Tiwar; **Mar :** Tivar; **Tam :** Madaipattai; **Tel** : Nallamanda, Mada.

Trade name : Bain.

Modern use : *Root* : aphrodisiac; *Bark* : astringent; *Unripe seed* : used as poultice for boils.

Phytography : Small to medium tree with smooth bark; leaves coriaceous, elliptic-oblong or ovate-oblong, very obtuse, glabrous, smooth, shining above, never glaucous, whitish beneath; flowers zygomorphic, yellow, sessile, in compact heads, expanded flower 6–8 mm in diameter; capsules above 2.5 cm, ovoid, compressed, beaked, green, wrinkled.

Phenology : *Flowering* : June–July; *Fruiting* : August.

Distribution : In marshy places and tidal creeks of India, Pakistan, Bangladesh, Sri Lanka and other countries.

Ecology and cultivation : Grows in mangrove swamps, along river flats, river slopes and ridge forests; wild.

Chemical content : *Plant* : Tannin.

Azadirachta indica A. Juss (Meliaceae)

Syn : *Melia azadirachta* L.

English names : Indian lilac, Margosa tree, Neem tree.

Sanskrit names : Arishta, Nimba.

Vernacular names : Asm : Mohanim; **Ben :** Nim; **Guj :** Jhada limbaelo, Limba; **Hin :** Nim; **Kan :** Olle bevu, Bevu, Nimban; **Kon :** Koddu nim, Kad limbu; **Mal :** Nimban, Veppu; **Mar :** Kadu limb, Balanta nimba; **Ori :** Nimba, Nimbu, Nima; **Pun :** Neeb, Nim; **San :** Bokom-dare; **Tam :** Vembu, Veepu, Veruttam; **Tel :** Vepa, Veppai.

Trade names : Margosa, Neem.

Traditional use : *SANTALS* and *RURAL FOLK OF INDIA* : use different parts of this plant in the following manner : (i) *Young branch* : used as tooth-brush, particularly in pyorrhoea; (ii) *Gum* : as a stimulant; (iii) *Leaf* : eat fried leaves as antidote to small pox, wash septic wounds with decoction of leaves; insecticide—commonly placed in clothes, papers and books to protect them from moths, etc., apply fume of burnt leaves in poultry and cattle sheds as mosquito-repellent; (iv) *Flower* : consume powder of flowers as anthelmintic, *dried flower* : as a tonic after fever under the name *Panchamrita*—a medicine prepared from flowers, fruits, leaves, bark and root of the tree.

ATHARVA VEDA : medicinally very important, *Juice* : cardiotonic and useful in tuberculosis, *Wind of this plant* : good for health; *CHARAKA SAMHITA* : useful in the diseases caused by over/inadequate eating; it purifies air, to protect a newborn child, the bed and clothes of the child as well as of the midwife should be disinfected by its *leaves*, and *young twigs* of the plant should be suspended on all the doors and windows of the labour room; *AGNI PURANA* : plant of great medicinal value; the plant, particularly the leaf, is beneficial to the patient of *kushtha* (leprosy), fever, pimple, wound, boil at anus; *Decoction of twigs* : for dropsy; according to this text, *leaves* of this plant should be chewed by a snake-bitten man for removal of the venom; *AYURVEDA* : a plant of significant medicinal importance; different parts of this plant are effective in different seasons as mentioned below : late spring and summer : *Bark of wood*; rainy season : *root-bark*; autumn, late winter and early spring : *leaf.* Use of different parts of the plant, for specific diseases, as prescribed by this system of medicine are as follows : (i) *Bark* : bitter tonic, astringent, antiperiodic, and used in dyspepsia, spermatorrhoea, liver-pain (bark+turmeric+emblica), persistent wound (*decoction of bark)*; (ii) *Leaf* : excessive urination associated with itching, blood sugar, vomiting, acidity, itching, jaundice, phlegm caused by heat, continuous low fever, intestinal worms; (iii) *Flower* : stomachic and tonic; (iv) *Berry* : purgative, emollient and anthelmintic; *(v) Seed-oil* : ringworm, wound in gum/buccal area, premature greying of hairs, rheumatism, skin diseases, external application is effective in birth control.

UNANI : traditional uses like Ayurveda.

HOMOEOPATHY : For control of mind, headache, and other ailments, eyes, ears, nose, face, mouth, throat, stomach, abdomen, stools, genito-urinary organs, chest, pulse, neck and back, sleep and dreams, fever, skin, etc.

Modern use : *Plant* : antilithic; *Tree-powder* : useful in white leprosy; *Bark* : (i) extract antibacterial against *E. coli*; (ii) aqueous extract contraceptive for male; (iii) *EtOh* (50%) *extract* : anticancer, antiviral, spasmogenic, beneficial to skin; *Leaf* : (i) antiviral, antineoplastic; (ii) alcoholic extract useful in eczema, ringworm and scabies; (iii) extract antifungal, anthelmintic—specially in ascariasis, mosquito larvicidal at 1% concentration, hypoglycaemic, antihyperglycaemic; *Flower* and *Fruit* : purgative, cure rheumatic pains; *Seed* : powdered kernel repellent against various insects, kernel extract larvicidal, pesticide, kernel is one of the constituents in hair compound, useful in preventing hairloss and treatment of dandruff; *Neem oil* or *margosa oil* : essential oil (i) from leaves and seeds antifungal, (ii) from flowers antibacterial, (iii) from seeds antimicrobial, antituberculosis, (iv) mosquito larvicidal and toxic to insectivorous fish and tadpoles, strong antifeedant against desert locust, (v) useful in diabetes, (vi) boiled with seeds of nuxvomica useful in eczema, (vii) used in skin diseases, wounds, ulcers and rheumatism, prevents propagation of malarial parasite within erythrocytes; *Nimbidin* : a chemical obtained from

Neem oil, effective in acute and chronic inflammation, psoriasis; it is analgesic, antipyretic and hypoglycaemic.

Phytography : Tree with a clear bole of 3–8 m and a girth of 2–3 m, bark moderately thick with oblique or longitudinal furrows on the outer surface, dark grey outside while reddish inside; though described as evergreen, it becomes deciduous in drier areas; leaves alternate, petiolate, imparipinnate, at times bipinnate, leaflets curved with two unequal longitudinal halves, slightly serrated, acute, glabrous; panicles axillary; flowers small, white, aromatic, pentamerous, complete, actinomorphic; fruits drupaceous berry with one seed, greenish yellow when ripe; seeds often enclosed in an aril, and winged.

Phenology : *Flowering* : March to May; *Fruiting* : June to August.

Distribution : Native of India; grows wild in the forests of Andhra Pradesh, Tamil Nadu and other parts of Deccan; widely planted all over India as roadside tree and also in crop-fields; villagers prefer to have it on the periphery of the hamlets; widely distributed in the Indo-Malayan region, in the open scrub forest of the dry zone of Myanmar and also found in tropical Africa.

Ecology and cultivation : Thrives best in drier climate with normal shade temperature not exceeding 40°C and rainfall varying from 45 to 112 cm/year; prefers black-cotton soil though grows on all kinds of soils.

Chemical contents : *Bark* : penta-nortriterpene, nimbin, 6-de-Ac-nimbin; *Trunkwood* : nimbolin A & B; *Heartwood* : sterols; *Leaf* : quercetin, β-sitosterol glucoside, n-hexacosanol, glucosides of quercetin and kaempferol; *Leaf-extract* : azadirachtin, meliantriol, nimbinene, 6-de-Acnimbinene and nimbadinol, petanortriterpene; *Fruit* : azadirone, azadiradione, azadirachtin, epoxyazadiradione, gedunin, salanin, vilasinin; *Seed* : azadirachtin, 6 new tetranortriterpenes; *Flower-oil* : quercetin-3-galatoside, kaempferol-3-glucoside, myricentin-pentoside; *Seed-oil* : meliantriol, nimbin, nimbandiol and its 6-OAc-derivatives, nimbinene and its 6-de-OAc derivatives, pentanortriterpene, tiglic acid, vepinin.

Remarks : The tree is well-suited for afforestation in dry situations. It is considered as a good purifier of air due to large leaf-area and is a good insect repellent.
In India, the tree is considered as sacred and as such not usually felled. The log is used in making deities of Jagannath temple of Puri and images of some other deities.
The timber is forbidden for constructional purposes by the BRIHATSAMHITA and MATSYA PURANA, though it seasons well and is durable in exposed situations. White ants do not attack the timber. The timber can be worked both by hand and machine and is suitable for carving, but does not take polish. It is usually used for making toys, boards, panels and agricultural implements.

Azima tetracantha Lam. (Salvadoraceae)

Sanskrit name : Kundali.

Vernacular names : Ben : Trikantagati; **Hin :** Katagur kamai; **Mar :** Sukkapat; **Tam :** Jeanka; **Tel :** Tella upi mulsangu.

Trade names : Katagur kamai, Kundali.

Traditional use : *Root* : diuretic, in rheumatism, dropsy; *Root-bark* : in rheumatism; *Juice of leaf* : relieves the cough of phthisis and asthma; *Bark* : expectorant.

Modern use : *Aerial parts* (50% EtOH extract) : spasmogenic.

Phytography : Rumbling spinous shrub; up to 1.5 (2.5) m; branchlets divaricate, 4-gonous; leaves elliptic, spine tipped; flowers cream, zygomorphic; fruits white, globose, berries 1- to 2-seeded, $\pm$ 6 mm in diameter.

Phenology : *Flowering* : throughout the year with a peak in February–April; *Fruiting* : April onwards.

Distribution : Found gregariously in the drier districts of South India near the coast, Orissa, Sundarbans; Sri Lanka and Myanmar.

Ecology and cultivation : Plains from the coast to 300 (500) m; very common in wastelands by the fences; ready coloniser; wild.

Chemical contents : *Root* : friedelin, glutinol, lupeol, β-sitosterol; *Leaf* : azimine, azearpine and all the chemicals present in root.

Baccaurea ramiflora Lour. (Euphorbiaceae)

Syn : *Baccaurea sapida* Muell. Arg., *B. wrayi* King ex Hook. f.

Vernacular names : **Asm** : Leteku; **Ben** : Latka, Latkan; **Hin** : Kataphal, Lutko; **Garo** : Dojuka; **Man** : Moktok; **Mik** : Damiyu-arong ; **Tri** : Bhubi.

Trade name : Latkan.

Traditional use : *HILL TRIBES OF NORTH-EAST INDIA* and *BENGAL* : *Bark* : in constipation.

Phytography : Evergreen tree, about 17 m high; stem pubescent, bark grey, with lenticels; leaves alternate, simple, petiole 2.5–3.7 cm, leafblade elliptic, rather membranous, 10–20 cm long; flowers unisexual, in separate racemes, barcteate, male florets about 0.3 cm in diameter, female florets 0.8–1.25 cm in length; fruits globose, yellow, puberulous, 1.8–2.5 cm in diameter, loculicidally dehiscent, each chamber 1-seeded; seeds orbicular, fleshy coat rose-coloured.

Phenology : *Flowering* : April–May; *Fruiting* : following May.

Distribution : Sub-Himalayan regions of Assam, Manipur, Tripura, Andamans; Bangladesh (Sylhet, Dhaka).

Ecology and cultivation : A plant of moist tropical forests, also grows in blazing cold climate; cultivated in domestic gardens.

Chemical contents : *Bark* : friedelin, epifriedelanol, Me-betulinate, β-sitosterol and an unknown ester.

Remarks : Fruits edible; Karbis use bark, lac and sometimes leaves to impart a chocolate-colour to cloth.

Bacopa monnieri (L.) Pennel (Scrophulariaceae)

(2n = 64, 68)

Syn : *Herpestis monniera* (L.) H.B. & K., *Moniera cuneifolia* Michx., *Gratiola monniera* L.

English name : Thyme leaved gratiola.

Sanskrit names : Brahmi, Nira brahmi.

Vernacular names : Ben : Adhabirni, Birmi shak, Barmi shak, Brahmi; **Hin :** Barambhi, Safed chammi; **Mal :** Nir bramhi; **Mar :** Bama; **Ori :** Urishnaparni; **Tam :** Nirbrami; **Tel :** Sambranichettu.

Trade name : Brahmi.

Traditional use : *ATHARVA VEDA* : strengthens body, improves quality of semen; *CHARAKA SAMHITA* : invigorating, life sustaining; *SUSHRUTA SAMHITA* : increases memory and lifespan; *AYURVEDA* : (a) *Plant* : bitter, diuretic, blood-purifier, invigorates sex; (b) *Plant-extract* : used in purification of blood, for anaemia, dermatitis, diabetes, filaria; (c) *Plant-juice* (along with ginger, sugar and bark extract of *Moringa oleifera*) : to children in stomach disorder; (d) *Leaves* (fried in *ghee)* : as a brain tonic, in nervous weakness, hysteria, epilepsy, insanity, anxiety neurosis, and to sharpen dull memory.

AGNI PURANA : *Plant* : poison-killer; *Plant-juice* : good for epilepsy.

UNANI : invigorating and good for cold and cough.

Modern use : *Plant-extract* : In experiments with mice, rat and dog, it has been proved to be tranquilizer, musculature relaxant, antispasmodic, anticancer; *Powdered dried leaf* : satisfactory results obtained in man in cases of asthenia, nervous breakdown, and other low adynamic conditions.

Phytography : Annual creeper, rooting at each node; stem soft, succulent, soft hairs present; branches 10–25 cm long; leaves alternate, simple, 1.25–1.8 cm long, sessile or subsessile, oval-shaped; flowers light blue or white, 0.8–1.5 cm long; capsules 2- grooved, valves separating from entire column, many-seeded.

Phenology : *Flowering* : April–June; *Fruiting* : June–December.

Distribution : Marshes throughout India, ascending up to 1300 m; Bangladesh, Pakistan and Sri Lanka.

Ecology and cultivation : Grows in damp or marshy areas, on the banks of ponds and canals, in the crop fields.

Chemical contents : *Plant* : Becoside A & B, betulic acid, betulinic acid, *d*-mannitol, stigmasterol, β-sitosterol, saponin, stigmastanol, hersaponin, monnierin, nicotine, luteoline and its glucosides.

Balanites aegyptica (L.) Delile (Balanitaceae)

(2n = 18)

Syn : *Balanites roxburghii* Planch, *B. aegyptica* var. *roxburghii* Duthie.

English names : Desert date, Soapberry tree, Thorn tree.

Sanskrit name : Ingudi.

Vernacular names : Ben and **Hin** : Hingan; **Mar** : Hinganbet; **Tam** : Nanjunda; **Tel** : Gari.

Trade name : Hingan.

Traditional use : *TRIBES OF PANCH MAHAL* (Maharashtra) and *RURAL FOLK*: *Bark, Leaf, Fruit* and *Seed* : as anthelmintic and purgative; *Fruit* : in boils, leucoderma and other skin diseases; *Fruit-pulp* : in whooping cough.

Modern use : *Plant* : cytotoxic and antitumour principle isolated; *Bark* : bactericide; *EtOH* (50%) *extract of bark* : spasmolytic; *Seed-kernel oil* : antimicrobial, useful in burns and ulcers, spermicidal.

Phytography : Spiny tree, about 6 m high; branches glabrous or puberulous and ending in very strong ascending spines; leaves alternate, bifoliate, ashy green, leaflets coriaceous; cymes axillary, 4- to 10-flowered; flowers green, velvety pubescent; fruits ovoid, about 5 cm in length, fleshy drupes, 1-seeded; seed oily but exalbuminous.

Phenology : *Flowering* : November, also during February to May; *Fruiting* : Winter.

Distribution : Bihar, Deccan, Gujarat, Uttar Pradesh (Kanpur) and Sikkim.

Ecology and cultivation : Xerophyte, found chiefly on black cotton soil; does not thrive in rocky areas.

Chemical contents : *Plant* : diosgenin and yamogenin; *Root* : steroidal sapogenin, diosgenin, yamogenin; *Stem-bark* : steroidal sapogenin, nitrogen glucoside, sesquiterpene, balanitol, bergapten, δ-marmesin; *Leaf* : 6 diosgenin glucosides; *Fruit* : steroidal sapogenin, yamogenin, 5 diosgenin glycerides, 6 diosgenin glucosides, diosgenin, balanitisins A, B, C, D & E; *Seed* : a saponin of diosgenin, non-edible fatty oil.

Remarks : Fruit-pulp is edible and used for cleansing silk and cotton. Fruit-shell is used in fire-cracker industry. Wood is chiefly used for making walking sticks and as fuel.

Balanophora polyandra Griff. (Balanophoraceae)

Traditional use : In asthma; the upright portion is used as aphrodisiac.

Modern use : *Plant-extract* : effective in cough.

Phytography : Root-parasite; glabrous fleshy herb, rootstock tuberous, lobed, warted with lenticels, leafless; peduncles bursting through the rootstock; flowers minute, unisexual, monoecious or dioecious, intermixed with clavate cellular bodies, male heads cylindric, female ovoid or oblong; fruits minute, crustaceous; seeds globose, adhering to the pericarp, endosperm oily.

Phenology : *Flowering* : September–October.

Distribution : Manipur, Meghalaya, Nagaland, Sikkim, West Bengal, Bihar, Orissa and Andhra Pradesh, up to 2000 m.

Ecology and cultivation : Root-parasite on trees; prefers damp and cool climate.

Chemical content : *Plant* : coniferin.

Remark : Plant is traditionally used in Thailand as antiasthmatic.

Baliospermum montanum (Willd.) Muell-Arg. (Euphorbiaceae)

(2n = 28)

Syn : *Baliospermum axillare* Blume, *Croton polyandrous* Roxb.

Sanskrit name : Danti(ka).

Vernacular names : Ben : Danti; **Hin :** Danti, Jangli jamalgota; **Mar :** Dantimul; **Mal :** Natadanti; **Tam :** Niradimultu; **Tel :** Nelajidi.

Trade name : Jangli jamalgota.

Traditional use : *TRIBES AND RURAL FOLK* : *Root* : cathartic and in dropsy, anasarca and jaundice; (ii) *Leaf-decoction* : in asthma; (iii) a preparation from *leaf* in abdominal tumours; (iv) *Seed-oil* : externally in rheumatism, as purgative and hydrogogue cathartic; (v) *Latex* : in rheumatism.

AGNI PURANA : (i) external application of the plant as ointment is beneficial to treat boil at anus; (ii) *Plant-decoction* : useful in dropsy.

Modern use : *EtOH extract of root* : showed *in vivo* activity in p-388 lymphocytic leukaemia and esters *in vitro* are anticancer.

Phytography : Erect subherbaceous leafy shrub, 1–2 m high; leaves alternate, simple, petiole stout, stipules glandular, lower ones 15–30 cm long while upper ones 5.0–7.5 cm in length, biglandular at base, rigid, ribs prominent; racemes 0.4 cm, may contain only male flowers or a few female flowers too; florets small, subsessile; capsules trilobed, 0.8–1.25 cm long; seeds mottled and endosperm oily.

Phenology : *Flowering* : mainly December–March.

Distribution : Throughout the sub-Himalayan tract from Kashmir (up to 1000 m) to Meghalaya, North Bengal, Bihar, Deccan peninsula; Bangladesh (Chittagong).

Ecology and cultivation : Usually undergrowth in damp places.

Chemical contents : *Root* : montanin, baliospermin, 12-deoxyphorbol and 12-deoxy-16-OH phorbol, 5 new phorbol esters and tigliane.

Adulterants : Seeds are sometimes substituted for the seeds of *Croton tiglium* (*jamalgota*); seed-oil resembles croton oil.

Ballota limbata Benth. (Lamiaceae)

Syn : *Otostegia limbata* Hook. f.

Vernacular names : Pun : Bhui, Phutkanda.

Trade name : Bhui.

Traditional use : *RURAL FOLK OF PUNJAB* : *Juice of leaf* : to children's gums and to ophthalmia in man and animals.

Phytography : Small, grey, spiny bush; branches tomentose, white terete; leaves subsessile, 2.5 cm long, elliptic-lanceolate, hoary on both surfaces; verticillasters axillary, lower bracts spinescent, upper dilated, pungent; flowers 0.4 cm; nutlets 0.5 cm, obovoid, smooth.

Distribution : Lower hills in rocky places west of the Jhelum to Salt range.

Ecology and cultivation : Xerophyte; wild.

Balsamodendron mukul Hook. ex Stocks (Burseraceae)

(2n = 26)

Syn : *Commiphora mukul* (Hook. ex Stocks) Engl.

English name : Indian bedellium tree.

Sanskrit name : Guggulu.

Vernacular names : Ben and **Hin** : Guggul.

Trade name : Gugal.

Traditional use : *SUSHRUTA SAMHITA* : useful in troubles of stomach, obesity, dropsy, bad smell of ear; *BAGBHATTA* : in respiratory troubles; *CHAKRADATTA* : in sciatica and rheumatism; *AYURVEDA* : in treatment of premature oldness, cold, cough, worms in alimentary system, inflammation of spleen, obesity, irregular menstruation, rheumatism, rheumatoid arthritis, muscular dystrophy, summer boil, irritation of hair follicles, sciatica; it is a ingredient of many Ayurvedic compounds; Ayurvedic drug *Myostal*, containing this, is useful in puerperal patients.

Modern use : *Oleoresin* : useful as anti-inflammatory, antirheumatic, hypocholesterimic, hypolipidaemic, dependable agent for hemiplegia and other artherosclerotic disorders like coronary insufficiency; increases fibrinolytic activity in ischaemic heart patients; effects of the following indigenous drugs have been accepted : (i) *Guggulipid* : in heart diseases, spondilitis and gout; (ii) *Septilin* : in rhinosinal infection and exodontia; (iii) *Biliarin* : in infective hepatitis; (iv) *Arogya-vardhini* : in acute viral hepatitis;
Resin : in indolent ulcers, as gargle in pyorrhea alveolaris, chronic tonsilitis, pharyngitis; *Fumes from burning gum* : inhalation is recommended in hay fever, chronic bronchitis, nasal catarrah, laringitis and phthisis; *Essential oil* : antibacterial.

Phytography : Tree, less than 2 m in height, branches slightly ascending; leaves alternate, 1–3 foliate, leaflets smooth and shining, toothed at the end; flowers nearly sessile, unisexual, brownish red; drupes red, containing two 2-celled stones.

Phenology : *Flowering* : Summer.

Distribution : Arid rocky hills of Rajasthan and Deccan.

Ecology and cultivation : Xerophyte; wild.

Chemical contents : *Oleoresin* : active constituents guggulsterol-I, -II & -III, myricyl alcohol, β-sitosterol, diterpenes, α-comphorene, cembrene A, cembrene alcohol, allycembrol (Mukulol), and long-chain aliphatic tetrols; *Flower* : flavanoids.

Remark : Resin-smoke is mosquito repellent.

Bambusa arundinacea Roxb. (Poaceae)

(2n = 72)

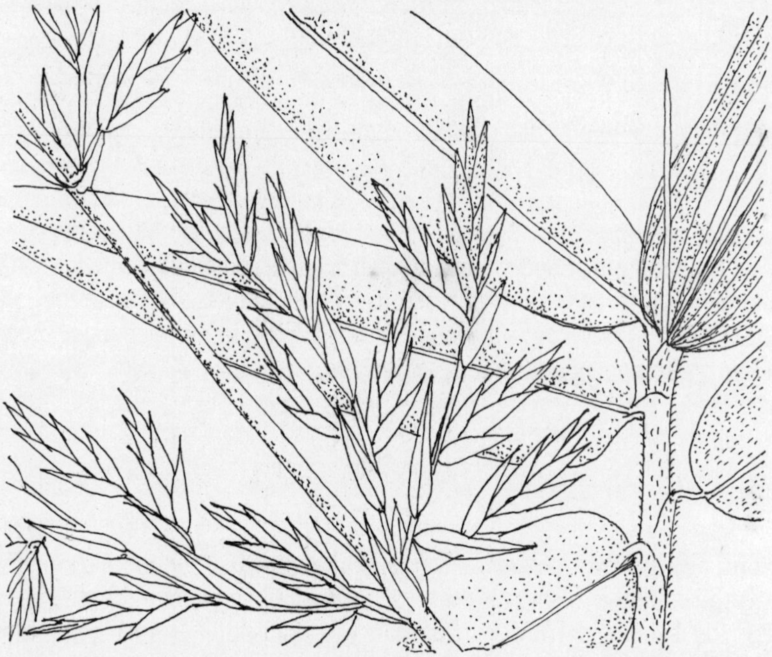

Syn : *Bambusa bambos* Druce

English names : Thorny bamboo, Bamboo.

Sanskrit names : Vamsha, Kichaka.

Vernacular names : Asm : Buah, Kotohag; **Ben :** Baanoh, Ketua, Kutuasi; **Guj :** Wans; **Hin :** Baans, Kattang; **Kan :** Biduru; **Mal :** Illi, Mungil, Venu; **Mar :** Mand gay; **Ori :** Baunsa; **Pun :** Magar; **Tam :** Kambu, Mangai, Mungil; **Tel :** Banguveduru, Pantevederu.

Trade names : Thorny bamboo, Kanta bansh.

Traditional use : This plant is mentioned in the *ATHARVA VEDA* as a medicinal one; *BHAVAPRAKASHA* : cooling the body with a hand-fan made of the leaves of this plant causes increase in bile and wind; *AYURVEDA* : cooling tonic and aphrodisiac, use of different parts is prescribed as under : (i) *Aerial parts* (stem and leaf) : as blood purifier, in leucoderma and inflammatory conditions; (ii) *Decoction of nodes* : for fertility control; (iii) *Aqueous extract of leaf* : bathing by the same is good for patients of piles; (iv) *Leaf-paste* : external application is beneficial in gout, wounds and for suppuration of boils; (v) *Root* : diluent, cures strangury, oedema; (vi) *Ash of the basal part of culm* and *thick root* : may be used along with sesame oil in treatment of baldness.
UNANI : (i) *Leaf-juice* : drinking the same with honey cures cough; (ii) *Ash of the basal part of culm* : if used as tooth-powder strengthens gum and makes the teeth shining; (iii) 'Tabasheer' (a siliceous secretion found in the culms) : good for asthma, cough, tuberculosis, leprosy and other debilitating diseases.

Phytography : Grows from a stout rootstock, stems woody, pointed, 24–30 m high and 15–17 cm in diameter, internodes may be 45 cm in length, hollow, walls 2.5–5.0 cm thick, nodes solid and prominent, lower nodes may have long horizontal shoots armed with 2–3 recurved spines, branched from the base, stem-sheathes 30–37 cm by 22–30 cm, coriaceous, young orange-yellow, often striped green or red, top rounded, edges plaited, young thickly clothed with golden hairs, blade triangular, outside glabrous, matted within; leaves alternate, up to 20.0 cm by 2.5 cm, tip sharp, stiff; flowers gregariously once in lifetime (25–50 years), inflorescence leafless, large, paniculate, rachis glabrous, shining, hard, flowers green and inconspicuous; caryopsis.

Distribution : Throughout the plains and lower hills of India, abundant in Assam, West Bengal, Orissa, hill forests of western and southern India, ascending up to 1000 m on the Nilgiris; Bangladesh, Myanmar, Sri Lanka.

Ecology and cultivation : Thrives best in monsoon forests, but becomes undershrub at temperate regions and at high altitudes; looks almost like common grass. The species is cultivated in the Ganga plain and in lower Himalaya. Since cultivation with the aid of seeds cannot be an annual occurrence, it is propagated usually by vegetative means in the following five methods :
(i) *Nodal cuttings* : From the time of Kautilya, uninodal cuttings of bamboo are being used for cultivation. Each cutting bears at least one short leaf-bearing branch and the open ends of the cuttings are treated with selected chemicals and manures, packed with soil and then sown; (ii) *Basal cane cuttings* : Normally, in the second year, cuttings are selected from the perimeter of a clump developed from not long wandering rhizomes for this purpose; (iii) *Rhizome cuttings* : When rhizomes grow like runners, they are cut into fragments and each fragment gives rise to a clump; (iv) *Clump division* : Full grown clumps are cut into several fragments; (v) *Layering* : New plantlets are produced by layering —this

method calls for better skill. When mature seeds are available, saplings can be raised from them too.

Chemical contents : *Plant* : Benzoic acid, oxalic acid, cyanogenic glucoside, lupeol, amyrin, β-amyrin, friedelin, tetraxerol, ferneol, 2-furaldehyde, lignin, hemicellulose, holocellulose, miliacin, glutinone, glutinol, cylindrin, crusgallin , cholin, betain, β-sitosterol, stigmasterol.

Remarks : Tribes and rural folk use *bamboo-knives* to severe umbilical chord. Bamboo leaves are used as medicine in several ailments of cow. Fresh young shoots are insecticidal and lethal to mosquito larvae. Culms are used in making several implements and the pulp for making paper.

Bambusa nutans Wall. ex Munoo (Poaceae)

Syn : *Bambusa falconeri* Munro.

Vernacular names : Asm : Bidhuli; **Ben :** Makla; **Hin :** Mala bansh; **Mik :** Chak.

Modern use : *EtOH* (50%) *extract of aerial parts* : anticancer and hypoglycaemic.

Phytography : Stem solitary from a creeping rhizome, height varies from 7–13 m and diameter from 3.7–7.5 cm, graceful, straight, nodes not much raised, often hairy, lower nodes rooting, internodes 37–45 cm long, thick-walled, white ringed below the nodes; stem-sheaths 15–22 cm, outside covered with black hairs; leaves alternate, linear-lanceolate, 15–30 cm by 2.5–3.7 cm, lower side white when dry, tip twisted, scabried, midrib and margins scabried, pellucid glands present, ligules long and hairy; inflorescence a stiff panicle with spiciform branches, spikelets in bracteate fascicles, many empty, fertile few 1.8–2.5 cm.

Distribution : Himachal Pradesh, Arunachal Pradesh, Sikkim and Meghalaya, ascending up to 1600 m; Bangladesh.

Ecology and cultivation : Wild; grows in tropical forests of lower Himalaya.

Bambusa tulda Roxb. (Poaceae)

Syn : *Bambusa macala* Ham. ex Wall., *B. trigyna* Roxb.

English names : Calcutta cane, Common bamboo of Bengal.

Vernacular names : Asm : Banh, Benjli; **Ben :** Bansh, Jowa; **Garo :** Wati; **Guj :** Kapura; **Hin :** Deobansh; **Kan :** Hebbidiru, Unde bidiru; **Kon :** Kondo; **Man :** Lep, Bleeng; **Mal :** Makor, Mulah; **Mar :** Bamboo; **Ori :** Baunsa; **Pun :** Bans; **Tam :** Mulai; **Tri :** Jowa.

Trade names : Deobansh, Bansh.

Traditional use : *HILL TRIBES* : *Plant* : in tetanus.

Phytography : Arborescent and gregarious in habit, stem 7–23 m high, 5–10 cm in diameter, nodes not swollen, internodes 30–60 cm long, walls 0.8–1.25 cm thick, white-ringed below nodes; stem-sheaths 15–22 cm by 15–25 cm, broad below, glabrous or hairy; leaves alternate, linear-oblong or lanceolate, 15–25 cm by 1.8–3.7 cm, base usually rounded, tip twisted, pellucid glands present, sheath glabrous, ligule narrow, obscure; inflorescence variable from leafless panicle to short leafy panicle, spikelets 2.5–7.5 cm, in interrupted clusters mixed with chaffy bracts, 4- to 6-flowered.

Distribution : Assam, West Bengal, Orissa, hills of Andhra Pradesh—ascending up to 600 m; Myanmar.

Ecology and cultivation : Wild in tropical monsoon forests; cultivated in plains of North India.
For cultivation : see *Bambusa arundinaceae*.

Bambusa vulgaris Schrad. ex Wendl. non Nees (Poaceae)

(2n = 72)

Syn : *Bambusa surinamensis* Rupr.

Vernacular name : San : Balkubans.

Traditional use : *SANTAL* : *Leaf* : in burns and wounds; *Paste of young leaf* : in diarrhoea.

Phytography : Arborescent; stem 7–17 m tall, 5–15 cm in diameter, green, yellow or striped, polished, nodes not raised or faintly raised, usually a ring of brown hairs present, internodes 25–45 cm long and thin-walled; stem-sheaths 15–25 cm by 17–22 cm, often streaked with yellow, thickly hairy above; leaves linear-lanceolate, tessellate by pellucid glands, hairy on both surfaces, ligule short and broad; panicles large, leafy, spikelets 1.50–2.0 cm, in bracteate clusters, clusters largest at nodes, lodicules three, winged and ciliate.

Distribution : Throughout tropical and hotter parts of India.

Chemical contents : *Shoot* : taxiphyllin, β-sitosterol, campesterol, stigmasterol.

Baphicanthus cusia Bremek. (Acanthaceae)

Syn : *Strobilanthus cusia* Imlay, *Strobilanthus flaccidifolius* Nees

English names : Assam indigo, Assam rum.

Vernacular names: Asm : Rampet, Rum; **Lus :** Tiny; **Man :** Khum; **Mik :** Bukangda.

Trade name : Rum.

Traditional use : HILL TRIBES : *Leaf* : astringent, diuretic and lithotriptic.

Phytography : Glabrous shrub; leaves opposite, simple, elliptic, acute at both ends, glabrous; panicles dense with whip-like and bracteate spikelets; flowers nearly 1.5 cm or more, tubular, bracteate; capsules 1.8 cm, base not constricted.

Distribution : In the lower hills of Mizoram, Manipur, Assam, North Bengal—ascending up to 1300 m; Bangladesh, Myanmar.

Ecology and cultivation : Grows in tropical rain forests.

Chemical contents : *Leaf (fresh juice)* : tryptanthrin.

Adulterant : *Strobilanthus secundus* T. Anders may be confused with this.

Barbarea verna (Mill.) Aschers (Brassicaceae)

Syn : *Barbarea praecox* R. Br., *Erysimum veernum* Mill.

English names : Beele-isie cress, Early winter cress, Scurvy cress.

Modern use : *Leaf* : antiscorbutic, detergent and pectoral.

Phytography : Erect, stiff-branched and leafy herb; stem angled and furrowed, green, 22–44 cm high; leaves simple, alternate, petiolate, lower leaves lyrate-pinnate while upper leaves ovate, entire or toothed; flowers yellow, petals with broad claw; pedicels of ripe fruits almost horizontal; fruits erect, stout, rigid.

Distribution : Native to Europe, introduced into the Nilgiris.

Ecology and cultivation : Plant of cold climate; the species usually grows as a cress, but it is not frequent.

Chemical contents : *Herb* : a glucoside and an enzyme; *Leaf* : rich in vitamin C (130 mg/100 g).

Adulterant : *Barbarea vulgaris* var. *taurica* Hook. f. & T. Anders is very similar to this herb.

Barbarea vulgaris R. Br. (Brassicaceae)

(2n = 14, 16)

Syn : *Barbarea vulgaris* var. *taurica* Hook. f. & T. Anders.

English names : Bitter winter cress, Hedge mustard, Yellow rocket.

Traditional use : *TRIBES OF HIMALAYA* : (i) *Plant* : anthelmintic, diuretic, stomachic; *Leaf* : in healing wounds.

Modern use : (i) *Herb* (pulverised) : for stimulating spermatogenesis; (ii) *Herb* (extract) : exhibits estrogen and toxic effect on experimental animals.

Phytography : Perennial branched herb, 22.5–45.0 cm high, erect, stiff; stem green, angled and furrowed; leaves alternate, young ones sessile while matured leaves petioled, simple, lyrate or pinnate, lower lobes auricled, stem-clasping; racemes many-flowered; flowers yellow, tetramerous, petals clawed; pods 1.6–3.7 cm long, narrow-linear, rigid, strongly keeled.

Distribution : Chiefly an European plant; found in temperate and subalpine Himalaya between 2000 and 3500 m; western Tibet, West Asia and Australia.

Ecology and cultivation : Grows in cold climate; wild.

Chemical contents : *Root* : singrin; *Leaf* : vitamin C, ascorbic acid.

Remarks : Exotic species, naturalised in a small area of India; deserves cultivation.

Barleria buxifolia L. (Acanthaceae)

Syn : *Dicranthus buxifolia* Oerst.

Traditional use : *Leaf* and *Root* : for cough and inflammations.

Modern use : *EtOH* (50%) *extract of plant* : spasmogenic.

Phytography : Small, prickly undershrub; branches hairy; leaves opposite, simple, petiolate, 0.8–1.6 cm long, broadly elliptic, hairy; flowers solitary, sessile or shortly pedicellate, bracteate and bracteolate, 2.5 cm in length, purple bluish or rose or white, showy; capsules 1.8 cm and 4-seeded.

Distribution : Southern part of Deccan peninsula, ascending up to 1300 m.

'*Barleria courtallica* Nees (Acanthaceae)

Sanskrit name : Chethasahacharam.

Vernacular name : Mar : Venkurunji.

Traditional use : RURAL PEOPLE OF MAHARASHTRA : (i) *Decoction of root* : in rheumatism and pneumonia; (ii) *Leaf* (boiled in oil) : in ear and eye complaints.

Phytography : Large shrub; leaves opposite, shortly petioled, obovate or elliptic, acuminate at both ends; spikes 2.5–7.5 cm, dense, pubescent; flowers bracteate and bracteolate, blue or yellow or white, 3.7–5.0 cm, pubescent; capsules 2.0–2.5 cm, 4-seeded; seeds 0.6 cm in diameter, silky.

Distribution : Konkan to Courtallum.

Ecology and cultivation : Prefers black soil and high temperature; wild.

Remarks : Plant of limited distribution; deserves conservation and cultivation.

Barleria cristata L. (Acanthaceae)

Syn : *Barleria ciliata* Roxb., *B. dichotoma* Roxb., *B. laciniata* Wall.

Vernacular name : Ben : Janti.

Trade name : Janti.

Traditional use : HILL TRIBES : *Root-decoction* : in anaemia; *Root* and *Leaf* : in cough and inflammations.

Modern use : *EtOH* (50%) *extract of plant* : spasmogenic and hypoglycaemic.

Phytography : Perennial, erect herb, branches yellow-hairy; leaves opposite, simple, petiolate, oblong or elliptic, 7.5–10.0 x 2.5 cm, yellow hairs on the lower surface; spikes panicled, dense; flowers bracteate and bracteolate, violet or purple-blue or white or rose, funnel-shaped; capsules 1.6 cm long, 4-seeded.

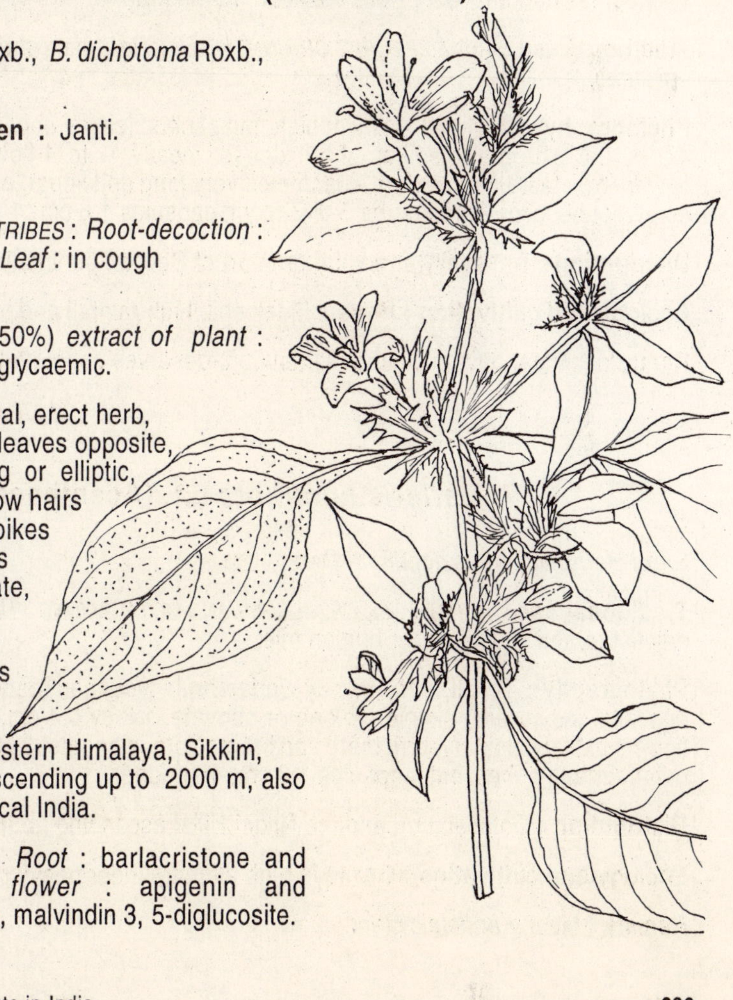

Distribution : Northwestern Himalaya, Sikkim, Meghalaya, Nilgiris—ascending up to 2000 m, also in the plains of subtropical India.

Chemical contents : *Root* : barlacristone and cristabarlone; *Violet flower* : apigenin and glucoronide, naringenin, malvindin 3, 5-diglucosite.

Barleria cuspidata Heyne ex Nees (Acanthaceae)

Modern use : *EtOH* (50%) *extract of plant*: Central Nervous System depressant.

Phytography : Prickly undershrub; leaves opposite, simple, 2.5–6.7 cm by 0.8 cm, elliptic, entire; flowers axillary, bracteate, tubular, 3.7 cm long, yellow; capsules 2.0–2.5 cm, two-seeded.

Distribution : Karnataka, Kerala—ascending up to 700 m.

Barleria longiflora L.f. (Acanthaceae)

Sanskrit name : Adyanda.

Vernacular names : Ori : Koilakha; **Tel** : Pinnagorta.

Traditional use : RURAL PEOPLE OF TINNEVELLY : *Root-decoction* : in dropsy, stone and stricture.

Phytography : Shrub, 60–120 cm high, pubescent; leaves opposite, petiole 0.6– 1.2 cm, 4.3 by 2.5 cm, ovate; spikes short, dense, mostly 1- to 4-flowered, softly pubescent; flowers bracteate, bracteolate, bracteoles very long and linear, outer surface of calyx and corolla pubescent, corolla tube 5.0–7.5 cm; capsules 1.6 cm, 4-seeded.

Distribution : Tinnevelly and southern part of Deccan peninsula.

Ecology and cultivation : Prefers black soil, high rainfall and temperature; wild.

Remarks : Plant of very limited distribution; deserves conservation and cultivation.

Barleria noctiflora L.f. (Acanthaceae)

Syn : *Barleriacanthus noctiflora* Oerst.

Traditional use : RURAL FOLK OF NILGIRIS AND OOTACAMUND : *Decoction of plant* : as an adjunct to and substitute for human milk.

Phytography : Small, very prickly undershrub; younger branches pubescent; leaves opposite, subsessile, simple, oblong or obovate, 1.8 by 0.4 cm, grey, young pubescent; flowers axillary, solitary, bracteate and bracteolate, elongate, 3 by 0.3 cm, white or yellow, outer surface pubescent; capsules 1.6 cm, 4-seeded.

Distribution : Ootacamund and the Nilgiri Hills, ascending up to 2500 m; Sri Lanka.

Ecology and cultivation : Grows in high altitude under heavy rainfall; wild.

Remark : Nearly *endemic* plant.

Barleria prionitis L. (Acanthaceae)

(2n = 30, 40)

Syn : *Barleria hystrix* L., *B. pubiflora* Benth., *Prionitis pubiflora* Miq.

Sanskrit names : Karunta(ka), Vajradanti.

Vernacular names : Ben : Kantajanti; **Hin :** Katsareya; **Kan :** Mullugorinda; **Mal :** Vetilakkunri; **Mar :** Koret, Korhanti, Vajradanti; **Ori :** Dasakerenta; **Tam :** Chemmulli, Varelmutti; **Tel :** Mundlagobbi.

Trade name : Kantajanti.

Traditional use : TRIBES OF NORTH-EAST INDIA : (i) *Root* : in boils and gland-swelling; (ii) *Leaf* : in cough; SANTAL : (i) *Leaf* : in toothache; TRIBES OF NORTH UTTAR PRADESH : (i) *Leaf* : in piles; TRIBES OF SAGAR DISTRICT (Madhya Pradesh) : (i) *Root* : to treat boil, gland-swellings, (ii) *Stem-bark* : in dropsy; TRIBS OF RAJASTHAN : (i) *Plant decoction* : in gum troubles, toothache, cough and tuberculosis; TRIBES OF RATAN MAHAL HILLS

(Gujarat) : *Leaf-infusion* : applied to soles of feet to strengthen them for tolerating extreme heat, cold and rough soil; SOME OTHER TRIBES : (i) *Plant* : as antiseptic; (ii) *Bark* : in anasarca; (iii) *Leaf-juice* : in urinary and paralytic affections, stomach disorders, catarrhal affections, lacerated sole, complaints of ear, rheumatism; (iv) *Leaf-juice with coconut oil* : on pimples, and (v) *Flowering-top* : as diuretic.
AGNI PURANA : a medicine for leprosy can be obtained by mixing the blue variety of this plant with powder of all the five parts of *Psoralea corylifolia* and decoction of *Acacia catechu*; AYURVEDA : (i) *Root* : febrifuge, decoction for toothache and mouth-wash; (ii) *Stem-bark* (fresh juice) : diaphoretic and expectorant; (iii) *Leaf* : bitter, applied with little honey or sugar in catarrhal affections of children which are accompanied by fever and much phlegm.

Modern use : *Plant* : in whoopig cough.

Phytography : Spiny shrub, reaching a height of 6.6–16.6 cm, very prickly; stem profusely branched, densely scabrid; leaves opposite, simple, petioles of the lower leaves may grow up to 2.5 cm; flowers axillary, bracteate, bracts foliaceous, yellow or orange or cream-coloured, funnel-shaped, may be 5 cm long; capsules 1.8–2.5 cm, 2-seeded.

Phenology : *Flowering* : Winter.

Distribution : Throughout the hotter parts of India; Bangladesh and Myanmar.

Ecology and cultivation : Xerophytic plant, well-adapted to mesophytic conditions; readily propagated by cuttings; often grown as a low hedge plant.

Chemical contents : *Plant* : β-sitosterol; *Stem* and *Leaf* : barlerin, Ac-barlerin, iridoids; *Flower* : scutellarcin-7-rhamnosylglucoside.

Barleria strigosa Willd. (Acanthaceae)

Syn : *Barleria caerulea* Roxb., *B. macrophylla* Heyne, *Pseudobarleria caerulea* Oerst.

Vernacular name : Ori : Koelekha.

Traditional use : TRIBES OF ORISSA : (i) *Root-decoction* : in cough, bronchitis, fever; (ii) *Leaf* : in skin diseases and pain in gums.

Phytography : Profusely branched herb, reaching 88 cm in height; branches fulvous—strigose upwards; leaves opposite, petiolate, simple, large—8.0 by 6.25 cm or bigger, widest towards base, ovate; spikes 2.5–7.5 cm, very dense, many-flowered; flowers blue, funnel-shaped, 3.7–5.0 cm, bracteate and bracteolate; capsules 1.8 cm, 4-seeded; seeds silky.

Distribution : Assam, Sikkim, West Bengal, Orissa, Bihar—ascending up to 1300 m.

Ecology and cultivation : Widely cultivated in India.

Chemical contents : *Plant* : β- and γ-sitosterol.

Barringtonia acutangula (L.) Gaertn. (Lecythidaceae)

Syn : *Stravadium acutangulum* Miers., *S. rubrum* Wall., *S. obtusangulum* Rhedii.

English name : Indian oak.

Sanskrit name : Dhatri.

Vernacular names : Ben : Hijal; **Mik :** Hijal; **San :** Hinjor.

Traditional use : MIKIR : *Branches* : in toothache; SANTAL : *Plant* : as an ingredient for medicine for cholera; TRIBES OF MAYURBHANJA : (i) *Bark* : in gum troubles, toothache; (ii) *Seed* : in liver troubles; TRIBES OF GORAKHPUR : *Bark* : in rheumatism.

AYURVEDA : (i) *Root* and *Leaf* : bitter tonic, (ii) *Powdered seed* : emetic, expectorant—specially for children; AGNI PURANA : (i) *Plant* : poison killer, yields a good medicine for the ailments caused by wind, beneficial for leprosy, checks fall of hair, and eating the *plant* during the period of menstruation can facilitate a lady to have a male child.
UNANI : Ingredient of 'Samandarphal'.

Modern use : *EtOH* (50%) *extract of root* : hypoglycaemic; *EtOH* (50%) *extract of stem-bark* : antiprotozoal.

Phytography : Evergreen tree, reaching 10–13 m in height and girth of 2 m; stembark rough dark brown, about 1.25 cm thick; leaves alternate, simple, short-petioled, crowded towards the ends of branches, elliptic, 12.5 by 5.0 cm; racemes terminal and lateral, elongated—often 33 cm, pendulous; flowers pink, 0.6 cm in length, funnel-shaped; fruits quadrangular, 2.5—3.7 cm long, broadest in the middle.

Distribution : Fairly common in Meghalaya, Assám, West Bengal, Bihar, Orissa, Madhya Pradesh and Deccan peninsula; Bangladesh, Myanmar and Sri Lanka.

Ecology and cultivation : Prefers moist climate, not found in mangrove forests.

Chemical contents : *Heartwood* : three triterpenoid sapogenols—barringtogenol B, barringtogenol C and barringtogenol D, triterpenic dicarboxylic acid, barringtonic acid, barrigenic acid; *Branchwood* : barringtogenol E, barrinic acid; *Leaf* : acutangulic acid, barringtogenic acid, oleanolic acid, tangulic acid, β-amyrin, β-sitosterol, stigmasterol and stigmasterol glucoside; *Fruit* : barringtogenol D, C, & B, barrigenic acid.

Remarks : Leaves are said to be edible; also used as fish-poison.

Barringtonia asiatica (L.) Kurz (Lecythidaceae)

Syn : *Barringtonia speciosa* J.R.& G. Forst., *Mammea asiatica* L.

English name : Queen of the shores.

Vernacular names: And : Cingola, Dod-da.

Traditional use : ANDAMANESE : (i) *Root* : for wounds, (ii) *Bark* : for yaws and to treat tuberculosis; (iii) *Fruit* : vermifuge, for yaws and wounds, (iv) *Seed* : for ringworm.

Phytography : evergreen tree, 10–17 m in height, glabrous; leaves alternate, sessile, simple, 37.5 by 17.5 cm, obovate; raceme short, erect; flowers bracteate—lower bracts leaf-like, white, 5 cm; fruits quadrangular or nearly ovoid, 7.5 cm in length and breadth, 1-seeded.

Distribution : Andamans, sea cost of southern part of Deccan peninsula.

Ecology and cultivation : Seaside tree ; wild.

Chemical content : *Fruit* : Bartogenic acid.

Barringtonia racemosa (L.) Roxb. (Lecythidaceae)

Syn : *Barringtonia alba* Miq., *Eugenia racemosa* L., *Butonica racemosa* Miers., *Stravadium album* DC.

Sanskrit name : Hijjala.

Vernacular names : Asm : Hidol; **Ben :** Hijal; **Hin :** Injar, Ijjal; **Kan :** Samudraphala; **Mal :** Katampiur; **Ori :** Hinjala; **Tam :** Callam, Kadapum; **Tel :** Kadapa, Samudrapandu.

Trade name : Ijjal.

Traditional use : *Root* : cooling, deobtruent; *Bark* and *Seed* : tonic, vermifuge, insecticidal.

Modern use : *Fruit* : in cough and asthma; *Kernel of fruit* (mixed with milk) : in jaundice and other bilious diseases; *Seed* : in colic and ophthalmia.

Phytography : Evergreen tree, nearly 17 m high, glabrous; leaves alternate, petioles 0.3–0.6 cm, simple, oblong, nearly 25.0 by 7.5 cm; racemes 30–45 cm, pendulous; flowers distant, broadly funnel-shaped, 0.9 cm; fruits ovoid, obscurely quadrangular below when ripe, 4.0 by 3.7 cm.

Distribution : Sundarbans, Karnataka, Andamans.

Ecology and cultivation : Halophyte; wild.

Chemical contents : *Bark* : ellagic acid and its mono-, di- and tri-OMe derivatives—mostly concentrated in resinous fraction; *Flower* : cyanidin- and delphenidin-3-sambubiosides.

Basella alba L. var. *rubra* Stewart (Basellaceae)

Syn : *Basella rubra* L.

English name : Indian spinach.

Sanskrit names : Upodaki, Putika.

Vernacular names : Asm : Purai; **Ben :** Puin shak; **Hin, Ori** and **Pun** : Poi; **Kan :** Kempa-basale, Kempu bayi basale; **Kon :** Bondival; **Mal :** Samparcheera; **Mar :** Mayak bhaji, Velbondi, Velgond; **Tam :** Sambarkeerai; **Tel :** Karubacochali.

Trade name : Poi shak.

Traditional use : *CHARAKA SAMHITA* : *Leaf* : ingredient of medicines for diarrhoea and piles; *SUSHRUTA SAMHITA* : *Leaf-juice* : in filaria, should be eaten and applied externally; *CHAKRADATTA* : (i) *Leaf-juice* : should be applied on pimples and tumours; (ii) *Leaf* : should be tied over them; *BANGASEN* : *Leaf-juice* : beneficial to cold of children and in constipation of pregnant women; *AYURVEDA* : (i) *Plant* : sweet, provokes cold, reduces the ailments caused by wind (*vata*) and bile (*pitta*), sedative and spermatopoietic, increases appetite, nutritive and useful in haemorrhagic troubles; (ii) *Leaf-juice* : beneficial for whooping cough, urticaria, gangrene; (iii) *Ash of leaf and stem* : useful for pyorrhea; (iv) *Seed-extract* (of red variety) : good in anaemia.

Modern use : *Herb* and *Leaf* : antiviral.

Phytography : Perennial, much-branched, twining herb, 2– 10 m long; stem fleshy but hard, contains mucilage; leaves alternate, petiolate, simple, fleshy, 7.5–12.5 cm, glabrous; spikes axillary, peduncled, simple or branched; flowers white or red or violet, bracteate, bracteolate; fruits pea-sized, red or white or black; seed subglobose, erect.

Phenology : *Flowering* and *Fruiting* : Winter.

Distribution : The red-leaved and green-leaved varieties are common in Assam, West Bengal, Orissa and South India while in Uttar Pradesh and Punjab, only the green-leaved variety is available.

Ecology and cultivation : Mesophyte; grown as pot-herb throughout India, except the hilly regions. The plant can be raised from roots or stem cuttings or seeds; the common practice is to raise from stem cuttings; space of 1 m is given between the saplings; often they are grown on stakes, and get ready for pickings in 80–90 days.

Chemical contents : *Plant* : proteins, calcium, iron, vitamins A, B_1 and B_2, fats and sterols; *Leaf* : polysaccharides, glycoprotein.

Bauhinia acuminata L. (Caesalpiniaceae)

Syn : *Bauhinia candida* Ait.

English name : Dwarf white bauhinia.

Sanskrit name : Jugmapatra.

Vernacular names : Asm and **Ben :** Kanchan; **Hin :** Kanchnar; **Mal :** Velutthamandrom; **Ori :** Kanchana; **Pun :** Kairal; **San :** Seeara; **Tam :** Vellaimandarai.

Trade name : Kanchan.

Traditional use : *SANTAL* : (i) *Root-decoction* : boiled with oil and applied to burns; (ii) *Stem-bark* : antidote to poison; (iii) *Decoction of bark* or *leaf* : in biliousness, bladder stone, leprosy and asthma.

Phytography : Erect shrub; branches obscurely downy and angular; leaves alternate, petiolate, simple, deeply cleft from the tip, 7.5–15.0 cm long; racemes axillary, short-peduncled, corymbose; flowers whitish, 3.7 cm long; pods 10.0–12.5 cm long, 1.5–1.8 cm wide, ribbed on each side, firm, glabrous, 8- to 12-seeded.

Distribution : Almost throughout India; Bangladesh.

Bauhinia phoenicia Heyne (Caesalpiniaceae)

Syn : *Bauhinia benthami* Bedd., *B. ruficarpa* Desv.

Traditional use : *RURAL FOLK OF KARNATAKA AND KERALA* : *Root* : in preparation of ointment against itches and abscesses.

Phytography : Extensive climber; branches slender, young ones covered with fine ferruginous hair; leaves alternate, petioled, simple, bi-cleft, subcoriaceous, silky on the ribs when young; racemes axillary or terminal, short-peduncled, few-flowered; flowers slightly cylindrical, glabrous, petals distinctly clawed; pods flat, 15 cm long and 2.5 cm wide, finely rusty, raised rib on either side.

Distribution : Moist parts of the Western Ghats of Karnataka and Kerala.

Ecology and cultivation : Wild.

Bauhinia purpurea L. (Caesalpiniaceae)

(2n = 28)

Syn : *Bauhinia triandra* Roxb., *B. coromandeliana* DC., *Phanera purpurea* Benth.

English names : Butterfly tree, Geranium tree, Mountain ebony, Orchid tree.

Vernacular names : Ben : Deva kanchan; **Hin :** Kaliar; **Kan :** Mandara , Ulipe ; **San :** Baper; **Tam :** Thirivathi; **Tel :** Devacanchanamu.

Trade name : Deva kanchana.

Traditional use : NAGA : (i) *Bark* : for tumour in stomach, (ii) *Flower* : in indigestion; GARO : (i) *Stem-extract* : rubbed on fractured bones and given to drink for setting the bones; TRIBES AROUND GUWAHATI : *Bark* : on wounds; TRIBES OF PURULIA (West Bengal) : (i) *Bark* : in small pox; SANTAL : (i) *Bark* and *Leaf* : in anasarca, convulsions, poisoning by *Datura*, dropsy, blackness in tongue, delirium febris, haemorrhages, pain, rheumatism, swellings, scorpion stings, snake and tiger bites; TRIBES OF DEHRA DUN (Uttar Pradesh) : (i) *Bark* : astringent; TRIBES OF TAMIL NADU : (i) *Leaf* : in jaundice.

Phytography : Middle-sized erect tree; branches stout, glabrescent; leaves alternate, petiolate, subcoriaceous, grey, glabrous, bi-cleft, cleavage reaching one-third or halfway down; corymbs axillary or terminal, short peduncled, few-flowered; flowers minutely bracteate, reddish or violet, 2.5 cm; pods 15–30 cm by 1.8–2.5 cm, flat, firm, glabrous, slightly decurved, 12- to 15-seeded.

Phenology : *Flowering* : Winter.

Distribution : Sub-Himalayan tract, ascending up to 1300 m, Assam, Meghalaya and western peninsula; Bangladesh (Chittagong) and Sri Lanka.

Chemical contents : *Flower* : astragalin, quercetin, isoquercetin, pelargonidin 3-glucoside, butein galactoside; *Seed* : trypsin and chymotrypsin inhibitors, two chalcone glycerides.

Remark : Santals eat leaves and fruits as vegetables.

Bauhinia racemosa Lamk. (Caesalpiniaceae)

Syn : *Bauhinia parviflora* Vahl, *B. spicata* Koenig, *B. timoriensis* Decaisne, *Pileostigma racemosa* Benth.

Vernacular names : Raj : Apta, Seta; **San :** Kaimu.

Traditional use : *TRIBES OF RAJASTHAN* : (i) *Bark* : in skin diseases, ulcers, glandular inflammations, goitre, (ii) *Leaf* : in worms, (iii) *Flower* : laxative, in cough, piles and haemorrhages.

Modern use : *EtOH* (50%) *extract of seed* : anticancer.

Phytography : Small, crooked, bushy, erect tree, branches drooping; leaves alternate, petiolate, simple, coriaceous, 3.7–5.0 cm long, deeply 2-cleft, dense grey hairs on the lower surface; racemes leaf-opposed or terminal, 15–30 cm by 1.25–2.5 cm, firm, glabrous, 12- to 20-seeded.

Distribution : Throughout India.

Chemical contents : *Stem-bark* : octacosane, β-amyrin and β-sitosterol.

Remark : Santals eat leaves and young fruits as vegetables.

Bauhinia semla Wunderlin (Caesalpiniaceae)

Syn : *Bauhinia retusa* Roxb., *B. roxburghiana* Voig., *Phanera retusa* Benth.

Vernacular name : San : Semla.

Traditional use : SANTAL : *Different parts* : in cholera, sores and snake bite.

Phytography : Erect; leaves alternate, simple, deeply cleft, rigidly coriaceous, width more than length; panicles terminal, corymbose, branches covered with silky grey hairs; flowers bracteate, bracteolate, yellow, 0.8 cm long, petals clawed; pods hard, flat, 12.5–15.0 cm by 3.0–3.7 cm, short-stalked, 6- to 8-seeded.

Distribution : Himachal Pradesh, Uttar Pradesh, ascending up to 1300 m.

Chemical contents : *Bark* : quercetin, glucoside and rutin; *Seed* : proteins.

Bauhinia tomentosa L. (Caesalpiniaceae)

(2n = 28)

Syn : *Bauhinia speciosa* Roxb.

English names : Bell bauhinia, Yellow bauhinia, St. Thomas tree.

Traditional use : TRIBES : (i) *Leaf* : an ingredient of a plaster applied to abscesses; (ii) *Seed* : tonic.

Phytography : Erect shrub; branches downy; leaves alternate, petiolate, simple, 2.5–5.0 cm in length, 2-cleft, incision less than half-way down from apex, hairless, coriaceous; flowers axillary, normally in pairs, short-stalked, bracteate, yellow, upper ones with red blotch on the face, 3.7–5.0 cm long; pods stalked, narrow, 1.25 cm wide, 10–12 cm long; 6- to 10-seeded.

Distribution : Southwards from Punjab; Sri Lanka.

Chemical contents : *Flower* : quercetin, isoquercetin, rutin; *Seed* : ebony oil (fatty oil), saponins, mucilage.

Remark : Commercial utilisation for 'rutin' may be explored.

Bauhinia vahliix Wt. & Arn. (Caesalpiniaceae)

Syn : *Phanera vahlii* Benth.

Vernacular names : Ben : Sehari; **Lod :** Jom-lar; **San :** Sihari-chop.

Traditional use : TRIBES OF TEHRI-GARHWAL : (i) *Fruit* : aphrodisiac, (ii) *Seed* : tonic and vermifuge; TRIBES OF BASTAR : (i) *Seed* : tonic; SANTAL : (i) *Plant* : in dysentery, stomachache, (ii) *Fruit* : to treat antifertility of women.

Phytography : Huge climber with many tendrils; branchlets hairy; leaves alternate, petiolate, simple, 2-cleft, cleavage goes down a quarter to one-third way from the apex,

7.5–45.0 cm long; racemes terminal, subcorymbose; flowers bracteate, 2.5 cm long; pods nearly 45 by 5 cm, rusty-downy, 7- to 12-seeded.

Distribution : Foot of central and eastern Himalaya, ascending up to 800 m, Bihar, Madhya Pradesh, Maharashtra and Karnataka.

Bauhinia variegata L. (Caesalpiniaceae)

Syn : *Bauhinia candida* Roxb., *Phanera variegata* Benth.

English names : Buddhist bauhinia, Camel's foot tree.

Vernacular names : Ben : Rakta kanchan; **Guj :** Kovidara; **Hin :** Khairwal; **Kan :** Bili mandar, Ulipe; **Kon :** Kudo, Tembri; **Mal :** Mandarum; **Ori :** Borara; **Pun :** Kachnar; **San :** Buruju-dare, Kouar; **Tam :** Tiruvatti; **Tel :** Adavimandara.

Trade name : Rakta kanchan.

Traditional use : *NAGA* : (i) *Bark* : in dysentery; *KUMAONI* : (i) *Bark* : in obesity, tumours, (ii) *Flower* : in tumours; *GARHWALI* : (i) *Bark* : in diarrhoea, indigestion, flatulence, skin diseases, snake bite, piles, malaria, (ii) *Flower* : in diarrhoea; *SANTAL* : (i) *Flower-bud powder* (with that of black pepper) : to women for regulating vaginal discharges; *TRIBES OF PURULIA* (West Bengal) : (i) *Flower* : in diarrhoea, intestinal worms; *TRIBES OF SAGAR* (Madhya Pradesh) : (i) *Bark* : in leprosy, scrofula, skin diseases, syphilis, ulcers; *TRIBES OF CHANDRAPUR* (Maharashtra) : *Leaf* : as laxative.
AYURVEDA : (i) *Root-decoction* : in dyspepsia, (ii) *Bark* : astringent, tonic, in skin diseases, wounds, leprosy, anthelmintic, blood purifier, useful in bone TB, (iii) *Flower-paste* : appetiser, (iv) *Flower-bud* (dry) : in piles, blood dysentery.

Modern use : *Extract of bark, bud* and *flower* : antibacterial; *Seed-extract* : agglutinates human blood.

Phytography : Tree; leaves alternate, petiolate, simple, 2-cleft, cleavage runs down the quarter of the length from the apex, broader; corymbs lateral, few-flowered; flowers pure white or white variegated with red and yellow, 3.7–5.0 cm long; pods 15–30 cm long, 1.8– 2.5 cm wide, flat, hard, 10- to 15-seeded.

Phenology : *Flowering* : February–April; *Fruiting* : starts in Summer, matures in the Rains.

Distribution : Himalayan foothills, Sikkim, ascending up to 1300 m; Myanmar.

Remark : Leaves, buds and fruits edible.

Beaumontia grandiflora (Roxb.) Wall. (Apocynaceae)

Syn : *Beaumontia longiflora* Lodd., *Echites grandiflora* Roxb.

English names : Easter-lily vine, Herald's trumpet, Nepal trumpet flower.

Vernacular name : Lep : Chomrik.

Modern use : *Seed* : cardiotonic glyceride detected.

Phytography : Evergreen climber; stem woody, very stout, rusty-pubescent; leaves opposite, petiolate, simple, 17.5–30.0 cm by 7.5–17.5 cm, upper side hairless, lower surface may be hairy, obovate-oblong, abruptly acuminate; cymes few- or many-flowered; flowers white, 7.5–12.5 cm long; follicles long, thick, woody; seeds 1.8 cm long, compressed, crowned with hairs.

Distribution : Indigenous to eastern Himalaya, ascending up to 1300 m and Bangladesh (Sylhet and Chittagong); grown in gardens throughout India, Bangladesh and Pakistan for its very attractive foliage and fragrant flowers.

Chemical contents : *Leaf* : quercetin, tamarixetin, n-hentriacontane, n-triacontane, palmitic acid, linoleic acid, ursolic acid, β-sitosterol and its glucosides; *Seed* : beaumontoside (cardiotonic), beauwalloside, oleandrin and wallinchoside; Fermented seeds of Indian plant contain 8 additional cardenolides.

Adulterant : *Beaumontia jerdoniana* is almost similar.

Beaumontia jerdoniana Wight (Apocynaceae)

Modern use : *Plant* : cardiotonic.

Phytography : Evergreen climber; stem stout, woody; leaves opposite, petiolate, membranous, nearly glabrous beneath; cymes few- or many-flowered; flowers white, bracteate, between campanulate and funnel-shaped, follicles 25.0 by 3.7 cm, cylindric.

Distribution : Occurs throughout the Western Ghats from Konkan to Kerala; also grown elsewhere as ornamental.

Chemical contents : *Plant* : cardiotonic glycerides.

Adulterant : Resembles very closely to *Beaumontia grandiflora.*

Beaumontia khasiana Hook. f. (Apocynaceae)

Modern use : *Plant* : cardiotonic.

Phytography : Evergreen climber; leaves opposite, petiolate, obovate or elliptic, lower surface tomentose, acuminate; cymes terminal, few- or many-flowered; flowers white, broadly campanulate, 5 cm long, throat rounded; pods long, thick, woody.

Distribution : Khasia Mountains (Meghalaya).

Chemical contents : *Plant* : cardiotnic glycerides.

Adulterant : Very similar to *Beaumontia multiflora* Teijsm & Binn.

Remark : An *endemic* plant; should be conserved and cultivated carefully.

Begonia josephii DC. (Begoniaceae)

Syn : *Begonia scutata* Wall.

Vernacular names : Gar : Derm; **Kha** : Jajew, Jabjew.

Traditional use : KHASI : *Bulb* : eaten raw in indigestion and stomachic.

Phytography : Bulbous herb, very variable in size and habit; rootstock of one or few bulbs; leaves alternate, simple, usually trilobed or orbicular with numerous lobes, 15 cm, usually hairless but may have a little hair, especially on the lower surface, stipulate—stipules deciduous; scape normally 30 cm, repeatedly dichotomously branched; flowers bracteate, bracts caducous, small, rose-coloured, unisexual, stamens monadelphous in male flowers; capsules 3-winged, one much longer than the others, 1.25 by 0.9 cm including wings.

Distribution : On roadsides near the Elephant Falls in Shillong (Meghalaya), moist places of Sikkim; Bhutan and East Nepal.

Ecology and cultivation : Plant of moist, cold climate; wild.

Adulterants : Often confused with *B. picta* and *B. scutata.*

Remark : *Endemic* to a small zone of the eastern Himalaya; should be protected and cultivated.

Begonia palmata D. Don var. *gamblei* Hara (Begoniaceae)

Syn : *Begonia laciniata* Roxb.

Vernacular name : Nag : Teisu.

Traditional use : NAGA : (i) *Plant* : diarrhoea, stomachic, (ii) *Root* (decoction) : fever, liver complaints, (iii) *Stem* (fresh) : toothache and other tooth troubles.

Phytography : Succulent herb; rootstock long, creeping; stem brown, tomentose, 15–60 cm; leaves alternate, stipules persistent, pubescent, petioles 3.7–10.0 cm long, brown, tomentose, simple, roundly ovate, 10–15 cm in diameter, more or less lobed, slightly serrated, upper surface pilose, pubescent on ribs; peduncles 7.5–15.0 cm, usually arises from one or two uppermost axils; flowers bracteate, bracts large, tomentose, unisexual, more or less glandular and pubescent on the back, nearly white to deep rose in colour; capsules 2.5 by 1.25 cm including wings, dehiscing by 4 lines.

Distribution : Sikkim, ascending up to 2300 m, Meghalaya; Nepal, Bangladesh (Chittagong) and Myanmar.

Ecology and cultivation : Subalpine plant; wild.

Remark : Plant of limited distribution; should be protected and cultivated.

Begonia picta Smith (Begoniaceae)

Syn : *Begonia echinata* Royle, *B. erosa* Wall.

Vernacular names : Mun : Luncliara; **Nag** : Longsilawa; **Ora** : Pakkan chatta.

Traditional use : MUNDA: *Leaf*: in colic and dysentery; NAGA: *Leaf* : in bristles on tongue, ulcer in mouth.

Phytography : Succulent herb, 15.0–37.5 cm high; rootstock of one or few tubers; stem hairless or obscurely hairy; leaves alternate, petioles 2.5 cm or little longer, stipules lanceolate and covered with hairs, leaf blade ovate, 7.5–12.5 cm by 5.0–7.5 cm, finely serrated, hairs on upper surface and only on veins on lower surface; peduncles hairy on upper side; flowers bracteate, bracts pubescent and persistent, pale rose, unisexual, stamens monadelphous in male flowers, in female flowers styles 3, each divided half-way up into two branches; capsules 2.5 cm, usually pubescent, may be glabrous, 3-winged, wings unequal, dehiscing by 3 faces.

Distribution : Khasia Mountains (Meghalaya), ascending up to 1300 m, Chotanagpur (Bihar) between 300 and 1400 m; Bhutan.

Ecology and cultivation : Subalpine plant; wild.

Chemical content : *Leaf* : acidic in taste.

Remark : Plant of limited distribution; should be conserved and cultivated.

Begonia roxburghii DC. (Begoniaceae)

Syn : *Bcgonia malabcrica* Roxb., *Diploclinium roxburghii* Miq.

Vernacular name : Asm and **Kha** : Kamchal.

Trade name : Kamchal.

Traditional use : *KHASI* : (i) *Root* : in bile, dysentery and fecal discharges; (ii) *Tuber* : boiled with fruits of *Solnea heterophylla* and decoction taken in fever.

Phytography : Succulent herb; roots fibrous; stem erect, 30–100 cm high, hairless when mature; leaves alternate, petioles 2.5–12.5 cm long, stipules lanceolate, deciduous, simple, 15–23 cm, hairs may be present only on veins, acuminate; cymes axillary, short, dichotomous, often few-flowered; flowers bracteate, unisexual, almost white; fruits pendent, thick-walled, succulent, finely dehiscent at the angles.

Distribution : Common in Assam and Meghalaya, ascending up to 1300 m, found in northern side of West Bengal; Nepal, Bangladesh and Myanmar.

Ecology and cultivation : Subalpine plant; wild.

Remark : Plant of limited distribution; should be protected and cultivated.

Begonia rubro-venia Hk. (Begoniaceae)

Syn : *Begonia barbata* Wall.

Vernacular name : Kha : Jahoksier.

Traditional use : *KHASI* : *Whole plant* and *root* : in diarrhoea and dysentery.

Phytography : Herb, 20–40 cm high; rootstock thick, aerial stem very short and minutely hairy; leaves alternate, petioles 2.5–5.0 cm long, stipules lanceolate, persistent, leaf blades simple, 7.5–15.0 cm, usually narrow, sometimes elliptic, small hairs present on veins, acuminate; inflorescence axillary, few-flowered; flowers bracteate, bracts persistent, white with rose-red veins, unisexual, stamens in male flowers monadelphous; capsules 1.25–1.5 by 2.5–3.1 cm including wings.

Distribution : Frequent in Meghalaya between 1000 and 2000 m, available in Sikkim; Bhutan.

Ecology and cultivation : Subalpine plant; wild.

Remark : Almost *endemic*; deserves protection and cultivation.

Belamcanda chinensis DC. (Iridaceae)

(2n = 32)

English names : Blackberry lily, Leopard lily.

Traditional use : *Rhizome* : acrid, antipyretic, alexipharmic, carminative, deobstruent and expectorant, used as blood purifier, in gonorrhoea, pulmonary complaints; *Stem* : chewed in food-poisoning.

Modern use : *Herb-extract* : useful in skin diseases and cosmetic preparations.

Phytography : Perennial herb; rootstock creeping, aerial stem erect; leaves alternate, ensiform, equitant, isobilateral; cymes invested with spathes, several-flowered; flowers stalked, bracteate, bisexual, trimerous, ovary inferior; capsules obovoid, membranous, loculicidal, leaving the seed-bearing axis persistent and free; seeds subglobose and shining.

Distribution : Native of China; wild in the Himalaya, ascending up to 2000 m.

Ecology and cultivation : Plant of cold climate; wild.

Chemical contents : *Rhizome* : isoflavone glucoside, shekanin—identical to tectoridin. Plants of China and Japan contain different chemicals.

Benincasa hispida (Thunb.) Cogn. (Cucurbitaceae)

(2n = 32)

Syn : *Benincasa cerifera* Savi.

English name : Ashgourd.

Sanskrit names : Kushmanda, Parvaphala, Phalashaka.

Vernacular names : Asm : Momora; **Ben :** Chaal kumraa, Chhaanchi kumra ; **Guj :** Kohola; **Hin :** Gol kaddu, Pethaa; **Kan :** Boodugumbala; **Kon :** Bhellpotri; **Mal :** Kumballanga, Kumpalam; **Mar :** Kohala; **Ori :** Pani kakharu; **Pun :** Petha; **Tam :** Kumpaly, Sambal poosani; **Tel :** Budaida gummadi, Pulla gummdi.

Trade names : Gol kaddu, Chaal kumraa.

Traditional use : TRIBAL : *Fruit-juice* : in jaundice.
SUSHRUTA SAMHITA : (i) *Young fruit* : beneficial in bile disorder, (ii) *Half-ripened fruit* : removes ailments caused by phlegm, (iii) *Ripe fruit* : useful in troubles of bowel and urinary system, chest troubles and perhaps in psychological imbalance; AYURVEDA : (i) *Ripe fruit* : in blood vomiting, pleurisy, tuberculosis, enlarged heart, constipation, colic pain, distension of stomach, urinary complaints and headache, (ii) *Seed* : to remove intestinal worms.
UNANI : The drug is called 'Petha'.

Modern use : (i) *Leaf* : purgative, juice or poultice in bruises; (ii) *Pollen* : allergenic, (iii) *Kernel of fruits* : in skin eruptions; (iv) *Seed* : ingredient of a drug for appendicitis, habitual consumption prevents hunger and prolongs life, (v) *EtOH* and *aqueous extract of seed* : anthelmintic, (vi) *Seed-ash* : in gonorrhoea.

Phytography : Large tendril climber, stem covered with soft white hairs, green, ribbed and furrowed, pith hollow in old stem, branched; leaves alternate, petioles 7.5–10.0 cm long, hairy, simple, slightly pentangular, dark green, 10–15 cm in diameter, covered with soft hairs; tendrils bifid, extra-axillary, green; flowers axillary, pedunculate, large, yellow or white, unisexual, monoecious; pepos large, 30–45 cm long, cylindric, fleshy, dark green, covered with white soft hairs; seeds many, oblong, compressed.

Phenology : *Flowering* and *Fruiting* : normally Rainy season, may be Autumn or Winter—in case of untimely crops.

Distribution : Cultivated in the plains of the entire Indian subcontinent.

Ecology and cultivation : Tropical plant; seeds are sown on the eve of rainy season; grows luxuriously during the rains, often the plant is raised on a bamboo platform or on the roofs of the huts. Untimely varieties are grown in autumn and winter.

Chemical contents : *Fruit* : n-triacontanol, mannitol, several aminoacids, glucose, rhamnose, lupeol and β-sitosterol.

Remark : Crops of rainy season only are medicinally effective.

Berberis aristata DC. (Berberidaceae)

(2n = 28)

Syn : *Berberis angustifolia* Roxb., *B. chitria* Ham., *B. tinctoria* Lesch.

English name : Indian barberry.

Sanskrit names : Darvihara, Daruharidra, Parjanyahari.

Vernacular names : Ben : Daruharidra; **Hin :** Chitra, Dar hald, Kashmal, Peel lakri, Rasaut; **Mar :** Daruhald.

Trade name : Daruharidra.

Traditional use : KUMAONI : *Root-bark* : in eye complaints and boil.
SHUKLA YAJURVEDA : cleans bile-gland, removes any disease caused by bile disorder; CHARAKA SAMHITA : cures bleeding piles, itching caused by bile, pus in earholes, healing up of wounds; SUSHRUTA SAMHITA : *Decoction of wood* : antidote of snake poison, effective

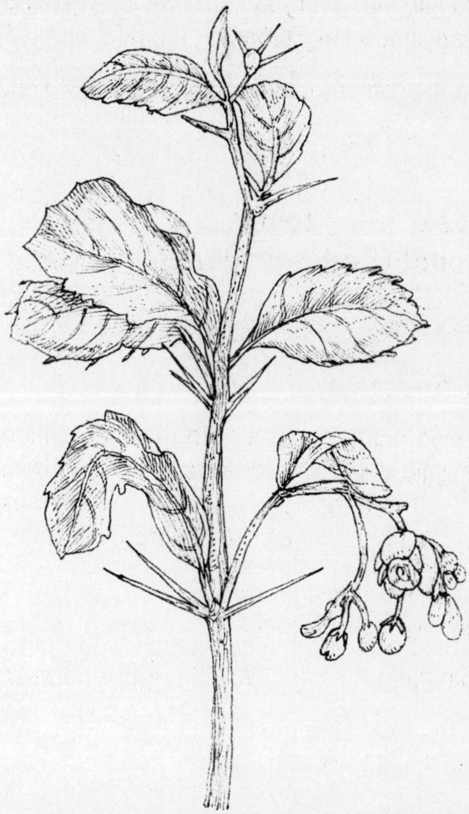

in gonorrhoea and eye troubles; BAGBHATTA : *Extract of wood* : useful in most types of common eye troubles, hydrocele and scrofula; AYURVEDA : (i) *Paste of root-bark* : as ointment for healing ulcers, (ii) *Root-bark extract* : for external application in diseased parts of eyelids, in chronic ophthalmia, (iii) *Tincture of root* : in intermittent fever, enlargement of spleen and liver associated with malaria, (v) 'Rasaut' *of root* (solid extract) : in stomatitis, leucorrhoea, (vi) *Paste of root* : in jaundice, (vii) *Paste of stem* : in pimples and freckles, oral ulcers, (viii) *Decoction of stem-bark* : in menorrhagia, leucorrhoea, healing of wounds;

UNANI: The drug called 'Rasaut Zarishk' : in enlargement of spleen and as emmenagogue.

Modern use : (i) *EtOH* (50%) *extract of root*: hypoglycaemic and anticancer, (ii) *Root* and *Root-bark* : gastro-irritant, (iii) *Berberine* (an alkaloid present in bark) : produces long-lasting fall in blood pressure of anaesthetised rabbits, useful in oriental sores, acute cholera, as sulphate used in cholera, diarrhoea, dysentery, eye complaints, tuberculosis, (iv) *Geriforte* (a drug containing root-extract) : in senile pruritus, antifatigue in rats; (v) *Vitafix* (a drug having the plant as an ingredient): in premature ejaculation; (vi) *Madhudarvyadhi* (an indigenous drug containing the plant) : in conjuctivitis, jaundice, (vii) *Stem-bark* : in liver complaints, (viii) *Fresh berries* : laxative, antiscorbutic, useful in piles, sores, eye diseases.

Phytography : Erect, spinous shrub, 2–6 m high, often gregarious; bark pale yellowish brown, deeply furrowed; leaves alternate, petiolate, evergreen or nearly so, obovate or oblong, acute or awned; racemes compound, often corymbose; flowers golden yellow with red pedicels, hermaphrodite; berries tapering and few-seeded.

Distribution : Temperate Himalaya between 2000 and 3500 m, Kulu to Kumaon, Nilgiri Hills; Bhutan and Sri Lanka (2000–2300 m).

Ecology and cultivation : Plant of temperate climate; wild.

Chemical contents : *Plant* : aromoline, berberine, karachine, oxyberberine, oxyacanthine, palmatine and taxilamine.

Adulterants : (i) Aerial stem of lighter yellow colour and having cork that detaches in long strips; (ii) roots and rhizomes of South American species of *Berberis*.

Remarks : Fruits are edible. *Semi-endemic* plant; deserves conservation and cultivation.

Berberis insignis Hook. f. & Thomas var. *shergaonensis* Ahrendt (Berberidaceae)

Vernacular name : Lep : Timburjhin.

Modern use : *EtOH* (50%) *extract of aerial parts* : diuretic.

Phytography : Erect, bushy shrub, 1.3–2.0 m high; leaves simple, alternate, short petiolate, spines at base, 7.5–17.0 cm long, elliptic or linear-lanceolate, shining on both surfaces, margins spinous toothed; flowers 3–20 in a fascicle, golden yellow; berries black, 0.8 cm long, uneven and irregularly reticulate on surface; seeds brownish black, obovoid.

Phenology : *Flowering* : May–August; *Fruiting* : August–October.

Distribution : Arunachal Pradesh, Sikkim, Darjeeling (West Bengal), between 2100–3300 m; Nepal, Bhutan and Tibet.

Ecology and cultivation : Grows in humid subalpine forests; wild.

Chemical contents : *Root* and *Stem-bark* : umbellatine.

Remark : Plant of limited distribution, should be conserved and cultivated.

Berberis jaeschkeana Sch. (Berberidaceae)

Vernacular name : Gar : Dam.

Trade name : Dam.

Traditional use : GARHWALI : *Root* : astringent, blood purifier, diuretic, useful in eye complaints, jaundice, menorrhagia, skin diseases.

Phytography : Shrub, up to 2 m in height; stem stout, angled, profusely branched; leaves whorled, subsessile, spines at base, narrowly obovoid or oblong, 1.0–2.5 cm by 0.4–1.0 cm, green, slightly lustrous above, pale beneath; inflorescence axillary, umbellate or subumbellate, 0.5–4.0 cm long including stalk, 3- to 8-flowered; flowers yellow, small, glands on petals; berry red, oblong, obscurely stylose.

Phenology : *Flowering* : May–July; *Fruiting* : July–September.

Distribution : Jammu and Kashmir, Himachal Pradesh, Uttar Pradesh, Sikkim; Nepal.

Ecology and cultivation : Forms dense patches on rocky or sandy slopes among boulders at 3000–5200 m; wild.

Remarks : *Endemic* to Himalaya, should be conserved and cultivated.

Berberis kumaoensis Sch. (Berberidaceae)

Vernacular name : Gar : Kingora.

Traditional use : *GARHWALI* : *Root*: astringent, blood purifier, diuretic, in eye complaints, jaundice, menorrhagia, skin diseases.

Phytography : Erect or prostrate shrub; stem glabrous, internodes 5–20 mm long; spine 3-fid, finely sulcate, 15–25 mm long; leaves subsessile, ovate, cuneate at base, semi-evergreen; flowers solitary, pedicels 5–9 cm long; prophylls concolorous, triangular, acuminate; berries red, ovoid, alternate and bent with short thick style at apex.

Phenology : *Flowering* and *Fruiting* : August–November.

Distribution : Garhwal Himalaya, between 3000 and 4500 m, Uttar Pradesh; Nepal.

Ecology and cultivation : Grows in temperate Himalaya; wild.

Berberis lycimum Royle (Berberidaceae)

(2n = 28)

Vernacular names : Gar : Kirmora, Chatroi; **Hin :** Kashmal; **Hill tribes of Uttar Kashi :** Chatroi, Kingora; **Mar :** Darhald; **Pun :** Kashmal.

Trade name : Darhald.

Traditional use : *TRIBES OF UTTAR KASHI* : *Root* : in eye complaints; *TRIBES OF SIWALIK* (Uttar Pradesh) : *Stem* : in malaria; *TRIBES OF LAHUL* (Punjab) : *Fruit* : in stomach disorder.

Modern use : *EtOH* (50%) *extract of root* : anticancer, antiprotozoal, antiviral and spasmolytic.

Phytography : Erect, rigid shrub, 1.25–2.5 m high, sometimes attaining a height of about 5 m; bark light grey or white, rough; leaves subsessile, spines at base, compound, leaflets with dentate margins and spines, bright green above; racemes simple or compound, drooping, longer than leaves; flowers yellow; berries violet, ovoid.

Phenology : *Flowering* : March–April; *Fruiting* : May–July.

Distribution : From Kashmir to Garhwal in clearances and along the road sides between 800 and 2700 m; Nepal, Pakistan.

Ecology and cultivation : Plant of cold climate; wild.

Chemical contents : *Root* : Berbamine, berbenine, berbericine, berbericinine and umbellatine.

Adulterants : *Berberis aristata* and *B. asiatica* are very similar to this.

Remark : *Endemic.*

Berberis pachyacantha Koehne (Berberidaceae)

Syn : *Berberis thunbergii* DC., *B. vulgaris* Auct. non L.

Modern use : *Plant-extract* : antibiotic and antiinflammatory.

Phytography : Deciduous shrub, around 3 m in height; stem dark red or pale yellow, subterete, spines present or absent; leaves alternate, distinctly petioled, spines at base, simple, obovate or elliptic or oblanceolate, 5–10 cm by 1.5–3.5 cm, many spinulose along margins; raceme or racemose, pendulous, simple, 15- to 20-flowered; flowers yellow, small; berries red, oblong or ovoid, nearly 8 mm long, compressed; seeds black, 6 by 2.5 mm.

Phenology : *Flowering* : May–August; *Fruiting* : July–November.

Distribution : Kashmir, Himachal Pradesh, Uttar Pradesh, ascending up to 3300 m; Nepal, Tibet, Pakistan, North Asia and temperate Europe.

Ecology and cultivation : Plant of cold climate; wild.

Chemical contents : *Root* : Magnoflorine and columbamine; *Different parts of the plant* : protoberberine.

Berberis petiolaris Wall. ex G. Don (Berberidaceae)

Vernacular name : Kum : Kilmora.

Trade name : Kilmora.

Modern use : *Root* : diuretic.

Phytography : Shrub, up to 6.5 m high; mature stem dark red to red-brown, terete, branches reddish or pale yellowish grey; leaves alternate, petioles 1–2 cm long, spines at base, simple, entire or spinulose or irregularly toothed, 4–8 cm by 1.4–3.0 cm, green on both sides, slightly leathery; racemes, sometimes umbellate, 8–15 mm long, glabrous; flowers yellow, approximately 6.5 by 5.0 mm, glands present; berries ellipsoid or oblong, 7.0 by 3.5 cm.

Phenology : *Flowering* and *Fruiting* : May–June.

Distribution : Garhwal Himalaya, Uttar Pradesh; Nepal.

Ecology and cultivation : Plant of cold climate; wild.

Chemical contents : *Root* : berbamine, berberine, berberine chloride, berbericine, berbericinine, palmitine, and a polar alkaloid.

Adulterants : Often confused with *Berberis pachyacantha* Koehne

Remark : *Endemic.*

Berberis pseudumbellata Parker (Berberidaceae)

Vernacular name : Jam : Pakkad.

Trade name : Pakkad.

Traditional use : RURAL FOLK OF RAJOURI (Jammu) : *Root* : in intestinal disorder, *Leaf* : in throat-ache.

Phytography : Shrub, 2–3 m high; stem dark or red-brown, finely sulcate to subterete; leaves alternate, short-petioled, spines at base, entire or inconspicuously spinose, oblong or obovate, 4 by 1.5 cm, dull grey-green above, grey pruinose beneath, deciduous;

umbellate or subumbellate, 3- to 15-flowered, pedicels stout, up to 15 mm long; flowers yellow, 5.5 by 3.0 mm, glands present; berries black to purple, obovoid or subglobose, 8–10 by 6–9 mm; seeds 6 mm long.

Phenology : *Flowering* : May–June; *Fruiting* : July–September.

Distribution : Jammu and Kashmir, Himachal Pradesh, Uttar Pradesh; Pakistan.

Ecology and cultivation : Plant of cold climate; wild.

Remark : *Endemic*; should be conserved and cultivated.

Berberis tinctoria Lesch. (Berberidaceae)

(2n = 5)

Vernacular name : Tribes of Nilgiris : Oosikala Thokku.

Traditional use : TRIBES OF NILGIRIS : *Root* : stomachic.

Phytography : Shrub, 3–5 m high; stem yellow, glabrous, very sulcate; leaves alternate, short-petioled, spines at base, simple, ovate, entire, rarely with 1 or 2 spinules, 1.5–3.5 cm by 0.7–1.7 cm, thick, dull above and white or greyish beneath, papillose; racemose, sometimes branched, drooping, 10- to 20-flowered; flowers yellow, 7 by 5 mm, glands present; berries blue or red, oblong, 10 by 5 mm.

Phenology : *Flowering* and *Fruiting* : February–June.

Distribution : The Nilgiri Hills (Tamil Nadu).

Ecology and cultivation : Wild.

Chemical contents : *Plant* : Berberine, Jatrordizine.

Adulterants : Often confused with *Berberis aristata*.

Remarks : Tribes use stem-bark for veterinary medicine and eat fruits. Plant of very small distribution; should be conserved and cultivated.

Berberis wallichiana DC. (Berberidaceae)

Vernacular name : Tribes of Subansiri (Arunachal Pradesh) : Tipetine.

Traditional use : TRIBES OF SUBANSIRI : *Plant* : in wounds.

Phytography : Shrub, up to 35 m high; stem glabrous, subangled, yellow; leaves almost opposite, subsessile, slender spines present at base, simple, oblong or lanceolate, thick, 4–12 cm by 1.2–2.8 cm, spinules along margins; flowers in axillary fascicles, yellow, 6 mm long, 7–8 mm across; berries black, ovoid, 8 by 4 mm.

Distribution : Arunachal Pradesh, Meghalaya; Nepal and Bhutan.

Ecology and cultivation : Grows in subtropical rain forest; wild.

Remark : *Endemic* to eastern Himalaya; should be conserved and cultivated.

Berchemia edgeworthii M. Laws (Rhamnaceae)

Syn : *Berchemia lineata* DC., *B. poiretiana* DC., *Rhamnus lineatus* L.

Vernacular names : Kum & **Gar :** Kameti; **Jau :** Angari.

Traditional use : Reported to be medicinal (Kirtikar/Basu, 587), considered a cure for fever in Lorali.

Modern use : Febrifuge.

Phytography : Diffuse shrub; branches subvoluble, branchlets puberulous; leaves alternate, simple, subsessile, stipules very small, setaceous, lamina ovate or suborbicular, 0.6–2.5 cm; flowers 2–3 together, fascicled in the axils or clustered at ends of branches, pedicellate, pedicels 0.6–0.9 cm, hermaphrodite, tubular; fruits blue, drupe, succulent, ovoid, 0.6 cm long; seed linear or oblong, albuminous.

Distribution : From Indus eastwards, Khazan valley between 1300 and 2300 m, Sikkim; Bhutan.

Ecology and cultivation : Plant of cold climate; wild.

Bergenia ciliata (Haw.) Stemb. (Saxifragaceae)

(2n = 34)

Syn : *Bergenia ligulata* Engl.

Sanskrit names : Ashmakabheda, Pashanbheda, Vatapatri.

Vernacular names : Hin : Pakharbhed; **Johari tribe :** Pashan-bhed; **Kum :** Gheepati; **Mon :** Tseikhanha; **Nag :** Jajew-maw-ramsong; **Tribes of Dharuchula** : Silpano, Silphora.

Traditional use : TRIBES OF DHARUCHULA (Uttar Pradesh) : *Root* : in boils, cuts, wounds, ophthalmia, kidney stone, urinary complaints; JOHARI : *Root* : in asthma, urinary troubles; KUMAONI : *Rhizome* : in fever and thirst; MONPA (Arunachal Pradesh) : *Leaf* : in boils, cuts, wounds; NAGA : (i) *Root* : in liver complaints, tuberculosis, (ii) *Leaf* : in boils, cuts, wounds; RURAL PEOPLE OF CENTRAL HIMALAYAN REGION : (i) *Plant* : in dizziness, headache, vertigo, (ii) *Leaf* : for dissolving kidney stones.
CHARAKA SAMHITA : useful in urinary complaints and stones; SUSHRUTA SAMHITA : in stones and sugars; CHAKRADATTA : urinary troubles and stones; RAJNIGHANTU : it is sweet

and bitter, useful in urinary diseases; BHAVAPRAKASHA : astringent, bitter and sweet, purifies the urinary bladder; AYURVEDA : *Leaf-juice* : in urinary troubles, cold, haemorrhagic diseases, distension of stomach and epilepsy
UNANI : in dissolving stones.

Modern use : (i) *Root* (alcoholic extract) : antilithic; (ii) *Root-bark* (acetone extract) : anti-inflammatory, cardiotoxic, central nervous system depressor, diuretic in mild doses but antidiuretic in high doses; (iii) *Cystone* (an indigenous drug containing the plant) : has been accepted as useful for urinary tract infections; (iv) *EtOH* (50%) *extract of rhizome* : spasmogenic, antiprotozoan and anticancer.

Phytography : A small plant; leaves about 2.5 cm in diameter, obovate to suborbicular, obtuse, tapering at base, glabrous on both sides, margin subentire and ciliate; flowers white to rose or purple, petals clawed.

Phenology : *Flowering* : Winter; *Fruiting* : Summer.

Distribution : Along the foothills of Himalaya, Khasia Hills (1300 m) to Kashmir (2300–3100 m); Bhutan.

Ecology and cultivation : Subtropical plant, grows closely appressed to rocky substratum; wild.

Chemical contents : *Root* : afzelachin, bergenin, β-sitosterol, carbon-glyceride, gallic.

Adulterants : *Kalanchoe pinnate* is often confused with this plant.

Remarks : People of Kumaon give root to cattle in diarrhoea and people of Meghalaya eat leaves as vegetable.

Bergenia purpurascens (Hk. f. & Th.) Engler (Saxifragaceae)

Syn : *Saxifraga purpurascens* Hk. f. & Th.

Vernacular name : Sik : Pashanved.

Traditional use : PEOPLE OF WEST SIKKIM : *Root* : in stomachache, to stop bleeding after childbirth.

Phytography : Herb; rootstock stout; leaves alternate, petiolate, a large sheath at the base of petiole, simple, obovate, 5.0–7.5 cm in diameter, glabrous, punctate beneath, margins undulate; scape corymbose, 15–45 cm high, pedicels very pubescent; flowers white or red, 1.8 cm long, tubular; fruit erect.

Distribution : Arunachal Pradesh, Sikkim—between 3000 and 5000 m.

Ecology and cultivation : Inhabits in cool temperate climate, especially alpine localities; wild.

Remarks : Plant of limited distribution; deserves protection and cultivation.

Bergenia stracheyi (Hk. f. & Th.) Engler (Saxifragaceae)

Syn : *Saxifraga stracheyi* Hk. f. & Th., *S. ciliata* Lindl. non Royle

Vernacular names : Gar : Silphari; **Kum** : Gheepati.

Traditional use : *GARHWALI* : *Root* : antiscorbutic, astringent, febrifuge, in ophthalmia and tonic; *KUMAONI* : *Root* : in cuts and wounds; *RURAL PEOPLE OF UTTAR PRADESH* : *Rhizome* : in menorrhagia, urinary diseases.

Phytography : Herb; rootstock stout; leaves alternate, petiolate, large sheath at the base of petiole, simple, obovate, glabrous on both surfaces, margin subentire, ciliate or crenate-serrated; scape corymbose, 15–45 cm high; flowers small, white or rosy or yellow, pedicels and calyx pubescent; fruits ovate-lanceolate.

Distribution : Kashmir to Kumaon from 2600 to 4600 m.

Ecology and cultivation : Plant of alpine localities; wild.

Bergia suffruticosa (Del.) Fenzl. (Elatinaceae)

(2n = 30)

Syn : *Bergia odorata* Edgew., *Lancretia suffruticosa* Del.

Vernacular names : Ratan Mahal Hills (Gujarat) **:** Kharsan; **Lad :** Runuund.

Traditional use : *TRIBES OF RATAN MAHAL HILLS* (Gujarat) : *Plant* : in scorpion bite; *RURAL PEOPLE OF GUJARAT* : *Leaf* : to sores as poultice; *LADAKH* : *Stem* : in dental care (as tooth brush).

Phytography : Woody, diffuse undershrub, 33 cm or less in height; bark aromatic, deciduous, papery; leaves opposite, sessile, simple, 0.2–0.8 cm, elliptic or ovate, thick, crenate or serrate; flowers axillary, fascicled, pedicelled 0.4 cm long; capsules 5-chambered; seeds many, minute, shining, slightly curved.

Distribution : Gujarat, Punjab, Ladakh; Pakistan (Sind).

Ecology and cultivation : Xerophytic; wild.

Beta vulgaris L. (Chenopodiaceae)

(2n = 18)

Syn : *Beta orientalis* Roth

English name : Sugar beet.

Vernacular names : Asm : Beet-palang; **Ben, Guj, Kan, Mal** and **Ori :** Beet; **Kon :** Betarab; **Hin** and **Pun :** Chukandar; **Tam :** Chakunda; **Tel :** Beetudumpa.

Modern use : *Pulp* : useful as blood plasma substitute; *Root-extract* : useful in vascular and metabolic disorders, shows antifertility activity; *Plant* : antimicrobial; *Leaf* : growth stimulator with pantothenic activity.

Phytography : Succulent herb, annual or perennial; root red, swollen, napiform; stem red, usually shortened, may grow up to 1 m, erect stem furrowed; leaves alternate, simple, base decurrent on petiole, yellowish green, margin wavy; spikes 15–45 cm long, slender; flowers minute, bisexual; seeds minute, horizontal, albuminous.

Distribution : Native of south Europe; cultivated in various parts of India.

Ecology and cultivation : A cool-weather crop but can be grown in warm climate; grows on all types of soils if friable enough for proper development of roots.

Chemical contents : *Root* : betanine, uronic acid, vitamin A, sugar, saponine, proteins; *Leaf* : caffeic acid, chlorogenic acid, ferulic acid, oxalic acid, volatile compounds; *Seed* : α-glucosidase.

Adulterant : It is used as an adulterant of *Armorachia lapathifolia* L.

Remark : Plant causes oxalate poisoning in cattle.

Betula utilis D. Don (Betulaceae)

Syn : *Betula bhojpatra* Wall.

English name : Birch tree.

Sanskrit name : Bhuurja patra.

Vernacular names : Ben and **Hin :** Bhujja patra.

Trade name : Bhujja patra.

Traditional use : JOHARI TRIBE : *Bark* : antiseptic, ear trouble, hysteria, jaundice; GARHWALI : Bark : wound; KUMAONI : *Resin* : burns, cuts, contraceptive.

Phytography : Tree, 13–17 m tall, dwarf at high altitudes; stem aromatic, pubescent, resinous; bark peeling in horizontal flakes; leaves alternate, petiolate, simple, 3.0–7.5 cm, ovate, pubescent when young, irregularly serrate, deciduous; male spikes pendulous, stipitate, male flowers bracteate and bracteolate; female spikes stout, female flowers bracteate; nuts margined or winged.

Distribution : Kashmir (2300–4000 m) to Sikkim (3000–4600 m); Bhutan, Tibet.

Ecology and cultivation : Plant of cold climate; wild.

Chemical contents : *Stem-bark* : betulin, lupeol, leucocyanidin, oleanolic acid, Ac-oleanolic acid, Karachic acid.

Adulterants : *Betula alnoides* is similar to this plant.

Remarks : Rural people of Bhagirathi Valley (Uttar Pradesh) use plant in treatment of disorders in alimentary canal of domestic animals; bark used for writing purposes in ancient India and still used by the Hindus for religious purposes.

Bidens bipinnata L. (Asteraceae)

(2n = 72)

Syn : *Bidens pilosa* L. var. *bipinnata* Hook. f.

English names : Beggar lice, Spanish needles.

Traditional use : RURAL PEOPLE : *Plant-juice* (warmed) : in conjuctivitis, earache and as styptic on wounds; *Root* and *Seed* : antiseptic, emmenagogue, expectorant, stimulant, in asthma.

Modern use : *Aqueous extract of stem, leaf* and *flower* : antibiotic against *Micrococcus pyogens* var. *aureus*, and yeast.

Phytography : Very variable erect herb, 15–60 cm high; stem glabrous or pubescent; leaves opposite, bipinnate, segments ovate or lanceolate or linear, entire or lobed or toothed; heads on long, stout peduncles, very variable in length, involucre of bracts with broad scarious margins; florets female and bisexual, yellow or white; achenes black, long, slender, narrowed from middle to top, awned and hooked.

Distribution : Throughout India, ascending the Himalaya and other mountains up to 2000 m.

Ecology and cultivation : Plant of warm climate; wild.

Chemical content : *Young plant* : rich in volatile oil with objectionable odour.

Remark : Cattle relish young plants, but consumption of the same may taint milk.

Bidens biternata (Lour.) Merrill & Sheriff ex Sheriff (Asteraceae)

Syn : *Bidens chinensis* Willd., *B. pilosa* auct. non-L.

Vernacular names : Guj : Samara kokadi; **Him :** Chorpushi ; **Hin** and **Pun :** Chirchitta; **Mun :** Huring sargujia-ba; **Ori :** Magha latenga.

Traditional use : *KHASI* and *JAIANTIA* : *Leaf* : in snake bite; *MIKIR* : *Leaf* : in headache; *ANGAMI NAGA* : *Leaf* : in cuts and wounds; *NAGA* : *Achene* : as appetiser; *TOTO* (West Bengal) : *Leaf* : in sores; *GARHWALI* : *Leaf* : in inflammation; *TRIBES OF SIWALIK* (Uttar Pradesh) : (i) *Leaf* and *Flower* together : in complaints of ear and eye, (ii) *Flower* : in diarrhoea; *TRIBES OF KANGRA VALLEY* (Himachal Pradesh) : *Plant* : in skin diseases, leprosy, ulcers; *TRIBES OF EASTERN RAJASTHAN* : (i) *Plant* : in eye troubles, (ii) *Root* : in toothache; *TRIBES OF NILGIRI HILLS* (Tamil Nadu) : *Leaf* and *Flower* (together) : in troubles of eye and ear.

Phytography : Erect herb, variable in size and character; height varies from 15 to 50 cm; stem pilose or pubescent; leaves opposite, bipinnate, segments lobed or toothed; heads on peduncles of various lengths; florets female and bisexual, may be neuter, yellow or white; achenes long, slender, barbed.

Distribution : Throughout India as weed, ascending up to 2000 m.

Ecology and cultivation : Plant of warm climate; wild.

Remarks : Rural people of West Sikkim and tribes of Hazaribagh (Bihar) use leaves as substitute for tea leaves. Khasis and Garos grow it in houses to ward off evil spirit. The barbed achenes often stick to garments.

Bidens cernua L. (Asteraceae)

English name : Burr marigold.

Traditional use : RURAL PEOPLE OF KASHMIR : *Essential oil* : as diuretic, laxative and for healing wounds.

Modern use : *Essential oil* : antimicrobial, inhibits gram-positive bacteria, dermophytes, certain yeast and mould fungi.

Phytography : Annual erect herb; stem stout, glabrous; leaves opposite, simple, 7.5–12.5 cm, serrated, base subconnate, sometimes auricled; heads 1.6–2.5 cm in diameter, outer involucre of bracts leafy, inner involucre of bracts broad, membranous, pale brown; florets bisexual or female or neuter, yellow or white; achenes very variable, cuneate, compressed, margins sometimes thickened, armed with short recurved spines and 1–4 awns.

Distribution : Kashmir to Chamba at 1500–2000 m.

Ecology and cultivation : Plant of cold climate; wild.

Chemical contents : *Herb* : ascorbic acid, carotene, five flavonoids, tannin; *Essential oil* : aromatic hydrocarbons, 1-Ph-hepta-1,3,5-triyne.

Bidens tripartita L. (Asteraceae)

(2n = 48)

Syn : *Bidens repens* Don, *B. trifida* Roxb.

English names : Burr marigold, Three-lobed butterburr.

Vernacular name : Tribes of Subansiri (Arunachal Pradesh) **:** Nikampusi.

Traditional use : TRIBES OF HIMALAYAN FOOTHILLS : (i) *Herb* : antiscorbutic, antiseptic, aperient, diaphoretic, diuretic, emmenagogue, febrifuge, mild narcotic, sedative; effective in respiratory and uterine haemorrhages, useful in dropsy, gout and haematuria; (ii) *Root* : aperient; (iii) *Seed* : diuretic, emmenagogue, expectorant, useful in stones of gall bladder and kidney.

Phytography : Stout annual herb; stem erect, 30–90 cm high, glabrous; leaves opposite, 7.5–12.5 cm, trilobed or tripartite or pinnatifid, segments lanceolate, toothed; head solitary, erect, peduncle long or short, discoid, 1.6 cm in diameter, outer involucre of bracts herbaceous, inner involucre of bracts oblong, margins scarious; florets bisexual or female or neuter, yellow or white; achenes 0.6 cm long, glabrous, armed with 2–3 very short awns.

Distribution : Western to central Himalaya between 1000 and 1600 m; Nepal.

Ecology and cultivation : Grows in marshes; wild.

Chemical contents : *Herb* : tannins, essential oil, flavonoids, luteolin and its seven glucosides, buteine glucoside, coumarins, isocoreopsin; *Leaf* : ascorbic acid; *Fruit* : fixed oil similar to sunflower oil.

Remark : People of Subansiri (Arunachal Pradesh) eat the leaves as vegetables.

Bieberstenia odora Steph. ex Fisch. (Biebersteiniaceae)

Syn : *Beiberstenia emodi* Jaub. & Spach.

Vernacular name : Lad : Drakspose.

Traditional use : LADAKHI : *Plant* : in cuts, wounds and peptic ulcers.

Phytography : Perennial, glandular, pubescent herb; rootstock densely tufted, branched, tuberous, holding a crown of leaves; leaves alternate, stipulate, linear, pinnatisect, leaflets small, broad, lobed; flowering stem a few cm high; inflorescence racemose or panicled; flowers yellow, 1.25 cm in diameter; fruit capsular, 5-lobed, crustaceous, wrinkled; seeds incurved on one side of fleshy albumen.

Distribution : Native of Asia, found in Ladakh at 5000–5600 m; alpine western Tibet.

Ecology and cultivation : Alpine plant; wild.

Adulterant : The plant looks like *Potentilla.*

Remark : *Endemic*; should be protected and cultivated.

Biophytum apodiscias Turez (Oxalidaceae)

Syn : *Oxalis petersii* Klotzsch, *Biophytum petersianum* Klotzsch, *Biophytum umbraculum* Welw.

Vernacular names : Mun : Janapid, Japid-sing, Durumbihir, Durum-sing, Durum-tasad; **Orn :** Lajauri.

Traditional use : ORAON : *Root* and *Leaf* : in insomnia.

Phytography : Annuai small herb; stem slender, unbranched, 2.5–5.0 cm high; leaves crowded forming a crown, petioles puberulous, sensitive, 2.5–3.7 cm, unipinnate, leaflets pale, slightly coriaceous; flowers red, 0.4 cm long and 0.6 cm across; capsules ovoid or oblong; seeds with scattered tubercles.

Distribution : Meghalaya, West Bengal, Bihar, Orissa, Andhra Pradesh, Tamil Nadu, Karnataka and Kerala.

Ecology and cultivation : Wild.

Biophytum reinwardtii Klotzsch (Oxalidaceae)

Syn : *Oxalis reinwardtii* Zucc.

Vernacular names : Mun : Japid-sing, Janapid, Durumbihir, Durum-sing; **Orn :** Lajauri.

Traditional use : MUNDA : *Herb decoction* : in fever; ORAON : *Root* and *Leaf* : in insomnia.

Phytography : Graceful little herb; stem simple, 7.5–25.0 cm high; leaves fascicled or almost whorled at the top of the stem, petiole swollen at base, pinnate, leaflets oblique,oblong, tip rounded, base glabrous; inflorescence terminal, umbellate; flowers small, yellow, rarely purple; capsules ovoid or oblong or subglobose, almost equalling the small sepals; seed spirally furrowed.

Phenology : *Flowering* and *Fruiting* : Rainy season and Winter.

Distribution : Kumaon to Arunachal Pradesh up to 1800 m, throughout hotter places of India and extending up to the Deccan peninsula.

Ecology and cultivation : Common in rocky jungles and damp places; wild.

Adulterant : It looks similar to *B. sensitivum.*

Biophytum sensitivum (L.) DC. (Oxalidaceae)

English name : Sensitive wood-sorrel.

Sanskrit name : Jhallapushpa.

Vernacular names : Asm : Bannarenga; **Ben :** Jhalai, Bannarenga; **Guj :** Jaharera, Zarer; **Hin :** Lajalu, Lakchana; **Kan :** Hora muni; **Kon :** Lojeri; **Mal :** Mukkuttii, Thindanazhi; **Mar :** Ladjari; **Mun :** Durum bihir; **San :** Jhapni.

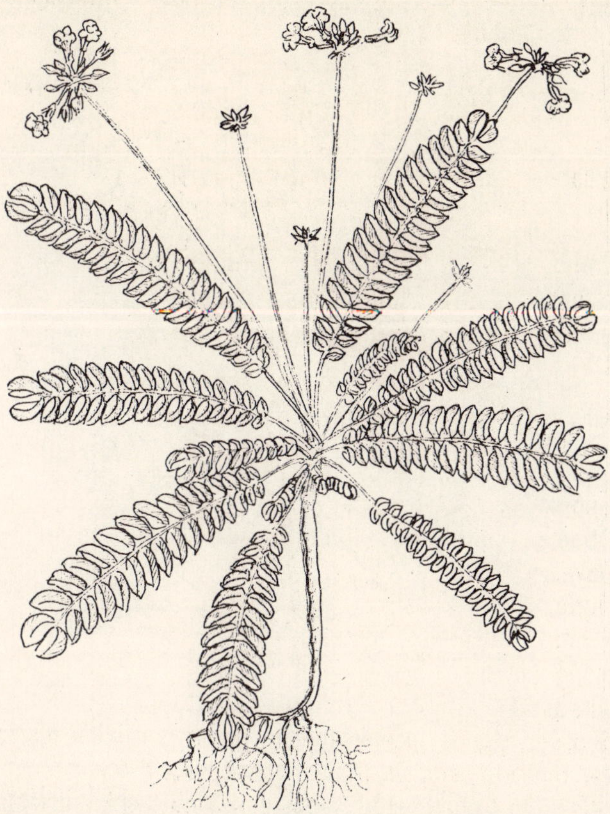

Traditional use : *KHASI* and *JAINTIA* : in headache, giddiness, fever; *BODO* : *Leaf* : as antiseptic, in bilious fever; *SANTAL* : (i) *Plant* : in convulsion, cramp, eye trouble (*Jhapni bai*), wounds with burning pain; (ii) *Root-paste* : burns, when limb(s)/body becomes cold as ice; *RURAL PEOPLE OF ORISSA* : *Plant* : antiexcitement; *TRIBES OF SAGAR DISTRICT* (Madhya Pradesh) : *Plant* : antiexcitement; *BHIL* : *Root* : to decrease sexual vigour; *TRIBES OF KHEDBRAHMA REGION* (Gujarat) : *Leaf* : in muscular pain, rheumatism; *TRIBES OF SAURASHTRA* (Gujarat) : *Leaf* : as tonic; *TRIBES OF RATAN MAHAL HILLS* (Gujarat) : *Dried leaf* : smoking with tobacco renders a man infertile, and excessive use of the same makes him impotent.

Modern use : *Leaf* (aqueous extract) : inhibits *Mycobacterium tuberculosis*; *Leaf* (saline extract): hypoglycaemic in rabbit.

Phytography : Annual herb, 7.5–10.0 cm high; stem simple; leaves sensitive, fascicled or almost whorled at the top of the stem, petioles swollen at the base, unipinnate, 5.0–12.5 cm long with 6–15 pairs of leaflets, leaflets oblique, oblong, terminal pair the largest; flowers shortly pedicelled, small, yellow or purple; capsules ovoid or oblong, covered by sepals; seeds with obliquely transverse tuberculed ridges.

Phenology : *Flowering* and *Fruiting* : Rainy season.

Distribution : Throughout hotter parts of India, ascending up to 2000 m in the Himalaya; Sri Lanka.

Ecology and cultivation : Xerophytic; wild.

Bischofia javanica Blume (Bischofiaceae)

Syn : *Bischofia oblongifolia* Dene, *Andrachne trifoliata* Roxb., *Stylodiscus trifoliatus* Bennet

English names : Bishop wood, Java cedar, West Indian cedar.

Vernacular names : Hin : Bhillar, Kain; **Mik** : Phangput-araung.

Traditional use : *RURAL PEOPLE* and *MIKIR* : *Plant* : in gastroenteritis; *Leaves* : in sores, eye diseases, toothache.

Modern use : *Seed-extract* : haemagglutinating.

Phytography : Round headed, more or less deciduous glabrous tree, 10–13 m high, bark smooth; leaves alternate, petiole 2.5–15 cm long, trifoliate, petiolules 1.25–1.8 cm, leaflets 7.5–12.5 cm, ovate to oblong-lanceolate, repand or toothed; racemes axillary or lateral, panicled, very slender; flowers minute, green, unisexual, males on short slender pedicels, females on stout pedicels; fruits globose, blue-black, smooth, pea-sized, on long thick pedicels, 3- to 4-chambered; seeds dark brown, shining, testa splitting longitudinally.

Distribution : Tropical Himalaya, from Kumaon eastwards, Assam southwards, Western Ghats, Nilgiri Hills (Tamil Nadu).

Ecology and cultivation : Prefers damp cold climate; wild.

Chemical contents : *Root* : β-amyrin, β-sitosterol, ursolic acid; *Stem-bark* : epifriedenol-OAc, friedelin, beutinic acid, β-sitosterol; *Leaf* : tartaric acid, vitamin C, ellagic acid, friedelin, friedelan-3α and its Ac-derivative, β-sitosterol.

Remarks : Mikirs extract a pink dye.
In Samoa, bark is taken for cough, while inner bark for burns, cracks on the soles of feet, diarrhoea in children, high fever and throat ailments.
Plant of medicinal importance and restricted distribution; deserves protection and cultivation.

Bistorta affinis Greene (Polygonaceae)

Syn : *Polygonum affinis* D. Don

Traditional use : *PEOPLE OF LAHUL* (Himachal Pradesh) : *Plant* : in cold, diarrhoea.

Phytography : Glabrous densely tufted perennial herb; rootstock woody, 10–20 cm long, branched, often tortuous, flowering stem simple, 10–30 cm high; leaves alternate but crowded on the top of rootstock, stipules tubular, 1.25–2.5 cm long, entire or split, brown, scarious when old, sessile or shortly petioled, simple, linear or elliptic or lanceolate, glaucous beneath; racemes peduncled, peduncles 7.5–20.0 cm long, stout, erect, flowers crowded; flowers suberect, rosy, 0.4–0.6 cm in diameter; nuts trigonous.

Distribution : Alpine and subalpine Himalaya from Jammu & Kashmir (2200–4600 m) to Kumaon (3600–4300 m); western Tibet, Nepal.

Ecology and cultivation : Plant of cold climate; wild.

Chemical contents : *Leaf* : flavone, isoaffinetin and some other flavonoids.

Remarks : People of Lahul eat the seeds.
Plant of limited distribution; deserves protection and cultivation.

Bistorta amplexicaulis Greene (Polygonaceae)

Syn : *Polygonum amplexicaule* D. Don

English name : Mountain fleece.

Vernacular names : **Kas :** Ajbar; **Gar :** Amli; **Uttar Pradesh :** Durpa tandar.

Trade name : Rootstock as *Anjubar.*

Traditional use : *PEOPLE OF KASHMIR* : *Root* : as haemostatic; *KUMAONI* : *Leaf* : as tonic; *PEOPLE OF BHAGIRATHI VALLEY* (in Uttarkashi Forest Division) : *Root* : in cough, dysentery.

Phytography : Glabrous densely tufted herb; rootstock stout, woody, branched, flowering stem 60-90 cm high, green, very slender, sparingly leafy; lower leaves 7.5–15.0 cm long, minutely papillose beneath, upper leaves amplexicaul, stipules tubular, 2.5–5.0 cm, narrow; racemes 5–15 cm long; flowers bracteate, red or white or greenish white, 0.4–0.6 cm in diameter; nuts included by the enlarged perianth.

Distribution : Subalpine and alpine Himalaya from Jammu & Kashmir to Sikkim between 1800 and 2400 m; Bhutan.

Ecology and cultivation : Plant of cold climate; wild.

Adulterant : It is used as a substitute for sarsaparilla.

Remarks : People of Kumaon use rootstock flakes as substitute for tea leaves.
In Tibetan system of medicine, powdered root is used as a diuretic, blood purifier, in post-natal complaints and to reduce biliousness.
The plant has been reported to be useful in oesophageal cancer.
Plant of significant medical information and limited distribution; deserves protection and cultivation.

Bistorta macrophylla Sojak (Polygonaceae)

Syn : *Polygonum macrophyllum* D. Don, *P. sphaerostachyum* Meissn.

Vernacular name : Gar : Kukhari.

Trade name : Kukhari.

Traditional use : *GARHWALI* : *Plant* : astringent.

Phytography : Perennial glabrous herb; rootstock woody, flowering stem simple, slender; leaves alternate, stipules tubular, petiole long, lamina simple, linear or oblong; spikes solitary, erect, stout, cylindric or capitate, densely flowered; flowers crimson or brilliantly coloured, drooping, minute.

Distribution : Garhwal to Sikkim between 3300 and 4600 m; western Tibet, Bhutan.

Ecology and cultivation : Plant of cold climate; wild.

Remark : Plant of limited distribution; deserves protection and cultivation.

Bixa orellana L. (Bixaceae)

(2n = 14, 16)

English names : Annatto, Arnato, Chinese monkey turmeric, Roucou, Saffron tree, Dye tree, Lipstick tree.

Vernacular names : Hin : Latkan; **Mun** : Rang-jo-daru; **Nay** (North Kerala) : Kungumapoovu; **Sad** : Rangphul, Rangphal.

Trade names : Annatto, Latkan.

Traditional use : *NAYADI* (North Kerala) : *Leaf* : in snake bite; *RURAL FOLK* : (i) *Twig* : in liver troubles, as emollient; (ii) *Leaf-decoction* : as a gargle for sore throat; (iii) *Leaf-infusion* : in dysentery; (iv) *Gummy substance obtained from pounded leaf* : as

diuretic, in gonorrhoea; (v) *Seed-pulp* : antidysenteric, digestive, diuretic, febrifuge, haemostatic, laxative, useful in epilepsy and skin diseases, applied to burns to prevent blisters and scars.

Modern use : *Root* (aqueous extract) : antispasmodic, hypotensive; *EtOH extract of defatted root-bark* : antimicrobial, spasmogenic, weak Central nervous System depressor; *Leaf* (HCl extract) : anticancer; *Seed-pulp* : useful in kidney diseases.

Phytography : Evergreen small bushy tree; leaves simple, stipules minute, petioles 5.0–7.5 cm long, slender, lamina 10–20 cm by 6.25–12.5 cm, cordate, glabrous, acuminate; panicles terminal; flowers purple or pink or white, 2.5–5.0 cm in diameter, bisexual; capsules scarlet, ovoid or subglobose, 3.7 cm, softly prickly, bivalved, pulp orange-red, many-seeded; seeds albuminous.

Phenology : *Flowering* : July–September; *Fruiting* : October–November.

Distribution : Native of tropical America; cultivated throughout India as ornamental.

Ecology and cultivation : Tropical plant.

Chemical contents : *Root* : triterpenoid, tomentosic acid and another unidentified one; *Leaf* : essential oil containing sesquiterpene and bixaghanene, 7-bisulphates of apigenin, luteolin, hypolaetin, flavonoids; *Seed* : bixin and isobixin.

Adulterants : Root is a substitute for saffron.

Remarks : In Columbia, bixa paste considered aphrodisiac. In Hawaii, dry wood used against fainting, mouth sores and impure blood.
Annatto dye, obtained from seeds, used in colouring food, cosmetics, pharmaceutical ointments and as mosquito repellent.

Blachia umbellata Baill. (Euphorbiaceae)

Syn : *Croton umbellatus* Willd., *Codiaewn umbellatum* Muell.

Modern use : *EtOH* (50%) *extract of aerial parts* : diuretic.

Phytography : Glabrous shrub; petiole 0.25–0.6 cm, lamina 10–17 cm, elliptic or sub-rhombic or oblong, broadest at or above the middle, cuspidately acuminate; male flowers umbelled on a capillary peduncle, peduncle 2.5–7.5 cm long, pedicels 0.4–0.8 cm,, buds subglobose, 0.6 cm in diameter, 1–4 female flower(s) on a rigid, erect, slender, 2.5–5.0 cm long peduncle, pedicels 0.3–1.25 cm long, thickened upwards, flowers 0.8 cm in diameter; capsules 1.6 cm in diameter, deeply trilobed, smooth.

Distribution : Kodagu (Coorg) and southwards; Sri Lanka.

Ecology and cultivation : Prefers seashore; wild.

Blepharis linariefolia Pers. (Acanthaceae)

Syn : *Blepharis sindica* Stocks ex T. Anders., *Acanthalium grossum* Wight

Vernacular name : Raj : Bhangari.

Trade name : Bhangari.

Traditional use : *PEOPLE OF RAJASTHAN DESERT* : *Seed* : tonic.

Modern use : *EtOH* (50%) *extract of flower* : anticancer.

Phytography : Rigid herb; stem 7.5–15.0 cm or shorter; leaves opposite or four in a whorl, simple, 2.5–5.0 cm, linear or narrowly oblong, pubescent, toothed or spinescent; heads few- or many-flowered, bracts 2.5 cm or more, ovate with divaricate lanceolate tips, prickly, hairs present on both sides, bracteoles 0.8 cm, linear, covered with hairs, spikes terminal or on suppressed axillary branches; flowers blue or pinkish, 1.25–1.6 cm, tubular, softly hairy; capsules shining brown, 0.8 cm, ellipsoid, compressed, 2-seeded; seeds 0.6 cm, echinate but the tufts of hairs have all their ends free.

Distribution : Desert of Rajasthan; Pakistan (Sind).

Ecology and cultivation : Xerophyte; wild.

Adulterant : May be confused with *Blepharis persica*.

Remarks : Plant of very limited distribution; deserves protection and cultivation.

Blepharis maderaspatensis (L.) Roth (Acanthaceae)

Syn : *Blepharis repens* (Vahl) Roth, *Acanthus repens* Vahl

Traditional use : IRULAR (of Tamil Nadu) : *Leaf-paste* : applied to forehead for curing headache.

Phytography : Perennial, wiry, prostrate herb; stem covered with rough hairs, branches 7.5–15.0 cm long; leaves usually four in a whorl, subsessile, simple, oblong, 1.25–2.5 cm, covered with hairs; heads terminal or axillary, mostly 1-flowered, many of outer bracts barren, bracteoles elliptic or oblong, bracts shorter than bracteoles,bracts and bracteoles bristle-ciliate; flowers pink or purple or nearly white, 0.6–0.8 cm; capsules shining glabrous, ellipsoid, compressed, 2-seeded; seeds densely covered with spines, numerous bundles of annular tubes connected at the base.

Distribution : Common in Tamil Nadu, available in Delhi and Chotanagpur; Sri Lanka.

Ecology and cultivation : Xerophyte; wild.

Adulterant : This has close similarities with *Blepharis boerhaaviaefolia*.

Blepharis persica (Burm. f.) Kuntze (Acanthaceae)

Syn : *Blepharis edulis* Pers., *Acanthus edulis* Forssk., *Acanthodium spicatum* Delile, *Ruellia persica* Burm.

Traditional use : RURAL PEOPLE OF PUNJAB : *Root* : diuretic, in urinary discharges, dysmenorrhoea; *Leaf* : in ascites, asthma, nasal haemorrhage, spleen complaints, throat inflammation, ulcer, wound, and as a purgative; *Seed* : in conjunctivitis, strangury and as deobstruent.
AYURVEDA : *Plant* : in impotency and rheumatism.

Phytography : Rigid herb; stem 30 cm or less, branched; leaves opposite or four in a whorl, simple, elliptic or oblong, often 1.25 cm broad, pubescent, spinescent; heads few- or many-flowered, bracts spinous, bracteoles linear, covered with hairs; flowers small; capsules 2-seeded.

Distribution : Punjab; Pakistan (Sind).

Ecology and cultivation : Xerophyte; wild.

Chemical contents : *Seed* : isocoumarin glucoside, blepharin, blepharigenin, β-sitosterol and a saponin.

Adulterants : Very similar to *Blepharis linariefolia.*

Remarks : Plant of limited distribution; should be protected and cultivated.

Blumea aurita DC. (Asteraceae)

Syn : *Laggera aurita* Sch.-Bip. ex C.B. Clarke, *Conyza aurita* L.f.

Vernacular names : Mun : Marang-puru, Soan-puru.

Traditional use : MUNDA : *Leaf* : to stop bleeding from a cut.

Modern use : *Essential oil* : inhibitory against *Escerchia coli* and *Klebsiella aerogenes.*

Phytography : Erect strongly aromatic (somewhat like turpentine) grey-green herb, 60–90 cm high; stem rather slender, viscidly hirsute pubescent or villous; leaves alternate, sessile or petioled, small, lobed or pinnatifid, covered with small hairs; heads in small lateral or terminal corymbs, 0.8–1.25 cm in diameter, involucre of bracts slender, soft, villous; florets pink, female or bisexual.

Phenology : *Flowering* and *Fruiting* : January–March.

Distribution : Throughout the plains of India; Pakistan (Sind), Bangladesh (Chittagong), Myanmar.

Ecology and cultivation : Grows in wastelands; wild.

Chemical contents : *Plant* : essential oil, laggerol, triterpenoids, alkanes, 2,3-di MeO-p-cymene.

Remarks : In treatment of rheumatic pain, strongly aromatic and sticky infusion of the plant is used in Angola.

Blumea balsamifera (L.) DC. (Asteraceae)

Syn : *Blumea densiflora* Hook. f., *Conyza balsamifera* L.

English names : Blumea camphor, Nagai camphor.

Vernacular name : Mik : Janiya.

Traditional use : MIKIR : *Leaf* : in stomach upset in infants due to defect in mother's milk. UNANI : ingredient of 'Kakronda'.

Modern use : *Plant* (alcoholic extract) : antibacterial against *Eschercchia coli.*

Phytography : Arboreous, aromatic (smelling strongly of camphor), herb, 10--20 cm high; stem tomentose or villous or silky woolly, corymbosely branched; leaves alternate, shortly petioled, simple, sometimes pinnatifid, coriaceous, elliptic or oblong or lanceolate, silky above; heads sessile or peduncled in rounded clusters or pyramidal panicle, 0.6–0.8 cm in diameter, involucre of bracts slender, puberulous; florets female or bisexual, corolla lobes glandular, achenes 10-ribbed, silky, pappus red.

Distribution : Tropical Himalaya, Sikkim (up to 1300 m), Assam, Meghalaya; Bangladesh (Chittagong), Myanmar, Nepal.

Ecology and cultivation : Plant of tropical rainforest; wild.

Chemical contents : *Leaf* : two quercetin derivatives.

Adulterant : It is used as an adulterant of *Cinnamomum camphora* (L.) Nees et Ebermaier.

Blumea densiflora DC. (Asteraceae)

Syn : *Conyza densiflora* Wall.

Traditional use : *TRIBES OF NORTH-EAST INDIA* : *Leaf* (fresh juice) : insect and mosquito repellent.

Phytography : Herb; stem stout, covered with thick white hairs; leaves alternate, petiole long-winged or appendaged, simple, 20–45 cm, elliptic-lanceolate, serrate or toothed, may be pinnatifid, undersurface very woolly white, upper surface puberulous; heads sessile in rounded clusters on a large branched panicle, 0.6 cm in diameter, involucre of bracts narrow, rather rigid; florets female or bisexual, corolla lobes of bisexual flowers hairy; achenes 10-ribbed, pubescent, pappus red.

Distribution : Sikkim, Assam, Meghalaya, Nagaland, up to 1300 m.

Ecology and cultivation : Plant of tropical rainforest; wild.

Adulterant : *Blumea balsamifera* is often confused with this species.

Blumea eriantha DC. (Asteraceae)

(2n = 20, 36)

Traditional use : RURAL PEOPLE OF KONKAN : *Leaf* (along with those of *Careya arborea* and *Vitex negundo*) : for fomentation.

Modern use : *Essential oil* : (i) *in vitro* : antibacterial, antifungal; (ii) *from leaves* : insecticide against larvae of *Culex fatigans* and *Culex pipiens*.

Phytography : Green herb, 30 cm high; stem very slender, pubescent or clothed with long hairs, dichotomously branched from base; leaves alternate, lower leaves petioled but upper leaves appear sessile, 2.5–7.5 cm, irregularly toothed; heads mostly on long slender peduncles of dichotomous cymes, rarely fascicled, 0.6–0.8 cm in diameter, peduncles and involucre of barcts covered with long silky hairs; florets female and bisexual; achenes, pappus white.

Distribution : Konkan.

Chemical contents : *Plant* and *Leaf* : essential oil ; *Flower* : erianthin; *Seed* : essential oil.

Remarks : *Endemic*; deserves protection and cultivation.

Blumea fistulosa (Roxb.) Kurz (Asteraceae)

Syn : *Blumea glomerata* DC., *Conyza fistulosa* Roxb.

Vernacular name : Mun : Bir nangula-ba.

Traditional use : PEOPLE OF ARAKU VALLEY (Andhra Pradesh) : *Leaf* : in boils, ulcers; MUNDA : *Plant* : in toothache, gum complaints; TRIBES OF SUBANSIRI (Arunachal Pradesh) : *Leaf* : in diarrhoea.

Modern use : *EtOH* (50%) *extract of plant* : diuretic; *Essential oil* : depressor of Central Nervous System.

Phytography : Erect herb, 30–60 cm high; stem slender, branched, pubescent or villous; leaves alternate, petioled, obovate or pinnatifid, sharply denticulate; heads in small sessile axillary clusters, purple, 0.6 cm in diameter, involucre of bracts villous; florets yellow; achenes 8- to 10-ribbed, glabrous or hairy,

Phenology : *Flowering* and *Fruiting* : January–May.

Distribution : Kumaon, Arunachal Pradesh, Sikkim, up to 1500 m, Assam, Bihar, Andhra Pradesh; Myanmar.

Ecology and cultivation : Wild.

Blumea lacera (Burm. f.) DC. (Asteraceae)

(2n = 18)

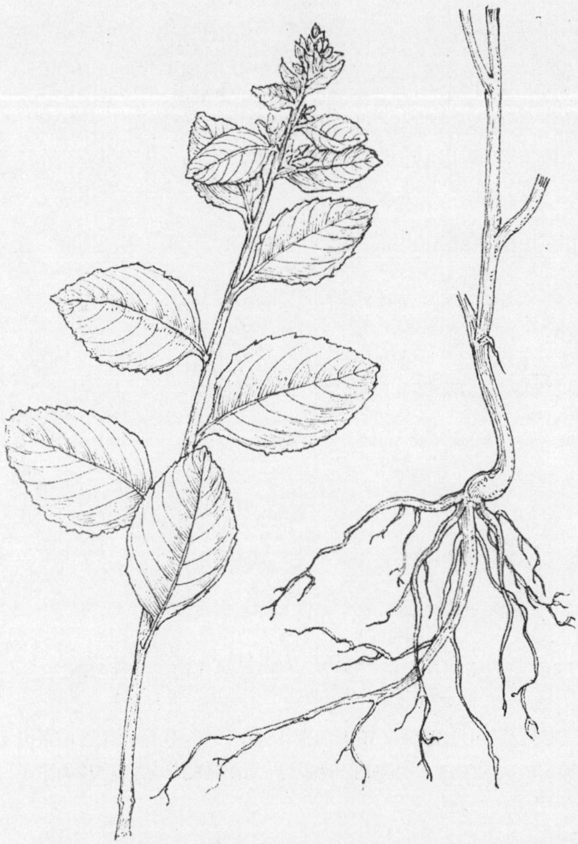

Syn : *Blumea leptoclada* Dalz. & Gibs, *Conyza lacera* Roxb.

Sanskrit names : Kukundara, Kukuradru.

Vernacular names : Asm : Kukur shuta; **Ben :** Bara kukshima, Kukur-sunga, Shealmotra; **Guj :** Kalara; **Hin :** Kukkurbanda, Kokranda; **Kan :** Gandhari gida; **Kon :** Burando; **Mar :** Burando, Nimurli; **Mun :** Marang kuru; **Ori :** Poka-sunga; **Pun :** Kokronda; **Tam :** Kattu mul-langi; **Tel :** Karn pogaku.

Traditional use : DANG (of Gujarat) : *Leaf* : in urinary complaints; PEOPLE OF DAHANU FOREST DIVISION (Maharashtra) : *Plant* : in bronchitis, fever; BHOXA (of Uttar Pradesh) : *Leaf* : in piles; MUNDA : *Leaf* : in burns, cuts, wounds, toe injury.

AYURVEDA : (i) *Whole plant* : in dyspepsia, itching and eye troubles; (ii) *Root* : bitter, alleviates dysentery, pain, purifies blood, useful in respiratory troubles and dropsy, increases amount of urine; (iii) *Root* along with *seed* of *Piper nigrum* : removes cholera; (iv) *Leaf* (fresh juice) : stops bleeding; (v) *Juice of leaf* along with *seed* of *Piper nigrum* : cures piles; (vi) *Flower* (aqueous extract) : in eye troubles.

Modern use : *Alcoholic extract of herb* : exhibit marked anti-inflammatory activity against carrageenin- and bradgkinin-induced inflammation in rats; *Leaf* (essential oil) : antimicrobial, antifungal and insect repellent.

Phytography : Slender, very variable weed with a strong turpentine or camphor odour, 45–60 cm high; stem erect, simple or branched, covered with hairs and glands, often grey in the more silky forms; leaves alternate, petioled, obovate, toothed; heads in short axillary cymes and collected into terminal panicles, 0.8 cm in diameter, involucre of bracts narrow, covered with hairs; florets female and bisexual, yellow; achenes nearly tetragonous and not ribbed.

Phenology : *Flowering* and *Fruiting* : Hot season.

Distribution : Throughout India, ascending up to 660 m; Sri Lanka.

Ecology and cultivation : Mesophyte; wild.

Chemical contents : *Herb* : essential oil containing Blumea camphor; *Aerial parts* (EtOH extract) : hentriacontane, hentriacontanol, α-amyrin, lupeol, β-sitosterol; *Root* and *Root-bark* : triterpene, sterols; *Leaf* : campesterol, coniferyl alcoholic derivatives, flavonoids, essential oil, cineol, 66 & δ-fenchone.

Remark : Tribes of Rajasthan use the plant as fish poison.

Blumea lanciniata (Roxb.) DC. (Asteraceae)

Syn : *Blumera lacera* var. *blumei* DC.

Vernacular name : Mun : Jugi ba.

Traditional use : GARHWALI : *Leaf* : in eczema and other skin diseases.

Modern use : *Essential oil* : antibacterial and antifungal.

Phytography : Strongly foetid, erect, green, stout herb, 60–120 cm high; stem pubescent or laxly tomentose, branched above; leaves alternate, petiolate, 7.5–15.0 cm, irregularly pinnatifid, lanciniate, scentless; heads many in open corymbs, 0.6–1.25 cm in diameter, involucre of bracts acuminate, villous; florets yellow, covered with hairs; achenes ribbed, silky.

Phenology : *Flowering* and *Fruiting* : March–December.

Distribution : Common in tropical Himalaya—Kumaon eastwards, Sikkim, Assam, Meghalaya, Bihar, West Bengal, found in Deccan peninsula.

Ecology and cultivation : Plant of tropical rainforest; wild.

Chemical content : *Plant* : essential oil.

Remark : Mundas use the plant as a remedy against mouth diseases of cattle.

Blumea lanceolaria (Roxb.) Druce (Asteraceae)

Syn : *Blumea myriocephala* DC., *B. longifolia* DC.

Vernacular name : Mik : Hanmoichu.

Traditional use : MIKIRS : *Leaf-decoction* (along with leaves of *B. balsamirera, B. sessiliflora* and *Mikania cordata*) : for bathing to cure body pain; *Leaf* : as poultice in rheumatism.

Modern use : *Plant-saponin* : haemolytic on human blood.

Phytography : Shrubby herb; stem thick, branched, glabrous or puberulous; leaves alternate, petiole may be winged, lamina 15–25 cm long, oblanceolate, glabrous; heads subsessile or clustered on the ascending branches of a pyramidal panicle, 0.6–0.8 cm in diameter, involucre of bracts tomentose; florets glabrous or glandular; achenes 10-ribbed, hairy, pappus red.

Distribution : Sikkim, Assam; Bangladesh (Sylhet, Chittagong), Myanmar.

Ecology and cultivation : Plant of tropical rain forest; wild.

Chemical contents : *Herb* : essential oil; *Root* and *Stem* : α-spinasterol.

Remark : In China, the plant is used as antibacterial, carminative and obstipation.

Blumea malcolmii Hook. f. (Asteraceae)

Traditional use : RURAL PEOPLE OF WESTERN DECCAN : *volatile oil from aerial parts* : antibacterial, attractant for house-fly, and insecticide.

Phytography : Stout herb; stem green, erect, densely clothed with silvery silky hairs; leaves alternate, sessile, leathery, 2.5–10.0 cm, obovate, closely jaggedly toothed; heads peduncled, peduncles covered with white silky hairs, fascicled towards the end of branches in open corymbs or panicles, 1.2–51.8 cm in diameter; achenes obscurely angled, covered with silky hairs, pappus white.

Distribution : Maharashtra, Karnataka, Kerala.

Ecology and cultivation : Wild.

Chemical content : *Aerial parts* : volatile oil.

Remark : Plant of limited distribution.

Blumea membranacea (DC.) Hook. f. (Asteraceae)

Syn : *Blumea hymenophylla* DC., *Conyza membranacea* Wall.

Vernacular name : Ori : Poksunga.

Traditional use : *RURAL PEOPLE OF ORISSA* : *Plant* : in body pain.

Modern use : *Essential oil from plant* : antibacterial, wide range in antifungal activity; *Leaf* (essential oil) : antibacterial.

Phytography : Tall herb; stem green, slender, much-branched, pubescent; leaves alternate, lower leaves petioled while upper leaves sessile and crowded, membranous, irregularly or sinuately toothed, softly pubescent; heads in open panicles, peduncles covered with soft spreading hairs and glands, 0.6 cm in diameter; florets female and bisexual, yellow, lobes glandular; achenes 8- to 10-ribbed, covered with spreading hairs, pappus white.

Distribution : Common throughout India.

Ecology and cultivation : Wild.

Chemical contents : *Herb* : sitosterol, essential oil; *Leaf* : essential oil.

Adulterant : Very similar to *Blumea virens*.

Blumea mollis (Don) Merr. (Asteraceae)

Syn : *Blumea wightiana* DC.

Vernacular names : Kan : Gobbusoppu; **Mun :** Ote jatang.

Traditional use : *GARHWALI* : *Leaf* : in skin diseases; *MUNDA* : *Boiled herb* : in diarrhoea.

Modern use : *Herb* (essential oil) : antibacterial, antifungal.

Phytography : Pleasantly aromatic (smell of hops) annual herb; stem green, erect, 30–60 cm high, covered with silky or woolly hairs; leaves alternate, petioled, simple, obovate, irregularly toothed or serrated; heads in dense cymes or panicles, peduncled, 0.6 cm in diameter; involucre of bracts narrow; florets purple, female and bisexual; achenes 4- to 5-angled but not ribbed, sparsely hairy.

Distribution : Common in the plains of India and outer Himalaya; Sri Lanka, Myanmar.

Ecology and cultivation : Mesophyte; wild.

Chemical contents : *Plant* : essential oil, stigmasterol.

Blumea oxyodonta DC. (Asteraceae)

Syn : *Conyza tenera* Wall.

Vernacular name : Mun : Huring karu.

Modern use : *EtOH* (50%) *extract of plant* : diuretic; *Leaf* (essential oil) : insecticide.

Phytography : Small prostrate decumbent herb; stem 10–25 cm, pubescent or villous or tomentose, branches spreading from base, dichotomously forked or simple; leaves alternate, petioled, upper leaves sessile, obovate, membranous, spinulose-toothed; heads on slender peduncles, solitary or few corymbose, involucre of bracts woolly or glabrate, 0.6 cm in diameter; florets pale yellow; achenes sparingly silky.

Phenology : *Flowering* and *Fruiting* : January–June.

Distribution : Garhwal, Kumaon, West Bengal, Bihar, Rajasthan, Deccan peninsula up to 1500 m; Myanmar.

Ecology and cultivation : Wild.

Remark : Mundas use the herb to kill fish.

Blumea riparia DC. (Asteraceae)

Syn : *Blumea chinensis* DC., *B. pubigera* Merr., *Conyza chinensis* L., *Conyza riparia* Blume

Traditional use : HILL TRIBES OF NORTH-EAST INDIA : *Decoction of leaf* : in beri-beri, colic.

Phytography : Scandent glabrous herb; branches terete, grooved, flexuous; leaves alternate, petioles 0.3–0.6 cm long, slender, lamina simple, elliptic or lanceolate, leathery, glandularly subserrate, acuminate; heads clustered on short distant branches of a long, terminal, pubescent panicle, 1.25 cm in diameter, outer bracts very short, leathery and obtuse, inner bracts long and thin; florets covered with hairs; achenes strongly 10-ribbed and covered with hairs, pappus white.

Distribution : Sikkim, Assam, Meghalaya, Andaman and Nicobar Islands; Bhutan.

Ecology and cultivation : Wild.

Blumeopsis flava Gagnep. (Asteraceae)

Syn : *Blumea flava* DC., *Laggera flava* Benth.

Vernacular name : Mun : Bocho tupuri ba.

Traditional use : *MUNDA* : *Plant* (with mustard oil) : against dropsy.

Phytography : Erect herb, 1.0–1.25 m high; stem slender, simple or corymbosely branched, glabrous or puberulous; leaves alternate, petioled, amplexicaul, oblong or cordate or hastate, 2.5–7.5 cm, irregularly toothed; heads few, corymbose, 0.6 cm in diameter, bracts multiseriate, narrow, often rigid; florets small, golden yellow; achenes minute, glabrous, pappus white.

Distribution : Kumaon, Sikkim—ascending up to 2000 m, Bihar, Golkunda Hills; Bhutan, Bangladesh, Myanmar.

Ecology and cultivation : Wild.

Boehmeria macrophylla Horn. (Urticaceae)

Syn : *Boehmeria platyphylla* D. Don

Vernacular names : Gar : Khaksha; **Him** : Samrali; **Hin :** Gargela; **Mun :** Najom jojo ara.

Trade name : Gargela

Traditional use : *TRIBES OF KANGRA VALLEY* (Himachal Pradesh) : *Leaf* (ground with black pepper) : in all types of eczema; *MUNDA* : *Root* : to stop vomiting.

Phytography : Evergreen, spreading, loose shrub, 2.0–3.30 m high; branches soft, glabrous; leaves mostly opposite, stipules usually free, petioles 2.5–12.5 cm, lamina broadly ovate or elliptic or orbicular, 7.5–22.5 cm, large toothed; monoecious or dioecious, flower clusters pisiform; fruit compressed, narrowed or beaked at the 4-teethed tip.

Phenology : *Flowering* : Rainy and Cold seasons; *Fruiting* : Cold and Hot seasons.

Distribution : Throughout India, ascending up to 1500 m.

Ecology and cultivation : Occurs in moist or shaded localities; wild.

Chemical contents : *Plant* : 3,4-di-OMe-w-(2'-piperidyl) acetophenone, cryptoleurine, secophenanthroquinolizidine; *Leaf* and *Twig* : oleanolic acid, pomolic acid, lupeol, 5α-stigmastane-3, 6-dione and β-sitosterol.

Boehmeria repens Nees & Mart. (Urticaceae)

Vernacular names : Asura (of Bihar) **:** Kokor polaba; **Mun :** Kokorpota ba.

Traditional use : ASURA : *Root* : in gout, paral, rheumatism; MUNDA : *Whole plant* : apply externally on swollen feet.

Phytography : Slender, much-branched, diffuse herb; branches 60–100 cm long; leaves alternate, petiolate, cordate; spikes long, monoecious or dioecious; flowers pink; fruits compressed.

Distribution : Bihar.

Ecology and cultivation : Wild.

Remarks : Plant of limited distribution; less studied taxonomically, phytochemically and pharmacologically.

Boehmeria rotundifolia Don (Urticaceae)

Syn : *Boehmeria caudigera* Wedd.

Modern use : *EtOH* (50%) *extract of aerial parts* : diuretic.

Phytography : Shrubby with soft glabrous branches; leaves alternate, petioled, broadly elliptic or ovate, tomentose pubescent or hispid—rarely smooth, large toothed–often confined to the upper parts; monoecious or dioecious, spikes usually longer than leaves; fruits compressed or angled.

Distribution : Sikkim, Meghalaya, Konkan; Sri Lanka.

Ecology and cultivation : Wild.

Remarks : Species of discontinuous distribution; deserves to be studied ecologically, phytochemically and should be conserved.

Boenninghausenia albiflora (Hk.) Reichb. ex Meissn. (Rutaceae)

Syn : *Ruta albiflora* Hook.

Vernacular names : Him : Pissumar; **Kum :** Upaniya jhar; **Mon** (of Arunachal Pradesh) **:** Ymari; **Sik :** Nukmma.

Traditional use : *TRIBES OF HIMACHAL PRADESH* : *Aerial Parts* : to kill lice, flee and other insects; *MONPA* : *Root* : paste applied to worm-infested wounds; *TRIBES OF WEST SIKKIM* : *Root* : in toothache.

Modern use : *Plant* : mild insecticide and insect repellent; *EtOH* (50%) *extract of plant* : hypoglycaemic and spasmogenic; *Essential oil from plant* and *leaf* : antibacterial, antifungal and dog-flea repellent.

Phytography : Perennial, slender erect branched herb, 33–66 cm high; stem glabrous or slightly pubescent; leaves alternate, petioles slender, lamina bi- or tri-pinnate, leaflets 0.6–1.8 cm, terminal one largest, petiolulate, obovate or obcordate, membranous, glaucous beneath; cymes terminal, panicled, leafy, many-flowered; flowers white, 0.6–1.25 cm in diameter, inclined or nodding; capsules 0.6–0.8 cm in diameter, few-seeded; seeds reniform, black, albumen fleshy.

Distribution : Kashmir to Arunachal Pradesh, Meghalaya, Mizoram—between 660 and 2600 m.

Ecology and cultivation : Plant of temperate and cold climate; wild.

Chemical contents : *Root* : 1-OH-N-Me-acridone, rutamarine, rutacridone, noracrony-cine, 1-OH-acridone, 1,7-di-OH-N-Ne-acridone; *Stem* : bergapten, xanthyletin, xanthotoxin, angenomalin, micropubscin, chalepin-OAc, octacosanoic acid, β-sitosterol and essential oil; *Leaf* : six coumarins, two isomeric coumarins, chalepensin and essential oil.

Remark : Roots and leaves are used as antiperiodic in China.

Boerhaavia chinensis (L.) Asch. & Schw. (Nyctaginaceae)

Syn : *Boerhaavia repanda* Willd.

Vernacular name : Mun : Piri kecho ara.

Traditional use : *ORAON* : *Root* : to ripen ulcers.

Modern use : *Root* : antileucorrhoea, antispermatorrhoea, anthelmintic.

Phytography : Diffusely branched herb; branches red, subscandent, glabrous or pubescent, 1–2 m long; leaves opposite, petioles slender, 1.25–3.7 cm long, lamina triangular or ovate or cordate, 2.5–7.5 cm long; umbels axillary and terminal, peduncled, peduncles 2.5–5 .0 cm, slender, 2- to 6-flowered; flowers pedicellate, pink, tubular; fruits covered with large glands, obscurely ribbed.

Phenology : *Flowering* and *Fruiting* : August–April.

Distribution : Throughout India as a common weed.

Ecology and cultivation : Wild.

Chemical contents : *Root* : alkaloids (0.1–0.15%)

Boerhaavia diffusa L. (Nyctaginaceae)

(2n = 26)

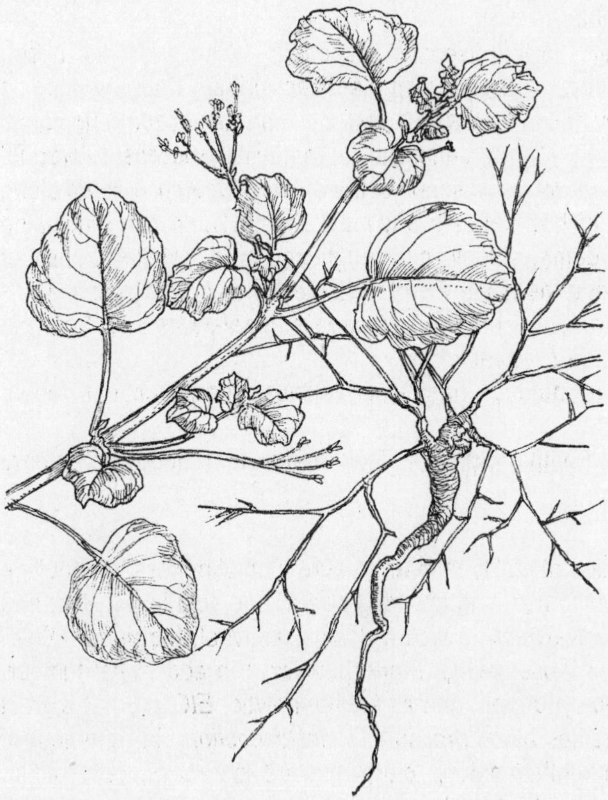

Syn : *Boerhaavia repens* L.

English name : Spreading hog weed.

Sanskrit names : Punarnavaa, Sathaagni.

Vernacular names : Asm : Pononua; **Ben :** Punarnaba, Purna; **Guj :** Ghetuli, Vakha khaparo; **Hin :** Biskhafra, Gadhapurva, Santh, Thikri; **Kan :** Balvadike, Gonajali, Rakta punarnava; **Kon :** Punarnavo; **Mal :** Tavalama, Thazhuthama; **Mar :** Punarnava, Vasu; **Mun :** Kecho ara; **Ori :** Ghodapuruni, Punarnava; **Orn :** Khapra arxa; **Pun :** Biskhapra, Itsit; **San :** Ohio-arak; **Tam :** Mukaratte-kirei, Kukkaratai, Uca; **Tel :** Punarnaba.

Traditional use : TRIBES OF INDIAN DESERT : *Root* : in eye complaints; TRIBES OF EASTERN RAJASTHAN : *Root* : in jaundice, *Leaf* : in child birth; RURAL PEOPLE OF GUJARAT : *Whole plant* : in liver complaints;

RURAL PEOPLE OF RATAN MAHAL HILLS (Gujarat) : *Root* : in fever; *VASAVA* (of Gujarat) : *Leaf*: in eye complaints; *TRIBES OF DAHANU FOREST* (Maharashtra) : *Leaf*: as blood purifier, in child birth, eye complaints; *RURAL PEOPLE OF KHED TALUKA* (Maharashtra) : *Root* : as blood purifier, in rheumatism; *BHIL*: *Root*: in dysentery; *TRIBES OF SAGAR DISTRICT* (Madhya Pradesh) : *Root*: in dropsy, gonorrhoea, heart diseases, kidney troubles, *Leaf*: in jaundice and other liver complaints; *RURAL PEOPLE OF DHOOMANOOR* and *CHAMPUKARAI* (Tamil Nadu) : *Root*: in wounds; *TRIBES OF ORISSA* : *Root* : in body heat, *Leaf* : in jaundice; *SANTAL* : *Plant*: in epilepsy, fistula, pain in abdomen, *Root* (pasted with 9 black peppers) : as laxative, *Seedpaste* : to pregnant ladies to hasten delivery; *TOTO* (of West Bengal) : *Root*: in jaundice; *RURAL PEOPLE OF PURULIA DISTRICT* (West Bengal) : *Root* : in stomach troubles; *KHASI* and *GARO* : *Leaf* : in rheumatism; *GARHWALI* : *Root* : in cough, eye complaints, heart diseases, piles, rheumatism, *Plant* : in liver troubles; *THARU* (of Uttar Pradesh) : *Leaf* : in eczema, itch; *RURAL PEOPLE OF DEHRA DUN* and *SIWALIK* (Uttar Pradesh) : *Root*: in asthma, jaundice; *RURAL FOLK OF DHARCHULA BLOCK* (Uttar Pradesh) : *Root* : in asthma, inflammation, urinary complaints; *RURAL FOLK OF DELHI* : *Plant* : in anaemia, heart disease; *RURAL PEOPLE OF KURUKSHETRA* (Haryana) : *Plant* : in cold, headache and menstrual complaints.

RIG VEDA : it is perennial, useful(?) in the diseases caused by water and *pitta* (biles); *SUSHRUTA SAMHITA* : useful in oedema, cough, skin diseases, urinary troubles, cardiac ailments, poisoning and in rejuvination; *MATSYA PURANA* : *Plant-decoction* : universal antidote of poison; *AGNI PURANA* : *Plant* : beneficial in removing abscess (*vidradhi*); *AYURVEDA* : *Root*: to remove thread worms, in sprain of the ankle joint; *Root* (pasted along with treated seeds of *Datura*) : in bites of rats and mad dogs; *Whole plant* including *root* : in urticaria, insomnia, oedema, calculi, old cough and cold, swelling, bites of honeybee; *Leaf*: in spermatorrhoea, oedema; *Flower* and *Seed* : as contraceptive.

SIDDHA : *Entire plant* : in preparation of the drug *Mukkiratai*; *Leaf-paste* and *Root* : in preparation of the drug *Caranai* and *Caranai ver*

UNANI : diuretic, effective in asthma, jaundice, dropsical swellings, worm affections, febrile affections and urethritis.

HOMOEOPATHY : beneficial for the treatment of head, liver, hands, feet, abdomen, urinary tract, cough, heart, etc.

Modern use : *Root* : for restoration of virility in man, to cure nightblindness, its poultice mixed with palm oil applied to boils, useful in lowering serum uric acid level; *Root* and *Leaf-ash* : in nightblindness; *Root-extract* : anticonvulsions, antiviral, febrifuge; *EtOH* (50%) *extract of root* and *plant* : spasmolytic; *Plant* (powder) : in abdominal tumour, cancer, renal diseases, nephrotic syndrome, and as antifibrinolytic; *EtOH* (50%) *extract of plant* : cardiotonic, relaxant, raises blood pressure; *Plant-decoction* : in treatment of nodules in the body; *Leaf* : anti-inflammatory, diuretic, used in dropsy, ophthalmia, muscular pain, gonorrhoea, to purify blood; one of the ingredients of the following drugs : (a) *Biliarin*—used in liver disorders, (b) *Styplon*—used in prevention of excessive bleeding in tonsillectomy.

Phytography : Diffuse herb; root fusiform, stout; rootstock woody, aerial branches reddish, 15–60 cm, glabrous or pubescent; leaves opposite, petiole as long as the blade or little shorter, simple, 1.25–5.0 cm, ovate or oblong or rounded, thick, usually green above and white beneath and borders red; small heads on axillary or terminal panicles, panicles with very slender branches; flowers minute, red; fruits 0.3 cm long, 5-ribbed, viscid, top rounded.

Phenology : *Flowering* and *Fruiting* : Rainy season and Winter.

Distribution : Throughout India, ascending the hot Himalayan valleys up to 2300 m; Bangladesh.

Ecology and cultivation : Common in grassy wet places; wild.

Chemical contents : *Plant* : β-sitosterol, punarnavine-1 & 2, myricyl alcohol, myristic acid, oxalic acid, punarnavoside, β-ecdysone, boeraavinones A, B and C, triacontanol, 5,7-dihydroxy-3',4'-dimethoxy-6, 8-dimethyl flavone; *Root* : hypoxanthine-9-*L*-arabinofuraxoside, hentriacontane, ursolic acid, β-sitosterol, aspartic acid, glutamic acid.

Remarks : Leaves are eaten as vegetables by Mundas, Irulars (of Tamil Nadu), tribes of Orissa, Mirzapur (Uttar Pradesh), Ratan Mahal Hills (Gujarat), Sagar (Madhya Pradesh), Purulia (West Bengal) and rural folk of many other parts of India.
Plant decoction, cooked with shea butter, is drunk in Ghana to relieve asthma. In Samoa, plant is used in dropsy.

Boerhaavia rubicunda Steud. (Nyctaginaceae)

Syn : *Boerhaavia elegans* Choisy

Traditional use : TRIBES OF RAJASTHAN : *Leaf* : as diuretic.

Phytography : Herb; rootstock woody, much divided, aerial branches erect, 33–66 cm, glaucous below; leaves opposite, sessile, few, 2.5–3.7 cm, linear or lanceolate or oblong, green, mottled with white when dry; cymes very large, excessively di- or trichotomously branched, peduncles and pedicels very long, capillary; flowers minute; fruits 0.25 cm long, narrowly ellipsoid, narrowed at both ends, ribbed, minute hairs present between ribs.

Distribution : Punjab, Rajasthan; Pakistan (Sind, Baluchistan).

Ecology and cultivation : Xerophyte; wild.

Remarks : Plant of limited distribution; deserves protection and cultivation.

Bombax ceiba L. (Bombacaceae)

(2n = 72, 92, 96)

Syn : *Bombax malabaricum* DC., *Salmalia malabarica* Schott & Endl.

English name : Red silk-cotton tree.

Sanskrit name : Shalmala (or -ee).

Vernacular names : Asm : Simalu; **Ben :** Shimool; **Guj :** Ratoshemalo, Shimlo; **Hin** and **Pun :** Shemal, Shimbal; **Kan :** Booruga, Gurla, Hatti, Kempubooruga; **Kon :** Kapxin; **Mal :** Mullilavu, Pula; **Mar :** Kate Savar, Shemuel, Shembal; **Mun :** Edel daru; **Ori :** Shimuli; **Orn :** Shimbali; **San :** Shimbar, Shimal-dare; **Tam :** Ilavam, Ilavu; **Tel:** Buraga, Konduburaga, Shalmali.

Trade name : Shimool.

Traditional use : MIKIR : *Root* : in cough, night pollution, urinary complaints; RURAL FOLK OF ASSAM : (i) *Root*: as aphrodisiac, in urinary complaints; (ii) *Gum* : as aphrodisiac; (iii) *Root* and *Gum together* : as tonic; (iv) *Leaf* : in infertility; TRIBES OF TIRAP DISTRICT (Arunachal Pradesh) : *Stem* : in night pollution, gonorrhoea, syphilis; TRIBES OF JALPAIGURI DISTRICT (West Bengal) : *Bark*: to restore vitality, in treatment of bone fracture; SANTAL : (i) *Seedling* : in low fever, continuous fever, rheumatism, pimples, chronic progredient sore, spermatorrhoea, haematuria; (ii) *Root of seedling/sapling* : in fistula, menorrhagia, body-pain during pregnancy, spider licking; (iii) *Fresh shoot* : in enlarged spleen, sores in legs (in connection with oedema), clavus, corns, syphilis, ulcer of the palate, rheumatoid ulcer, leprosy, as antidote of snake bite; (iv) *Stem-bark* : in cholera, intercostal neuralgia, pleurisy, pneumonia, stings of scorpions, centipedes, spiders, etc; (v) *Gum* : in scabies; (vi) *Knots on stem* : in bleeding gums, toothache and caries, sores in mouth, small pox; (vii) *Pith of stem* : in osteomyelitis, atrophy, emaciation or cachexy; (viii) *Prickles* : in headache; (ix) *Flower* (paste) : on boils for suppuration; ORAON : *Bark* and *Gum* : as astringent in diarrhoea; RURAL PEOPLE OF RANCHI and HAZARIBAGH DISTRICTS(Bihar) : *Root* : in abdominal pain, boils, dysentery; ASUR (of Bihar) : (i) *Bark* : in urinary complaints; (ii) *Leaf* : in anaemia; TRIBES OF ORISSA : (i) *Root* : in burns; (ii) *Seed* : in chicken pox, small pox; BHOXA (of Uttar Pradesh) : *Fruit* : in dysentery; GARHWALI: (i) *Root*: in diarrhoea, dysentery, impotency; (ii) *Stem-bark*: in skin diseases; (iii) *Gum* : in asthma; (iv) *Fruit* : in snake bite; RURAL PEOPLE OF DEHRA DUN and SHIWALIK : (i) *Root* : as tonic; (ii) *Stem-bark* : in dysentery; (iii) *Gum* : in dyarrhoea, dysentery; BHIL : *Flower*: in sex weakness; TRIBES OF DAHANU FOREST (Maharashtra) : (ii) *Root* : as tonic, in impotency; (ii) *Gum* : in debility; TRIBES OF KHANDALA (Maharashtra) : *Bark* : in dysentery, sex weakness; RURAL FOLK OF CHANDRAPURA DISTRICT (Maharashtra) : *Bark* : in blood dysentery; TRIBES OF MOUNT ABU (Rajasthan) : *Bark* : as tonic.

CHARAKA and SUSHRUTA have mentioned the usefulness of this plant; AYURVEDA : *Root of sapling* : in insufficiency of sperm, sex weakness, gonorrhoea; *Root-bark* : in menorrhagia; *Stem-bark* (paste) : in boils; *Gum* : in diarrhoea, dysentery, eczema, emaciation, old cough; *Petiole* (paste) : in dysentery, dehydration; *Leaf-blade* : in swelling of glands; *Dry flower* (powdered) : constipation in children, haemorrhagic disease (*raktapitta*); menstrual complaints, troubles in spleen; BRAHMAVAIVARTA PURANA : prohibits the use of this plant in brushing the teeth.

Modern use : *Seed* (hot aqueous extract) : moderate oxytocic.

Phytography : Tall tree, growing up to 45 m in height and 12 m in girth; stem tall, covered with sharp conical prickles, base buttressed, branches horizontal and in whorls of 5–7; leaves petiolate, palmately compound with 5–7 leaflets, leaflets petiolulate, lanceolate, 10–20 cm; flowers 10–12.5 cm across, red, stamens monadelphous; capsules oblong, hard, woody, 10–12 cm long, dehiscent, many-seeded; seeds glabrous, covered with silky wool.

Phenology : *Flowering* : January–May; *Fruiting* : March–May and December–March.

Distribution : Native of India and Myanmar, found throughout the plains of India.

Ecology and cultivation : Mesophyte; wild, some efforts have started to bring it under cultivation.

Chemical contents : *Root* : triacontanol, β-sitosterol and a glyceride; *Root-bark* : a naphthoquinone compound, lupeol, β-sitosterol, lactone, phenolics, four sesquiterpenes; *Stem-bark* : β-sitosterol, mucilage, lupeol; *Fresh petals* : essential oil, hentriacontane, hentriacontanol, kaempferol, quercetin, β-sitosterol and its glucoside, glucosides of pelargonidin and cyanidin; *Seed* : hexacosanol, tocopherol, terpenes.

Remarks : In Kwangton and Taiwan, plant is used in dysentery, enteritis, hepatitis, lymphoadenoma, menorrhagia.
Mundas and Oraons eat young root, calyx and flower as vegetables.

Borassus flabellifer L. (Arecaceae)

(2n = 36)

Syn : *Borassus flabelliformis* Murr., *B. dichotomous* White

English names : Barb tree, Palmyra palm, Toddy palm.

Sanskrit name : Trina raaj.

Vernacular names : **Asm** and **Ben** : Taal; **Guj :** Tad; **Hin :** Tari; **Kan** : Tale mara, Tati nungu; **Kon** : Tadmadd; **Mal :** Karimp-ana, Pana, Talam; **Mar** : Tada; **Mun** : Rola daru; **Ori** : Tala; **Pun :** Tar; **San** : Tar, Tardare; **Tam** : Talam, Panai; **Tel** : Tali, Thadi; **Urd :** Tad.

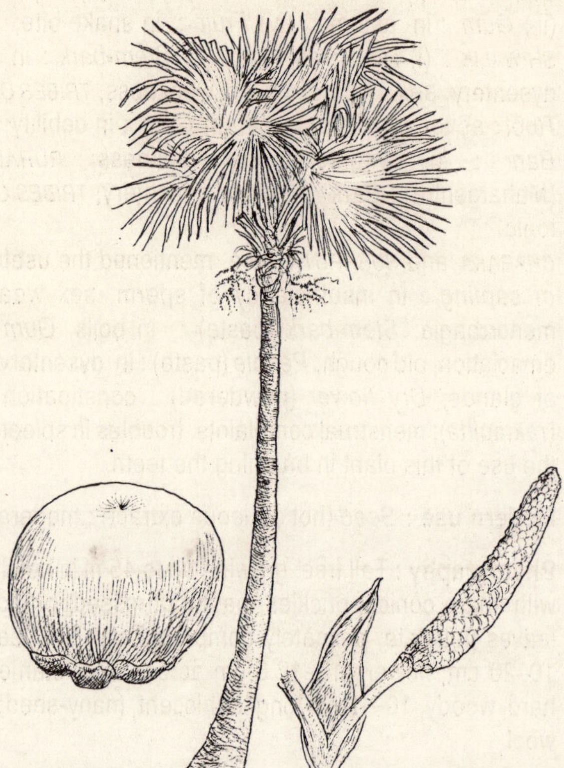

Trade name : Taal.

Traditional use : BHIL : *Wood-pulp* : as abortifacient; SANTAL : (i) *Root of seedling* : in spider-lick; (ii) *Bulb of seedling* : in epilepsy, haemiplagia, (iii) *Pith* : in menorrhagia; (iv) *Juice of leaf-stalk* : in earache, headache, heat stroke, nausea, vomiting; (v) *Male flower* : in adenitis fulosa colli, burns, clavus, corns, sores, syphilis, perforating ulcer of palate, (vi) *Ash* (obtained by burning *male inflorescence*) along with powder of 21 black peppers and fresh cow milk (2:1:1) : as contraceptive; TRIBES OF DAHSAN VALLEY : *Inflorescence* : in rheumatism;
ATHARVA VEDA : Sap of the plant (*Sura*) gives strength and energy; SUSHRUTA SAMHITA : *sap of the plant*, *kernel of the fruit*, *young leaf* and *root* have medicinal properties, *paste of young root* is diuretic; CHARAKA SAMHITA : *Kernel of young fruit* (boiled in *ghee*) : useful in urinary troubles; AGNI PURANA : oil obtained from this plant ensures luxurious growth of hair and makes the hair black; BRAHMAVAIVARTA PURANA : consumption of ripe fruit helps to eliminate bile (*pitta*) and wind (*vayu*), but drinking of water immediately after eating the fruit gives rise to deadly bile; AYURVEDA : *Sap of the plant* : cooling, strengthening, refreshing, excitant and diuretic; *Fruit-kernel* : nutritive, diuretic, useful in skin diseases; *Toddy-poultice* (made by powder of kernel and rice) : heals carbuncle, gangrenous

wound; *Ash of male inflorescence* : in acidity.
SIDDHA : *palmyra jaggery/palmyra sugar candy* : in preparation of drug *Panai.*

Modern use : *EtOH* (50%) *extract of fruit* : diuretic; *Fruit-jaggery* (*gur*) : energy providing food for convalescents, laxative, useful in anaemia; *Gur-solution* : excellent food for typhoid patients, laxative; *Sugar* : antibilous, alterative; *Palm -candy* : in coughs, pulmonary affections, laxative for children.

Phytography : Tall, dioecious palm, attaining a height of 20–23 m and a girth of 1–2 m; trunk stout, unarmed, very rarely branching, often swollen above the middle; 30–40 leaves form a crown at the top; leaves alternate (spiral), petioles 0.6–1.2 m long, wide, armed with spines at margins, ligules short, leaf blade palmately fan-shaped, plicately multifid, 2–3 m in diameter, segments 60–120 cm, linear, bifid, margins spinulose; spadix of male and female florets more than 1 m long, very stout, peduncles sheathed with open spathes, female spadix sparingly branched; male florets small, mixed with scaly bracts, female florets 2.5 cm in diameter, globose, perianth fleshy, greatly accrescent; fruits large, 20 cm in diameter, subglobose, drupe with 1–3 obcordate fibrous pyrenes, brown when young, black when mature; seeds oblong, top trilobed, testa adherent to the pyrene, consists of a soft, sweet, gelatinous pulp with a little liquid in them.

Phenology : *Flowering* : March–May; *Fruiting* : May–September.

Distribution : Native of tropical Africa; found in many of the comparatively dry zones along the coasts of Deccan peninsula, Orissa, Bihar, West Bengal; Bangladesh, Myanmar, Sri Lanka.

Ecology and cultivation : Grows well in sea coast, also occurs in fields and barren lands; wild and cultivated.
Plants develop from seeds. The seedlings are first planted in nurseries in bamboo or earthen pots. After sprouting, they are transplanted to pits. The distance between the pits should be from 3.5 to 5.5 m depending upon the nature of the soil. Nearly 1000 saplings can be sown in an acre of land. In the earlier stages, the underground portion of the stem grows in thickness. Usually, more than 15 years are required for proper growth and development of the aerial portion to assume the characteristic features.

Chemical contents : *Plant* : Sucrose, butyric acid.

Remarks : Santals use crushed leafstalk in haemorrhagic septicaemia of animals.
Fruits, seeds, *nira* (the fresh sap), toddy (fermented nira), palm-jaggery and sugar are consumed by Irular, Santal and modernised people.
The plant is used as a special symbol in sculpture on Vishnu temples.
In Kampuchea, flowers and roots are used as sinapism in tumours of uterus. All parts of the plant are used in diverse ways.

Borreria articularis (L.f) Will. (Rubiaceae)

(2n = 56)

Syn : *Borreria hispida* (L.) Schum., *Spermacoce hispida* L.

Vernacular names : Hin : Madanaghanti; **Mun :** Hada pota; **Orn** : Satigithia; **San :** Pitua arak, Kucha.

Traditional use : GARO : *Leaf* : in stomach complaints; PEOPLE OF DARJEELING HILLS (West Bengal) : (i) *Root* : as appetiser; (ii) *Leaf* : in toothache; (iii) *Seed* : as stimulant; SANTAL : (i) *Plant* : in bodyache, clavus, corns, dysentery, local acute swelling (*Bonga Khoda*), spleen complaints; (ii) *Root* : antidote to scorpion stings and snake bite, in blindness, conjunctivitis, gum troubles, otis media suppurative, toothache; (iii) *Leaf* :

conjunctivitis, earache, inflammation of eyes, intermittent fever; *MUNDA* : *Leaf* : in diarrhoea; *PEOPLE OF ORISSA* : *Leaf* : in eye troubles; *TRIBES OF KHEDBRAHMA* (Gujarat) : *Leaf* : in scabies, skin diseases, ulcers, wounds.

Modern use : *EtOH* (50%) *extract of plant* : diuretic.

Phytography : Procumbent herb, scabrid hirsute or hispid; branches stout, usually square, 15–35 cm, ascending; leaves opposite, stipules connate with petioles into broad tube, margins bristled, leafblades 1.25–3.7 cm by 0.8–1.6 cm, obovate or spathulate or elliptic, margins flat or wavy, rigidly coriaceous, pale when dry, very scabried above; flowers 4–6 in a whorl, blue or white, 0.6–1.25 cm, bracteate; capsules hispid or pubescent, rounded at the base; seeds very variable, narrow or broad.

Phenology : *Flowering* : July–October; *Fruiting* : September–December.

Distribution : Throughout India, from Meghalaya and Assam to Himachal Pradesh, ascending up to 1000 m; Sri Lanka.

Ecology and cultivation : Wild.

Chemical contents : *Plant* : β-sitosterol, ursolic acid, δ-mannitol; *Seed* : isorhamnetin.

Remarks : Leaves are eaten as vegetable by Mundas and tribes of Andhra Pradesh. Natives of Manbhum and Hazaribagh Districts of Bihar consume the fruits.

Boswellia serrata Roxb. ex Colebr. (Burseraceae)

(2n = 44)

Syn : *Boswellia glabra* Roxb.

English names : Indian olibanum tree, White dammar.

Sanskrit name : Shallaki.

Vernacular names : Ben : Lubarn, Salai; **Gon :** Salai; **Hin :** Salai, Salga, Salhe; **Kon :** Chitta; **Mar :** Salai, Salphullai; **Mun :** Salga daru; **Orn :** Salhai mann; **San :** Salea, Salga, Salhai, Salai-dare; **Tam :** Gigilu, Kindrikam, Kungli, Madara; **Tel :** Anduga, Anduku, Parangi.

Trade name : Salai.

Traditional use : *SANTAL* : (i) *Stem-bark* (paste) : in epilepsy, leucoderma, pleurisy, as antidote of snake bite; (ii) *Resin* : in diabetes; *ASUR* : *Stem-bark* : as antiseptic; *ORAON* : *Bark* : in wounds; *TRIBES OF ABUJH-MARH* (Madhya Pradesh) : *Seed* : in snake bite; *TRIBES OF BASTAR* (Madhya Pradesh) : (i) *Stem-bark* : in snake bite; (ii) *Leaf* : in wounds;

TRIBES OF VARANASI (Uttar Pradesh) : (i) *Stem-bark* : in cold, cough; (ii) *Resin* : in stomach troubles; TRIBES OF MIRZAPUR (Uttar Pradesh) : *Stem-bark* : in eye troubles, toothache; TRIBES OF MOUNT ABU (Rajasthan) : *Stem-bark* : in diarrhoea, dysentery; TRIBES OF KHEDBRAHMA (Gujarat) : (i) *Stem-bark* : in cold, cough; (ii) *Resin* : in skin eruptions, sores; (iii) *Leaf* : in eye troubles; TRIBES OF SAURASHTRA (Gujarat) : *Resin* : in diarrhoea, dysentery, skin eruptions, sores; TRIBES OF EAST GODAVARI (Andhra Pradesh) : (i) *Resin* : in pulmonary diseases, sores; *(ii) Seed* : for growth of hairs.

MATSYA PURANA : *Paste of Plant* : steriliser of clothing, bed, furniture, coat of armour, ornaments, canopy, *camara* (fan made of furs and wools). K. L. Dey (1896–1984) states, "It is used externally in the form of ointment as a rubefacient and stimulant application to boils, carbuncles, etc. It has been occasionally given as an astringent and diaphoretic. It may be given in bronchitis, chronic laryngitis and bonchorrhoea".

UNANI : Ingredient of 'Kundwi'.

Modern use : *EtOH* (50%) *extract of root* : anticancer; *EtOH* (50%) *extract of stem* and *fruit* : hypoglycaemic; *Bark* (*EtOH extract*) : active against potato virus X; *Gum-resin* : anti-inflammatory, anti-atherosclerotic and anti-arthritic in animals; *Non-phenolic fraction of gum-resin* : marked sedative and analgesic in rats, produce marked and long lasting hypotensive in anaesthetised dogs; *Essential oil from gum-resin* : antibacterial and antifungal.

Phytography : Balsamiferous tree, nearly 7 m high; bark reddish or grey, peeling off in flakes; leaves alternate or nearly opposite, petiolate, imparipinnate with opposite sessile leaflets, very variable in size, shape and degree of pubescence, deciduous; racemes axillary or terminal, shorter than leaves; flowers small, white; drupes trigonous, containing three 1-seeded pyrenes; seeds pendulous, compressed.

Phenology : *Flowering* : January–March; *Fruiting* : May–June.

Distribution : Almost throughout India.

Ecology and cultivation : Very common in jungles; wild.

Chemical contents : *Bark* : β-sitosterol; *Gum-resin* : serratol, β-boswellic acid,-Ac-β-bosewellic acid, 11-Keto-boswellic acid, Ac-11-Keto-boswellic acid and some other acids.

Adulterant : Oleo gum-resin of this plant is often used as a substitute of *guggulu.*

Remarks : Tribes of Abujh-Marh use resin in ulcers of cattle.
Oraons and Santals hang branches and leaves in cattle- and poultry-sheds for keeping flies away; Oleogum-resin is used as incense.
Santals and other tribes of Hazaribagh District eat fruits and seeds.
Bhils consider it as auspicious and give wooden armlets to the brides.

Brachiaria mutica Stapf (Poaceae)

Syn : *Panicum muticum* Forssk.

English name : Water grass.

Vernacular names : Ben : Nardul; **Kan :** Mauritius hullu; **Mal :** Lombopullu; **Tam :** Niradipul.

Trade name : Nardul.

Modern use : *Grass* : cynogenic and slightly estrogenic.

Phytography : Stem ascending, 2.0–2.6 m high, stout, nodes softly bearded, rarely glabrescent; leaves erect, 15–25 cm by 1.25–1.6 cm, sheaths glabrous or hairy, ligule short, ciliate; panicles erect, deltoid, glabrous, 7.5–12.5 cm, rachis stiff, spikes alternate, rather distant, stout, simple or branched, lower ones 5–10 cm, upper gradually shorter, spikelets green or purplish, turgid.

Distribution : Naturalised in West Bengal, cultivated throughout India; Sri Lanka.

Ecology and cultivation : Grows in marshes.

Adulterant : Very similar to *Panicum molle.*

Brachyactis royeli Wend. (Asteraceae)

Syn : *Brachyactis umbrosa* Benth.

Vernacular name : Lad : Sathi.

Traditional use : LADAKHI : *Plant* : in rheumatism.

Phytography : Small, annual, glandular and villous herb, 10–20 cm high, branched from base, branches ascending; leaves alternate, petiolate, obovate, lobulate, 1.25–1.8 cm long; heads many, axillary, sessile or long peduncled, bluish or rosy, 0.8 cm in diameter, heterogamous; florets minute; achenes narrow, 0.2 cm long, sparingly silky, pappus scanty, reddish.

Distribution : Kashmir at 2200–4000 m; Tibet.

Ecology and cultivation : Plant of cold climate; wild.

Remarks : Plant of very limited distribution; deserves protection and cultivation.

Brassica alba (L.) Boiss. (Brassicaceae)

Syn : *Sinapis alba* L.

Sanskrit names : Gaura sarshapa, Shveta sarshapa.

Vernacular names : Ben : Sada sarisha, Shveta rai.

Trade names : Sveta rai, Sada sarisha.

Traditional use : SUSHRUTA SAMHITA : *Seed* (pasted along with seed of *Pongamia pinnata* and cow dung) : in *urustambha* (stiffness and swelling of legs); *Seed* (pasted with cow's urine) : in epilepsy and lunacy; *Seed-paste* : in pin-pricks and nervous disability, dyspepsia, indigestion, *vatarakta* (vitiated blood), insufficient menstrual flow, opium poisoning; AGNI PURANA : plant of great medicinal importance.

Phytography : Annual herb, 60 cm high; stem erect, usually hispid; leaves alternate, petiolate, pinnatisect, pubescent; flowers yellow, tetramerous, petals clawed; pods 2.5 cm long, stalked, cylindric, beak long and flattened, valves and base of the beak white with hispid hairs.

Phenology : *Flowering* and *Fruiting* : Winter.

Distribution : Cultivated in Punjab and Uttar Pradesh.

Ecology and cultivation : Cultivated.

Chemical contents : *Seed* : p-OH-benzyl isothiocyanate, sinapine, sinalbin, bound phenolic acids.

Remark : Leaves are eaten as spinach.

Brassica campestris L. (Brassicaceae)

(2n = 20,22,44)

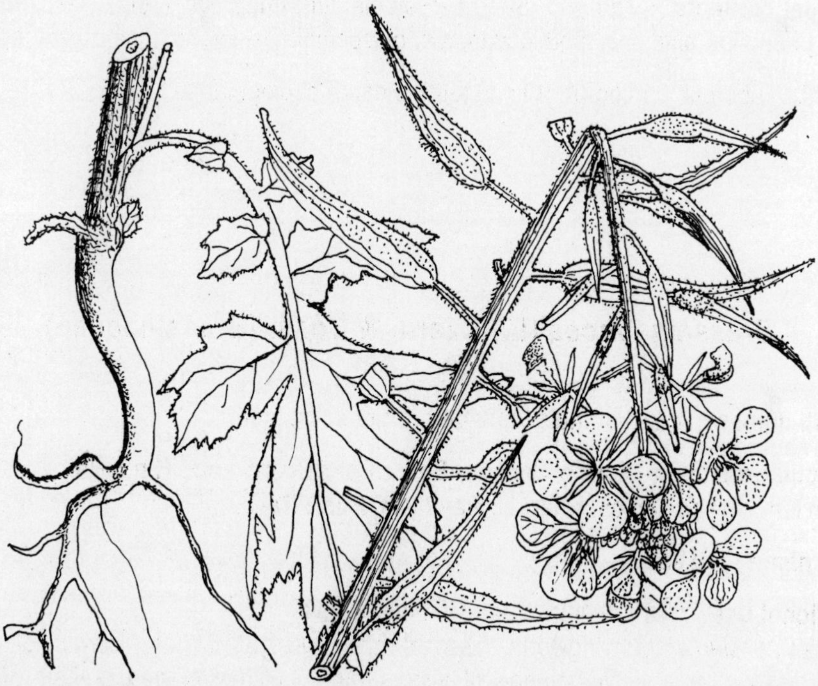

Syn : *Brassica rapa* L.

English names : Field mustard, Indian colza, Rape, Turnip.

Vernacular names : Ben : Shalgom; **Hin :** Chukandar.

Trade name : Chukandar.

Traditional use : *TRIBES OF MEGHALAYA AND ASSAM* : *Seed* : in muscular pain; *TRIBES OF TIRAP DISTRICT* (Arunachal Pradesh) : *Plant* : in migraine; *GARHWALI* : *Seed* : in skin diseases; *RURAL PEOPLE OF KASHMIR* : *Seed-oil* : in boils; *SANTAL* : *Turnip* (ground) : in bronchitis, chill, cold, leprosy, pneumonia; *Plant* : in tenacious cough; *Seed* : in scabies.

Modern use : *Stem* : antibacterial; *Fresh leaves* (ground) : applied to abscess; *Seed extract* : antifungal.

Phytography : Annual herb; root tuberous or fusiform; stem erect, stout, simple or branched, 30–100 cm high; leaves alternate, petioled, large, more or less pinnatifid; raceme or corymb; flowers yellow, pedicellate, tetramerous; siliquas 3.7–7.5 cm, glabrous; seeds small, pale or dark, smooth.

Distribution : Cultivated throughout India.

Ecology and cultivation : Winter-season crop.

Chemical contents : *Turnip* : S-Me-L-cysteine sulfoxide; *Stem* and *Leaf* : minerals; *Seed* : phenolics, sinapine, isothiocyanates, glucosinolates, cyano compounds, goitrin.

Remark : Tuberous portion (root and stem) is eaten as vegetable.

Brassica juncea (L.) Czern. & Coss. (Brassicaceae)

(2n = 36)

English names : Brown or Indian mustard, Chinese mustard.

Vernacular names : Asm : Sarisha; **Ben :** Rai sarisha, Lahi; **Hin :** Rai, Sarson rai, Sarson lahi, Gohna sarson, Khar rai, Rava lahi; **Pun :** Raya.

Trade name : Rai sarson.

Traditional use : *MANY TRIBES* : *Seed* : in boils, ringworm
CHARAKA SAMHITA : in ringworm, headache; *SUSHRUTA SAMHITA* : medicinal plant; *ATRISAMHITA* : removes imbalance of wind (*vayu*), bile (*pitta*) and phlegm (*kapha*), appetiser; *AYURVEDA* : *Seed-oil* : in filaria, wounds in gum, mumps, skin diseases; *Seed powder* : in rheumatism, sinusitis; *Seed-paste* : in menstrual complaints.

Phytography : Tall, erect, annual herb; stem 1–2 m high, slender, branched, usually glabrous; leaves alternate, petiolate, 15–30 cm, sometimes pinnatifid, upper ones lanceolate; racemes long; flowers bright yellow, tetramerous, bisexual, petals clawed; siliqua 1.25 cm, linear-lanceolate, beak straight, flattened; seeds small, dark, rugose.

Phenology : *Flowering* and *Fruiting* : Winter.

Distribution : Introduced from China into northeastern India; abundantly cultivated in Punjab, Rajasthan, Uttar Pradesh, Madhya Pradesh, Bihar, Orissa, West Bengal and Assam; Pakistan.

Ecology and cultivation : Winter crop; grows best in medium or heavy soils. May be cultivated as a single crop or as a mixed crop with wheat, barley and gram. Mixed cropping is common in Uttar Pradesh. In mixed cropping, seeds are usually grown in parallel rows, 1.2–1.8 m apart, alternating with the main crop. In single cropping, seeds are broadcast in the entire field. Broadcasting of seeds may be resorted to in mixed cropping too. The crop matures in about 110–160 days. It is harvested in February and March.

Chemical contents : *Seed* : two antithiamine compounds, flavonol glycerides, p-OH-benzoic acid.

Adulterants : Seeds of wild mustard or charlock [*Brassica kaber* (D.C.) Wheeler], common rape seeds (*Brassica napus* L.) and the seeds of rocket or *Eruca sativa* Lam are used as adulterants.

Brassica napus L. (Brassicaceae)

(2n = 38, 76)

English names : Colza, Indian rape.

Sanskrit name : Sarshapa.

Vernacular names : Asm : Sariha; **Ben :** Sarisha; **Guj :** Sarassava; **Hindi :** Sarson; **Kan :** Sasiva; **Mal :** Katuku; **Mar :** Rai sarson; **Tam :** Kadugu; **Tel :** Avelu.

Trade names : Sarson, Sarisha.

Modern use : *Rapeseed-oil* : nutritive, effects cardiac tissue, ingredient of a compound used for treating osteoarticular rheumatism and collagenic patients; *Bran-extract* : anticoagulant.

Phytography : Annual herb, 30–100 cm high; stem erect, stout, simple or branched, glabrous or slightly hispid; root fusiform; leaves alternate, petiolate, thick, bluish green, more or less pinnatifid, glaucous and glabrous; corymbs axillary; flowers large, bright yellow, tetramerous, petals clawed, deciduous before the corymbs lengthens; pods thin, 3.7–7.5 cm long torulose, in the early stage spread out on the fruit axis and ultimately droop down; seeds small, pale or dark.

Phenology : *Flowering* : Winter; *Fruiting* : April–May.

Distribution : Introduced from Japan into Punjab; now also cultivated in Uttar Pradesh.

Chemical contents : *Stem* and *Leaf* : barbarin, goitrin, napoleiferin; *Flower* : brassicoside; *Pollen* : brassinolide; *Rapeseed-oil* : triterpenoids, brassicasterol, tocopherol, brassidic acid, thioglucosides, 7-oxosterols, sterols, dihydrosterols, sinapine, polyphenols.

Remarks : Leaves are highly appreciated as sweet and delicious *shag* during winter and early summer; used as fodder for cattle in scarcity.

Brassica nigra (L.) Koch. (Brassicaceae)

(2n = 16)

Syn : *Sinapis nigra* L.

English names : Black mustard, True mustard.

Vernacular names : Hin : Banarasi rai, Kali rai; **Kan :** Karisaasiva; **Tam :** Sirukadagu; **Tribes of Araku Valley :** Avallumokka.

Traditional use : *TRIBES OF ARAKU VALLEY* (Andhra Pradesh) : *Seed-oil* : in rheumatism. *SIDDHA* : *Seed* : in preparation of the drug *Katuḳu*.

Modern use : *Seed-oil* : rubefacient, liniment, an ingredient of medicines of pleurisy and pneumonia.

Phytography : Annual herb, 60–200 cm high; stem rigid, much branched, more or less hispid; leaves alternate, petiolate, 10–20 cm long, covered with hairs, lower leaves lyrate while upper ones small and entire; corymbose raceme; flowers small, 0.8–1.2 cm in diameter, bright yellow, tetramerous, sepals spreading; pods terete or subtetragonal,

0.6–1.25 cm long, subulate, torulose, small beaked, valves keeled, 3- to 10- seeded; seeds small, oblong, mucilaginous.

Distribution : Common weed, also grown as a garden crop in Tamil Nadu, Andhra Pradesh, Uttar Pradesh, Punjab; Tibet.

Ecology and cultivation : Cold-climate crop; wild and cultivated.

Chemical content : *Plant* : glucosinoles.

Remarks : Indians use the seeds in pickles and curries. Black mustard is ground with white mustard for preparing table mustard and medicinal preparations like mustard bran and mustard flour.
In Europe, it is used for making lubricant, soap.
Plants, if consumed by livestock, may cause severe diarrhoea, gastroenteritis, irritation of mouth, pain and salivation in the animals.

Brassica oleracea L. var. *botrytis* L. sub. var. *cauliflora* DC. (Brassicaceae)

(2n = 18, 36, 72)

English name : Cauliflower.

Vernacular names : Asm : Phool kabi; **Ben :** Phool kapi, **Guj :** Phool Kobi,Phoovar; **Hin** : Phool gobi; **Kan :** Kukesu; **Mal :** Cauliflower; **Mar :** Phoolkobi; **Ori :** Phoola kobi; **Pun :** Phool gobhi; **Tam** : Gospoovu; **Tel** : Phoolgobi, Purugobi.

Trade names : Phool gobi, Phool kapi.

Modern use : *Leaf* : on blistered surfaces, in gout, rheumatism, piles, skin diseases; *Cauliflower* : anti-inflammatory, anti-tussive, cardiotonic, useful in biliousness and urinary diseases.

Phytography : Annual herb; stem stout, short; leaves alternate, petioles small, lamina large, whitish green, upper part inwardly curved, midrib white, thick strong; corymbose raceme terminal; flowers very small, white, pedicellate, form an almost continuous solid white or creamy mass of succulent tissue extending up to 30 cm in diameter; pods sessile, cylindric, beak either absent or conical.

Phenology : *Flowering* : Winter.

Distribution : Introduced form United Kingdom; cultivated in India; Bangladesh.

Ecology and cultivation : Plant of cold climate.
Plants are raised from seeds grown in nursery beds. The seedlings are transplanted when

4- to 6-weeks-old. The plant requires very heavy manuring and continuous protection from fungi and insects. It is better not to expose the inflorescence to direct sunlight, at least in the earlier stages. The crop matures in 12 weeks.

Chemical contents : *Root-peels* : 2-Ph-Et-isothiocyanate; *Green leaf* : calcium, phosphorus; *Inflorescence* (cauliflower) : S-Me-L-cysteine sulfoxide, vitamin U (S-Me-methiodine), folic acid.

Remark : Young leaf and cauliflower are relished as vegetables.

Brassica oleracea L. var. *capitata* Linn. f. (Brassicaceae)

(2n = 18, 36, 72)

English name : Cabbage.

Vernacular name : Asm : Bandha kabi; **Ben :** Bandha kapi; **Guj :** Kobij; **Hin :** Bandh gobhi, Pat gobi; **Kan :** Yelekosu; **Mal :** Mattakese; **Mar :** Kobi; **Ori :** Bandha kobi; **Pun :** Bandhagobhi; **Tam :** Muttaikose; **Tel :** Akugobi.

Trade name : Bandha gobi.

Modern use : *Cabbage* : in Pb and Hg poisoning; *Juice* : in ulcer of stomach and duodenum, atherosclerosis; *Seedling* (extract) : fungistatic.

Phytography : Grown as annual; stem stout, short, apex wide; leaves alternate, subpetiolate, simple, large, green, sometimes tinged with red pigments, margin wavy, overarches the apical bud forming a big ball-like structure; panicled raceme; flowers very small, yellowish white, tetramerous; pods sessile, subtetragonal, beak absent or conical.

Distribution : Introduced from Europe and cultivated in the hills and plains of India; Bangladesh and Sri Lanka.

Ecology and cultivation : Grows well in cool climate on sandy loam soil; may grow on clay or silt loam soil; does not grow well in highly acidic soil, the optimum pH is between 5.5 and 6.5.
The seeds are sown in raised nursery beds in August–September or in September–October. The seedling transplantation method is like cauliflower. It requires continuous supply of moisture.

Chemical contents : *Root* and *Leaf* : glucobrassicin; *Cabbage-head* : indole-3-acetonitrile, allyl isothiocyanate, vitamins A, B, B_2, B_6, C and U (S-Me-methionine), S-Me-L-cysteine sulfoxide, folate, mevalonic acid, polyphenols, minerals and goitrogens; *Seed* : glucosamine.

Remarks : One of the oldest cultivated vegetables; also used by the ancient Romans and Greeks.

Breynia retusa Alston (Euphorbiaceae)

(2n = 26)

Syn : *Breynia patens* Benth., *Phyllanthus patens* Roxb.

Vernacular names : Tot : Mimaising; **Ori :** Raktatrichuli.

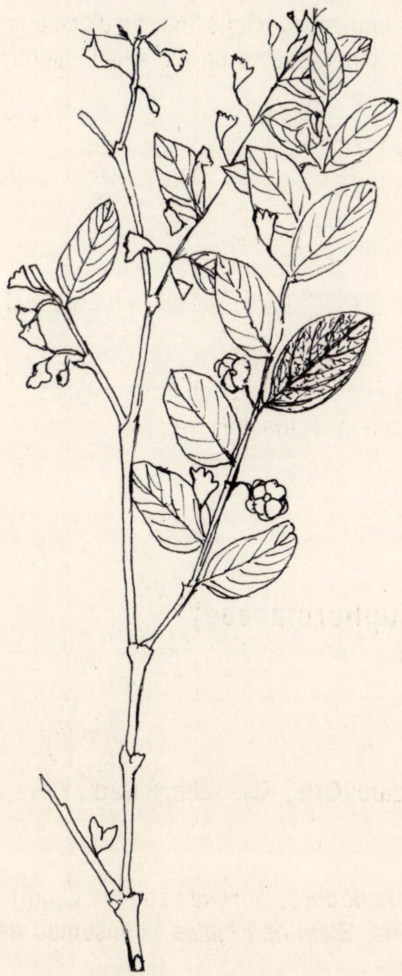

Traditional use : TOTO (of West Bengal) : *Plant* : in eye complaints; TRIBES OF ORISSA : *Bark-paste* : in inflammation; TRIBES OF MADHYA PRADESH : *Stem* : in eye complaints. AYURVEDA : this is an ingredient of the drug *Ayapon*, which checks bleeding after medical termination of pregnancy.

Modern use : *Plant* : in dysmenorrhoea, constitiuent of many medicines; *Plant-extract* (along with extract of *Leptadenia reticulata*) : effective galactogogue.

Phytography : Erect tree, 1–2 m high, branches spreading; leaves distichous, petiolate, stipules minute, lamina simple, 1.25–2.5 cm long, rather glaucous beneath, veins raised; flowers axillary, minute, monoecious, stalked, male flower 0.15–0.2 cm, drooping, female flower 0.25– 1.25 cm broad.

Distribution : Assam, Meghalaya, Manipur, Orissa, Deccan peninsula; Nepal, Bangladesh (Sylhet, Chittagong), Myanmar, Sri Lanka.

Ecology and cultivation : Plant of tropical rainforest; wild.

Remark : Indigenous drug, 'Aloes compound', useful for infertile brood-mares, is made by this plant.

Breynia vitis-idaea Burm. f. (Euphorbiaceae)

(2n = 56)

Syn : *Breynia rhamnoides* Muell. Arg.

English name : Indian snowberry.

Vernacular name : Mun : Huring rangphui.

Traditional use : *TRIBES OF CHOTANAGPUR* : *Plant* : in diarrhoea.

Modern use : *Plant* (petroleum extract) : poor sedative.

Phytography : Shrub or small tree, 1.3–1.6 m high; branches horizontal, flexuous, bifarious; leaves alternate,, stipules minute, petiole 0.25 cm long, lamina 1.2–1.5 cm long, membranous, elliptic or ovate, or resembling pinnate leaves, pale beneath, dark brown or black when dry; flowers on short stalks, 0.15–0.2 cm long, monoecious, may be dioecious; berries bright red, globose, 0.6 cm in diameter, succulent, seated on the small, slightly enlarged calyx; seeds 0.3 cm long.

Phenology : *Flowering* : March–December; *Fruiting* : March.

Distribution : Throughout tropical India; Sri Lanka.

Ecology and cultivation : Wild.

Chemical contents : *Leaf* : triacontane, penta- and triacontanic acids, ceryl alcohol, lanosterol and cholesterol.

Remarks : Tribes of Chotanagpur use fruits to colour combs and such other articles. In Philippines, decoction of root is used as mouthwash in toothache.

Bridelia airy-shawii Li (Euphorbiaceae)

(2n = 26, 28)

Syn : *Bridelia retusa* Spreng.

Vernacular names : Hin : Kaja; **Mun :** Karaka daru; **Orn :** Kiss xochol; **Sad :** Kasa.

Trade name : Kaja.

Traditional use : *TRIBES OF CHOTANAGPUR* : *Stem-bark decoction* (with country liquor) : to prevent pregnancy; *Stem-bark paste* : to wounds; *Stem-bark juice* : consumed as antidote to snake bite.

Modern use : *EtOH* (50%) *extract of stem-bark* : anticancer, antiviral and hypotensive.

Phytography : Deciduous tree, 17 – 20 m high, may be shrubby, conical prickles on the trunk; leaves alternate, petioles 0.6 – 1.2 cm long, lamina 7.5 – 12.5 cm, may be larger, glabrous above, glabrous or glaucous or finely pubescent beneath; flowers small, dioecious, greenish, axillary or in long spikes, bracts small, villous; fruits globose, size of a pea, purple-black, cocci dehiscing.

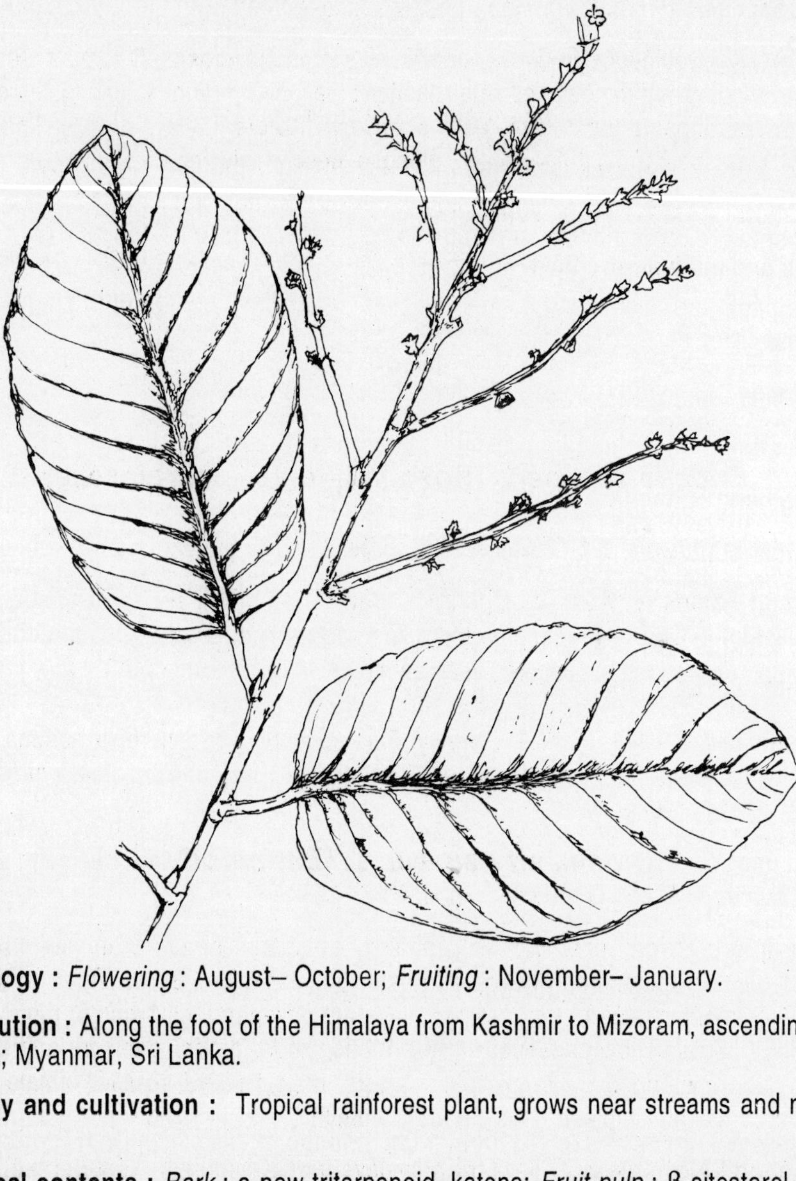

Phenology : *Flowering* : August– October; *Fruiting* : November– January.

Distribution : Along the foot of the Himalaya from Kashmir to Mizoram, ascending up to 1200 m; Myanmar, Sri Lanka.

Ecology and cultivation : Tropical rainforest plant, grows near streams and nullahs; wild.

Chemical contents : *Bark* : a new triterpenoid, ketone; *Fruit-pulp* : β-sitosterol, ellagic acid, gallic acid; *Seed* : ellagic acid.

Bridellia crenulata Roxb. (Euphorbiaceae)

(2n = 26,28)

Syn : *Bridelia retusa* Spreng. var. *roxburghiana*

Vernacular name : Ori : Kosigacha.

Traditional use : *TRIBES OF ORISSA* : *Stem-bark* : in antifertility.

Phytography : Deciduous tree, may be shrubby; branchlets densely tomentose; leaves alternate, stipules minute, petioles small, lamina small, membranous, oblong, tip rounded, lower surface densely tomentose, pale when dry; flowers axillary or in spikes, small, greenish, dioecious; fruits purple-black, globose, cocci dehiscing.

Distribution : Orissa, Deccan peninsula.

Ecology and cultivation : Wild.

Bridelia scandens (Roxb.) Willd. (Euphorbiaceae)

Syn : *Bridelia stipularis* Bl., *Cluytia scandens* Roxb.

Vernacular names : Asm : Mouhilika; **Ben :** Harinhara; **Hin :** Kangiabel; **Kan :** Bisalballi; **Mal :** Cheruka; **Mar :** Asana; **Ori :** Gaurkarsi; **Meg :** U-mei-tong-krong.

Trade names : Kangiabel, Harinhara.

Traditional use : *TRIBES*: *Plant*: in pleurisy; *Bark-decoction*: to children in asthma, cough, fever, and as gargle for sores in mouth; *Tender leaf* : in jaundice; *Leaf-emulsion* : in anaemia due to pregnancy.

Modern use : *EtOH* (50%) *extract of aerial parts* : hypotensive; *Seed*: haemagglutinating against human A, B and O red cells.

Phytography : Large, more or less scandent, evergreen shrub; branches straight or flexuous, usually tomentose; leaves alternate, petioles 0.6–0.8 cm, stout, tomentose, lamina 5–20 cm by 1.25–12.5 cm,elliptic, undulate or repand, glabrous or pendulous or puberulous above, subcoriaceous; flowers monoecious, green, often in long panicled spikes, often subtended by long stipular bracts, male flowers sessile, female flowers pedicelled; fruits bluish black, 1.25 cm long, smooth.

Distribution : Along the foothills of the Himalaya from Kashmir to West Bengal, Deccan peninsula; Bhutan.

Ecology and cultivation : Plant of tropical rainforest; wild.

Chemical contents : *Root* : taraxerone; *Bark* : β-sitosterol, friedelin.

Remark : Tribes of hilly regions of Maharashtra, Goa and Kerala (Cannanore District) eat the fruits.

Bridelia squamosa (Lamk.) Gaertn. (Euphorbiaceae)

(2n = 26, 28)

Syn : *Bridelia retusa* Juss.

Vernacular name : Hin : Khaja.

Trade name : Khaja.

Traditional use : *TRIBES OF ORISSA* : *Bark* : in antifertility; *Leaf and Fruit* : in stomachache; *TRIBES OF BIHAR* : *Bark* : in antifertility, tuberculosis; *TRIBES OF MADHYA PRADESH* : *Bark* : in antifertility, rheumatism; *TRIBES OF CHANDRAPURA DISTRICT* (Maharashtra) : *Fruit* : as laxative; *TRIBES OF DAHANU FOREST* (Maharashtra) : *Bark* : in debility, pain in waist, rheumatism; *TRIBES OF SALSETTE ISLAND* (near Bombay) : *Bark* : as tonic; *DANG* (of Gujarat) : *Bark* : in diarrhoea, earache, snake bite, wounds; *TRIBES OF BANSWARA DISTRICT* (Rajasthan) : *Bark* : in cuts; *TRIBES OF DEHRA DUN* and *SIWALIK* (Uttar Pradesh) : *Root* and *Bark* : as astringent, *Fruit* : as laxative; *TRIBES OF KANGRA VALLEY* (Himachal Pradesh) : *Bark* : in arthritis, rheumatism.

Modern use : *EtOH* (50%) *extract of aerial parts* : antiviral.

Phytography : Large, more or less scandent, evergreen shrub; leaves alternate, petiole 0.6–0.8 cm, simple, 5–20 cm by 1.2–12.0 cm; spikes long and panicled; flowers green, monoecious, male flowers sessile, female flowers pedicelled; fruits bluish black, ± 1.25 cm.

Distribution : Throughout the hotter parts of India, along the foothills of Himalaya from Kashmir to West Bengal, in the forests around Gorakhpur, Chotanagpur, Madhya Pradesh and the Western Ghats; Bhutan, Sri Lanka.

Ecology and cultivation : Plant of tropical rainforest; wild.

Remarks : Tribes of Santal Parganas use the bark in veterinary medicine and leaf as fodder. Tribes of Uttar Pradesh, Maharashtra and Goa eat the fruits.

Brucea mollis Wall. ex Kurz (Simaroubaceae)

Vernacular name : Asm : Koinine.

Trade name : Koinine.

Traditional use : MIKIR : *Fruit*: in malaria; *Seed*: in malaria, stomach complaints.

Modern use : *EtOH* (50%) *extract of aerial parts* : anticancer.

Phytography : Bitter shrub; leaves alternate, very large, unequally pinnate, leaflets 7.5 by 5.0 cm, pubescent to nearly glabrous; panicles axillary, rachis pubescent or nearly glabrous, secondary cymes often many-flowered; flowers very small, tetramerous; drupes ovoid, 0.6–0.8 cm, brown, glabrous, scarcely reticulated, 1-seeded; seed exalbuminous.

Distribution : Subtropical eastern Himalaya, Sikkim, Assam; Bhutan.

Ecology and cultivation : Subtropical plant, grows between an altitude of 1000 and 2000 m; wild.

Bruguiera gymnorrhiza (L.) Lamk. (Rhizophoraceae)

Syn : *Bruguiera eriopetala* Hensl., *B. conjugata* (L.) Merrill, *Rhizophora gymnorrhiza* Roxb.

Vernacular names : Ben : Kankara; **Ori :** Kekar; **Tam :** Stgapukokandan; **Tel :** Thuddaponna.

Trade name : Kankara.

Modern use : *EtOH* (50%) *extract of aerial parts* : hypotensive.

Phytography : Large, evergreen tree; leaves opposite with interpetiolar oblong deciduous stipules, petiolate, lamina 7.5–15.0 cm by 3.7–5.6 cm, oblong, coriaceous, slightly acuminate; flowers axillary, solitary, peduncles shorter than petioles, complete, about 2.5 cm in diameter, coriaceous, calyx and corrolla 10–14 cleft, ovary inferior; fruits coriaceous, indehiscent, crowned with calyx limb, 1-chambered, 1-seeded; seeds exalbuminous, germination viviparous.

Distribution : Littoral and especially estuarian forests of mainland of India and the Andamans; Sri Lanka.

Ecology and cultivation : Mangrove plant; wild.

Chemical contents : *Stem* and *Bark* : brugierol, isobrugierol; *Leaf* : an alcohol, gymnorhizol, sitosterol.

Remark : Tribes of the Andaman and Nicobar Islands eat the fruits.

Bruguiera sexangula Poir. (Rhizophoraceae)

Syn : *Bruguiera eriopetala* Wight & Arn. ex Am. Hensl.

Traditional use : *TRIBES OF DECCAN* : *Root* and *Leaf* : in burns.

Modern use : *Bark-extract* : antitumour against Sarcoma 180 and Lewis lung carcinoma.

Phytography : Tree; leaves opposite, stipules interpetiolar, petiolate, lamina coriaceous, ovate or oblong, acuminate at both ends; peduncles axillary and 1-flowered; flowers large, coriaceous, calyx about 10-cleft and about 2.5 cm long, petals with margins densely hirsute with silvery hairs from base to apex, ovary inferior, 3- to 4-chambered; fruits coriaceous, indehiscent; germination viviparous.

Distribution : Tidal shores of Orissa, Andhra Pradesh, Tamil Nadu, Kerala; Sri Lanka.

Ecology and cultivation : Mangrove plant; wild.

Chemical contents : *Stem-bark* : toxic alkaloids including brugine, tropine and its esters.

Buchanania axillaris (Desv.) Ramam. (Anacardiaceae)

Syn : *Buchanania angustifolia* Roxb., *Mangifera axillaris* Lamk.

English names : Bughanan's mango, Cuddapah almond.

Vernacular names : Hin : Piyala; **Kan :** Murligidda; **Mal :** Chaara paruppu; **Tam :** Saraparuppu; **Tel :** Pedda sara; **Tribes of Bastar :** Tolemorli.

Trade name : Piyala.

Traditional use : *TRIBAL* : *Plant-gum* (dissolved in goat milk, mixed with flour of *Eleusine coracana*, boiled to paste) : on joints to alleviate pain.
UNANI : An ingredient of 'Habb-us-Samena'.

Modern use : *EtOH* (50%) *extract of aerial parts* : Central Nervous System depressor.

Phytography : Tree; stem glabrous, branchlets stout; leaves alternate, petiole 1.8-3.0 cm, slender, simple, 10–15 cm by 3.7–5.0 cm, linear-oblong or elliptic or elliptic-lanceolate, thinly coriaceous, glabrous, upper surface shining, tip obtuse or rounded; panicles axillary and terminal, 10–15 cm long, peduncles and branches slender, spreading, glabrous; flowers sessile or nearly so, clustered, small, white, complete, apocarpous; drupes obliquely spherical, 1.25 cm in diameter.

Distribution : Madhya Pradesh, Konkan and southwards; northern part of Sri Lanka.

Ecology and cultivation : Common in dry deciduous forests and in open scrubs; wild.

Remarks : Tribes of Bastar eat the fruits.
Plant of limited area, deserves protection and cultivation.

Buchanania lanzan Spr. (Anacardiaceae)

Syn : *Buchanania latifolia* Roxb.

Sanskrit names : Priyala, Piyala, Rajavriksha.

Vernacular names : Ben : Piyal; **Gon :** Accr, Char, Erka, Morlichettu; **Hin :** Chironji, Kath-bhilawa; **Mun :** Char-daru, Tarob daru; **Orn :** Kitti; **Sad :** Piyar; **San :** Tarop.

Trade name : Chironjee.

Traditional use : TRIBES OF PURULIA DISTRICT (West Bengal) : *Root* : in venereal diseases; SANTAL : (i) *Root* : in bronchitis, consumption, dry cough; (ii) *Bark* : in burns, cuts, inflammation of gum, cholera, dysuria, snake bite, sting of scorpion and other insects, wounds; (iii) *Bark-juice* : in fever; (iv) *Leaf* : in burns; *MUNDA* : *Bark* : in wounds; *ORAON* : *Leaf* : in dysentery; *ASUR* : *Bark* : in snake bite; *TRIBES OF ABUJH-MARH AREA* (Madhya Pradesh) : (i) *Bark* : in skin diseases; (ii) *Leaf* : in wound; (iii) *Seed* : in pimples and as tonic; *TRIBES OF BASTAR* (Madhya Pradesh) : *Leaf* : in wounds; *TRIBES OF CHANDRAPUR A* (Maharashtra) : *Leaf* : in wounds; *VASAVA* (of Gujarat) : *Bark* : in snake bite; *TRIBES OF SIWALIK AND DEHRA DUN REGIONS* (Uttar Pradesh) : (i) *Gum* : in diarrhoea; (ii) *Leaf* : in wounds; *TRIBES OF KANGRA VALLEY* (Himachal Pradesh) : *Seed-oil* : in skin diseases. *MATSYA PURANA* : plant of great medicinal value; *AGNI PURANA* : *Plant-extract* (along with honey and some other plants : in urinary obstruction; *BRAHMAVAIVARTA PURANA* : teeth must not be brushed with twigs of this plant; *AYURVEDA* : medicinal plant. *SIDDHA* : *Kernel* in preparation of the drug called 'Caraiapparuppu'.

Modern use : *EtOH* (50%) *extract of aerial parts* : anticancer; *Leaf* : general tonic and cardiotonic; *Kernel-oil* : in skin diseases caused by defects in blood.

Phytography : Subdeciduous tree, 13–17 m high and up to 1.3 m in girth; bark thick, dark grey, divided into small rectangular plates resembling crocodile hide; leaves alternate, petiolate, simple, 15–25 cm in length, very coriaceous or hard; panicles axillary and terminal, pyramidal, branches stout; flowers sessile, greenish white, 0.6 cm in diameter, complete; drupes black, subglobose, slightly compressed, 1.25 cm long; stone hard, bi-valved, kernel has a pleasant sweetish acidic flavour.

Phenology : *Flowering* : January–March; *Fruiting* : April–May.

Distribution : Native of India, found throughout the hot and drier parts of India, foothills of Himalaya, ascending up to 500 m; Myanmar.

Ecology and cultivation : Common in forests; wild.

Chemical content : *Kernel* : rich in oil.

Remarks : Different tribes eat the fruits. Flesh is palatable and kernels are a good substitute for almonds; used for confectionery under the name Chironjee; seed-oil is used as *ghee*.

Buddleja asiatica Lour. (Buddlejaceae)

(2n = 30, 38)

Syn : *Buddleja neemda* Buch. Ham.

English name : White butterfly bush.

Vernacular name : Mon : Agia-chita.

Traditional use : *MONPA* (of Arunachal Pradesh) : *Leaf* : for fomentation on inflammation.

Modern use : *EtOH* (50%) *extract of leaf* : hypotensive; *Essential oil* from *bud* and *flower* : antibacterial.

Phytography : Evergreen shrub, usually 1–2 m high, rarely 5 m; leaves opposite, petioles 0.6 cm, stipules narrow, lamina 12.5 by 3.0 cm, lanceolate, usually narrowed at the base, entire or crenate or toothed, acuminate, lower surface grey or white, tomentose or pubescent; spikes axillary and terminal, often panicled, 5–20 cm long, grey, tomentose or pubescent; flowers urn-shaped, 0.4 cm, sweet-scented; capsule 0.6 cm, ellipsoid.

Phenology : *Flowering* : December–February; *Fruiting* : February– March.

Distribution : Throughout India, ascending up to 2300 m.

Ecology and cultivation : Wild.

Chemical contents : *Flower* : linarin, quercetin, essential oil.

Remark : Monpas prepare edible drink with leaves.

Bupleurum falcatum L. (Apiaceae)

(2n = 16, 28)

Syn : *Bupleurum imaicolum* A. Kerner

English name : Hare's ears.

Vernacular name : Mun : Bir yambiri.

Traditional use : MUNDA : *Plant* (crushed) : in headache; TRIBALS OF SIWALIK : *Root* : in abdominal inflammation, fever, liver complaints.

Modern use : *Root* : analgesic, antipyretic, antitussive, anti-inflammatory, haemolytic.

Phytography : Erect herb, 30–130 cm high, not much branched, corymbose upwards; leaves half-amplexicaul or sessile, long, acuminate; yellow flowers in umbels, bracts and bracteoles present; fruits brown, 0.4–0.5 cm, ridge distinct.

Phenology : *Flowering* and *Fruiting* : September–January.

Distribution : From Kashmir to Meghalaya between 1000 and 4000 m altitude, high hills of Chotanagpur; Bhutan.

Ecology and cultivation : Plant of cold climate; usually wild; efforts are on to cultivate.

Chemical contents : *Root* : saikosaponins, essential oil, α- and β-spinasterol, stigmasterol, stigmastenol, bupleurumol, ribitol, β-sitosterol, coumarins, polyacetylenic compounds; *Aerial parts* : saponins; *Leaf* and *Flower* : rutin.

Remarks : Chemical contents differ in the varieties of the species. Cultivation and breeding affect yield of chemicals.

The plant (known as *Chai-hu* in Chinese) is an important ingredient of various Chinese medicines meant to resolve tightness and resistance syndrome at the coastal margin that might be related to inflammation of diaphragm and enlargement of liver. Modern experiments extend support to the Chinese medicines.

Tribes of North-East India eat root as vegetable.

Bupleurum hamiltonii Balakr. (Apiaceae)

Syn : *Bepleurum tenue* Buch. ex D. Don non Salisb.

Traditional use : KASHMIRI : *Root* : as antipyretic, diaphoretic and in liver complaints.

Phytography : Erect herb, 30–90 cm high, much-branched upwards; lower leaves linear, upper leaves oblong, distant; umbels usually terminal; flowers very small, bracts and bracteoles present; fruits broadly oblong, 0.25–0.3 cm long, yellowish brown, ridges prominent, furrows 1-vittate.

Distribution : Kashmir to Sikkim from 300 to 2700 m, common in western Himalaya and in the lower hills.

Ecology and cultivation : Plant of cold climate; wild.

Remark : The species occurs in a limited area, deserves attention.

Bupleurum lanceolatum Wall. ex DC. (Apiaceae)

Vernacular name : Kas : Sarpabuti.

Trade name : Sarpabuti.

Traditional use : KASHMIRI : *Fresh juice of herb* mixed with *Piper longum* : as remedy for snake bite.

Phytography : Perennial erect herb, 30–150 cm high, simple or with spreading horizontal branches; lower leaves long-petioled, lanceolate, upper leaves ovate, 7.5–12.5 cm by 0.6–7.5 cm, acuminate; small yellow flowers in umbels, bracts and bracteoles may be absent; fruits 0.4–0.5 cm, ridges distinct in imperfectly ripe fruits, small obsolete in the fully ripe, carpophore entire; seeds terete.

Distribution : North-West Himalaya between 1300 and 3000 m; Nepal.

Ecology and cultivation : Plant of cold climate; wild.

Remark : The species of very limited area, should be conserved and cultivated.

Bupleurum longicaule Wall. ex DC. (Apiaceae)

Syn : *Bupleurum rupestre* Edgw.

Vernacular name : Gar : Kala-ajwain.

Trade name : Kala-ajwain.

Traditional use : GARHWALI : *Root* : in liver complaints.

Phytography : Erect herb, 20–60 cm high, corymbose upwards, sometimes branching from base; leaves linear or lanceolate; flowers in umbels, black, bracts and bracteoles present; fruits oblong, 0.3–0.4 cm long, ridged and furrowed.

Distribution : Kashmir to Sikkim between 2600 and 4000 m.

Ecology and cultivation : Plant of cold climate; wild.

Remark : The species grows in a limited area, should be conserved and cultivated.

Bupleurum longicaulie Wall. ex DC. var. *himalayense* (Kl.) Cl. (Apiaceae)

Vernacular name : Lad : Sah-kuchak.

Traditional use : *LADAKHI* : *Plant* : in colic, gastrointestinal disorders, hepatitis and renal complaints.

Phytography : Erect herb, 20–60 cm high, corymbose upwards, sometimes branching from base; upper leaves ovate-lanceolate; small umbels, bracts and bracteoles present, bracteoles very wide.

Distribution : Ladakh, frequent at 4000 m altitude.

Ecology and cultivation : Plant of cold climate; wild.

Remark : Species of very limited region, should be conserved and cultivated.

Bupleurum wightii Mukherjee (Apiaceae)

Syn : *Bupleurum mucronatum* W. & A. non Brouss. ex Spreng.

English name : Common hare's ear.

Vernacular names : Mun : Tasad ependung; **Orn :** Latenga.

Traditional use : *MUNDA* : *Crushed herb* : in pain in legs and arms.

Phytography : Tall, much-branched herb, up to 2 m high; upper leaves linear or linear-oblong, mucronate, narrowed at base; umbels much panicled; flowers small, yellow, bracts prominent and lanceolate; fruits with prominent ridges, furrows 2- or 1-vittate.

Phenology : *Flowering* and *Fruiting* : October–January.

Distribution : Orissa, Bihar, Madhya Pradesh, Maharashtra, South India, plentiful between 1700 and 2600 m; Sri Lanka.

Ecology and cultivation : Grows in hot lower hills; wild.

Bursera serrata Colebr. (Burseraceae)

Syn : *Limonia pentagyna* Roxb.

Vernacular names : **Mun** : Kandeordaru; **Sad** : Kanror.

Traditional use : *SAN* : *Plant* : in cholera; *TRIBES OF CHOTANAGPUR* : *Twigs* : in dyspepsia.

Phytography : Balsamiferous evergreen tree; leaves alternate, imparipinnate, rarely 1-foliate, 15–30 cm; panicles axillary, lax, shorter than leaves; flowers very small, greenish, hermaphrodite, pubescent externally; drupes globose or ovoid.

Phenology : *Flowering* : April–May.

Distribution : Meghalaya, West Bengal, Bihar; Bangladesh (Chittagong).

Ecology and cultivation : Grows on small hills; wild.

Remark : Tribes of Purulia (West Bengal) eat the fruits.

Butea buteiformis (Voigt.) Grier (Fabaceae)

Syn : *Butea minor* Buch.Ham.

Vernacular name : **Nag** : Fu.

Trade name : Fu.

Traditional use : *NAGA* : *Seed-paste* (taken with water) : to kill intestinal worms.

Phytography : Climber; branches robust, young ones clothed with grey or brown silky hairs; leaves alternate, petioled, trifoliate, leaflets 15–37 cm long, leathery, scabrous above, finely silky beneath; racemes 7.5–30.0 cm long; flowers dull red, 2.5 cm long, papilionaceous, densely velvety on the outside; pods subsessile, 7.0–8.5 cm long.

Distribution : Eastern Himalaya; Nepal, Bangladesh (Sylhet).

Ecology and cultivation : Grows on tropical hills; wild.

Remark : Species of very limited area; should be conserved, studied and cultivated.

Butea monosperma (Lamk.) Taub. (Fabaceae)

(2n = 18, 32)

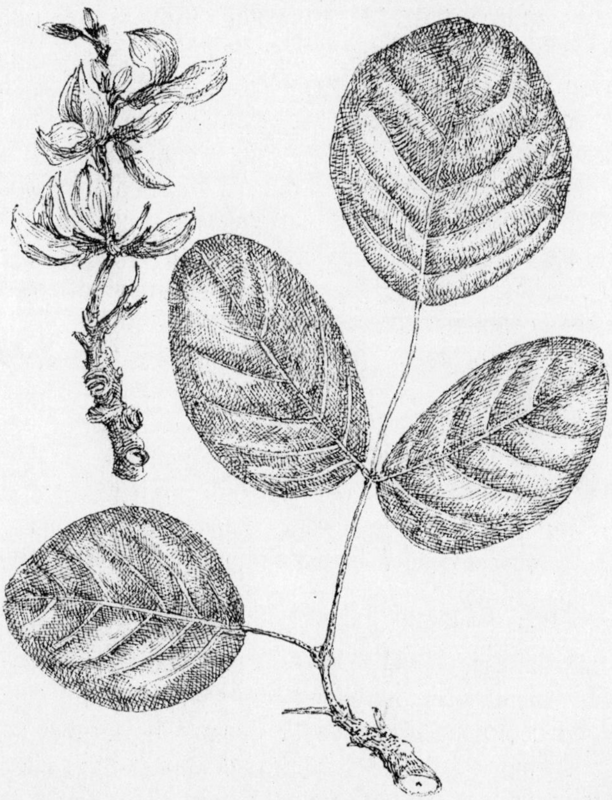

Syn : *Butea frondosa* Roxb., *Erythrina monosperma* Lamk.

English names : Flame of the forest, Parrot tree.

Sanskrit name : Palasha.

Vernacular names : Asm and **Ben :** Palash; **Guj :** Khakar; **Hin :** Dhak; **Kan :** Muttugathoras, Muttugu-mara; **Mal :** Chamath, Khakar; **Mar :** Palash; **Mun :** Murud ba; **Orn :** Murka; **Ori :** Palasa; **Pun :** Dhak; **Sad :** Palas, Paras; **Tam :** Samithu, Palasam; **Tel :** Moduga, Palashamu.

Trade names : Palasha, Dhak.

Traditional use : *KHASI* and *GARO* : *Leaf* : in delirium; *TRIBES OF PURULIA* (West Bengal) : *Seed* : in ascaris; *TRIBES OF MAYURBHANJA* (Orissa) : *Seed* : as contraceptive; *TRIBES OF SANTAL PARGANAS* (Bihar) : *Root* : in tuberculosis; *TRIBES OF VARANASI* (Uttar Pradesh) : *Leaf* : in boils; *Seed* : as vermifuge; *TRIBES OF MIRZAPUR* (Uttar Pradesh) : *Bark* : in dysentery; *Gum* : in diarrhoea, dysentery; *TRIBES OF SIWALIK* (Uttar

Pradesh) : *Gum* : as tonic; BHAT : *Seed* : as abortifacient; BHOXA : *Bark* : in bone fracture, *Gum* : in piles, urinary complaints; GARHWALI : *Leaf* : in boil, inflammation, *Flower* : in diarrhoea, dysentery, pimples, *Seed* : as anthelmintic; THARU : *Gum* : as diuretic, *Seed* : as cooling agent; FOLKS OF DELHI : *Gum* : as astringent, *Flower* : as aphrodisiac, astringent, diuretic, *Seed* : as anthelmintic; FOLKS OF KURUKSHETRA (Haryana) : *Flower* : in stomachache; DANG : *Bark* : in diarrhoea; TRIBES OF RATAN MAHAL HILLS (Gujarat) : *Flower* : in eye complaints; KORKU (of Maharashtra) : *Flower* : in dysentery; TRIBES OF KHANDLA (Maharashtra) : *Flower* : in dog bite, urinary complaints; TRIBES OF CHANDRAPURA (Maharashtra) : *Leaf* : in skin diseases; TRIBES OF JHABUA (Madhya Pradesh) : *Root* : in dog bite; TRIBES OF SAGAR (Madhya Pradesh) : *Leaf* : as vermifuge, *Flower* : in diabetes, diarrhoea, piles; TRIBES OF EAST GODAVARI (Andhra Pradesh) : *Gum* : in diarrhoea; TRIBES OF NILGIRI (Tamil Nadu) : *Bark* : as haemostatic, in wounds, *Flower* : in eye complaints; TRIBES OF KANNANORE (Kerala) : *Flower* : in antifertility.

ATHARVA VEDA : *Extract of stem* : beneficial for sperms and helps securing conception; CHARAKA SAMHITA : *Stem-extract* : useful in leprosy, piles, gastroenteritis and menorrhagia; SUSHRUTA SAMHITA : useful in diseases caused by *vayu* (wind), *Seed* : effective against intestinal worms; AYURVEDA : *Bark* : useful against snake venom, wounds, indigestion, gastroenteritis, fever, tuberculosis, *Gum* : astringent, beneficial to children and women, *Leaf* : astringent, sex stimulant, useful in intestinal worms, dyspepsia, piles, menorrhagia, pimples, wounds in mouth/throat, *Flower* : diuretic, sex stimulant, helps menstruation, useful in gastroenteritis, *Seed* : useful against intestinal worms.

SIDDHA : *Flower-juice* : used in preparation of the medicine *Murukkam, Seed* and *Kernel* : in *Palacu*

UNANI : Ingredient of the medicine called 'Dhak(tesu)' and 'Samaghke Dhak'.

Modern use : *Plant* (alcoholic extract : produces persistent vasodepression in cats, shows activity against earthworms; *Bark* : insecticide against house flies; *Alcohol extract of bark* : inhibitory against *E. coli* and *Micrococcus pyogenes* var. *aureus*; *Gum* : solution applied to check conception; *Root* (bark) : aphrodisiac, analgesic, anthelmintic, useful in elephantiasis, applied in sprue, piles, ulcers, tumours and dropsy; *EtOH* (50%) *extract of leaf* : spasmogenic; *Flower* : effective in leprosy, gout; *Alcoholic extract* : antiestrogenic in mice; *Aqueous extract* : anti-implantation in rats; along with *Hygrophila auriculata* leaf and root taken with milk to cure leucorrhoea; *Seed* (freshly powdered) : effective against Ascaris; *Extract* (*in vitro*) : anthelmintic against *Asacridia galli* worms; *finely powdered* along with *Acorus calamus* rhizome or mixed with juice of *Cyperus rotundus* rhizome : cures delirium; *Saline extract* : agglutinates erythrocytes of animals; *Hot alcoholic extract* : anti-implantation and anti-ovulatory in animals.

Phytography : Erect tree, reaching a height of 13–17 m; young stem clothed with grey or brown silky hairs; leaves alternate, petioles 7.5–15.0 cm long, trifoliate, leaflets 10–20 cm long, leathery, lower surface covered with dense silky hairs; racemes 15 cm long; flowers bright orange-red, thickly clothed on the outside with silvery small hairs; pods 15–20 cm by 3.7–5.0 cm, narrowed suddenly into a stalk longer than calyx.

Phenology : *Flowering* : February–April; *Fruiting* : May–July.

Distribution : Plains of India, ascending up to 1300 m; Myanmar.

Ecology and cultivation : Mesophyte; wild.

Chemical contents : *Plant* : flavonoids, glucosides, butin, butrin, isobutrin, palastrin; *Flower* : butrin, coriopsin, monospermoside, sulphurein, chalcones; *Seed* : palasonin, *Seed oil* : d-lactone of n-heneicosanoic acid, monospermine, new phytolectin.

Remarks : An important tree for lac cultivation, but the lac produced on it is of inferior quality. Bark yields fibre, wood yields timber of poor quality; stem-bark used as fish poison by tribes of South Rajasthan. Plates and bowls are made by stitching the leaves by the tribes of Purulia and Saurashtra. Flowers yield a yellow dye of little permanency.
Flowers are eaten as vegetables by tribes of Manbhum and Hazaribagh Districts of Bihar while fruits by Garhwalis.
Tree is sacred to the Hindus and Buddhists. Flower is an essential item of *Saraswati Puja*.

Butea parviflora Roxb. (Fabaceae)

Syn : *Spatholobus roxburghii* Benth.

Vernacular name : Mun : Bandu nanri.

Traditional use : *GARO* : *Bark* : in toothache, gum trouble, *Stem* : in dysentery; *KHASI* : *Bark* : in toothache and gum trouble; *TRIBES OF PURULIA* (West Bengal) : *Root* : in venereal diseases; *SANTAL* : *Root* : in diarrhoea, mucus in urine, pain, *Bark* : in cholera, haematuria, madness, stomachache; *TRIBES OF MADHYA PRADESH* : *Root* : in rheumatic pain.

Phytography : Robust evergreen climber; trunk attaining 1 m in girth, branches densely grey-downy; leaves alternate, trifoliate, petioles 7.5–15.0 cm, leaflets 25 cm long, subcoriaceous, dull green; panicles reaching 30–45 cm, branches very downy; flowers bright red or cream-coloured or white, 7.5 cm long; pods 7.5–10.0 cm long, 2.5 cm broad, clothed with ferruginous hairs, narrowed suddenly into a stalk 1.25 cm long.

Phenology : *Flowering* : August–December; *Fruiting* : February–March.

Distribution : Plains of India; Sri Lanka.

Ecology and cultivation : Wild.

Chemical contents : *Seed* : β-sitosterol, freidelin, oil.

Remarks : Santals use bark to treat rinderpest of cattle; tribes use this as fish poison. It yields good fibre for rope.
Lac insect is reared on it.

Butea superba Roxb. (Fabaeae)

(2n = 18)

English name : Creeping butea.

Vernacular names : Ben : Latano palash; **Hin :** Palas lata, Kimsuka; **Mun :** Lara murud.

Trade name : Palas lata.

Traditional use : *TRIBES OF MEDINIPUR* (West Bengal) : *Flower* : in antifertility; *SANTAL* : *Root* : in gravel, *Bark* : in continuous fever, dysuria; *TRIBES OF MAYURBHANJA* (Orissa) : *Bark* : in cuts.

Modern use : *Root* : cardiotonic, hypotensive, rejuvenating; *Root-bark* : aphrodisiac; *Stem-bark* : used in tonics, elixirs; *Seed* : anthelmintic, haemagglutinating against human A, B, O red cells, sedative; *Seed-oil* : anthelmintic, hypotensive.

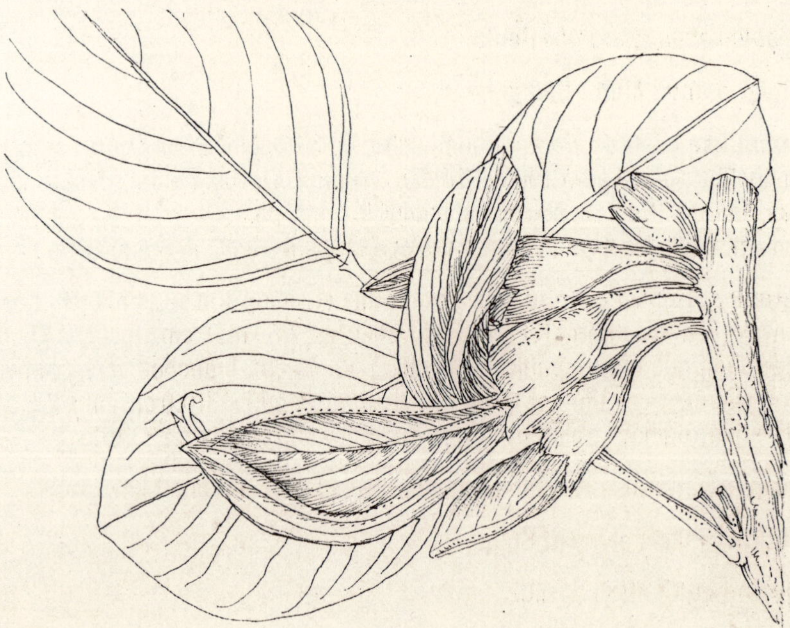

Phytography : Large woody climber; leaves alternate, trifoliate, leaflets attaining 45 cm in length; racemes 30 cm long; flowers gorgeous orange-scarlet, may be 50 cm long, borne in great profusion on the leafless branches.

Phenology : *Flowering* : March–April; *Fruiting* : June–July.

Distribution : Forests of West Bengal, Bihar, Orissa and Karnataka.

Ecology and cultivation : Tropical forest plant; wild.

Chemical contents : *Root* : estrogenic substance, poisonous substance.

Remarks : Korkus of Maharashtra use Root in ulcers of animals and also active against fungus—*Helminthosporium sativum.*

Cadaba fruticosa (L.) Druce (Capparidaceae)

(2n = 18)

Syn : *Cadaba indica* Lam., *C. farinosa* Forssk., *Cleoma fruticosa* L.

Vernacular names : **Hin** : Kodhab; **Kan** : Chegaviche; **Mal** : Kattakatti; **Mar** : Habab, Viludi; **Tam** : Kattagatti, Vili; **Tel** : Adanmorinika.

Trade names : Farsa, Balambo.

Traditional use : *TRIBES OF DAHANU FOREST* (Maharashtra) : *Leaf* : in gastrointestinal troubles, urinary complaints and as vermicide; *Leaf* and *Root* : purgative, anthelmintic, antispasmodic, deobstruent, emmenagogue, aparient, prescribed in uterine obstructions; *Leaf* : used as poultice for sores.

Phytography : Straggling, much-branched, wiry shrub; leaves simple, ovate or oblong, 0.6–1.2 cm; corymbs terminal, few- flowered; flowers greenish white, 2.5 cm in diameter, bracteate; fruits cylindric, 1.2–3.7 cm.

Distribution : Punjab, Rajasthan, Madhya Pradesh, Gujarat, dry parts of Maharashtra, Karnataka; Pakistan (Sind).

Ecology and cultivation : Found in arid zones; xerophytic in nature; wild.

Chemical contents : *Leaf* : cadabine, alkaloids; *Pod* : dilactone, cadabalone.

Remark : In Africa, infusion of leaves inhaled to clear head from colds.

Caesalpinia crista L. (Caesalpiniaceae)

(2n = 24)

Syn : *Caesalpinia bonducella* Fleming

English names : Malacca bean, Fever nut, Bonduc nut.

Sanskrit names : Kuberakshmi, Putikaranja.

Vernaculars names : Ben : Nata, Natakaranja; **Hin :** Karanju, Nathamala, Katkaranja; **Lod :** Kata-karanj, Janum-karonjo; **Mal :** Kazanchik kuru; **Mar :** Gajaga; **Mun :** Nata; **San :** Bagui; **Tam :** Kazhi-chikay; **Tel :** Gacha-chakaya.

Trade name : Karanju.

Traditional use : LODHA : (i) *Root-powder* : in c-onsti- tutional disorder; (ii) *Paste of twig* (with lime water) : in intestinal worms; (iii) *Leaf-powder* (with decoction of long pepper) : as tonic; (iv) *Seed-powder* : in dyspepsia, colic pain; MUNDA : *Seed -oil* : in facial paralysis; SANTAL : (i) *Seed-powder* : in malarial fever; (ii) *Seed-oil* : in hydrocele.
UNANI : an ingredient of 'Karanjwa'.
HOMOEOPATHY : for control of mind, for the treatment of different ailments of head, eyes, nose, mouth, tongue, abdomen, liver, spleen and fever.

Modern use : *Kernel* : antiperiodic, anthelmintic, bitter tonic, febrifuge and used in fever; *Seed* : for the treatment of hydrocele.

Phytography : An extensively climbing or scrambling shrub covered with short, straight prickles, those on the leaf rachis reflexes; branchlets fulvous hairy; leaves simple or large persistent compound or pinnatifid, stipules foliaceous, pinna 6–8 pairs; flowers pale yellow; young pods softly echinate, old broad-oblong, prickly, dehiscent, 1- to 2-seeded.

Phenology : *Flowering* : August–September; *Fruiting* : December–April.

Distribution : Throughout the hotter parts of India; Bangladesh and Pakistan; up to an elevation of 1000 m.

Ecology and cultivation : Grows in dry and semiarid regions, usually in open hedges; evergreen.

Chemical contents : *Seed* : bitter substance phytosterinin, bonducin, saponin, fatty oil 20–24% starch, sucrose, 2-phytosterols; bitter amorphous glycoside bonducin isolated from the oil; bitter principle ineffective against bird malaria.

Caesalpinia digyna Rottl. (Caesalpiniaceae)

English name : Teripods.

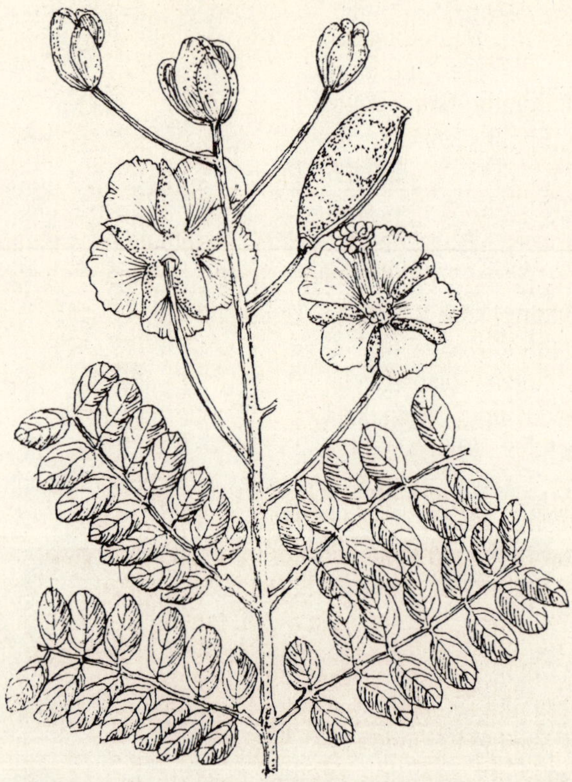

Vernacular names : Asm : Heri; **Ben :** Umul kuchi; **Hin :** Vakeri mul; **Lod :** Gilo; **Mar :** Vakeri-mula; **Ori :** Gilo; **Tel :** Nune gacea.

Trade name : Tari (trade name of the pods).

Traditional use : *LODHA* : (i) *Root-decoction* : against alcoholic toxicity, (ii) *Root-powder* (with leaf juice of bamboo) : in diabetes.

Modern use : *EtOH* (50%) *extract of root* : active *in vitro* against *Mycobacterium tuberculosis* in guinea pigs; *Plant* : a minor constituent of indigenous drug 'Geriforte'.

Phytography : Large, branched climbing shrub, armed with recurved prickles; leaves pinnately compound, pinnae 8 or 10 or 12, minutely pubescent or fulvous hairy; flowers pale yellow; pods oblong, thick, fleshy, fruiting pedicels slender; seeds oblong, dark brown or black.

Phenology : *Flowering* : July–October; *Fruiting* : February–April.

Distribution : Andamans, West Bengal, Assam and the adjoining states; Bangladesh, Myanmar, Sri Lanka.

Ecology and cultivation : Grows in the damper places, prefers well-drained sandy soil; common in bushes.

Chemical contents : Tannin 53.8–59.9% in pod-cases, but not in seeds; bitter substance bonducin, saponin.

Caesalpinia nuga (L.) Ait. (Caesalpiniaceae)

(2n = 24)

Vernacular names : Ben : Singrilata; **Mal :** Kakamullu; **Tel :** Mulutiga.

Trade name : Singrilata.

Traditional use : *Root* : diuretic, tonic, useful in gravel and stone in bladder as well as lithotriptic.

Phytography : Large prickly climber with prickles on the branches; leaf pinnately compound, pinnae 3 pairs, leaflets usually only 2 pairs on each pinna; flowers yellow, fragrant in panicled racemes; pods compressed, woody; seeds large, brown, flattened ellipsoid .

Phenology : *Flowering* : January–April; *Fruiting* : May.

Distribution : Karnataka, Orissa, Sundarbans; Bangladesh (Chittagong, Sylhet, Khulna, Barisal).

Ecology and cultivation : A littoral species.

Remark : Pulped fruit and stem : yield a fish poison.

Caesalpinia sappan L. (Caesalpiniaceae)

(2n = 24)

English names : Japan wood tree, Sappan wood tree.

Sanskrit names : Pattanga, Pattranjaka.

Vernacular names : Ben and **Hin :** Bakam, Patang; **Kan :** Pattanga; **Mal :** Sappannam; **Mar :** Patang; **Tam :** Patungam; **Tel :** Bakamu.

Trade names : Patang, Sappan wood tree.

Traditional use : *Decoction of wood* : considered as emmenagogue, useful in diarrhoea and dysentery, given internally in certain skin diseases.

Modern use : *EtOH* (50%) *extract of stem* : semen coagulant, anticancer; *Heartwood* (MeOH extract) : anti-inflammatory in rats; *Essential oil from leaf* : antimicrobial.

Phytography : Small tree with small and few or no prickle; leaf pinnately compound, pinnae 8–12 pairs, leaflets close, sessile, oblong, oblique and falcate, 10–12 pairs; flowers yellow, racemes panicled; pods prominently beaked, polished, indehiscent; seeds 3-4.

Phenology : *Flowering* : Rainy season.

Distribution : Tamil Nadu, Kerala, Karnataka, Andhra Pradesh and West Bengal, usually as a cultivated hedge plant.

Chemical contents : *Heartwood* : α-amyrin, amino acids, β-sitosterol, brasilin essential oil; *Leaf* : 0.6–25% essential oil, δ, α phellandrene as chief constituent; oscimine also present; *Pod-cases* and *Bark* : tannin.

Caesalpinia sepiaria Roxb. (Caesalpiniaceae)

Vernacular names : Hin : Arlu, Kando, Relu, Uri; **Mal :** Inna; **Mar :** Chillur; **Ori :** Gilo; **Tam :** Kadiyindu; **Tel :** Gaddakorinda.

Trade names : Relu, Kandu.

Traditional use : *Root* : purgative; *Leaf* : emmenagogue, laxative, applied to burns.

Phytography : Extensive shrub, rambling and climbing by means of the recurved prickles on the stem and leaf rachis; branchlets and leaf rhachises and panicle rusty pubescent or subtomentose; leaf pinnately compound, pinnae 5–10 pairs, leaflets 6–12 pairs; flowers showy, bright, sulphur-yellow; pods oblong, dry, compressed, tipped with a straight, hardened style, fruiting pedicels stout, woody.

Phenology : *Flowering* : December--May; *Fruiting* : February.

Distribution : Throughout India; Bangladesh and Pakistan.

Ecology and cultivation : Planted in gardens as hedge plant.

Cajanas cajan (L.) Millsp. (Fabaceae)

(2n = 22, 44, 66)

Syn : *Cajanas indicus* Spreng., *Cytisus cajan* L.

English names : Cango pea, Cajan pea, Pigeon pea, Red gram.

Sanskrit names : Adhaki, Tuvari, Tuvarika.

Vernacular names : Ben : Arhar; **Hin** and **Mar** : Arhar, Tur, Tuver; **Kan** : Togare; **Lod** : Rahri; **Mal** : Thuvara; **San** : Arki; **Tam** : Thovaray.

Trade name : Arhar.

Traditional use : LODHA : (i) *Leaf* and *Seed* : made into paste, warmed and applied to check excessive secretion of milk, (ii) *Leaf -decoction* : for washing septic wounds; MUNDA : (i) *Leaf-paste* (fresh) : in leprosy, (ii) *Tender leaves* : to chew for curing apthae and spongy gum;- SANTAL : *Leaf-juice* (fresh) : in amoebiosis.

Phytography : Annual or perennial shrub, 0.6–3.6 m high with grooved silky branches; leaves trifoliate, leaflets oblong, acute, entire, densely silky beneath and indistinctly gland-dotted, stipules small, filiform; flowers borne in axillary corymbose racemes, reddish yellow; pods linear, straight, narrowed at both ends, densely glandular, pilose with oblique depressions between seeds.

Phenology : *Flowering* and *Fruiting* : June–March.

Distribution : Cultivated throughout India, Bangladesh and Pakistan as pulse crop; sometimes found as feral.

Ecology and cultivation : Plants grow in tropical/subtropical climate but cannot tolerate frost; can be cultivated up to an elevation of zoom; pH 5–7 are best.

Chemical contents : *Seed* : two globulins—cajanin and concajanin; extraction of active urease; peptisation of red gram proteins and their characterisation by electrophoresis.

Calamintha vulgaris (L.) Druce (Lamiaceae)

(2n = 20)

Syn : *Calamintha clinopodium* Benth., *Clinopodium vulgare* L., *Melissa clinopodium* Benth.

English name : Wild basil.

Vernacular names : Ben and **Hin :** Ban tulsi.

Trade name : Ban tulsi.

Traditional use : GARHWALI : *Plant* : as astringent, carminative and in heart troubles.

Modern use : *Aquatic extract of aerial parts* : Central Nervous System depressant and sedative in mice.

Phytography : Erect herb; rootstock woody, stoloniferous, stem up to 1 m high, slender, branches few; leaves opposite, remote, 2.5–5.0 cm, ovate, subcrenate; verticillasters axillary and terminal, 2.0–2.5 cm in diameter, depressed; flowers bilabiate, purple, 2.0–2.5 cm, hairy; nutlets minute, subglobose, smooth, dry.

Distribution : Western temperate Himalaya from Kashmir to Kumaon, between 1300 and 4000 m.

Ecology and cultivation : Plant of cold climate; wild.

Chemical contents : *Rhizome* : stachyrose.

Calamus rotang L. (Arecaceae)

English names : Common rattan, Rattan.

Sanskrit name : Vetra.

Vernacular names : Ben : Bet, Chachibet; **Hin :** Bet, Bemt, Chachibet; **Mal :** Cural; **Tam :** Cural, Pirambu; **Tel :** Pemu.

Trade names : Chachibet, Rattan.

Traditional use : *Root* : astringent, acrid, alexiteric, antidysenteric, anti-inflammatory, bitter, cooling, expectorant, diuretic, febrifuge and tonic; beneficial in vitiated conditions of *pitta* (bile), bronchitis, burning sensation, cough, erysipelas, leprosy, skin diseases, general debility and some other ailments; *Leaf* : astringent, laxative, useful in skin diseases; *Seed* : astringent, sour, depurative, pruritus; *Sprouts* : thermogenic, useful in stomach ailments.

Phytography : Scandent, slender shrub with cylindrical, uniformly thick yellowish white stem armed with short flat spines on the leaf-sheaths; leaf-sheaths glabrous, hollowed below, ochreate, short truncate, margin with small straight or recurved spines, laterally compressed, rachis flat near the base and smooth above; leaflets trigonous upwards, many, equidistant, one-ribbed, ribs bearing scattered bristles; flowers unisexual, male spadix very long, flagelliferous, spathe elongate, tubular, female flowers scattered along the slender branches of the spadix; fruits on minute perianths, subglobose, mucronate, scales many, pale yellow in vertical series.

Phenology : *Flowering* : with a peak during March–April; *Fruiting* : throughout the year.

Distribution : Central and South India; Bangladesh and Pakistan; in arid tracts up to 450 m.

Ecology and cultivation : Plains from the coast; scrub jungles; occasional clumps.

Calendula officinalis L. (Asteraceae)

(2n = 28, 32)

English name : Pot marigold.

Vernacular name : Pun : Zergul.

Trade name : Marigold.

Traditional use : *Plant* : astringent, carminative, stimulant, home remedy for jaundice, vulnerary.

Modern use : *EtOH* (50%) *extract of plant* : antiprotozoal, spasmolytic; *Aqueous extract of plant* : antitumour against Crocker sarcoma 180 in mice; *Leaf* : resolvent, diaphoretic; *Inflorescence* : in the treatment of ulcers (internal and external) and exposed sores; *Flowers* : stimulant, antiseptic, emmenagogue; an ointment made of flowers is considered to be useful in skin diseases; *Flower-extract* : antimicrobial.
HOMOEOPATHY : Extensively used in Homoeopathic system; in Europe, mother tincture is used in amenorrhia and after dilution for pains and bruises.

Phytography : More or less hairy, annual, 30–80 cm high; leaves thickish, oblong to oblong-ovate, 50 mm to 1.20 cm long, entire, remotely denticulate, more or less clasping; heads solitary on stout stalk, showy, 40–80 mm across, flat spreading rays white, yellow to deep yellow, closing at night.

Phenology : Blooming the whole season in cycles; flowers collected mostly in August.

Distribution : Wild in Punjab; Pakistan (Sind, Peshawar) and arid zones; cultivated in many parts of India and Bangladesh as an ornamental.

Ecology and cultivation : Cultivated widely as a garden ornamental; thrives best in moderately rich garden soil.

Chemical contents : *Roots* : eight triterpenoid glycerides, calenduloside A & B; *Plant* : rich in Mn and carotene, flavonoids, oleanolic acid glycerides, saponins, saponosides; *Leaves* and *Seedlings* : sterols, polyprenyl quinones, α- tocopherol; *Flowers* : triterpenoid alcohols, sterols, glycerides of oleonolic acid, volatile oil, n-tricosane, glyceroids; *Extract of inflorescence* : caryophyllin, carotenoids, a compound similar to pyrethins; *Seeds* : haemagglutinins.

Adulterants : Capitula used as an adulterant of saffron. Ligulate corollas of florets of the marigold, which are often coloured with methyl orange and are sometimes known as feminell and Chinese safflower.

Remarks : Dried ligulate floret is used in preparing homoeopathic drug. Not more than 2% of foreign organic matter; it yields not more than 2% of acid-soluble ash; the drug has somewhat an aromatic odour and a distinctly bitter taste.

Callicarpa arborea Roxb. (Verbenaceae)

Sanskrit name : Bagodi.

Vernacular names : Asm : Khoja; **Ben :** Kojo; **Gar :** Ghiwala.

Traditional use : *Bark* : carminative, aromatic, bitter and tonic; *Decoction of bark* : applied to cutaneous diseases.

Phytography : A small or medium-sized tree with compressed, 4-angled, densely tomentose branchlets; leaves ovate to ovate-lanceolate, densely stellate tomentose below; lilac-purple flowers small, in dichotomous corymbose cymes; drupes purple.

Phenology : *Flowering* : October, April–June; *Fruiting* : August, November, December.

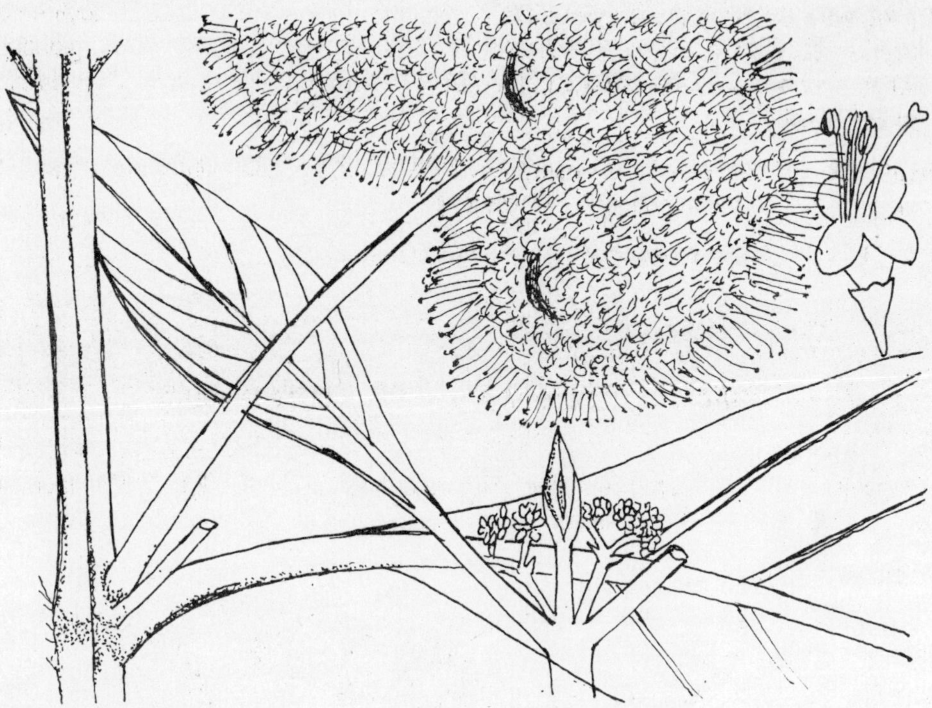

Distribution : Assam, Meghalaya, in lower hills of Kumaon and Sikkim, Uttar Pradesh, in some parts of West Bengal, South India, especially in Andhra Pradesh, Karnataka; Bangladesh.

Ecology and cultivation : In damper places and hills of the drier regions.

Chemical contents : *Bark*: Me-betulinate, baurenol, β-sitosterol-OAc; *Leaf*: β-sitosterol, maslinic acid, oleanolic acid, ursolic acid, lupeol-OAc, β-amyrin-OAc, *Heartwood* : β-sitosterol, oleanolic acid.

Callicarpa cana L. (Verbenaceae)

Syn : *Callicarpa candicans* Hochr.

Vernacular name : Ben : Arusha.

Traditional use : *Root*, *Leaf* and *Bark* : in skin diseases; *Leaf* : remedy for abdominal troubles, for poulticing wounds and boils.

Phytography : A large shrub; young shoots stellately hairy; leaves opposite, toothed or subentire; flowers in axillary cymes, shorter than the leaves; drupe small, globose, pyrenes 4 or fewer by suppression, 1-seeded; seeds shortly oblong; berries purple, ultimately black.

Distribution : A native of Malaya peninsula, often cultivated; occasionally naturalised in some parts of Bangladesh including Chittagong.

Remark : Leaves are used as fish poison.

Callicarpa macrophylla Vahl (Verbenaceae)

(2n = 34)

Vernacular names : Ben : Mathara; **Hin** and **Kum :** Daya; **Mal :** Chimpompil; **Pun :** Sumali; **San :** Buddhighassic.

Trade name : Daya.

Traditional use : *Leaf* (heated) : applied to joints for relieving rheumatic pain; *Oil from root* : aromatic, stomachic.

Phytography : A stout shrub; branches, leaves, (beneath) and inflorescence densely stellately woolly; leaves 1.25–2.50 cm long, elliptic, rarely ovate-lanceolate, long acuminate, crenate or crenate-dentate; flowers rose-coloured in dense dichotomous cyme; drupe 3–6 cm, white.

Phenology : *Flowering* : August–November; *Fruiting* : October–December.

Distribution : Western Himalaya from Kashmir eastwards up to 2000 m, upper Ganga plains, Bengal, Assam; Bangladesh, Pakistan (Hazara).

Ecology and cultivation : Evergreen, edge of forests, steep slopes and ghats, open jungles and wastelands.

Chemical contents : *Root* : an aromatic oil, *Aerial parts* : 2-tetracyclic diterpenoids, calliterpenone and its mono-OAc; *Leaf* : β-sitosterol, calliterpenone and its mono-OAc, luteolin, apigenin and its 7-glucuronides, ursolic acid, crategolic acid; *Seed* : calliterpenone and its -OAc.

Callicarpa tomentosa (L.) Murr. (Verbenaceae)

(2n = 130, 136)

Syn : *Tomex tomentosa* L., *Callicarpa lanata* L., *C. wallichiana* Walp

Vernacular names : Ben : Massandari; **Hin :** Bastra; **Kan :** Ardi; **Mal :** Nallapompil; **Mar :** Aisar; **Tam :** Verrilaippati, Vettilaipattai.

Traditional use : *Root* : in cutaneous affections; *Decoction of bark* : in skin diseases, hepatic obstructions and fever; *Leaf* (boiled in milk) : as a wash for aphthae of the mouth.

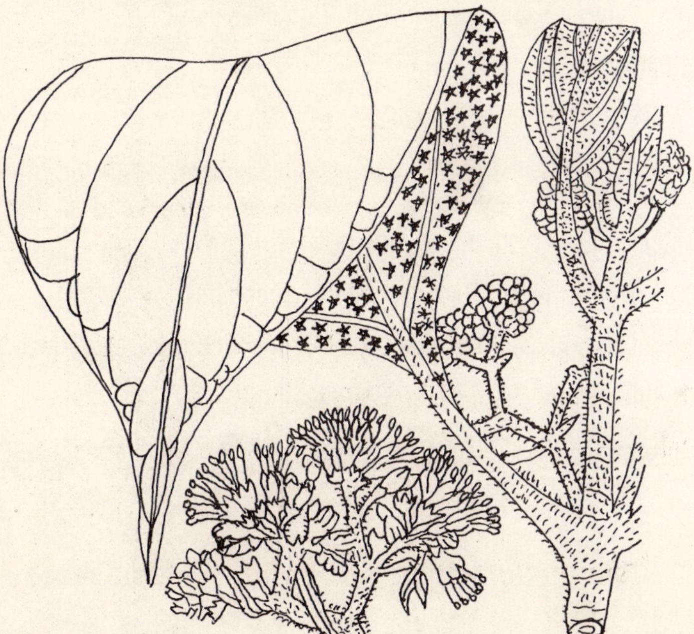

Phytography : A shrub or small tree, ± 10 m long; leaves broadly elliptic-ovate, coriaceous, lateral nerves ± 8 pairs, pubescent along nerves above, tomentose below, petiole up to 9 cm long; inflorescence axillary and terminal, umbellate cyme, bracts linear, deciduous, bractlets fulvous-tomentose; flowers 4-merous, subsessile; drupes globose, 4 mm, of four 1-celled pyrenes, red on ripening; seeds oblong.

Phenology : *Flowering* : with a peak during March–April; *Fruiting* : throughout the year.

Distribution : Occurring in the Western Ghats, Maharashtra and some parts of Karnataka; Sri Lanka.

Remark : Tribes of Kerala use root, bark and leaf as masticatory.

Calonycton aculeatum House (Convolvulaceae)

(2n = 30)

Syn : *Ipomaea bona-nox* L., *Convolvulus aculeata* L.

English name : Moonflower.

Sanskrit name : Chandrakanti.

Vernacular names : Ben and **Hin :** Dudhiakalmi; **Mal :** Mandavalli; **Mar :** Gulchandani; **Ori :** Kunjalata; **Tam :** Naganamukkari; **Tel :** Panditivankayya.

Trade name : Dudhikalmi.

Traditional use : *Different plant parts* : in snake bite.

Phytography : Glabrous or nearly so, with milky juice, stem 3–6 cm long; leavess cordate to subreniform or hastate, angular or trilobed, acute, long petioled; flowers fragrant, usually closing in the morning, sometimes remaining open till noon.

Phenology : *Flowering* : from Rainy season to December.

Distribution : An ornamental herb cultivated throughout India; Bangladesh and Pakistan.

Ecology and cultivation : Sometimes found as feral.

Chemical content : Resin-like substance separated from the plant.

Calophyllum apetalum Willd. (Clausiaceae)

Syn : *Calophyllum wightianum* T. Anders, *C. spurium* Choisy

English name : Poonspar of Travancore.

Vernacular names : Guj : Sarpuna; **Kan :** Irai; **Mal :** Churupinnai; **Mar :** Bobbi; **Tam :** Sirubinnai.

Traditional use : *Seed-oil* : used in leprosy and skin infections; *Resin* : antiphlogistic and anodyne.

Phytography : Middle-sized tree with yellow bark; young shoots 4-gonous, glabrous; leaves coriaceous, obovate-oblong, cuneate, rounded at the apex, often emarginate; flowers ± 12 mm in diameter, boat-shaped, caducous; fruit 16 mm long, ellipsoid, smooth, apicute, red when ripe.

Phenology : *Flowering* : Winter.

Distribution : Western Ghats—Maharashtra, Karnataka, Kerala and West Coast.

Ecology and cultivation : Common, mostly along the banks of rivers.

Chemical contents : *Seed* : yields fixed oil; *Kernel of seed* : yields 45-50% of oil with bitter taste.

Calophyllum elatum Bedd. (Clausiaceae)

Syn : *Calophyllum tomentosum* T. Anders, *C. angustifolia* Dalz. et Gibs.

English names : Poonspar tree, Sirpoon tree.

Vernacular names : Hin : Sirpon; **Kan :** Viri; **Mal :** Kattupunna; **Mar :** Pun, Nagani; **Tam :** Kattupinnai.

Trade name : Poonspan.

Traditional use : *Gum* : slightly astringent.

Phytography : Big tree, young shoots, buds and panicles clothed with rusty tomentum, young branches quadrangular; leaves coriaceous, oblong-lanceolate, acuminate, acute at the base, glabrous, shining; flowers 13–18 mm in diameter, in terminal panicle from the axil of the upper leaves, slightly tomentose; fruits obliquely ovoid, pointed, smooth.

Phenology : *Flowering* : March–May.

Distribution : Gujarat, Maharashtra, Karnataka and Kerala, ascending up to 1725 m.

Chemical content : *Seed-kernel* : yield 70% oil.

Calophyllum inophyllum L. (Clausiaceae)

(2n = 22)

English names : Alexandrian laurel, Borneo mahogany, Sirpoon tree.

Sanskrit names : Nagchampa, Punnaga.

Vernacular names : Ben and **Hin :** Sultan champa; **Kan :** Vuma, Honne, Wima, Hona, Pinokae; **Mal :** Punna; **Mar :** Undi, Surangi; **Ori :** Poonang; **Tam :** Punnagam; **Tel :** Pouna.

Trade name : Poon.

Traditional use : *Pounded bark* : in orchitis; *Juice of bark* : purgative; *Bark-decoction* : employed for indolent ulcers; *Resin* : emetic and purgative; *Oil* : in rheumatism and for skin diseases.

Modern use : *Aromatic resin* : possess emetic and purgative properties.

Phytography : An exceedingly handsome, moderate-sized tree with large glabrous, oblong, ovate, obtuse or emarginate monocot-like leaves; leaves 1–1.50 cm, very shining on both surfaces; racemes axillary, drooping; flowers white, fragrant, ± 20 cm in diameter; drupe globose, yellow, pulpy when matured, ± 25 mm in diameter.

Phenology : *Flowering* : May–June and again in October–November; *Fruiting* : October–November and April–May.

Distribution : Distributed on the sea shores of India, particularly of Orissa, Karnataka, Maharashtra, Andamans; Bangladesh, Pakistan. Also cultivated as an ornamental tree.

Ecology and cultivation : Indigenous to the coasts; cultivated.

Chemical contents : *Bark* : 11.9% tannin; *Leaf* : saponin and hydrocyanic acid; *Seedcoat* (hard testa) : leucocyanidin; *Kernel* : yield 50–73% oil.

Remarks : Plant suitable for avenue planting. Leaf used as fish poison.

Calotropis gigantea (L.) R. Br. ex Ait. (Asclepiadaceae)

(2n = 22)

Syn : *Asclepias gigantea* Willd.

Sanskrit name : Arka.

Vernacular names : Ben : Akanda; **Hin :** Ak, Akona, Madar; **Guj :** Akado; **Kan :** Arkagida, Palati, Palti; **Lod :** Orakka kulatoy, Shwet akanda; **Mal :** Erikku; **Mar :** Rui,

Akanda; **Mun** : Palati, Patla dudha; **Orn** : Akuan, Ark; **Sad** : Gadesa; **San** : Parkha; **Tam** : Arkkam; **Tel** : Jilledu.

Trade name : Akanda.

Traditional use : *MIRI* : (i) *Root-bark* : in dysentery, (ii) *Latex* : as purgative, (iii) *Leaf* : in burns; *NAGA* : *Leaf* : in bodyache, skin diseases; *LODHA* : (i) *Root-powder* (with country liquor) : in leprotic wounds, (ii) *Root-decoction* (with paste of long peppers) : in leucorrhoea, (iii) *Leaf* (fomented) : in enlarged liver; *ORAON* : *Root* and *Latex* : in body pain, dropsy, as a purgative, febrifuge; *SANTAL* : (i) *Root* : in anasarca, ailments of spleen, carbuncle, cholera, convulsions, convulsions with tongue drawn in, dropsy, hemiplagia, hydrocele, lumbago, muscular pain, neuralgia, orchitis, osteomyelitis, palsy, pleurisy, pneumonia, post-natal complaints, protracted labour, rabies, rheumatism, ringworm, scabies, sores, strangulation of intestine, syphilis, (ii) *Bark* : in cholera, epilepsy, small pox, (iii) *Latex* : in boils, chest pain, dysentery, kala-azar, stings, toothache; *THARU*: *Leaf*: in swelling; *KUMAONI*: (i) *Latex* : in leprosy, (ii) *Leaf-juice* : in complaints of ear; *TRIBES OF RAJASTHAN* : (i) *Latex* : on wounds, (ii) *Leaf* : in intestinal worms, swelling; *DANG* : *Leaf-juice* : in complaints of ear; *VASAVA* : *Latex* : in stomachache; *BHIL* : *Latex* : in swellings; *TRIBES OF UPPER GODAVARI DISTRICT* (Andhra Pradesh) : (i) *Root* : in dysentery, elephantiasis, (ii) *Leaf* : in complaints of eye; *IRULAR* : *Latex* : as abortifacient.

HOMOEOPATHY : diaphoretic, expectorant, emetic and valuable remedy for skin diseases; aqueous and alcoholic extract caused marked contraction of isolated rat uterus; the contractions was not altered by atropine but were blocked by papaverine.

Modern use : *Plant* : shows anticancer properties.

Phytography : A stout, hoary, tomentose shrub or occasionally a small tree; leaves large, more or less oblong-ovate, 1–2 cm with a cordate, often amplexicaul base; flowers handsome, reddish, light purple or lilac or white or nearly blue; follicles 75 mm –1.0 cm, recurved, obtuse; seeds many with comas.

Phenology : *Flowering* : December–July; *Fruiting* : February–June.

Distribution : Distributed throughout India; Bangladesh and Pakistan.

Ecology and cultivation : Frequent in open waste ground, along railway tracks, exposed situations, sometimes subgregarious; favours arid conditions, cannot stand waterlogging; wild.

Chemical contents : *Plant* : bitter resin akundarin, calotropin; *Latex* : strong irritant to skin, 0.45% uscharin, 0.15% calotoxin, β-calotropeol, β-amyrin and calcium oxalate, *Root-bark* : β-amyrin, 2-isomeric crystalline alcohol, giganteol and iso-giganteol.

Adulterant : It is a substitute for ipecacuanha.

Remarks : Latex is used by Lodhas to treat septic wounds of cattle and by Santals in rinderpest. Ethnic communities of eastern India use leaves to treat swelling of legs of domestic animals.
Mundas use latex to kill crows.
Bow strings and fishing nets are made with fibres of the plant.
Lodhas believe that the white-flowered variety, when planted near a house, protects the inmates from epidemic diseases like cholera and small pox.
The Hindus consider the species as sacred, and use the flowers to worship Lord Shiva.
A bachelor, according to religious belief, should marry this plant before marrying a widow.
In Sri Lanka, fresh leaves are used as poultice.
In China, fresh leaves are rubbed over skin of elephants to cure *Keserayer* disease.

Calotropis procera (Ait.) R. Br. (Asclepiadaceae)
(2n = 22, 26)

Sanskrit name : Alarka.

Vernacular names : Ben : Akanda; **Hin :** Akanda, Madar; **Lod :** Akon; **Mar :** Mandara; **Mun :** Kula toa; **Pun :** Ak; **San :** Arak; **Tam :** Vellurukku.

Traditional use : LODHA : *Latex* : in toothache, in extracting thorn or spine—inserted deeply in any muscle, in fever caused by spleen disorder; MUNDA : (i) *Latex* : to mothers lacking milk, (ii) *Leaf* (fomented) : against abdominal pain; ORAON : *Latex* : on rheumatic swellings; SANTAL : *Leaf* (fomented) : to swellings due to sprain.

Modern use : Same as *Calotropis gigantea.*

Phytography : A shrub resembling *Calotropis gigantea* but not over 1.20 m high; leaves opposite, simple, oblong, cordate, usually suddenly, often sharply acute or submucronate; peduncles often paired, tomentose, and often branched; flowers about

12 mm, having lobes erect, white with purple blotch on the upper half, coronal process with a purple tinge, obtuse, muticous tip and fleshy upturned white base; follicles recurved, somewhat sausage-shaped, outer and inner coats with fibrous tissue between; seeds ovate, ± 6 mm, smooth with long comas.

Phenology : *Flowering* and *Fruiting* : most of the year.

Distribution : More or less throughout India in warm, dry places in Rajasthan, Uttar Pradesh, Bihar, Punjab, Western, Central and South India; Pakistan (Sind, Punjab and NW Frontier Province).

Ecology and cultivation : Wastelands and arid zones.

Chemical contents : *Stalk* and *Leaf* : calotropin, calotropagenin; *Latex* : contains five cardiac steroid glucosides with the same aglycone, calotropagenin, for which a structure is being postulated.

Remarks : Lodhas and Santals use fresh latex as piscicide and to kill mad dogs. Tribes and rural people of India make pillows with seed floss, get the same warmed by sun for using against pain on neck region.

Caltha palustris L. (Ranunculaceae)

(2n = 28, 32, 52, 60)

English name : Marsh marigold.

Vernacular names : Pun : Baringu, Mumiri, Mamiri.

Trade name : Mumiri.

Traditional use : *Root* : acrid, poisonous.

Modern use : *Plant* (EtOH extract) : spermicidal activity in human semen and used in fertility control, Central Nervous System depressant.

Phytography : Succulent, perennial herb; stem hollow, ± 60 cm high; leaves cordate, 0.5–1.5 cm wide; flowers bright yellow, 2–5 cm broad.

Phenology : *Flowering* : April–June.

Distribution : Western temperate Himalaya, Kashmir from ± 2200 to 3500 m; NW Frontier Province of Pakistan, .

Chemical contents : *Root* : helleborin and veratrin; *Whole plant* : saponin.

Calycopteris floribunda (Roxb.) Lam. (Combretaceae)

(2n = 48)

Sanskrit names : Susavi, Shvetadhataki.

Vernacular names : Hin : Kokkarai, Kokaray; **Kan :** Marasadaboli; **Mal :** Pullani, Varalli; **Mar :** Ukshi; **Ori :** Kukundi, Dhonoti; **Tam :** Minnarkoti; **Tel :** Adivijama, Bandimurugudu, Muruguditige.

Trade name : Kokaray.

Traditional use : *Root* : useful in snake bite; *Leaf*: bitter, astringent, laxative, anthelmintic, depurative and febrifuge; also beneficial in intestinal worms, colic, dysentery, leprosy, malarial fever, ulcers as well as vomiting; *Flower* : as poultice for headache; *Fruit* : useful in jaundice, pruritus and skin diseases.

Phytography : Scandent, woody shrub with slender, brown-streaked, diffused branches, occasionally twining around supports and storing abundant water; leaves simple, opposite, elliptic, acute or acuminate, surface tomentose, main nerves 5–8 pairs; flowers in terminal panicles, yellowish green, apetalous; fruits 5-ribbed, one-seeded.

Phenology : *Flowering* : February–March; *Fruiting* : April onwards.

Distribution : In deciduous forests up to an elevation of 750 m, as well as along the watercourses and wastelands in Madhya Pradesh, Orissa, Assam; Bangladesh (Chittagong).

Ecology and cultivation : Foothills, wastelands and near watercourses, semi-evergreen straggler with arching branches/lianes; globose leafgalls characteristic.

Chemical contents : *Plant* : Calycopterin. *Leaf* : 7% tannin.

Camellia sinensis (L.) O. Kuntze (Theaceae)

(2n = 30)

Syn : *Camellia theifera* Griff., *C. thea* Link

English names : Tea, Chinese tea.

Sanskrit names : Shyamaparni, Chaha.

Vernacular names : Ben : Cha; **Hin :** Chay, Chayeh; **Mal :** Teyila; **Tam :** Teyilai; **Tel :** Theyaku.

Trade names : Chaye, Cha.

Traditional use : *Leaf* : Astringent, appetiser, bitter, carminative, diuretic, diaphoretic, detergent, digestive, resolvent, thermogenic and nervine tonic; also useful in hyperdipsia, hemicrania, cordiodynia, ophthalmia, haemorrhoids, abdominal disorders, fever, fatigue, inflammations and strangury.

Phytography : Evergreen shrub, 9–13 m high (60–150 cm high owing to pruning); leaves simple, alternate, elliptic-ovate or lanceolate with serrate margins, usually glabrous, leathery; flowers white, fragrant, solitary or 2–4 together; fruits depressed; capsules 3-cornered, 3-seeded.

Phenology : *Fruiting* : throughout the year.

Distribution : Throughout India, cultivated at high altitudes up to 3500 m, Assam and hilly countries to the east and south of it—Brahmaputra, Surma Valleys, Darjeeling, Jalpaiguri, Tripura, Kangra, Kumaon, Malabar, Coimbatore and the Nilgiris are the principal places for tea growing.

Ecology and cultivation : Prefers slopes of hills, damp atmosphere; cultivated in specific gardens.

Chemical contents : Distinctive characters of tea beverage due to caffeine, polyphenols, essential oil. *Leaf* : contain carotene, riboflavin, nicotinic acid, pantothenic acid, ascorbic acid, malic acid, oxalic acid, kaempferol, quercetin, theophylline, theobromine, xanthine, hypoxanthine, adenine, gums, dextrins and inositol.

Adulterants : Chicory and many others.

Cananga odorata (Lamk.) Hook. f. & Thoms. (Annonaceae)

(2n = 16)

Syn : *Canangium odoratum* L.

English names : Inang-inang tree, Ylang-ylang.

Vernacular names : Kan : Apurvachampaca; **Tam** : Maladi, Karumugai, **Tel** : Chettu sampangi.

Traditional use : *Flower* : a pomade prepared is rubbed on the skin to guard against skin diseases; *Oil* : used as an application in cephalalgia , ophthalmia and gout.

Phytography : Tree of medium or large size with somewhat drooping branches; leaves oblong-ovate, 1.5–2.0 cm long, acuminate with sharp point, glossy above, slightly pubescent beneath; flowers many, very fragrant, greenish to yellowish, narrow, hanging; fruits $\pm$ 25 mm, greenish, oblong-cylindric.

Distribution : Cultivated in gardens throughout India; Pakistan and Bangladesh .

Ecology and cultivation : Introduced; planted for fragrant flowers.

Chemical contents : *Flower* : essential oil.

Canarium commune L. (Burseraceae)

English names : Java almond, Kinari nut.

Vernacular names : Hin : Jangli badam; **Kan :** Kagil mara; **Mal :** Karaichingasi; **Mar :** Jangli badam.

Trade name : Jangli badam.

Traditional use : *Resin* : in the form of ointment applied to indolent ulcers; substitute for *Mistura amygdalae*; *Fruit* : laxative.

Modern use : *Volatile oil* : antiseptic against fungi and bacteria.

Phytography : Balsamiferous tree, extremities tawny puberulous or glabrate; leaves alternate, stipules auricled, often deciduous, leafblade imparipinnate, leaflets 7–9, ovate, acuminate, glabrous, leaves of flowering branches 23.5–45.0 cm; panicles terminal, puberulous with spreading successively shorter lateral branches; flowers bracteate, polygamous, variable in size; drupes ellipsoidal, subtrigonous with a bony 1- to 3-celled stone.

Distribution : Indigenous to Malaysia and now cultivated extensively in different parts of South India, particularly in Kerala.

Chemical contents : *Essential oil from oleoresin* : 34% anethole and a small quantity of terpenes; *Seed* : moisture, protein, fat and ash.

Remark : In Cambodia, tubers are considered diaphoretic, styptic.

Canavalia gladiata (Jacq.) DC. (Fabaceae)

(2n = 22)

Syn : *Canavalia ensiformis* sensu Baker non DC., *Dolichos gladiata* Jacq.

English names : Jack bean, Sword bean, Broad bean.

Sanskrit names : Asisimbi, Masimbi.

Vernacular names : Ben : Makhan shim; **Hin :** Lalkudusumpal, Khadsampal, Barasim, Makhan sim; **Kan :** Shembi avari, Tumbekai; **Mal :** Valvara, Valpayar, Valamara; **Mar :** Pandhri abai; **Tam :** Segapputampaltai, Vellai tambattai; **Tel :** Vella tamma.

Trade names : Barasem, Makhan sem.

Traditional use : *Fruit* : appetiser, astringent, cooling, digestive, sweet and vulnerary, useful in anorexia, burning sensation, dyspepsia, hyperdipsia, wounds, ulcers and vitiated conditions of *kapha* and *pitta*.

Modern use : *Aerial parts* (50% EtOH extract) : diuretic.

Phytography : Stout perennial twiner with glabrous stem and branches; leaves compound, leaflets membranous, acute or acuminate, apiculate, base rounded, glabrous in both sides; flowers large, white or lilac in lax few-flowered curved axillary racemes; fruits large, sword-shaped flattened pods; seeds reddish brown or white, strongly compressed.

Phenology : *Flowering* : October, January and July–September; *Fruiting* (*Pods*) : throughout the year.

Distribution : Throughout India; Bangladesh and Pakistan.

Ecology and cultivation : Plains from the coast to 300 m, very common in the scrub jungles and thickets; abundant where ground has been disturbed.

Chemical contents : *Fruit*: Cystin, tyrosin, tryptophan, and alkaloids.

Remark : Nomenclature is very puzzling, as has been pointed out by Wight and Arn. It requires immediate revision.

Canna indica L. (Cannaceae)

(2n = 18)

Syn : *Canna orientalis* Rose., *C. indica* L. var. *orientalis* Rose. (including *C. indica* L. of Watt.).

English names : Canna, Indian shot.

Sanskrit names : Devakili, Sarvajaya.

Vernacular names : Ben and **Hin :** Sarvajaya; **Kan :** Hudingana; **Lod :** Kodaimulo-ba, Rokto-sora; **Mal :** Kattuvalla; **Mar :** Devakeli; **Mun :** Kodai-muli ba; **Tam :** Kalavalai, Puvalai; **Tel :** Krishnamatara.

Trade names : Sarvajaya, Canna.

Traditional use : MUNDA : *Rhizome* (paste) : on bone fracture; SANTAL : (i) *Rhizome* : (paste) : on ringworm (ii) *Rhizome* (decoction) : in irregular menstruation.

Modern use : *Aerial parts* (50% EtOH extract) : Hypotensaive.

Phytography : Herb, rhizome creeping, stem glabrous, erect; leaves ovate-elliptic or oblong, 30–50 by 10–15 cm, chartaceous, glabrous, base tapering, entire, acuminate, nerves close; panicles up to 20 cm, bracts ovate or narrowly linear, 4–7 cm long; flowers paired, up to 10 cm long, yellow, orange or red; capsules 2–4 by 1–2 cm, warty; seeds globose, up to 8 mm.

Phenology : *Flowering* : throughout the year.

Distribution : Cultivated throughout India, Bangladesh and Pakistan for its ornamental flowers and foliage.

Ecology and cultivation : Hills, generally as ferals, by plantations, near habitations.

Remark : Stalk cut into pieces and boiled with rice-water and pepper, given to cattle as antidote to effects caused by eating poisonous grasses.

Cannabis sativa L. (Cannabinaceae)

(2n = 20)

English names : Cannabis, Indian hemp (true and soft), Marijuana.

Sanskrit names : Bhang, Ganjika.

Vernacular names : Ben, Hin and **Mar :** Bhang, Charas, Ganja, Siddhi; **Guj :** Bhang, Ganja, Charas, Jia, Siddi; **Kan :** Bhangi; **Pun :** Bhang; **Tam :** Bhang, Ganja; **Tel :** Ganzai.

Trade names : Hemp, Ganja.

Traditional use : KHASI AND GARO : *Leaf* : in skin diseases; TRIBES OF PURULIA (West Bengal) : *Leaf* : as narcotic; BHOXA (of Uttar Pradesh) : *Leaf* : in piles; KUMAONI : (i) *Leaf* : as anthelmintic, in cuts, piles, (ii) *Leaf* and *Fruit* : as sedative; GARHWALI : (i) *Leaf* : in skin diseases; (ii) *Fruit* : as laxative; TRIBES OF DEHRA DUN (Uttar Pradesh) : (i) *Bark* : in hydrocele; TRIBES OF SAGAR (Madhya Pradesh) : (i) *Leaf* : as appetiser, in bowel complaints, gonorrhoea; (ii) *Leaf* and *Fruit* : as nervine stimulant;

TRIBES OF BANSWARA (Rajasthan) : *Leaf* : in piles; *TRIBES OF KURUKSHETRA* (Haryana) : *Leaf* : in bronchitis, piles.
AYURVEDA : Plant used as tonic, intoxicant, stomachic, antiseptic, analgesic, narcotic, sedative and in anodyne asthma; it acts upon the nervous system, producing first excitement accompanied by hallucinations and afterwards lethargy and sleep. It is used as a sedative in mania and hysteria.
UNANI : ingredient of 'Qinnab (Bhang)'.
HOMOEOPATHY : Dispelling of illusion, oblivion, mental weakness, heart diseases, headache, urinary troubles, whooping cough, female diseases and gonorrhoea.

Modern use : Isolation of a sedative and antibacterial principle distinct from unknown constituents of hemp. Bactericidal effect observed on gram-positive microorganism up to a dilution of 1 : 1,50,000.

Phytography : Hemp is a tall annual herb, 1.2–4.8 m high with erect angular stem; leaves palmately divided; flowers greenish, dioecious, rarely monoecious, male flowers borne in long drooping panicles and female flowers in short axillary spikes; fruit ovate, seed-like achene.

Distribution : Naturalised in the sub-Himalayan tract and abundant in wastelands from Pakistan to West Bengal, Bihar and South India; Bangladesh.

Ecology and cultivation : Largely cultivated in temperate countries for its strong fibres; seed is sown in August and seedlings are planted out in September on a carefully tilled and well-manured field.

Chemical contents : Sp. Gr. 0.9285, Sap. val. 90–193, Iod no. 150–166. Unsap. 1–1.3% cannabinol, pseudo-cannabinol, cannabinin.

Remarks : *Bhang* is composed of dried leaves and flowering shoots of male and female plants, both cultivated and wild. *Ganja* consists of dried, unfertilised female inflorescence. *Ganja* and *Charas* are two forms in which drug is used for smoking, whereas *Bhang* is taken in the form of a drink.
Dried pistillate plants (flowering top) are used medicinally for smoking. Unlawful in many countries, much used by Fakirs, Sadhus and poorer section of the people in Africa and Asia. Used by criminals to drug people; for this purpose it is mixed with black Datura and sugar. Excessive smoking is physically and mentally injurious, causing moral weakness and depravity. *Bhang, siddi* or *patti* are the dried leaves used for preparation of a drink.

Canscora decussata Schult. (Gentianaceae)

(2n = 64)

Sanskrit name : Shankhapushpi.

Vernacular names : Ben : Dankuni; **Hin :** Shankhaphuli; **Mar :** Sansvel.

Trade name : Shankhaphuli.

Traditional use : *TRIBES OF SANTAL PARGANAS* : *Stem* : in stomachache, as blood purifier; *Plant* : laxative, alterative and nerve tonic; *Fresh juice of plant* : prescribed in insanity, epilepsy and nervous debility.

Modern use : *Root* : anti-inflammatory, Central Nervous System depressant; *Whole plant* : fertility control by spermicidal activity in human semen.

Phytography : Herb, 15 to 35 cm; stem narrowly 4-winged; leaves elliptic-lanceolate, 2–3 by 1.0–1.5 cm, 3-nerved, base rounded to acute, apex gradually acute; cymes dichasial, axillary or terminal, peduncles 2.5 cm, pedicel 1 cm, winged; flowers white; capsules obscurely two-valved, 1.5 by 0.4 cm,; seeds angular.

Phenology : *Flowering* : November–March; *Fruiting* : December onwards.

Distribution : Occurring mainly on the walls of the plains and throughout India, Pakistan and Bangladesh up to an elevation of 1500 m and growing in moist situations.

Ecology and cultivation : Plains up to 1500 m, often in clumps of moist places like exposed slopes, wayside puddles with the rains.

Canthium dicoccum (Gaertn.) T. & B. (Rubiaceae)
(2n = 44)

Syn : *Canthium didymum* Gaertn., *Plectronia didyma* Kurz, *Psydrax dicoccum* Gaertner.

Sanskrit name : Garbhagoja.

Vernacular names : Kan : Hatteranike; **Mar :** Arsul, Varsinghi; **Tam :** Imburuttan; **Tel :** Nalla balasu.

Traditional use : *RURAL FOLK OF DARJEELING* : *Bark* : in fever; *TRIBES OF ORISSA* : *Root* : in boils, diarrhoea.

Phytography : Densely foliaceous evergreen unarmed tree, ± 14 m, tender parts pubescent, crown spreading, compact; leaves dark green, shining, elliptic-ovate or oblanceolate, 8–12 by 4–7 cm, stipules lanceolate; cymes laxly dichashial or densely umbellate; flowers cream-coloured; drupes ellipsoid-globose, 0.6–0.8 cm across.

Phenology : *Flowering* : December–March, July–September; *Fruiting* : throughout the year.

Distribution : Meghalaya, Sikkim, Arunachal Pradesh, all plain districts of Tamil Nadu, on both sides of the peninsula.

Ecology and cultivation : Grows on hills between 500 and 1200 m, on the slopes; less common in scrub jungles in the plains.

Chemical contents : *Bark* : α-amyrin, canthic acid, oleanolic acid, quinovaic acid, Ac-quinovaic acid, sapogenin, scopoletin, sitosterol; *Leaf* : δ-mannitol, 3-epibetulin, flavonoids, tannins.

Canthium parviflorum Lamk. (Rubiaceae)
(2n = 44)

Syn : *Plectronia pa* (Lamk.) Bedd.

Sanskrit name : Gangekuri.

Vernacular names : Mar : Kirma, Kadbar, Kirni; **Tam :** Karai, Mullukaarai, Sengaraarai, Theravai, Theranai; **Tel :** Karai, Balasu.

Traditional use : *TRIBES OF ORISSA* : (i) *Root* : in swelling of neck; (ii) *Fruit* : in headache.

Phytography : Armed shrub up to 5 m; spines up to 4 cm, supra-axillary; leaves 2–4 by 1.5–2.5 cm, elliptic-ovate to obovate, chartaceous, lateral nerves 4–5, glabrous, base alternate, apex subacute, petioles up to 0.7 cm, stipules linear; flowers in decussate lax cymes, up to 1 cm, greenish; drupes globose, orange on ripening, 1 cm across, pyrenes furrowed, pedicel up to 1 cm.

Phenology : *Flowering* : May–August; *Fruiting* : July onward.

Distribution : Peninsular India, Coromandal coast and Karnataka towards Sri Lanka.

Ecology and cultivation : Plains from the coast; scrub jungles up to 750 (1000) m, common; wild.

Capparis decidua (Forssk.) Edgew (Capparidaceae)

(2n = 44)

Syn : *Capparis aphylla* Roth, *Sodada decidua* Forssk.

Sanskrit names : Gudhapatra, Karira.

Vernacular names : Guj : Ker; **Hin** : Karer, Kurrel, Karira; **Kan :** Chippu; **Mal :** Kari, Ker; **Pun :** Delha, Karil; **Tam :** Sengam; **Tel :** Kariramu.

Trade name : Karira.

Traditional use : *TRIBES OF INDIAN DESERT* : *Stem* : in tooth troubles including pyorrhoea; *TRIBES OF GUJARAT* : *Leaf* : as appetiser, in cardiac problems, tooth troubles; *TRIBES OF BANDA DISTRICT* : (i) *Stem* : in tooth troubles, (ii) *Fruit* : in intestinal worms; *TRIBES OF DAHSAN VALLEY* : *Stem* : in rheumatism and tooth troubles.
AYURVEDA : paste of coal from wood is applied to muscular injuries; *Fruit* and *Seed* : used in cholera, dysentery and urinary purulent discharges.
UNANI : *Shoot* (along with shoot of *Peganum hurmala*) : used against antifertility.

Phytography : Leafless, much-branched, dense shrub; thorns in pairs, straight; glabrous, linear, pungent leaves found on very young shoots; flowers in simple corymbs on short lateral shoots, red-brown, 2.5 cm in diameter; sepals ciliate; fruits 1.25–1.9 cm, glabrous, long-beaked.

Distribution : Gujarat, Kutch, Rajasthan, Punjab, upper Ganga plains, Madhya Pradesh and in many parts of Tamil Nadu, Karnataka and Andhra Pradesh; Pakistan (Punjab, Sind, Baluchistan).

Chemical contents : *Root* : Capparine, cappariline, capparinine; *Roo-tbark* : n-pantacosane, n-tricontanol, β-sitosterol, l-satachydrine; *Flower* and *Fruit-husk* : phthalic acid, l-stachydrine; *Seed* : n-triacontanol, β-sitosterol, β-carotene, glucocapparin.

Remarks : Bhils use stem and leaves in treatment of stomach troubles of domestic animals. Rural folks of Delhi, Rajasthan and Saurashtra (Gujarat) eat leaves and fruits as vegetables during famine.

Capparis grandis L.f. (Capparidaceae)

(2n = 42)

Vernacular names : Guj : Dhuti; **Kan :** Revapi; **Mal :** Waghutty; **Mar :** Kuntel, Puchaonda; **Tam :** Nakkulinjan, Turatta; **Tel :** Oridonda.

Traditional use : An infusion of *Bark* and *Leaf* is used for swelling and eruptions.

Phytography : A perennial shrub; leaves simple, petioled, usually with spines or bristle-like stipules; flowers white, often bracteate; berry many-seeded, globose.

Distribution : Rajasthan, Gujarat, Maharashtra, Karnataka, Andhra Pradesh and some other parts of South India; also found in the drier regions of Pakistan.

Ecology and cultivation : Grows in dried hard soil, especially in the arid zones; wild.

Chemical content : *Seed* : yield a fatty oil used as an illuminant.

Capparis moonii Wight (Capparidaceae)

Vernacular name : Kor : Waghati.

Traditional use : KORKU (of Maharashtra) : *Fruit* : in cough, spitting, weakness in tuberculosis.

Modern use : *EtOH* (50%) *extract of aerial parts* : causes depression on Central Nervous System.

Phytography : Large woody climber, branches glabrous, thorns recurved; leaves simple, oblong, 7.5–15.0 cm by 3.7–5.6 cm, leathery, shining above, pale beneath; corymbs terminal, 6- to 12-flowered; flowers white, 10.0–12.5 cm in diameter, pedicels 2.5–5.0 cm, stout, gynophore 7.5–10.0 cm long; fruits subglobose or obovoid, 5–10 cm in diameter, many- seeded; seeds large like bean.

Distribution : Konkan, ascending up to 1000 m.

Ecology and cultivation : Prefers hot climate; wild.

Capparis sepiaria L. (Capparidaceae)

(2n = 40,42)

Sanskrit name : Kakadani.

Vernacular names : Ben : Kaliakara; **Hin :** Kanthaii.

Traditional use : TRIBES OF MOUNT ABU (Rajasthan) : (i) *Plant* : in swelling; (ii) *Leaf* : in skin diseases; ETHNIC COMMUNITIES OF KHEDABRAHMA (Gujarat) : *Root* : in earache, mumps; TRIBES OF DAHANU FOREST (Maharashtra) : *Root* and *Stem-bark* : in dropsy, gout; TRIBES OF DHARMAPURI FOREST (Tamil Nadu) : *Seed* : as antidote to snake bite.

AYURVEDA : (i) *Plant* : considered as pungent, bitter, effective rejuvenator, appetiser and preventer of aging; (ii) *Decoction of plant* : used in cutaneous affections, as febrifuge and tonic.

SIDDHA SYSTEM : *Bark* used to prepare *Cenkattarip pattai.*

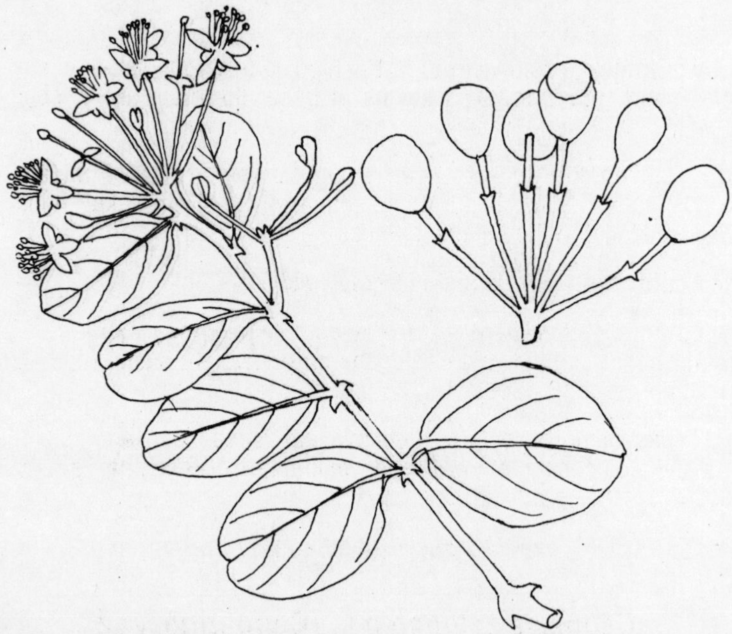

Phytography : A large climbing shrub, thorns curved; leaves elliptic, 1.9–3.2 cm by 1.2–1.9 cm, petioles 0.2 cm; umbels simple; flowers white, 0.8–1.2 cm in diameter, pedicels short; fruits borne in clusters, pisiform, black.

Phenology : *Flowering* : Summer; *Fruiting* : Rainy season.

Distribution : Dry places throughout India; Pakistan, Sri Lanka and Myanmar.

Ecology and cultivation : Prefers dry climate; wild.

Chemical contents : *Leaf* : taraxasterol, β-sitosterol, α & β-amyrin, erythrodiol.

Capparis spinosa L. (Capparidaceae)

(2n = 24, 38)

English name : Caper bush.

Vernacular names : Hin : Kabra; **Kum :** Bussar; **Pun :** Kabra; **Tel :** Kokilakshamu.

Trade name : Kabra.

Traditional use : *TRIBES OF NORTH-EAST INDIA* : (i) *Bark* : in paralysis, toothache; (ii) *Root* and *Bark* : in rheumatism. Ingredient of indigenous drug 'Geriforte'.
UNANI : componenet of 'Kabar'.

Modern use : *Plant-juice* : Drug for acute viral hepatitis, major ingredient of 'Liv-52', useful in cirrhosis of liver.

Phytography : Straggling spiny shrub, ± 1 m high; leaves roundish or ovate, 25–50 mm long, petioles short with 2 reflexed spines at base; flowers axillary, solitary on thick peduncles, white.

Distribution : Native to the Mediterranean region; distributed to NW Himalaya, Rajasthan, plains of Indus and Jhelum, Salt Range, Maharashtra, Karnataka, Andhra Pradesh; Pakistan (Baluchistan and the adjacent places).

Ecology and cultivation : Prefers dry climate; wild.

Chemical contents : *Root-bark*, *Cortex* and *Leaf* (aqueous extract) : stachydrine; *Flower-bud* : glucoside, rutin, 40% pentosans on dry weight basis, rutic acid, a volatile emetic constituent, saponin; *Seed* : 34–36% of an oil.

Remarks : Tribes of northwestern Himalaya eat leaves as vegetable; hill tribes of Maharashtra and Goa consume flower-bud, while tribes of Rajasthan eat the fruits.

Capparis zeylanica L. (Capparidaceae)

(2n = 40)

Syn : *Capparis horrida* L. f.

Sanskrit names : Ahimsra, Karambha.

Vernacular names : Ben : Kalokera; **Hin :** Ardanda, Arnanda, Gitoran; **Kan :** Tottula; **Lod :** Maru-janam; **Mar :** Govindi; **Mun :** Gaterna, Mari-janum; **Orn :** Mari-punp; **Pun :** His, Karvila; **Sad** and **San :** Bagnai; **Tam :** Adondai, Sudhuthorati, Morandaan; **Tel :** Adonda.

Trade names : Arnanda, Ardando.

Traditional use : TRIBES OF PURULIA (West Bengal) : *Root* and *Fruit* : as antidote to snake bite; LODHA : (i) *Root-bark* (pasted with long peppers) : in small pox, (ii) *Roo- bark* (paste) : to cure swelling of testicles; MUNDA : (i) *Root* (pounded) : rub as curative of aching body, leg and foot, (ii) *Leaf* (crushed) : in small pox; ORAON : *Stem-bark* (decoction) : on cholera; THARU (of Uttar Pradesh) : *Root-bark* : in boils, swelling of testicles; RURAL FOLKS OF DEHRA DUN and SIWALIK region : (i) *Root-bark* : as anthelmintic, in cholera; (ii) *Leaf* : in piles; TRIBES OF GORAKHPUR (Uttar Pradesh) : *Fruit* : in snake bite; TRIBES OF MOUNT ABU (Rajasthan) : *Leaf* : in boils, piles; SANTAL : use different parts in colic, hemiplagia, neuralgia, pleurisy, pneumonia, sores. AYURVEDA : antirheumatic and anti-inflammatory.

Modern use : *EtOH* (50%) *extract of aerial parts* : spasmolytic.

Phytography : Straggler, 4–8 m; branchlets stellate tomentose or glabrescent; leaves ovate, elliptic or lanceolate, rarely obovate, 2.0–6.5 by 1.5–5.0 cm, coriaceous, apex obtuse or retuse, strongly mucronate, petiole up to 8 mm, prickles/thorns recurved; flowers in serial clusters on fresh shoots, 4 cm across, pedicels 1 cm; berry globose, 3.5 cm; seeds many, 5 mm.

Phenology : *Flowering* : February–April; *Fruiting* : Rainy season.

Distribution : Throughout major parts of India; Bangladesh and some parts of Pakistan.

Ecology and cultivation : Grows in moist habitat; wild.

Chemical contents : *Leaf* and *Seed* : α-amyrin, a fixed oil, glucocapparin, n-tricontane.

Remark : Fruits are eaten by the tribes of Santal Parganas, hilly regions of Poona and the people of Salsette Island (near Mumbai).

Capsella bursa-pastoris (L.) Medic. (Brassicaceae)

(2n = 16)

English name : Shepherd's purse.

Traditional use : *RURAL FOLK OF DEHRA DUN AND SIWALIK REGION* : *Plant* : in blood pressure, diarrhoea and dropsy.

Modern use : *Plant* : haemostatic, good substitute for ergot of rye to arrest bleeding from the kidney, lungs and uterus; *Leaf* (extract) : show antifertility activity in animals.

Phytography : Small-branched annual herb, height varies from 15 to 40 cm, glabrous or hairy; leaves simple, alternate, usually pinnatifid, rarely entire; flowers in racemes, white, 0.25 cm in diameter, cruciform; pods 0.6–0.8 cm; seeds many, oblong, punctate.

Distribution : A weed of cultivation throughout temperate India, especially in northwestern Himalaya.

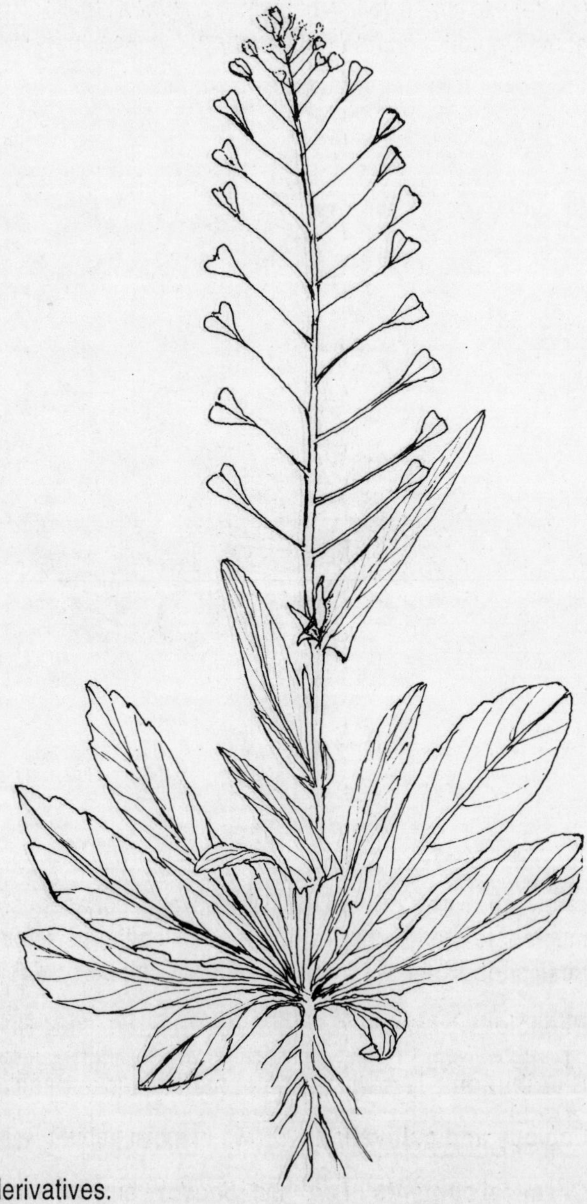

Chemical contents : *Plant* : rutin, hesperidin, choline, histamine, sinigrin, stigmasterol, fumaric acid, essential oil; *Dried plant* yields saponin, an alkaloid bursin, a fatty oil (35%); *Seed* : quercetin derivatives.

Remarks : Used in Chinese medicine in ailments of eyes and also in dysentery. *Herba bursae-pastoris* was used as diuretic, febrifuge and haemostatic.

Capsicum frutescens L. (Solanaceae)

(2n = 12, 22, 48)

Syn : *Capsicum annuum* L., *C. baccatum* L.

English names : Green pepper chilli, Cayenne pepper.

Sanskrit names : Katuvirah, Raktamaricha.

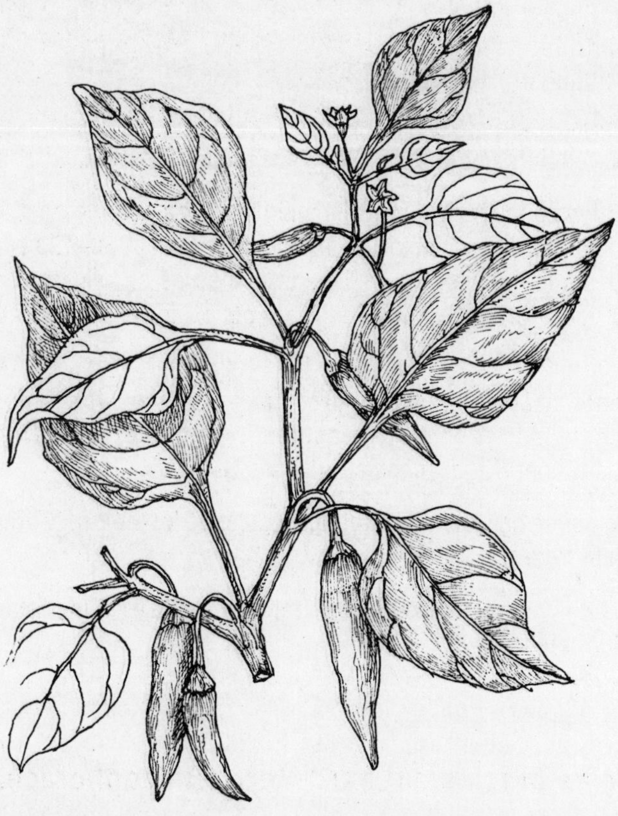

Vernacular names : Ben : Lanka, Marich, Lalmarich; **Hin :** Lal mirch, Gachmarich; **Kan :** Mensina kai; **Mal :** Mulaku; **Tam :** Milagay; **Tel :** Mirapa kaya.

Trade names : Lal mirch, Marich (dry chilli of commerce).

Traditional use : Fruits extremely pungent and given internally as powerful stimulant and carminative, especially in flatulent dyspepsia; *Tincture* of capsicum sometimes added to gurgles for pharyngitis and relaxed throat troubles; externally employed for the treatment of lumbago, neuralgia and rheumatism; in the form of ointment, applied to painful joints in rheumatoid arthritis, hoarseness and in yellow fever, in diarrhoea and snake bite; also to rouse appetite.

In *SIDDHA SYSTEM*, fruit is used to prepare the drug named *Milakai.*
HOMOEOPATHY : good medicine for the constriction of nose, cheek, throat, dysentery, earache, asthma, respiratory troubles and gonorrhoea.

Phytography : A perennial shrub; inflorescence of several flowers; fruits small, clustered, erect, conical, pointed, 2–3 cm long berries, young fruits green, becoming red on ripening, extremely pungent.

Phenology : In india, *kharif* crop is planted in July–September and *rabi* crop in February–March; *Flowering* : with a peak in January–May; *Fruiting* : throughout the year—green through yellow to red.

Distribution : An important condiment crop; native to West Indies and tropical America; extensively cultivated throughout India; Bangladesh and Pakistan.

Ecology and cultivation : Chilli is grown almost from sea-level to an altitude of 1500 m in tropical and subtropical conditions with an annual rainfall of 60–130 cm. In low-rainfall tracts or when grown in hot weather, chilli is cultivated as an irrigated crop; the rainfed crop does well on deep, fertile, well-drained, black cotton soil and somewhat heavy clayey loam; propagated by seeds.

Chemical contents : *Leaf* : luteolin glycerides; *Fruit* : quercetin derivatives, apiin, luteolin 7-glucoside carotenoid, capsanthene, carotene, capsorubin, oleoresin, capsaicin(e), capsicidine; *Seed* : capsicoside, sterols.

Adulterants : Japanese chillies; these are distinguished by their very bright, crimson red colour and freedom from stalk.

Remark : The plant is not mentioned in any Sanskrit treatise on medicine; introduced in India at some later date.

Carallia brachiata (Lour.) Merr. (Rhizophoraceae)

Syn : *Carallia interrima* DC.

Vernacular names : Asm : Kanthikera; **Ben :** Kierpa, **Hin :** Maniawya; **Kan :** Andipunaru; **Mal :** Varanga; **Mar :** Panasi, Punschi; **Ori :** Manj; **Tel :** Karalli.

Trade name : Carollina wood.

Traditional use : *MIRI* (of Assam) : *Fruit* : in stomach complaints.

Phytography : Small tree with shining, leathery, oblong, elliptic or obovate, shortly suddenly obtuse or obtusely acuminate leaves and stout dense di- to tri-chotomous cymes of a small greenish flowers with inconspicuous white erose petals; fruits ± 6 mm.

Phenology : *Flowering* : December–April; *Fruiting :* Rainy season.

Distribution : Eastern Himalaya, Assam, West Bengal, Maharashtra, South India, Andamans; Bangladesh and Myanmar.

Caralluma edulis Benth. ex Hook. f. (Asclepiadaceae)

(2n = 22)

Sanskrit name : Dugdha.

Vernacular names : Mar : Pippa, Pippu; **Pun :** Chung.

Traditional use : *Plant* : used as a cooling, alterative and anthelmintic, in leprosy and diseases of blood.

Phytography : Erect herb; stem and branches subterete, 4-grooved; leaves linear, acute; flowers 1–4 together from nodes of the stem, subdivided to the base, glabrous, companulate, purple inside.

Phenology : *Flowering* : September.

Distribution : Punjab, Peninsular India; Pakistan (Baluchistan, Sind).

Remark : The plant is eaten as vegetable.

Cardamine pratensis L. (Brassicaceae)

(2n = 16, 20, 30, 36, 76)

Syn : *Cardamine laxotemonoides* Schultz.

English names : Cuckoo flower, Ladies' smock.

Trade name : Cuckoo flower.

Traditional use : *Plant* : stimulant, diaphoretic and diuretic; *Leaf* : pungent and antiscorbutic; *Flowering top* : reported to be used in epilepsy and other nervous affections.

Modern use : *Plant-juice* : against tumours.

Phytography : Slender, erect, glabrous perennial herb, 3–7 cm high with attractive flowers in spring and early summer; leaves pinnately divided, small, oblong or rounded;

flowers white or rose-purple, bright and pretty, ± 13 cm long in a terminal erect corymb-like raceme, sometimes double; fruit ± 25 cm long.

Phenology : *Flowering* : Spring and Summer.

Distribution : Kashmir; western Tibet.

Chemical contents : *Flower* : Kaempferol glyceride; *Seed* : myronic acid and an oil similar to mustard oil.

Adulterant : Sometimes used as a substitute for Cress.

Remark : Emergency food plant.

Cardiospermum halicacabum L. (Sapindaceae)

(2n = 22)

English names : Ballon vine, Heartwood, Heartseed.

Sanskrit name : Karnasphota.

Vernacular names : Ben : Lataphatkari, Sibjhul; **Guj :** Karolio; **Hin :** Kanphuti; **Lod :** Merom-med-nari; **Mar :** Kapalphodi; **Tam :** Mudukottan; **Tel :** Buddakakara.

Trade names : Kanphuti, Sibjhul, Ballon vine.

Traditional use : LODHA : (i) *Plant-paste* (in cow's milk) : in swelling of legs, (ii) *Root* (decoction) : as emetic, (iii) *Root-paste* (with paste of long pepper and oil of *Schleichera oleosa*) : in gout, (iv) *Leaf-juice* (with paste of long pepper) : in diabetes; MUNDA : (i) *Leaf-juice* : inhale in headache; (ii) *Dried seed-powder* : as nervine; ORAON : *Leaf-juice* (fresh) : in eye diseases.
UNANI : used in the preparation of 'Habb-ul-Qilqatil'.

Modern use : *Plant* (50% EtOH extract) : hypotensive, spasmolytic; *Seed-oil* : blood presure depressant, *in vitro* antibacterial, hypotensive.

Phytography : Annual wiry herb, thinly pubescent or nearly glabrous with acuminate leaflets; flowers white, very minute; capsules depressed, pyriform; tomentose seeds globose, up to 5 mm.

Phenology : *Flowering* : July–February; *Fruiting* : August–March.

Distribution : Occurring throughout the plains and in lower elevations (up to 1200 m) of India; Bangladesh and Pakistan.

Ecology and cultivation : Mainly in wastelands; plains to the lower slopes (usually 400–500 m); optimum size and abundance on disturbed ground; ready colonizer.

Chemical contents : *Plant* : saponin; *Seed* : cyanogenic glyceride.

Careya arborea Roxb. (Lecythidaceae)

(2n = 26)

English name : Tummy wood.

Sanskrit name : Kumbhi.

Vernacular names : Ben and **Hin :** Kumbhi; **Guj :** Kumbi; **Kan :** Kaval, Doddala, Asanda; **Lod :** Kumbi-daru, Kumbhi; **Mal :** Alam, Pelu; **Mun :** Asanda-daru, Kumbhir; **Sad :** Kumbi; **San :** Khubati; **Tam :** Ayma; **Tel :** Araya, Duddipa.

Trade names : Kumbhi, Kumbi.

Traditional use : LODHA : (i) *Root-paste* : in body pain, (ii) *Root-bark decoction* (with long pepper) : in fever, (iii) *Stem-bark powder* (paste with honey) : to children in cold and cough; SANTAL : *Stem-bark* : (paste with margosa) : in leucoderma

Phytography : Medium-sized tree with large ovate, ovate-oblong leaves clustered at the end of the branchlets, old leaves often red/purple; flowers large, white and pink, in dense spikes, fruits large globose green, crowded with calyx tube, deciduous.

Phenology : *Flowering* : March–May; *Fruiting* : July.

Distribution : Sub-Himalayan tract occurring throughout India up to an elevation of 1700 m.

Chemical contents : *Bark* : lupeol, betulin; *Leaf* : ellagic acid, hexacosanol, quercetin, β-sitosterol, α-spinasterol taraxerol and its OAc, valoneic acid, careaborin, β-amyrin; *Seed* : α-spinasterol, α-spinasterone, baringtogenol-C, careyagenol-E.

Remarks : Only plant named by William Roxburgh in honour of William Carey.
Lodhas use fresh stem bark decoction for washing septic wounds of cattle, and keep a dry fruit in a room as snake repellant.
Oraons use powdered stem, root and leaf for poisoning fish.

Carica papaya L. (Caricaceae)

(2n = 18, 36)

English names : Papaya, Papaw tree.

Vernacular names : Ben : Pepe, Pappaiya, Papeya; **Guj :** Papayi; **Hin :** Papaya, Papita; **Kan :** Parangi mara; **Lod :** Papaw; **Mar :** Papaya; **Mun, Orn** and **Sadri :** Papita; **Tam :** Pappali, Pappayi; **Tel :** Boppayi.

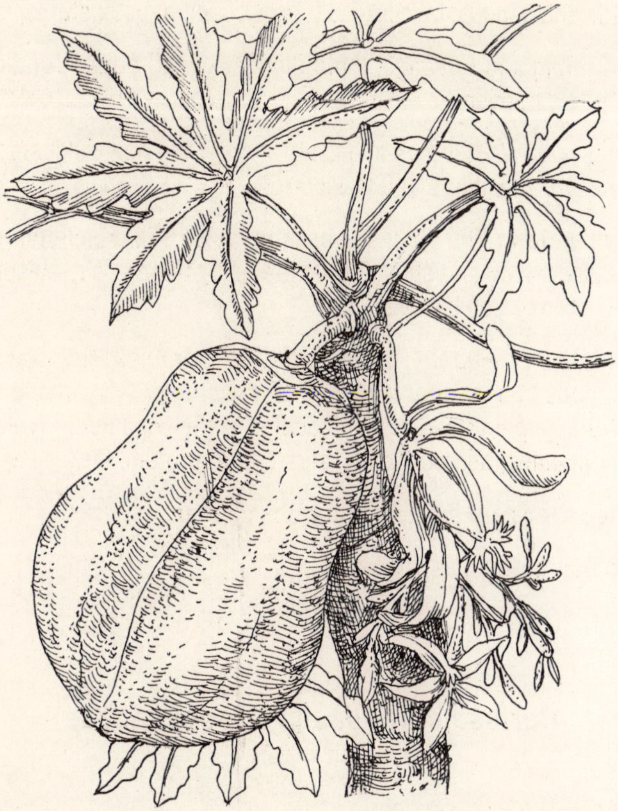

Trade names : Papaya, Pepe.

Traditional use : *Milky latex* obtained from green immature fruit is a valuable commodity of commerce for its protein-digesting property; *Milky juice of unripe fruit* used as a cosmetic to remove freckles and other blemishes from the skin; anthelmintic, particularly effective in the expulsion of lumbrici; *Ripe fruit* : stomachic, carminative and diuretic; *Seed* : vermifuge, emmenagogue, used to quench thirst.
UNANI : an ingredient for the preparation of 'Papita desi'.
HOMOEOPATHY : for preparation of medicines for different stomach ailments.

Modern use : *Papain* is also an important ingredient in digestive medicines. Papain used in the preparation of cholera medicine and in the cure of diphtheria, bleeding piles, ulcers, enlarged spleen, *carpeine*, an alkaloid occurring in green parts and seeds, is reported to have been used as heart-stimulant; it is an amoebicide, particularly for amoebic dysentery.

Phytography : A rapidly growing small tropical American plant containing milky latex; leaves in a terminal cluster, alternate, exstipulate and usually palmately lobed; flowers pentamerous, staminate, pistillate and hermaphrodite, racemose, pistillate and hermaphrodite with a superior one spuriously 5-celled ovary containing numerous ovules; fruit a berry; seeds albuminous; laticiferous tubers occur in roots, stems, leaves and fruits.

Phenology : *Flowering* and *Fruiting* : throughout the year, peak in end of Rainy season and beginning of Cold season.

Distribution : Native of West Indies and Central America, now cultivated throughout the greater part of India; Bangladesh and Pakistan.

Ecology and cultivation : The plant thrives best in rich soil of uniform texture, having a depth of 2 m; medium black and alluvial soils are also suitable; clayey soils are not good; papaya is usually propagated by seed.

Chemical contents : *Papain* is obtained commercially by slashing the fruits; approximate composition of the raw fruit is moisture 92%, pectic substance 1%, carbohydrates 4.5%, papain 0.5–1.0%; in addition to the constituents referred above, the dry fruit is rich in vitamins, ascorbic acid, β-carotene, protein, fatty oil and volatile oil.

Adulterant : Dried seeds are adulterant of black pepper.

Remark : Green fruit is eaten as vegetable.

Carissa carandas L. (Apocynaceae)

(2n = 22)

English names : Caranda, Karanda.

Sanskrit name : Karamarda.

Vernacular names : Ben : Karamcha; **Guj** : Karamarda; **Hin** : Karaunda; **Kan** : Karekayi; **Mar** : Karavanda; **Tam** : Kalakkay; **Tel** : Vaka.

Trade names : Karaunda, Karamcha.

Traditional use : *Root* : bitter, stomachic, anthelmintic; *Decoction of leaf* : given at commencement of remittent fever; *Unripe fruit* : antiscorbutic, astringent; *Ripe fruit* : cooling, acid.

Phytography : An evergreen, diffuse and spiny shrub with rigid, dichotomous branches and pairs of varicate very strong thorns ± 25–50 mm; leaves 35–80 mm, broadly elliptic-oblong or oblong with rounded or very obtuse; flowers in rather lax, usually binate, terminally produced cyme, pedicels often slender; berry ellipsoid, 1.25–1.75 cm, 4 or more seeded.

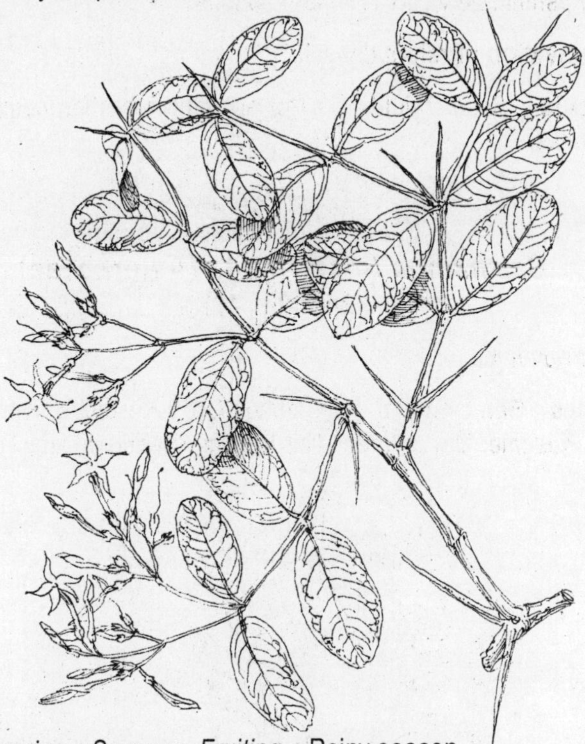

Phenology : *Flowering* : Summer; *Fruiting* : Rainy season.

Distribution : Throughout India; Pakistan and Bangladesh.

Ecology and cultivation : Very commonly cultivated for its fruits.

Chemical contents : *Root* : 4 cardioactive compounds, carissone, β-sitosterol and water-soluble polyglyceride fraction, carindone, carinol; *Fruit* : ascorbic acid (915 mg/100g).

Carthamus lanatus L. (Asteraceae)

(2n = 22)

English name : Saffron thistle.

Modern use : *Plant* : anthelmintic, febrifuge and sudorific.

Phytography : Thistle-like herb; leaves alternate, entire, rigid, unarmed or spinulose-serrate; head inflorescence, generally homogeneous; flower bisexual, orange-red; achene obovoid with pappus.

Distribution : Kashmir, between 1750 and 2000 m.

Ecology and cultivation : Prefers dry soil.

Chemical contents : *Seed* : yields a fixed oil with properties comparable to that of *Carthamus tinctorius* L.

Carthamus tinctorius L. (Asteraceae)

(2n = 24, 48)

Sanskrit name : Kusumba.

Vernacular names : Ben : Kusum, Kusumphul; **Guj :** Kusumbo; **Hin :** Kusum, Karrah; **Kan :** Kusambe, Kusume; **Mar :** Kardai, Kurdi; **Tam :** Senduriakam; **Tel :** Kusumba.

Trade names : Kusumphul, Kusum.

Traditional use : *Capitulum* : laxative and diaphoretic, used in jaundice; *Flower* : laxative and used as diaphoretic, in jaundice, stimulant, sedative, emmenagogue; *Seed* : purgative, diuretic, tonic, used in rheumatism; *Oil* : charred and used for healing sores and in rheumatism.
Unani : in the preparation of 'Qurtum'.

Modern use : *Flower* : used in preparation of medicines for polio myelitis, paralysis of brain and spine, other nerve disorders, heart diseases; *Corolla-extract* : effective in coronary occlusion in animals; *Seed* (50% EtOH extract) : spasmolytic; *Seed-oil* : lowers blood cholesterol in human, increase plasma insulin.

Phytography : An annual herb, 30–60 cm in height; leaves ovate, spiny, toothed; leafy bracted heads of tubular flowers, entire or broken, orange-red to brownish red tubular florets, up to 40 mm long when entire, consists of an oblong ovary with tubular corolla, style terminating in a bifid stigma.

Phenology : *Flowering* and *Fruiting* : Autumn and Winter.

Distribution : Cultivated throughout the greater part of India, mainly Andhra Pradesh, Karnataka, Maharashtra, Madhya Pradesh; Bangladesh.

Ecology and cultivation : Dry, winter crop in the plains; cultivated both as a rainfed and irrigated crop; being drought-resistant, it could be grown even on poor sandy soil.

Chemical contents : *Aerial parts* : acetylenic compounds; *Flower* : yellow pigment safflomin A; *Fruit* : alkaloids; *Seed* : matairesinol monoglucoside, aellobioside, 1,8,11,-14-heptadecatetraene, polyacetylenes, tracheloside, steroid glyceride, oil.

Adulterants : The tubular florets of *Carthamus tinctorius*, commonly known as the American saffron or safflower, have long been substituted for saffron and *Calendula officinalis.*

Remark : Yellow colouring matter, available from flower, is used as dye.

Carum bulbocastanum Koch. (Apiaceae)

(2n = 22)

Syn : *Bunium persicum* (Boiss.) Fedts.

English names : Black caraway, Earth nut.

Vernacular names : Ben : Kalo zira; **Hin :** Kali jira, Shah zira; **Kan :** Shimejeerige; **Kas :** Gunium; **Lad :** Umbhu; **Mal :** Shima jirakam; **Tam :** Shemai shiragam, Pilappu; **Tel :** Sima jilakara.

Trade name : Kalo jira.

Traditional use : *Fruit*: used for similar purposes as those or *Carum carvi*.

Modern use : *Fruit*: stomachic, carminative, lactagogue.

Phytography : Perennial herb, up to 1 m in height; branchlets glabrous; leaves bi- to tri-pinnate, leaflets deeply partite, chartaceous, glabrous, sheathing at the base, margins membranous; umbel compound; flowers 2 mm across, white; fruits ovoid, laterally compressed, carpophore, entire.

Distribution : Kashmir; Pakistan (Baluchistan), above 2000–3000 m.

Ecology and cultivation : Introduced; hills above 2000 m.

Chemical contents : *Fruit*: yields 2% essential oil containing 18% aldehydes.

Carum carvi L. (Apiaceae)

(2n = 20, 22)

Syn : *Apium carvi* Crantz, *Seseli carvi* Lam.

English name : Caraway.

Sanskrit name : Sushavi.

Vernacular names : Ben : Jira; **Hin :** Shiajira; **Kas :** Gunyum; **Mar :** Wilayati zirah; **Pun :** Zira siah; **Tam :** Shimai shembur; **Tel :** Shimaisapu.

Trade names : Shia jira, Jira.

Traditional use : *Fruit* : stomachic, carminative, lactagogue and occasionally used in flatulent colic; *Caraway-oil*: used to impart flavour to medicines and to correct their nauseating effects; *Caraway water* : useful remedy in flatulent colic of infants.
UNANI: an ingredient for 'Zeera Siyah'.

Modern use : *Seed-extract* : agglutinate human blood cells; *Seed* (alcoholic extract) : antispasmodic; *Essential oil* (from seed): antibacterial, antifungal.

Phytography : A biennial herb producing during the first season a cluster of bi- and tri-pinnate radical leaves; during the second year, there arises a slender stem bearing alternate, pinnate to bipinnate leaves, the terminal segments of which are quite narrow and pointed; these cauline leaves have large sheathing brown petioles; long-stalked compound umbel; flowers white, calyx teeth small, petals ovate and notched, stylopod conical; fruits oblong, laterally compressed, dark brown, cremocarps.

Phenology : *Flowering* and *Fruiting* : Winter; occasionally in Summer also.

Distribution : Native to Europe and West Indies and is now wild in the North Himalayan region; cultivated in the plains as a cold-season crop and in the hills of Kashmir, Kumaon, Garhwal, Chamba and Pakistan at an altitude of 3000–4000 m as a cold-season crop.

Ecology and cultivation : Caraway is a biennial plant and is usually grown on a loamy soil; seed is sown in March–April and harvest is reaped in the summer of the following year at the end of August.

Chemical contents : *Leaf* : flavonoids; *Fruit* : abscisic acid, herniarin, lipids, umbelliferone, scopoletin; *Essential oil* : carvone, menthone, pulegone, trans- & cis-carveol.

Adulterants : Caraway may be adulterated by exhausted fruits with admixture of small stems and rays of the umbels and by Indian Dill; fruits of *Nigella sativa* (Ranunculaceae) are sometimes known as black-caraway.

Caryota urens L. (Arecaceae)

(2n = 32)

English names : Fishtail palm, Indian sago palm, Toddy palm, Wine palm.

Sanskrit names : Dhoajavriksha, Dirgha, Mada.

Vernacular names : Asm : Baraflawar; **Guj :** Shankarjata, Shivajata; **Hin :** Mari; **Kan :** Bagani; **Lod :** Salpa; **Mal :** Anapana; **Mar :** Birlimad, Bherlimad, Kittul; **Mun :** Sabudana-daru; **Orn :** Salpa; **Ori :** Salapa; **San :** Salapa; **Tam :** Irambanai, Tippili; **Tel :** Jilugujattu.

Trade name : Mari.

Traditional use : KARBI (Mikir of Assam) : *Seed* : as masticatory; LODHA : *Ash of old leaf* (with honey) : in tympanitis; ORAON : *Inflorescence* (fermented juice) : in hypochondriasis.

Modern use : *Nut* : acrid, cooling to allay thirst and fatigue; in case of hemicrania, it is applied in head; *Freshly drawn toddy* : laxative.

Phytography : Unbranched, robust, monocarpic palm; stem up to 2 m long, up to 60 cm wide near base, light brown in colour, surface smooth; leaves ± 5 m long, palmately compound, ultimate leaflets one-nerved, leaf-sheath broad, semiwoody, margins converge and form two elevated confluent lines on the upper side; spadix terminal and axillary, about 4 m long, pendulous; prophyll and bracts semiwoody; male flowers prominently 2-bracteate, female flowers with two scale-like bracts, the terminal inflorescence and 2–3 inflorescences below the terminal bear fruits; fruits globose, ± 1.8 cm in diameter, apiculate, 2-seeded, testa adherent; the palm reaches maturity in about 15 years and once started flowering, the tree takes its beautiful shape.

Distribution : West and East coast of India and cultivated in Punjab, Andhra Pradesh, north Canara, Nilgiris, Bihar, Orissa, West Bengal, Arunachal Pradesh, Assam, Manipur; Bangladesh, Pakistan, Nepal, Myanmar.

Ecology and cultivation : Common in forest areas; widely cultivated in large gardens and parks for landscape beautification.

Chemical contents : *Fresh juice of palm* : contains 13.6% sucrose and a trace of reducing sugar, while fermented juice of toddy contains 1% reducing sugar, 3–4.5% alcohol and 0.3% acetic acid.

Remarks : Fibres obtained from leaf-sheath are used. Sago is obtained from the pith of stem. Ethnic communities and rural people prepare a spirituous drink, *toddy*, from the juice of inflorescence.

Casearia tomentosa Roxb. (Samydaceae)

Syn : *Casearria elliptic* Willd.

Vernacular names : Ben : Maun; **Hin :** Churchu, Chilla; **Kan :** Biliyubina, Rore; **Lod :** Bidi-daru, Cara farel, Churchu-daru; **Mal :** Anakkarana, **Mun :** Bir-churchu, Churchu-daru; **Orn :** Biri; **Ori :** Benimanj, Kokra; **San :** Chorcho; **Tam :** Kadichai kutti; **Tel :** Chilakadieddi.

Trade names : Chilla, Churchu.

Traditional use : LODHA : (i) *Root-paste* (with kusum oil) : in leucoderma, (ii) *Root-powder* : as hallucinogen, (iii) *Stem-bark* (powder) : to smoke in dropsy; MUNDA : *Fresh leaf-paste* (with lime) : in ringworm; SANTAL : *Root-bark decoction* (with paste of long peppers) in diabetes; TRIBES OF BASTAR (Madhya Pradesh) : *Root-bark* : as tonic in anaemia.

Modern use : *Aerial parts* (50% EtOH extract) : antiviral, hypotensive and spasmolytic.

Phytography : A small tree or shrub with pubescent or tomentose twigs; leaves simple, oblong or smaller one somewhat ovate or elliptic, pubescent, especially on ribs beneath; flowers axillary on the new shoot, green, very small; capsule soft green, axillary and from leafless axils, oblong, 6-angular; seeds arillate.

Phenology : *Flowering* : January–May; *Fruiting* : February–May.

Distribution : Upper Ganga plains, westwards to Chenab, ascending up to ± 1500 m, Garhwal Himalaya, Kumaon, Sikkim, Karnataka, Orissa and some other places.

Ecology and cultivation : Very common in waste ground, forest areas, river valleys, laterite and cotton soils and the like.

Chemical contents : *Bark* : tannin 10%.

Remark : Dried leaf-powder and crushed fruits are used for poisoning and benumbing fish, respectively.

Casearia zeylanica (Gaertn.) Thied. (Samydaceae)

Syn : *Casearia esculentum* Roxb.

Sanskrit name : Bhurigandha.

Vernacular names : Kan : Doddahanise; **Mal :** Malampa vattu, Vella kunnan, Pannimurangam; **Mar :** Mori; **Tam :** Kottargovai, Kakkaipilai.

Traditional use : *Root* : astringent, cathartic, promotes action of liver; *Decoction of root* : used in diabetes and piles.

Phytography : Shrub or small tree; branches glabrous; leaves alternate, petioled, undivided, entire or obscurely crenate, broad, lanceolate, 7.5–15.0 cm by 2.5–5.0 cm, thick, subcoriaceous; flowers small, greenish yellow, clustered in the axils, pedicels less than 7.5 cm, surrounded by scales, jointed at the base; capsules succulent, globoid or ovoid, somewhat triangular or six-ribbed when dry; seeds many, angular or obovoid, with a fleshy coloured aril.

Distribution : Maharashtra, Karnataka, other places of west coast; Sri Lanka, Myanmar.

Ecology and cultivation : Plant of warmer climate, found mostly in wastelands and scrub jungles; wild.

Remark : Tribes of North-East India eat shoot and leaves as vegetables.

Cassia absus L. (Caesalpinaceae)

(2n = 26, 28, 56)

Vernacular names : Guj : Chimar; **Hin :** Chaksu; **Mal :** Karinkolla; **Mun :** Piri masuri-ara; **Tam :** Mulaippal virai; **Tel :** Chanupala vittulu.

Trade name : Chaksu.

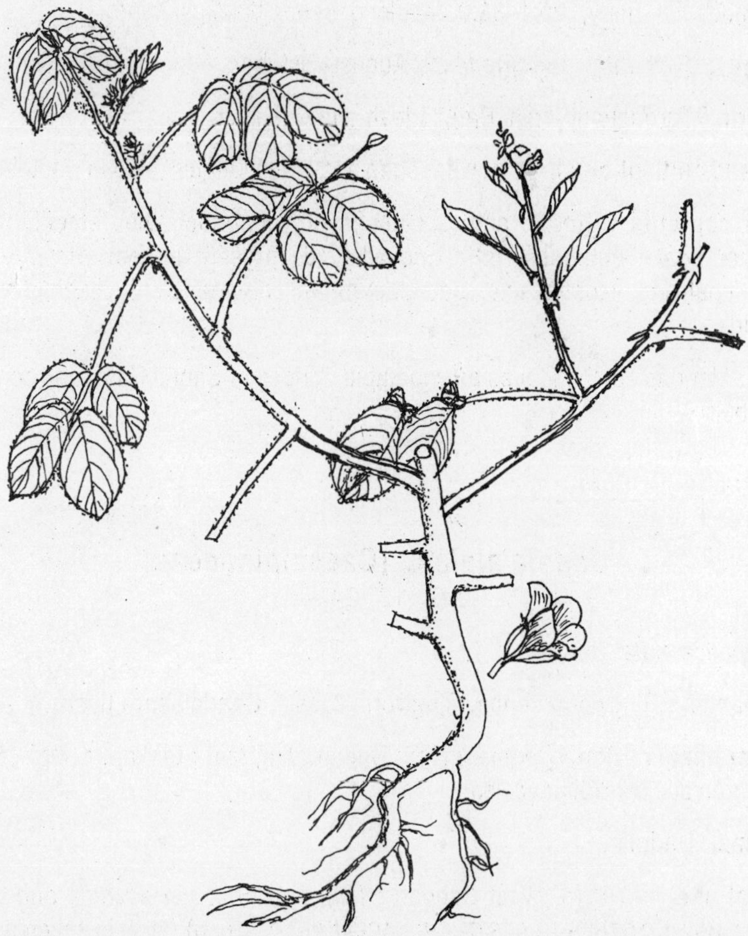

Traditional use : *TRIBES OF ALMORAH* (Uttar Pradesh) : (i) *Leaf* : in cough; (ii) *Seed* : in ophthalmia; *TRIBES OF SIWALIK* (Uttar Pradesh) : (i) *Leaf* : in cough, (ii) *Seed* : in skin diseases; *TRIBES OF SAURASHTRA* (Gujarat) : *Leaf* : in wounds; *TRIBES OF KHEDBRAHMA REGION* (Gujarat) : *Seed* : as cooling agent, in eye infections; *TRIBES OF NILGIRIS* (Tamil Nadu) : (i) *Leaf* : in healing wounds; (ii) *Seed* : in ophthalmia.
UNANI : it is an ingredient for the drug 'Chaksu'.

Modern use : *Seed* : astringent, cathartic, in ringworm, skin affections, conjunctivitis and ophthalmia.
Chaksine, the alkaloid from the seeds, in small doses depresses medullary centres; in larger doses depresses other regions of the brain, has depressant action on intestinal and other involuntary muscles.

Phytography : Erect, viscuous, hairy herb, 22.5–45.0 cm high with two pairs of ovate or rhombic-ovate leaflets; terminal raceme of small yellow or red flowers with only 4 perfect stamens; pods flat, hairy, 25.1–35.0 mm long; seeds black, shining.

Phenology : *Flowering* and *Fruiting*: August–October.

Distribution : Throughout India; Bangladesh and Pakistan.

Ecology and cultivation : In somewhat open and stony jungles, also in wastelands; wild.

Chemical contents : *Root* : anthroquinones chrysophenol, aleo-emodin, chaksine, isochaksine; *Leaf* : quercetin, rutin and alkaloids present in root; *Seed* : sitosterol glyceride, chaksine, isochaksine, galactomannan compound, mucilage, flavonoides, triterpenoids.

Remarks : Mundas eat the leaves as vegetable. Tribes of Santal Parganas consume the seeds as pulse.

Cassia alata L. (Caesalpinaceae)

(2n = 12, 24, 28)

Syn : *Herpetica alata* Raf.

English names : Ringworm senna, Ringworm cassia, Candalabara bush.

Vernacular names : Ben : Dadmari; **Hin :** Dadmurdan; **Mal :** Elakajam; **Mar :** Dadmardana; **Tam :** Anjali; **Tel :** Simayavisa.

Trade name : Dadmari.

Traditional use : TOTO (of West Bengal) : *Leaf* and *Fruit* : in eczema and other skin diseases; NAGA AND OTHER TRIBES OF ARUNACHAL PRADESH : (i) *Leaf*: in eczema and other skin diseases; (ii) *Decoction of leaf* and *flower*: used internally in bronchitis and asthma and for washing eczematous patches.

Modern use : *Infusion of root* : for rheumatism; *EtOH* (50%) *extract of aerial parts* : anti-inflammatory, Central Nervous System depressor, diuretic; *Plant-extract* : anti-inflammation, antitumour; *Leaf-extract* : antifungal; *Leaf-juice* : in eczema; *Seed*: used as vermifuge.

Phytography : Small bushy tree or big shrub, about 2.8 m high; leaves 30–90 cm long, 4–16 pairs of broad, long, obtuse leaflets, 50–175 mm long, glandless; flowers golden yellow, 25 mm across, in large spike-like racemes, fertile stamens 7; pods flat, linear, 1.0–1.5 cm long with 4 broad crenulate wings.

Phenology : *Flowering* : Rainy season; *Fruiting* : Winter.

Distribution : Throughout West Bengal and Tamil Nadu; Bangladesh, pantropic.

Ecology and cultivation : Wastelands, mainly in rainy season.

Chemical contents : *Root* : quinone pigments; *Plant* : aloe-emodin, chrysophanol, emodin, rhein, santhone cassiollin (pinselin); *Leaf* : free aleo-emodin, kaempferol, volatile oil, anthraquinones and their derivatives, *Fruit* : anthraquinones.

Remarks : *Plant* : poisonous to livestock and fish.

In Samoa, bark is used in skin diseases and leaf as poultice and purgative. Plant is also used medicinally in Spain.

Cassia angustifolia Vahl (Caesalpiniaceae)

Syn : *Cassia medicinalis* Bisch.

English name : Indian or Tinnevelly senna.

Sanskrit names : Bhumiari, Bhupadma.

Vernacular names : **Ben** : Sonamukhi, Sonpat; **Guj** : Nat-ki-sanna; **Hin** : sena; **Kan** : Nilavarika; **Mal** : Nila vaka; **Mar** : Sonamukhi; **Tam** : Nila virai; **Tel** : Nila tengades.

Trade name : Sonamukhi.

Traditional use : *Leaf* and *Fruit* : used as laxative, cathartic, vermifuge and purgative, especially useful in habitual costiveness. *UNANI* : ingredients for the preparation of 'Sanna', 'Itrifal Ustukhudus'—used for purging.

Modern Use : Standardised tablets useful against constipation in patients in post partum period; no effect on breast-fed baby if taken by mother.

Phytography : A short, low-growing shrub, with whitish stem, leaves stipulate, paripinnate, with pale yellowish leaflets, frequently bordered and usually more abruptly pointed; large yellow flowers borne in axillary racemes; pods narrowly elliptical or reniform, 8-seeded.

Phenology : *Flowering* : with a peak in Winter; *Fruiting* : throughout the year.

Distribution : Native of Sudan and Arabia; now cultivated in different parts of Tamil Nadu, Andhra Pradesh and Karnataka.

Ecology and cultivation : Cultivated on irrigated land.

Chemical contents : *Leaf*: Glucoside, sennoside C, kaempferin, arthraquinone, essential oil, chrysophanic acid, isorhamnetin Ca-oxalate in 12% leaves; kaempferol, rhein and small quantity of emodin, aloe, aloe-emodin or emodin glucoside, occurrence of oxymethyl-anthraquinone also reported; *Pod* : glycerides of rhein, chrysophanic acid, sennosides A & B, anthraquinone glycerides.

Adulterants : Tinnevelly senna is usually free from adulterants, but it is found that some consignments of Indian senna have been adulterated with the leaflets of *Cassia auriculata.*

Cassia auriculata L. (Caesalpiniaceae)

(2n = 14, 16, 28)

English names : Avaram, Tanner's cassia.

Vernacular names : Ben : Tarevar; **Guj :** Awal; **Hin :** Tarevar; **Kan :** Kakki; **Mal :** Avara; **Mar :** Tarwad; **Tam :** Avaram; **Tel :** Tangedu.

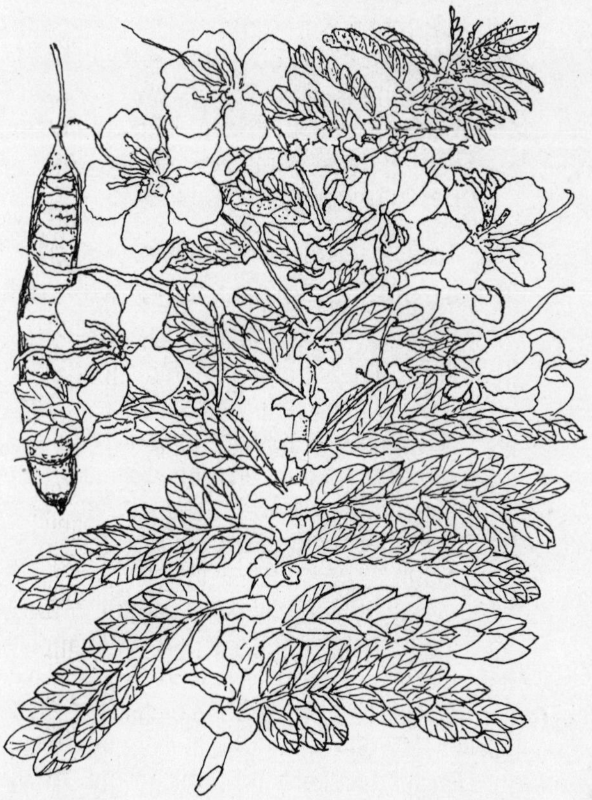

Trade name : Tarevar.

Traditional use : *Bark* and *Root* : astringent; *Root* : used in skin diseases; *Leaf* and *Fruit* : anthelmintic; *Seed* : in ophthalmia and conjunctivitis, in diabetes and chylous urine.

Modern use : *Seed* : in diabetes.

Phytography : A much-branched handsome shrub, 1.0–3.25 m high with hairy twigs; leaves pinnately compound, leaflets 8–12 pairs, elliptic, obtuse, mucronate, hairy, easily recognised by the large, broad, foliaceous, obliquely cordate stipules on the young leaves; flowers large, yellow, in subterminal corymbs; pods 0.75–1.25 cm by 1.75 cm, flat, obtuse at each end; seeds 6–10.

Phenology : *Flowering* : October–January; *Fruiting* : January– March.

Distribution : Wild in dry areas of Madhya Pradesh, Uttar Pradesh, Maharashtra, Gujarat, Rajasthan; Pakistan (Sind, NWFP and others).

Ecology and cultivation : Xerophytic in nature; wild in dry regions, cultivated in other areas; prefers cottony soil.

Chemical contents : *Bark* : contains tannin 15.2–19.1%.

Adulterant : Substitute for senna.

Cassia fistula L. (Caesalpiniaceae)

(2n = 24, 26, 28)

English names : Indian laburnum, Golden shower, Purging cassia.

Sanskrit names : Aragvadha, Avaghataka, Karnikara, Kritmala, Rajavriksha, Sampaka, Suvarnaka.

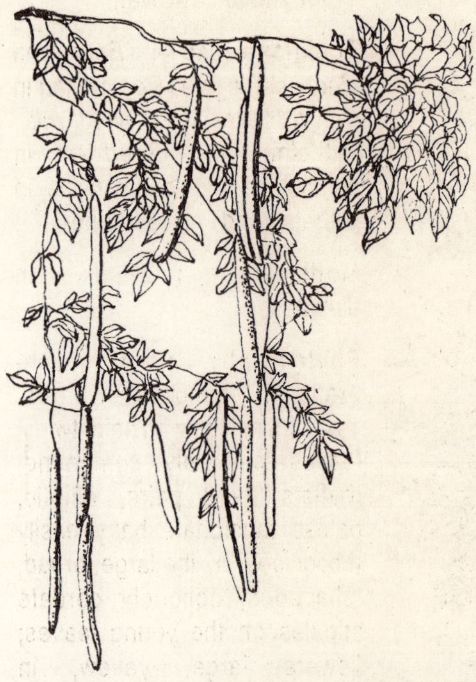

Vernacular names : Ben : Sundali, Banarlathi; **Guj :** Garmala; **Hin :** Amaltas; **Kan :** Kakki; **Lod :** Bondar-lasi, Hari, Punden; **Mar :** Bhava; **Mun :** Hari; **Orn :** Sonarkhi, Funan; **Sad :** Bandar lathi; Bandar lauri; **San :** Badar, Sodal, Bero; **Tam :** Konnei; **Tel :** Rela.

Trade names : Amaltas, Banarlathi.

Traditional use : GARO (of Meghalaya) : *Fruit* : as purgative; BODO (of Assam) : (i) *Root* : in skin diseases including leprosy; (ii) *Leaf* and *Seed* : as purgative; RURAL FOLK OF GUAHATI : *Root* : in skin diseases, as purgative; RURAL PEOPLE OF SIKKIM : *Fruit* : as purgative; RURAL FOLK OF PURULIA (West Bengal) : (i) *Gum* : in dysentery; (ii) *Fruit-pulp* : in constipation; TRIBES OF SANTAL PARGANAS : (i) *Leaf* : for blood purification; (ii) *Seed* : in antifertility, gastric troubles, as cooling agent; RURAL FOLK OF ORISSA : *Fruit* : in pimples; THARU (of Uttar Pradesh) : (i) *Fruit* : in liver complaints; (ii) *Seed* : in constipation, jaundice; RURAL FOLK OF MIRZAPUR (Uttar Pradesh) : *Root* : in snake bite; BHOXA (of Uttar Pradesh) : *Fruit* : in dysentery and indigestion; GARHWALI : (i) *Leaf* : as antiseptic; (ii) *Seed* : as purgative; RURAL FOLK OF RAJOURI (Jammu and Kashmir) : *Seed* : as purgative and in throat troubles; RURAL PEOPLE OF RAJASTHAN : *Fruit* : in constipation, snake bite and swelling; DANG (of Gujarat) : (i) *Bark* : in blindness; (ii) *Fruit* : in snake bite; RURAL PEOPLE OF SAURASHTRA (Gujarat) : *Fruit* : as tonic, purgative, in chest infection; TRIBES OF KOLHAPUR (Maharashtra) : (i) *Bark* : in stomachache; (ii) *Leaf* : in ringworm; RURAL FOLK OF KHED TALUKA (Maharashtra) : *Root* and *Fruit* : as purgative; TRIBES OF KANGRA RANGE (Madhya Pradesh) : *Leaf* and *Fruit* : in abdominal pain; TRIBES OF ABUJH MARH AREA (Madhya Pradesh) : (i) *Root* : in skin diseases; (ii) *Stem-bark* and *Fruit* : in scorpion sting and snake bite; TRIBES OF SAGAR (Madhya Pradesh) : (i) *Leaf* : in ringworm, rheumatism, deopsical swelling; (ii) *Fruit* : in pimples and as purgative.

CHARAKA, *SUSHRUTA*, *BAGBHATTA* and *CHAKRADUTTA* have extolled the medicinal value of different parts of this plant. It is useful in treatment of fever, heart troubles, high blood pressure, constipation, liver troubles, scrofula, fistula, wounds and leprosy.
In *SIDDHA* system of medicine, the *fruit-pulp* is a constituent of the drug named *Charakkonrai puli*.
In the *UNANI* system, the *fruit-pulp* is used in treatment of constipation and other stomach troubles.

Modern use : *EtOH* (50%) *extract of stem-bark* and *pod* : anticancer, antiviral, hypoglycaemic; *Leaf* : in ringworm; *Flower* : in fungal infection; *Fruit* : abortifacient; *Fruit-pulp* : in chest and heart diseases; *Dried fruit* : as purgative, laxative for habitual constipation.

Phytography : A tree cultivated as an ornamental for attractive yellow blossoms in pendent racemes, attaining a height of about 15 m with grey bark and pinnate leaves with 3–7 pairs of ovate leaflets; fruit elongated, reddish brown to black, cylindrical lomentum with numerous compartments, each containing a flattened, ovoid, shiny, reddish brown seed embedded in a brownish hardened pulp.

Phenology : *Flowering* : May–June; *Fruiting* : Autumn mainly.

Distribution : Throughout India; Bangladesh and Pakistan.

Ecology and cultivation : Prefers laterite soil; often planted as roadside tree.

Chemical contents : *Root-bark* : tannin, phlobaphenes and oxyanthraquinone substances; *Leaf* : anthraquinone derivatives and very little tannin; *Fruit-pulp* : rhein, small quantity of volatile oil, 3 waxy substances and a resin substance.

Adulterants : Pods of *Cassia grandis* L. (horse cassia); pods of *C. moschata* H.B. & K.

Remarks : Lodhas use sweet pulp around seeds with paste of *Hemidesmus indicus* and molasses in tongue sore of cattle and seed-powder with lime in swollen throat of cows. Santals administer fruit-powder with mustard oil and powdered turmeric to cows in treatment of intestinal worms.

Cassia mimosoides L. (Caesalpiniaceae)

(2n = 16, 32)

Vernacular names : **Mun :** Merai; **Orn :** Jhirn; **San :** Patwa-ghas; **Tel :** Nelaponna.

Traditional use : *ORAON* : *Root* and *Leaf* : given in spasms of stomach and headache.

Modern use : *Leaf-* and *seed-extracts* : anthelmintic, fungitoxic against *Helminthosporium oryzae*.

Phytography : Prostrate herb; branchlets densely hispid; leaves pinnate, 1.5–4.5 cm, leaflets 25–40 pairs, oblong, elliptic, inequilateral, chartaceous, glabrescent, base and apex obtuse, stipules auriculate, lanceolate; flowers axillary, solitary, yellow, 6 mm across; pods short, stipitate, flat, strap-shaped, compressed; seeds ellipsoid, 3 mm.

Phenology : *Flowering* : October–February; *Fruiting* : throughout the year.

Distribution : Throughout India; Pakistan and Bangladesh.

Ecology and cultivation : Grows on hills, 900–1600 m, on base slopes, in pasture lands, perennially moist places; less in the plains, to the coast, on the floor of scrub jungles sheltered from grazing.

Chemical contents : *Root* : physcion; *Leaf* : emodin and its glycerides, emodic acid, luteolin-7-glucoside, physcion.

Remarks : Decoction of leaves is used in Japan as substitute for tea.
Seed is a minor constituent of indigenous drug 'Geriforte'.
In Samoa, leaves are used to treat asthma, stomach trouble and typhoid.

Cassia obtusa Roxb. (Caesalpiniaceae)

Syn : *Cassia obovata* (L.) Collad., *C. burmani* Wight

English names : Country senna, Dog senna, Italian senna.

Sanskrit name : Bhutalapota.

Vernacular names : Guj : Suratisonamukhi; **Hin :** Chottataroda; **Mal :** Serruvanni; **Mar :** Bhuitarwar; **Tam :** Kattunilavirai.

Trade name : Deshi senna.

Traditional use : *Leaf* : purgative.

Phytography : Leaves broadly obovate, apex abruptly tapering, venation pinnate, distinct, constituents similar to those of senna; pods distinctly curved.

Distribution : South India, Gujarat, Rajasthan, Punjab; Pakistan (Sind, Baluchistan).

Chemical contents : Oxymenthyl-anthraquinone isolated from leaflets and pods.

Adulterant : Used as an adulterant of Tennevelly senna.

Cassia occidentalis L. (Caesalpiniaceae)

(2n = 26, 28)

Syn : *Senna occidentalis* L.

English names : Coffee senna, Negro coffee.

Sanskrit name : Kasamarda.

Vernacular names : Ben : Kalkasunda, Kalkasundi; **Hin :** Kasondi; **Mal :** Natramkatara; **Mun :** Marang chakonda; **Orn :** Koha chakonda; **Sad :** Barka chakonda; **Tam :** Nattamakatarai; **Tel :** Kasinda.

Trade names : Kasondi, Kalkasundi.

Traditional use : *Plant* : febrifuge, purgative, diuretic, tonic; *Root* : in snake bite; *Root, Leaf* and *Seed* : purgative; *Leaf* and *Seed* : used externally in skin diseases, antiperiodic.

Modern use : One of the ingredients of 'Liv-52', used in hepatitis; useful as stomachic, in dropsy, fever, rheumatism, ringworm, eczema.

Phytography : A small shrub, 1–2 m high; branchlets glabrescent; leaves pinnate, up to 15 by 6.5 cm, leaflets 4 or 5 pair, opposite, oblong, lanceolate, petioles up to 4 cm; racemes axillary and terminal; flowers 2 cm across, yellow; pod short, stipitate, slightly falcate, flat, compressed; seeds many, ovoid, longitudinal.

Phenology : *Flowering* and *Fruiting* : throughout the year.

Distribution : Pantropic, throughout India; Bangladesh and Pakistan.

Ecology and cultivation : Mesophyte; wild.

Chemical contents : *Plant* : xanthone casiolline, islandicin, helminthosporin, xanthorin; *Root* : emodol, physcion, phytosterol; *Leaf* : dianthronic heteroside, anthraquinone pigments; *Pod* : apigenin sennosides; *Seed* : anthraquinone glycerides, anthraquinones, N-Me-morpholine, polysaccharides, tannic acid, fatty acid, mucilage.

Adulterant : Seeds are used as substitute for coffee.

Cassia pumila Lamk. (Caesalpiniaceae)

(2n = 28)

Syn : *Senna prostrata* Roxb.

Vernacular names : Kan : Nelatagache; **Mal :** Sarmal; **Mun :** Ote masuria, Piri masuria, Seta kurti; **Sad :** Chotka aura; **Tel :** Nallajiluga.

Trade name : Sarmal.

Modern use : *Plant* (50% EtOH extract) : Central Nervous System depressant, diuretic, spasmolytic.

Phytography : Prostrate or decumbent herb; branchlets down, villose; leaves pinnate, 2.4–5.0 cm, leaflets 10–20 pairs, oblong-elliptic, glabrous above, sparsely villose below, nerves prominent, base cuneate, margin ciliate, apex obtuse, mucronate, petiole up to 4 mm with a basal stipitate gland, stipules lanceolate, up to 4 mm; flowers axillary, solitary or in a cluster, yellow, 1 cm across; pod short, stipitate, compressed.

Phenology : *Flowering* and *Fruiting* : throughout the year.

Distribution : India; Bangladesh and Pakistan.

Ecology and cultivation : Very common in wastelands and pastures, grows on bare slopes up to 800 m; wild.

Remark : In Indo-China, seeds are given as purgative.

Cessia sophera L. (Caesalpiniaceae)

(2n = 24, 28)

English name : Sophera senna.

Sanskrit name : Kasamarda.

Vernacular names : Ben : Kalkasunda; **Hin :** Kasundi, Kasunda; **Lod :** Kasunda-ba; **Mal :** Pounantakara; **Mar :** Kasudi; **Tam :** Sularai; **Tel :** Kondakashinda.

Trade name : Kasundi.

Traditional use : *Plant* : febrifuge, purgative, diuretic, tonic; *Decoction of plant :* used in acute bronchitis; *Root* : in snake bite — (properties similar to *C. occidentalis*); *Leaf :* used externally in ringworm; *Root*, *Leaf* and *Seed* : purgative, cathartic; *Leaf* and *Seed* : used externally in skin diseases, antiperiodic.

Modern use : *Plant* (50% EtOH extract) : spasmolytic.

Phytography : A small shrub, 2.4–3.25 cm high; leaves with 6–12 pairs of lanceolate, finely acute or acuminate leaflets; flowers 25–75 mm diameter in 1 shore axillary and terminal panicle; pods subterate or terate, covered; seeds compressed, black.

Phenology : *Flowering* : August–September; *Fruiting* : November–December generally; but found also in March–April.

Distribution : Throughout India; Bangladesh and Pakistan.

Ecology and cultivation : Usually in hedges and wastelands in wild condition.

Chemical contents : *Plant* : emodin and chrysophanic acid and a new flavonol glucoside; *Leaf* : new flavonol-8-c-glyceride; *Flower* : anthraquinone, flavonol glyceride.

Remarks : Lodhas take fried seed-powder as coffee.
In Samoa, bark is used for skin diseases.

Cassia surattensis Burm. f. (Caesalpiniaceae)

(2n = 28, 32)

Syn : *Cassia glauca* Lamk.

Vernacular names : Mal : Vellatakara; **Mar :** Mothatarvad; **Tam :** Kovalai; **Tel :** Kondatantemu.

Traditional use : *Bark* and *Leaf* : in diabetes and gonorrhoea.

Modern use : *Aerial parts* (50% EtOH extract) : Central Nervous Systen depressant, diuretic.

Phytography : A small tree or shrub, grows up to 10 m in height; branchlets sparsely pubescent; leaves 10–15 cm, leaflets 7–9 pairs, oblong, elliptic, 1.5–4.5 by 1.2 cm, chartaceous, apex obtuse, petiole up to 3 cm, glandular; stipules caducuous; flowers 4 cm across, yellow; pods long, stipitate, flat, thin, compressed.

Phenology : *Flowering* and *Fruiting* (Pods) : throughout the year.

Distribution : Throughout India; Pakistan and Bangladesh.

Ecology and cultivation : Usually grown as hedge in the gardens; doubtfully wild.

Chemical contents : Glucosides, chrysophanic acid.

Cassia tora L. (Caesalpiniaceae)

(2n = 24, 26, 28, 52)

Syn : *Cassia obtusifolia* L.

English names : Stinking cassia, Sickle senna.

Sanskrit names : Chakramarda, Dadmari.

Vernacular names : Ben : Chakunda; **Guj :** Kovariya; **Hin :** Pamaar, Chakunda; **Lod :** Jomaikain; **Mar :** Takla; **San :** Bheda-deren; **Tam :** Tagarai; **Tel :** Tantemu.

Trade name : Chakunda.

Traditional use : *LODHA* : (i) *Root-bark paste* (with albumen of hen's egg) : in bone fracture, (ii) *Seed-powder* (with lime) : in ringworm and other cutaneous diseases; *SANTAL* : (i) *Leaf-paste* : in eczema, (ii) *Leaf-decoction* (with paste of black peppers) : as laxative; *ETHNIC COMMUNITIES OF HAZARIBAGH DISTRICT* : *Root* : in fever; *ETHNIC COMMUNITIES OF VARANASI DISTRICT* : *Leaf* : in blindness; *THARU* : *Leaf* : in boils; *GARHWALI* : *Leaf* : as antiseptic, in boils; *KUMAONI* : *Seed* : in cold; *ETHNIC COMMUNITIES OF BASTAR* : (i) *Leaf* : in cuts, (ii) *Seed* : in eczema ; *ETHNIC COMMUNITIES OF DAHANU FOREST DIVISION* : (i) *Leaf* : in itching, (ii) *Seed* : in eczema, ringworm; *ETHNIC COMMUNITIES OF KHED TALUKA* : *Root* : in ringworm; *ETHNIC COMMUNITIES OF CHANDRAPURA* : *Leaf* : in cuts; *DHANG* : *Root* : in abnormal child birth.

Modern use : *Plant* (50% EtOH extract) : antiviral, diuretic, spasmolytic, active *in vivo* against P 388 lymphocytic leukemia in mice; *Seed* : antibacterial; *Seed* (50% EtOH extract) : spasmolytic.

Phytography : A small weed, subshrubby, up to 75–150 cm; branchlets glabrous; leaves up to 4 cm, leaflets 3 pairs, obovate, 1.5–3.0 cm by 1.0–1.5 cm, thin, coriaceous, glabrous above, pubescent below, base oblique, cuneate, rotund, margin entire, apex obtuse, apiculate, petiole up to 2 cm; racemes terminal, corymbose, bracteoles two, linear; flowers 1.5 cm across, golden yellow; pods short, stipitate, flat, compressed, up to 9 mm, 4–6 mm across; seeds 20–30, oblong, longitudinal.

Phenology : *Flowering* and *Fruiting* (Pods) : Throughout the year, peak in September to January.

Distribution : Throughout India; Pakistan and Bangladesh.

Ecology and cultivation : Plains from the coast, in low-lying places, river banks, other moist places like fallow fields, abundant in waste places, up to 1000–1400 m.

Chemical contents : *Root* : anthraquinons, β-sitosterol; *Stem* and *Leaf* : β-sitosterol, δ-mannitol, myricyl alcohol; *Gum* : kaolin, lactone, suspending agent for calomel; *Pod* : sennocide rich concentrates; *Seed* : anthraquinones, anthraquinone glucosides, gluco obtusifolin, glucoaurantioobtusin, rubrofusarin glyceride, torachrysone,chrisophenol-β-gentiobioside, chrysophanic acid-9-anthrone, questin, torosachrysone, naphthalenic lactones, isotoralactone, cassiialactone, seed oil.

Adulterant : Seeds are substitutes for coffee.

Remarks : Ethnic communities of Rajasthan use seeds as galactogogue in cattle. Leaf is eaten as vegetable by ethnic communities of Salsette Island, Maharashtra, Goa, Saurashtra, Purulia (West Bengal) and Irulars of Tamil Nadu. Santals and tribes of Rajasthan consume fruits as vegetable.

Cassytha filiformis L. (Cassythaceae)

(2n = 24, 48)

Sanskrit name : Akashavalli.

Vernacular names : Ben : Akasbel; **Hin :** Amarbeli; **Kan :** Akashaballi, Beluballi; **Mal :** Akashavalli; **Mar :** Amarvela; **Mun :** Harihar-ronar; **Tam :** Erumaikkotan; **Tel :** Nulutega.

Traditional use : *Plant* : tonic, alterative, in bilious affections, chronic dysentery, urethiritis and skin diseases, insecticidal; powdered and mixed with gingili oil, it is used as hair tonic; mixed with butter and ginger, it is used for cleansing invertebrate ulcers; *Juice of plant* : mixed with sugar, considered as specific in inflamed eyes.

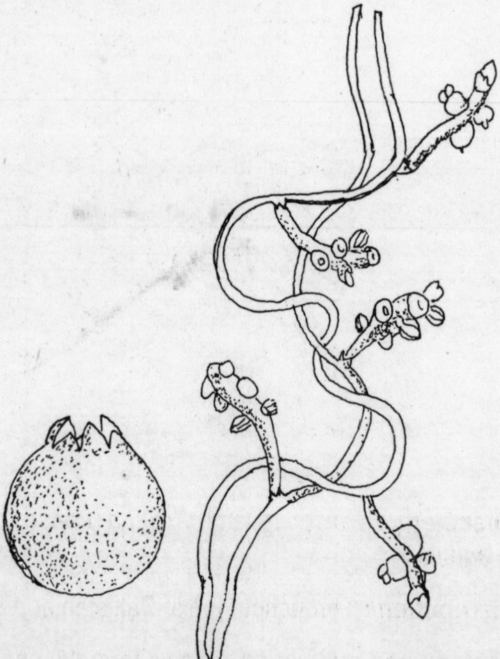

Phytography : A parasitic leafless twiner; stem pubescent or glabrous; spikes pubescent from the axils of scale-leaves; flowers bracteate, sessile, white, outer perianth lobes small, orbicular; berry very small.

Phenology : *Flowering* and *Fruiting* : most of the year.

Distribution : Throughout the greater part of India, especially near the coast; Bangladesh and Pakistan.

Ecology and cultivation : *Cassytha* and *Cuscuta* are seldom found in the same locality; generally found in the scrub jungles on *Carissa*, *Holarrhena*, *Ziziphus* and Sal trees.

Chemical contents : *Stem* : cassyfiline, cassythidine and its -OMe derivative, cassythine, nantenine.

Remarks : Plant contains an alkaloid laurotetanine, which produces cramps and, in large doses, even death; dulcital isolated from the methyl alcohol extract of the plant.
In Exuma and Long Islands, plant is used as aphrodisiac, in rheumatism and waist pain.

Castanea sativa Mill. (Fagaceae)

(2n = 22, 24)

Syn : *Castanea vulgaris* Lam., *C. vesea* Gaertn.

English names : Sweet chestnut, European chestnut, Spanish/Eurasian chestnut.

Trade name : Sweet chestnut.

Traditional use : *Aqueous infusion of leaf* : tonic, astringent, effective in paroxysmal cough and irritable conditions of the respiratory organs.

Modern use : *Stem-bark* (50% EtOH extract) : antiviral, spasmolytic; *Nut-extract* : may be used as platelet inhibitors in thrombosis and aetherosclerosis.

Phytography : Tall tree, up to ± 34 m; buds large, leaves oblong-lanceolate, 1.2–2.7 cm long, acute to acuminate, narrowed or rounded at base, coarsely serrate with spreading teeth and marked rounded sinuses between, tomentose beneath when young, becoming more or less glabrous, dark green above, light green beneath; nut over 25 mm wide.

Distribution : Cultivated in many parts of the Himalaya, in Punjab, Darjeeling and Meghalaya; Pakistan (NW Frontier Province).

Chemical contents : *Wood* : quercetin, a triterpenic acid, vanillin; *Leaf* : betulin, β-sitosterol, fatty acids, flavanoids, lupeol, 3-0-p-coumarylquinic acid, ursolic acid.

Adulterants : Substitute for coffee; leaves are sometimes substituted for witch hazel, *Hamamelis virginiana* L. (Fam. Hamamelidaceae) and tobacco.

Casuarina equisetifolia L. (Casuarinaceae)

(2n = 18, 24)

Syn : *Casuarina muricata* Roxb., *C. litorea* L.

English names : Beetwood, Casuarina, She oak, Swamp oak, Bull oak.

Vernacular names : Ben : Jhau; **Guj :** Vilayatisaru; **Hin :** Jangli jhau; **Kan :** Chabaku, Mara, Sarve; **Mal :** Chavukku; **Mar :** Suru, Vilayatisaro; **Tam :** Savukku; **Tel :** Saruku.

Trade name : Jhau.

Traditional use : *RURAL PEOPLE OF TEJPUR* (Assam) : *Bark* : astringent, useful in diarrhoea and dysentery; *Decoction of leaf* : used in colic pain.

Phytography : A tall and handsome tree with drooping branches; very slender sulcate branchlets with whorls of 6–8 scale-like leaves; male spikes usually numerous at the end of the same branches on which the females are borne lower down, slender, cylindric or fusiform, bracts subulate with pubescent margins; female head in flowers, ellipsoid; fruiting head having a general superficial resemblance to that of a coniferous tree.

Phenology : *Flowering* : May; *Fruiting* : June.

Distribution : East side of Bay of Bengal, from Chittagong (Bangladesh) southwards; cultivated elsewhere in India, Bangladesh and Pakistan.

Ecology and cultivation : Often planted on sandy soil and found as avenue tree.

Chemical contents : *Bark* : polyols, alicyclic and amino acids, casuarin, polyphenols; *Leaf* : kaempferol and its glycerides, isoquercitin and glycerides of quercetin.

Remarks : In Philippines, decoction of bark is used as emmenagogue, and infusion of young stem as diuretic.

In Tonga, infusion of bark is used in constipation, cough, nervous disorders, stomachache and ulcers.

Catharanthus roseus (L.) G. Don (Apocynaceae)

(2n = 16, 32)

Syn : *Lochnera rosea* (L.) Spach, *Vinca rosea* L.

English names : Madagascar periwinkle, Old maid, Red periwinkle.

Vernacular names : Ben : Nayantara; **Hin :** Sada sawagon; **Lod :** Swet chandu; **Mal :** Ushamalari; **Mar :** Sadaphul; **Ori :** Ainskati; **Pun :** Rattanjot; **Tam :** Sudukadu Mallikai; **Tel :** Billaganneru.

Trade name : Nayantara.

Traditional use : BODO : (i) *Plant* : in cancer, diabetes, (ii) *Leaf* : in menorrhagia; LODHA : (i) *Root-paste* : in septic wounds, (ii) *Root-decoction* (with paste of long peppers) : in fever, (iii) *Leaf-juice* : in blood dysentery, (iv) *Leaf-decoction* : to babies in gripping pain; SANTAL : (i) *Latex* : in scabies, (ii) *Seed-powder* (with decoction of black pepper) : in epilepsy; ETHNIC COMMUNITIES OF EAST GODAVARI DISTRICT : *Root* : in cancerous wounds.

Modern use : *Plant-extract* : antimitotic; *Root* (alkaloids) : in cancer, and as emetic, hypotensive, sedative and antiviral.

Phytography : A small herb or subshrub, up to 75 cm high; leaves elliptic-ovate to oblong, 4–10 by 2–4 cm, glabrous to puberulous, base acute or cuneate, apex obtusely apiculate, lateral nerves 10–12 pairs, petiole 1.0–1.5 cm; flowers in axillary or terminal cymes; solitary or paired, shortly pedicellate, pink or white or white with pink or yellow ring in orifice region; mericarps 3–4 by 0.3 cm, puberulous.

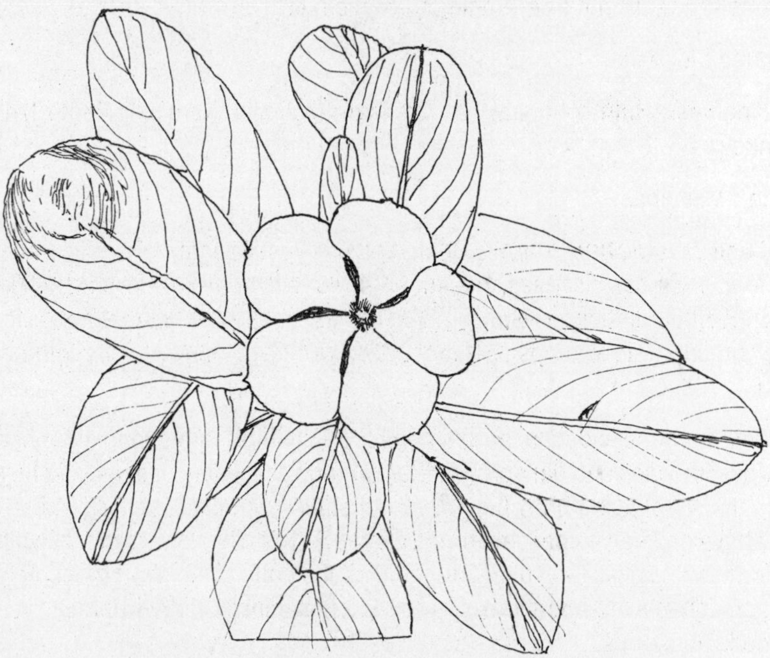

Phenology : *Flowering* and *Fruiting* : throughout the year.

Distribution : A native of West Indies; commonly grown in gardens throughout India; Bangladesh and Pakistan.

Ecology and cultivation : Plains from the coasts, in wastelands, fallow fields, less on the hills 800–1400 m, also widely cultivated.

Chemical contelts : *Root-bark* : vincaline I & II; *Root* : vinblastine or vinleukoblastine (VLB), vincristine or vinleurocristine (VCR), ursolic acid, oleanolic acid, ajmalicine, alstonine; *Stem* : vincarodine, vincoline, vinamidine, leurocolombine, vincathicine, vincubine; *Leaf* : leurosine, vindoline, catharanthine, lochnerine, tetrahydroalstonine, roseoside, essential oil; *Seed* : vincedine, vincedicine, tabersonine.

Catunaregum spinosa (Thunb.) Tirvengadum (Rubiaceae)

Syn : *Xeromorphis spinosa* (Thunb.) Keay, *Randia dumetorum* (Retz.) Lam.

English names : Emetic nut tree, Common emetic nut.

Sanskrit name : Madana.

Vernacular names : Hin : Mainphal, Madan; **Mal :** Malanka, Karaculli; **Tam :** Madkarai, Marukkalankey.

Trade name : Mainphal.

Traditional use : *ETHNIC COMMUNITIES OF MEDINIPUR*(West Bengal) : *Fruit* : in amoebiosis; *SANTAL* : *Root* : in ear diseases, malaria; *ETHNIC COMMUNITIES OF BASTAR* (Madhya Pradesh) : (i) *Root* : in gonorrhoea, (ii) *Fruit* : as emetic, *KORKU* : *Fruit* : in boils; *DANG* : *Stem-bark* : in wounds; *ETHNIC COMMUNITIES OF HARYANA* : *Stem-bark* : in rheumatism.

Phytography : Large, deciduous, armed shrub, ± 9 cm high with strong straight spines, horizontal rigid branches and dark brown or grey rough scaly bark; leaves simple, usually fascicled on the suppressed branches, obovate, obtuse, wrinkled, shining, more or less pubescent above and nerves beneath, main nerves 6–10 pairs; flowers at first white, then turning yellow, scented, solitary, or 2–3 together at the ends of the short branchlets; fruits globose or broadly ovoid, smooth or obscurely longitudinally ribbed berries, yellow on ripening; seeds numerous.

Phenology : *Flowering* : two peaks around January and July; *Fruiting* : throughout the year.

Distribution : India; Bangladesh and Pakistan.

Ecology and cultivation : Common in foothills, scrub jungles; lower dry deciduous slopes up to 500 m.

Cayratia carnosa (Wall.) Gagnep. (Vitaceae)

(2n = 80, 98, 120)

Syn : *Cayrata trifolia* (L.) Domin., *Vitis carnosa* Wall., *Cissus carnosa* Lam.

English name : Fox grape.

Vernacular names : Asm : Ghepetalot; **Ben :** Amallata, Bundal; **Guj :** Khat, Khadumdu; **Hin :** Amalbel, Ramchana; **Kan :** Heggoli; **Lod :** Baiong, Didhi mukurinari; **Mal :** Sorivalli; **Mar :** Ambat bel; **Mun :** Goal; **Orn :** Dabir; **Pun :** Arnalbel; **Tam :** Tumans; **Tel :** Kurudinne, Kamputige.

Trade name : Amalbel.

Traditional use : LODHA : *Stem-paste* : for healing bone fracture; DANG : *Root* : in anaemia; RURAL PEOPLE OF DELHI : (i) *Root* : in snake bite, (ii) *Leaf* : in boils.

Modern Use : *Aerial parts* (50% EtOH extract) : Central Nervous System depressant.

Phytography : Vine up to 4 m high, densely pubescent; leaves alternate, 3- or 5-foliate, 7–9 cm, leaflets ovate, suborbicular, 4–6 cm by 3–4 cm, thin, coriaceous, pubescent, base obtuse, margin dentate, stipules 2, triangular, cymes umbellate or corymbose; flowers 5 mm across, greenish white; berry ovoid, ± 5 mm, pink when ripe; seeds 2, pyriform.

Phenology : *Flowering* : with a peak during July–October; *Fruiting* : throughout the year.

Distribution : Tropical Himalaya and the hotter parts of India; Pakistan and Bangladesh.

Ecology and cultivation : Plains from the coast, on the wayside, fences, common in deeper forests, bushes, thickets; wild.

Chemical contents : *Stem*, *Root* and *Leaf* : contain hydrocyanic acid.

Remarks : Leaves and seeds are used in the treatment of yoke sores on the neck of bullocks.
Tribes of Saurashtra and Goa consume leaves and fruits as vegetables.

Cayratia pedata (Wall.) Gagnep. (Vitaceae)

(2n = 40, 72, 80)

Syn : *Vitis pedata* Vahl ex Wall., *Columella pedata* Lour., *Cissus pedata* Lam.

Sanskrit names : Godhapadi, Suvaha.

Vernacular names : Asm : Tusamphordoukha; **Ben :** Goalilata; **Mal :** Vellutasorivalli, Tripadi; **Mar :** Gorpadvel; **Ori :** Pittapotalo; **Tam :** Kattuppirandai; **Tel :** Edakulamandulam ari, Gummaditige.

Trade names : Gadhapati, Goalilata.

Traditional use : *Leaf* : astringent, refrigerant, used for ulcers; *Leaf-decoction* : used for checking uterine reflexes.

Phytography : A large, weak climber with young branches and pedately 7-foliate leaves softly hairy; leaves alternate, palmately compound, leaflets ± 1–2 cm, strongly veined; flowers tetramerous, small, grey or white in large subcorymbose cymes as long as the petiole; fruit subglobose, 2- to 4-seeded, rather dry.

Phenology : *Flowering* : August–September; *Fruiting* : Winter.

Distribution : Assam, West Bengal, Maharashtra, Karnataka and some other places.

Ecology and cultivation : Prefers humid conditions; plains from the coast (scarce), on thickets as straggler, roadsides, etc., but abundant on the hills 1000–1400 m, where it becomes a liane on forest trees.

Cedrela toona Roxb. ex Rottl. & Willd. (Meliaceae)

(2n = 52, 56)

Syn : *Toona ciliata* M.I.Roem., *T. febrifuge* King

English names : Indian Mahogany, Cedrela tree, Red cedar, Toon, Moulmein cedar.

Sanskrit names : Toona, Toonaka, Toonika, Kacchapa, Kanta, Kutheraka, Nandivriksha.

Vernacular names : Asm : Foma; **Ben :** Nandivriksha; **Hin :** Toon; **Kan :** Mandurike; **Mal :** Malarveppu; **Man :** Tairel; **Mar :** Tuni; **Mun :** Katangari, Katanga-daru; **Orn :** Tund; **Pun :** Khuding; **Tam :** Santhanavembu, Tunumaram; **Tel :** Nandichettu.

Trade name : Toon.

Traditional use : RABHA (of West Bengal) : *Leaf* : in bronchitis; SANTAL : *Stem* : in gastric troubles; GARHWALI : (i) *Bark* : in dysentery, fever, as antiseptic, (ii) *Fruit* : as antiseptic.

RAJANIGHANTU : it is pungent, bitter, nutritive, has sweet aroma and spermatopoietic property; used in burning sensation, dyserasia, headache and leucoderma.

AYURVEDA : (i) *Bark* : in purifying blood, dysuria, chronic dysentery, it is antiperiodic, astringent, tonic and as external application for ulcers, (ii) *Flower* : emmenagogue, useful in menstrual disorders.

Phytography : Tree, up to 25 m; branchlets tomentose; leaves even-pinnate, up to 25 x 12 cm, leaflets 5–12 pairs, subopposite, oblong-lanceolate, thin, coriaceous, glabrous, petiole up to 5 cm; panicles terminal or axillary, 25 x 10 cm, peduncle 2 cm, pedicel up to 2 mm; flowers pentamerous, bisexual, white; capsules septifragal, 3 x 1.5 cm, 5-valved; seed 25–30, oblong, 2 cm, thin-winged at both ends.

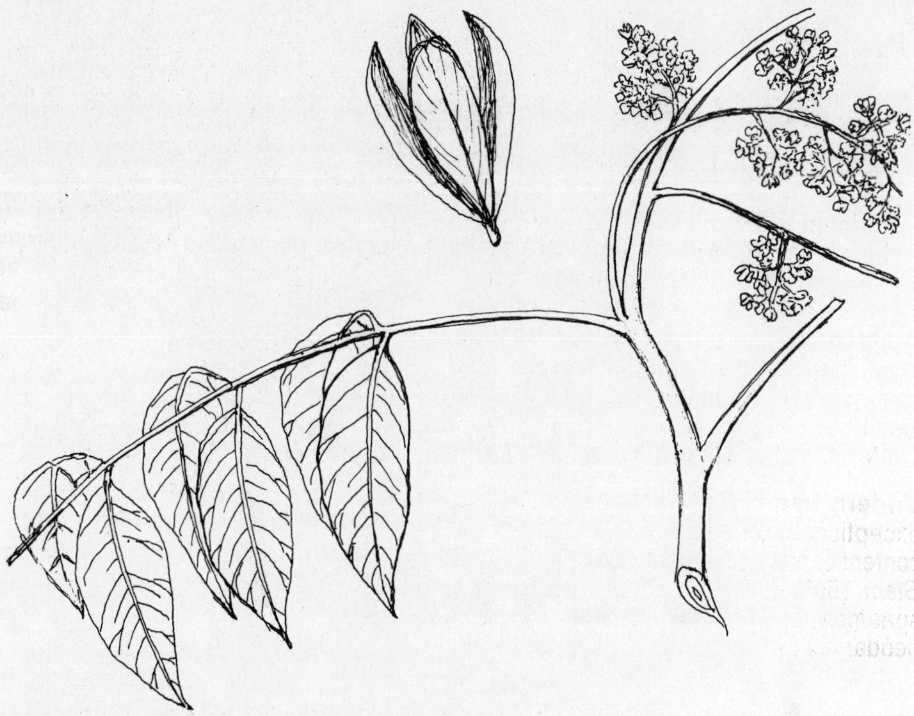

Phenology : *Flowering* : February–April; *Fruiting* : June–August.

Distribution : Occurring in hilly tracts and planted as avenue trees in India; Pakistan and Bangladesh (Chittagong).

Ecology and cultivation : Dense, foliaceous, deciduous tree with a spreading crown; leaf-fall in December; new foliage in January, red, rapidly turning to brown.

Chemical contents : *Bark* : toonacilin, β-seco-tetranortriterpenoid and its 6-acetoxy derivatives, 23-(R,S)- and 21-(R,S)-hydroxy toonacilids; *Wood* : alloaromadendrene, cadalene, cadinene, cadinol, calacorene, calamene, copaenecubenol, fernesene, geranylgeraniol, leucicyanidin, ledol, siderin,T-murolol, cedrelone, β-sitosterol; *Flower* : flavone glucoside, nyctanthin, *Seed* : bargapten, β-sitosterol, cedrelone, and 1,2-dihydro-cedrelone.

Remarks : One of the most valuable timber trees, also a good avenue tree.

Cedrus deodara (Roxb.) Loud. (Pinaceae)

Syn : *Cedrus libani* Berrel var. *deodara* Hook. f., *Pinus deodara* Roxb., *Abies deodara* Lindley

English names : Himalayan cedar, Cederwood, Deodar.

Sanskrit name : Devadaru.

Vernacular names : Hin : Deodar, Diar, Paludar; **Kas :** Deodar.

Trade name : Deodar.

Traditional use : GARHWALI : *Resin* : anthelmintic; ETHNIC COMMUNITIES OF UTTARKASHI : *Stem* : in rheumatism, ulcer. AYURVEDA : *Bark* : astringent, febrifuge; *Wood* (decoction) : in diarrhoea, dysentery, fever, urinary disorders; *Oleoresin* and dark-coloured oil (turpentine) : diaphoretic, emetic.
UNANI : an ingredient of the drug 'Deodara'.

Modern use : *Bark* (aqueous extract) : exceptionally effective in reducing sugar content of diabetic patient's urine and blood; *Stem* (50% EtOH extract) : anticancer, spasmolytic; *Wood-oil* (cederwood oil or deodar oil) : *in vitro* antibacterial, fungitoxic.

Phytography : Tall, evergreen tree, $\pm$50 m high with spreading branches and pendulous branchlets, branchlets densely pubescent; bark thick, fissured at places; wood aromatic, oily; leaves needle-shaped, 25–50 mm long, as thick as broad, dark bluish green; cones reddish brown, 80 mm–1.25 cm long and 50–80 mm across.

Phenology : Cones fall apart at maturity.

Distribution : Distributed in northwest Himalaya from Kashmir to Garhwal; forests of deodar occur in Kulu, Kashmir, Chamba, Tehri Garhwal, Almora, Bushahar, Simla, Chakrata and Mussourie hill stations.

Ecology and cultivation : Prefers granite belts and extreme cold climates, typically gregarious, often forms dense forests.

Chemical contents : *Stem-bark* : deodarin, dihydroflavonol, flavonoids, β-sitosterol, quercetin, taxifolin; *Wood* : cederin, deodorin, cedrinoxide, dihydromyricetin, kaempferol glucoside, polyphenolic lignoids; *Hexane extract of wood* : allohimachalol, himadarol, centdorol, isocentdarol; *Leaf* : vitamin C; *Pollen-grains* : several abietic acid derivatives, naringenin, β- sitosterol-β-D-glucoside.

Ceiba pentandra (L.) Gaertn. (Bombacaceae)

Syn : *Eriodendron antractuosum* DC., *Bombax pentandrum* L.

English names : White silk cotton tree, True kapok tree.

Sanskrit names : Kuta shalmali, Sweta shalmali.

Vernacular names : Ben : Shwet simul; **Hin :** Safed simul, Hattiam, Kattan; **Kan :** Buraga, Bili buraga; **Mal :** Ilavu, Mullilavu; **Mar :** Salmali, Pandhari; **Tam :** Ilavum; **Tel :** Tella buraga.

Trade names : Safed shimul, Shwet shimul.

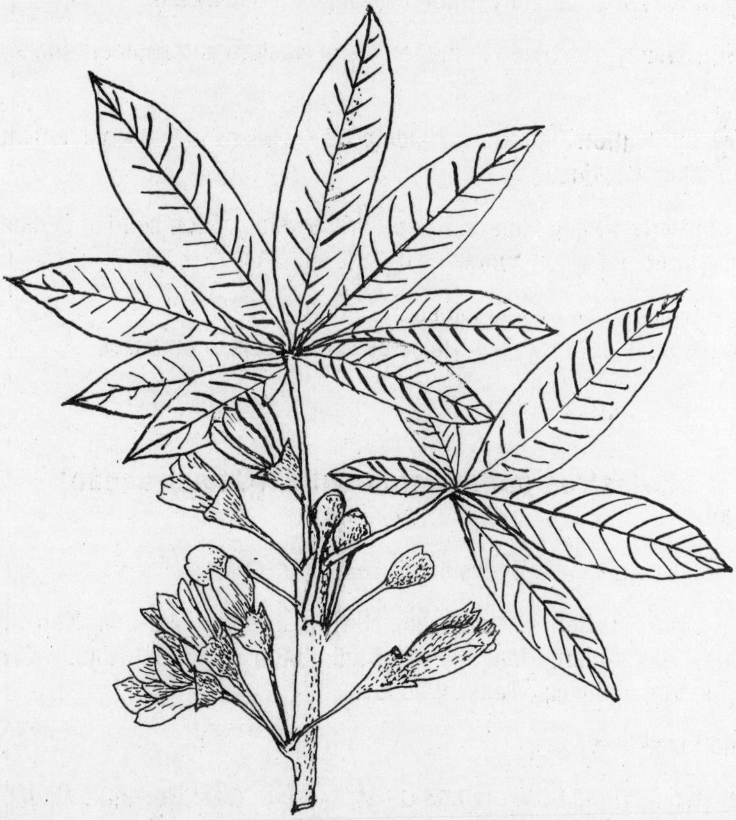

Traditional use : *BHAVAPRAKASH NIGHANTU* : it is astringent, bitter, hot, and useful in the diseases caused by phlegm, bile, fat, abdominal lump, spleen, poison, flatulence and mental disorders.

AYURVEDA : (i) *Root* : diuretic, increases the quantity of sperm, tonic; (ii) *Bark* : diuretic, emetic, removes fever; (iii) *Gum* : tonic, light purgative, useful in different types of stomach trouble; (iv) *Young leaf* : emolient, useful in gonorrhoea; (v) *Flower* (decoction) : mild purgative, useful in post-delivery menstruation; (vi) *Fruit* (unripe) : anodyne, astringent, increases retention power, useful in headache; (vii) *Seed* (oil) : unctous.

Modern use : *Aerial parts* (50% EtOH extract) : Central Nervous System depressant and diuretic.

Phytography : Tree, up to 30 m; branches horizontal, in whorls of 3, clearly tiered; leaves 5- to 9-foliate, 6–20 cm, leaflets usually oblanceolate, elliptic or oblong, thin, coriaceous, 12- to 15-nerved, prominent, margin entire; flowers white, clustered at the end of the branchlets, pedicel 4.5 cm; capsule ellipsoid to fusiform, indehiscent, valves with silky fibres; seeds many, subglobose.

Phenology : *Flowering* : January–April; *Fruiting* : March onward.

Distribution : Widely distributed in the forests of western and southern India, especially in hotter areas.

Ecology and cultivation : Introduced; planted in the plains to the coast; optimum size on hills above 800 m; deciduous.

Chemical contents : *Plant* : linarin; *Seed* : resembles cotton seed in composition, but contains little or no gossypol; contains 20–25% oil.

Remarks : Egyptians use root in diabetes.
Toys are made from the wood; seed floss used in pillows, mattresses.

Celastrus paniculatus Willd. (Celastraceae)

(2n = 46)

Sanskrit names : Agniphala, Jyotishmati, Tejovati.

Vernacular names : Ben : Malkangni; **Hin :** Malkangni, Malkunji; **Kan :** Kariganne; **Lod :** Ara-ba-kuiri, Kujari; **Mal :** Palulavam; **Mun :** Ara-ba, Kujuri; **Orn :** Kujur; **Sad :** Munjini; **San :** Munjui; **Tam :** Valuvuvai.

Trade name : Malkangi.

Traditional use : *ETHNIC COMMUNITIES OF JALPAIGURI* (West Bengal) : *Bark* : in swollen veins; *LODHA* : (i) *Root-bark* : in wounds, (ii) *Root-paste* : as abortifacient, (iii) *Seed-oil* : in rheumatism, scabies; *MUNDA* : (i) *Root* (crushed) : in pneumonia, (ii) *Leaf* (fresh) : in headache, (iii) *Leaf* (paste) : in pneumonia; *ORAON* : (i) *Fresh leaf-juice* (with common salt) : as antidote to toxic effect of alcohol consumption, (ii) *Seed-oil* : to babies for development and strengthening of skull bones; *SANTAL* : (i) *Bark* : in gastric complaints, (ii) *Seed* : in gout, rheumatism, (iii) *Seed-oil* : in scabies, leucoderma, bodyache, beriberi;

ASUR : *Bark* : as abotifacient; ETHNIC COMMUNITIES OF MAYURBHANJ (Orissa) : (i) *Root* : in sex diseases, (ii) *Fruit* : in skin diseases, (iii) *Seed* : in scabies; BHIL : *Seed-oil* : in eczema; ETHNIC COMMUNITIES OF BASTAR (Madhya Pradesh) : *Root* and *Leaf* : in headache; ETHNIC COMMUNITIES OF MAHARASHTRA : *Bark* : in wounds; DANG : *Bark* : in bronchitis; ETHNIC COMMUNITIES OF MT. ABU (Rajasthan) : *Seed* : in paralysis, for improvement of memory; ETHNIC COMMUNITIES OF HARYANA : *Seed* : in digestive complaints, rheumatism, as stimulant; ETHNIC COMMUNITIES OF RAJOURI (Jammu) : *Seed* : in backache; GARHWALI : *Seed* : in cold, fever, leprosy; THARU : *Seed* : as hair tonic.

ATHARVA VEDA (*Vaidyakakalpa*) : *Seed-oil* : Tonic, enhance memory and life span. CHARAKA SAMHITA : *Seed-oil* : beneficial in the diseases caused by phlegm. SUSHRUTA SAMHITA : *Seed-oil* : useful in the diseases caused by vitiated phlegm and fat, worms, skin diseases. BHAVAPRAKASHA : astringent, bitter, tonic, emetic, gastric stimulant, enhances intelligence and memory, useful in the diseases caused by deranged phlegm and wind. AYURVEDA : (i) *Bark* : abortifacient, (ii) *Leaf* : appetiser, emetic, mild laxative, sex invigorating, useful in fever, (iii) *Seed* : bitter, cough expectorant, memory promoter, useful in diseases caused by wind, (iv) *Seed-oil* : useful in beriberi, oedema, rubefacient. UNANI : an ingredient of 'Malkangunj'.

Modern use : *Plant* : minor ingredient of indigenous drug 'Geriforte'; *Aerial parts* (50% EtOH extract) : antiviral; *Leaf* : emmenagogue; *Seed-oil* : tranquilizing, sedative, useful in hysteria, vasoconstrictor and spasmolytic in rat's uterus; *Resin* (from seed-oil) : lowers blood pressure in dogs.

Phytography : Large climbing shrub, unarmed straggler/liane up to 12 m, branchlets puberulous; leaves alternate, ovate to orbicular, thin, coriaceous, base obtuse to subacute, margin dentate, petiole up to 3 cm; panicles terminal; flowers polygamous, greenish white; capsules 0.6–1.0 cm; seeds 6, ellipsoid to ovoid, up to 5 mm.

Phenology : *Flowering* : March–May, sometimes up to September; *Fruiting* : September.

Distribution : Distributed in the sub-Himalayan tract from the Jhelum eastwards up to 2000 m in different hilly parts of India; Pakistan and Bangladesh.

Ecology and cultivation : Extensive straggler on liane on tree; hills, 750–1400 m along the forest borders; very common.

Chemical contents : *Seed*, *Seedcoat* and *Aril* : yield fatty acids; *Distillation of seed* : yields a black ampyreumatic oil, Oleum nigrum, which has stimulating action followed by diaphoresis unattended by exhaustion; alkaloid, glucoside, colouring matter; seeds also yield brownish oil with unpleasant taste; *Seed-fat* and *Seedcoat-fat* : contain formic acid, acetic and benzoic acids.

Celosia argentea L. var. *argentea* (Amaranthaceae)

(2n = 36, 72)

English name : Quail grass.

Vernacular names : Ben : Shwet moragphul; **Guj :** Lapadi, Lambadi; **Hin :** Safed murgha, Sarwari; **Mar :** Kurdu, Kuradu; **Mun :** Sirgiti-ara; **Orn :** Kimar-xa; **Pun :** Sarwali, Sarpankha, Sil; **Sad :** Siliari; **Tel :** Gurugu, Panchechettu.

Trade names : Safed murgha, Shwet moragphul.

Traditional use : *GARHWALI* : *Seed* : as purgative; *ETHNIC COMMUNITIES OF DEHRA DUN AND SIWALIK* : *Seed* : in blood purification, sores in mouth; *ETHNIC COMMUNITIES OF EASTERN RAJASTHAN* : *Seed* : in blood purification, diarrhoea.
AYURVEDA : *Seed* : aphrodisiac, useful in diarrhoea, diseases of the eye.

Modern use : *Plant* (50% EtOH extract) : antiprotozoal, spasmolytic.

Phytography : Common weed; herb up to 1–2 m; leaves elliptic- ovate below, linear-lanceolate above, 7–13 x 0.5–3 cm, base cuneate to attenuate, apex acute, petiole up to 1 cm, spikes dense, cylindric, very showy, flowers 4 mm across, pink, gradually turning white; bract lanceolate, 6 mm, scarious; capsules globose, reticulate.

Phenology : *Flowering* and *Fruiting* : Winter.

Distribution : Cosmopolitan, probably originating in Africa; common weed in cultivated fields, hedges, riverbanks and open places throughout India, Pakistan and Bangladesh; occasionally ascending up to 2000 m in hilly regions.

Ecology and cultivation : Hills in arable lands, as a weed of cultivation; conspicuous in harvested field; less common in plains.

Chemical contents : *Plant* : 10-coumaroyl-betanicin-5-0-glucuronosylglucoside, 10-feruloyl-isobetanidine-5-0-glucuronosylglucoside.

Celtis philippensis Blance var. *wightii* (Planch.) Soep. (Ulmaceae)

Syn : *Celtis wightii* Planchon.

Vernacular names : Mal : Mannalli; **Tam :** Vakkanai; **Tel :** Kaka-mushtii.

Traditional use : Wood scraped fine and mixed with lemon juice taken internally as a blood purifier in itch and cutaneous eruptions and is used as a fumigator and in headache.

Phytography : Tree, ± 15 m in height; leaves elliptic to oblong lanceolate, 9–14 cm by 3–7 cm, base rounded, apex acute, 3-nerved from base, nerves running almost up to the apex of the leaf, flattened above; stipules elliptic-ovate, acute; panicles short, many-flowered; drupe ellipsoid, epicarp leathery, coriaceous, beaked when young.

Phenology : *Flowering* : with a peak in December–April; *Fruiting* : February onwards, red when ripe.

Distribution : Hills of South India, Andamans and Sri Lanka.

Ecology and cultivation : Foothills up to 120 m; very common in forests; densely foliaceous tree; size increasing with increasing altitude.

Celtis timorensis Spanoghe (Ulmaceae)

(2n = 20)

Syn : *Celtis cinnamomea* Lindl., *C. trinervia* Beddome

Vernacular names : Mal : Putan; **Ori** : Gundukambhara; **Tam** : Pinari.

Traditional use : *Wood* : scraped and mixed with lemon juice taken internally as a blood purifier in itch and cutaneous eruptions and is used as fumigator and in headache; used for nervous disorders.

Phytography : Deciduous tree, up to 25 m high; bark whitish outside; tender part pubescent; leaves chartaceous or (sub)coriaceous, base rounded, entire or serrate along the upper half, acute to shortly acuminate, nerves 1–2 pairs, pubescent along the nerves on both sides; petiole up to 0.5 cm; inflorescence racemose, ± 15-flowered; flowers greenish, 3–5 mm across; drupes ovoid-globose, 0.5 cm across.

Phenology : *Flowering* : March–April on leafless branchlets, followed by new foliage; *Fruiting* : April onwards.

Distribution : Sikkim-Himalaya, West Bengal, Madhya Pradesh and South India.

Ecology and cultivation : Hill above 1200 m; common; leaf fall January–February.

Chemical content : *Wood* : skatole.

Cenchrus biflorus Roxb. (Poaceae)

Syn : *Cenchrus barbetus* Schum., *C. catharticus* Del.

Vernacular names : Mar : Baront; **Raj :** Bharut.

Trade name : Bharut.

Traditional use : *Decoction of fruit* : diuretic and pectoral.

Phytography : Herb, culm up to 40 cm in height and up to 2.5 mm wide; leafsheaths 3–8.5 cm, scabrid on nerves, ligule up to 1.5 mm, blades oblong, folded; panicles up to 8 cm; involucre turbinate, bristle rigid, apex retrorsely barbellate; spikelets 3–4 mm; caryopsis oblong, apex beaked.

Distribution : Punjab, Uttar Pradesh, Bihar and some other semi-arid zones; Pakistan.

Ecology and cultivation : Prefers dry soil.

Remark : Ethnic communities of arid zones prepare bread with seed-flour.

Centaurea cyanus L. (Asteraceae)

(2n = 20, 24)

English names : Bachelor's button, Blue bottle, Cornflower.

Traditional use : *Plant* : remedy for disorders of nervous system; also used for many eye ailments; *Flower* : tonic, stimulant and emmenagogue; *Florets* : astringent and soothing in conjunctivitis; used in the preparation of Eau de Casse-lunettes, a French preparation for weak eyes.
UNANI : ingredient of 'Behman safaid".

Modern use : *Seed-extract* : antifungal.

Phytography : An ornamental herb, slender branching annual; ± 30–60 cm high, woolly white at least when young; leaves linear, 75–150 mm long, entire or the lower toothed, sometimes pinnatifid; heads naked on slender peduncles, ± 25–40 mm across; flowers blue, pink, purple or white fringed with short scarious blackish teeth; pappus about equalling achene.

Distribution : North-West India—in cultivated places and in corn fields.

Ecology and cultivation : Grows in arid climate.

Chemical contents : *Flower* : n-nonacosane, triterpane, steroidal compounds.

Centaurium roxburghii (G. Don) Druce (Gentianaceae)

(2n = 56)

Syn : *Erythraea roxburghii* G. Don

Vernacular names : Ben : Gima, Girm; **Guj :** Jangli, Khiatu; **Hin :** Barik chirayata, Kheta chirayata; **Mal :** Thuporuppenpullu; **Mar :** Lantak, Kadavinai; **San :** Gadasigirk.

Trade name : Kheta chirayata.

Traditional use : Bitter, tonic, stomachic, substitute for *chiretta*, in fever, febrifuge.

Phytography : Annual herb; leaves opposite, sessile or clasping, entire; flowers usually pink in spikes.

Distribution : Throughout India; Bangladesh and Pakistan; ascending up to 700 m.

Ecology and cultivation : Common in cultivated land; wild.

Adulterant : Substitute for *chiretta.*

Centella asiatica (L.) Urban (Apiaceae)

(2n = 18, 22)

Syn : *Hydrocotyle asiatica* L.

English name : Asiatic pennywort.

Sanskrit names : Mandukaparni, Tvastri.

Vernacular names : Ben : Thankuni; **Hin :** Brahma manduki, Brahmi; **Mar :** Kerivana; **Tam :** Vallarai; **Tel :** Brahmi.

Trade names : Thankuri, Thankuni.

Traditional use : *NAGA* : *Leaf* : in dysentery, cholera; *KHASI* and *JAINTIA* : *Leaf* : in cough, diarrhoea; *RURAL FOLK OF ASSAM* : *Leaf* : in blood dysentery, syphilis; *TOTO* (of West Bengal) : *Leaf* : in gastric disorders; *RURAL FOLK OF ORISSA* : *Leaf* : in dysentery; *THARU* (of Uttar Pradesh) : *ROOT* : as diuretic; *RURAL FOLK OF PITHORAGARH* (Uttar Pradesh) : *Whole plant* : as anthelmintic; *KUMAONI* : *Leaf* : in fever and as brain tonic; *GARHWALI* : *Whole plant* : as brain tonic, *Leaf* : in eczema, eye complaints, fever, tuberculosis; *TRIBES OF KANGRA VALLEY* (Himachal Pradesh) : *Leaf* : in blood dysentery, as brain tonic; *TRIBES OF RAJASTHAN* : *Whole plant* : as brain tonic, *Fruit* : in blood dysentery, nervine disorder, respiratory troubles, skin diseases; *BHIL* : *Plant* : in urinary complaints; *TRIBES OF MADHYA PRADESH* : *Plant* : as brain tonic, in skin diseases; *TRIBES OF THE NILGIRIS* (Tamil Nadu) : *Leaf* : as cooling agent; *NAYADI* (of Kerala) : *Leaf* : in dysentery, nervine disorder.
ATHARVAVEDA : this plant is mentioned as brain tonic. *AYURVEDA* : *Leaf* : in dysentery, fever, wounds, as brain tonic.
SIDDHA SYSTEM : *Leaf* : used to prepare the drug named *Vallari*.
HOMOEOPATHY : used to treat ascariasis, eczema, elephantiasis, granular cervicitis and ulcer of womb.

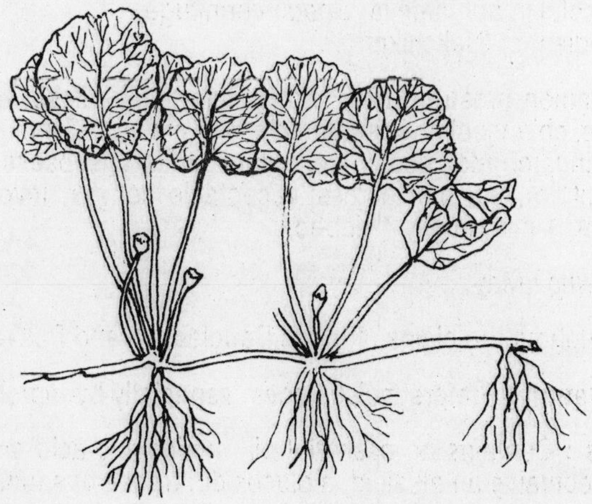

Modern use : *EtOH* (50%) *extract of plant* : antiprotozoal, spasmolytic and mild sedative.

Phytography : A prostrate herb rooting at nodes; young leaves and petioles villous, leaves orbicular, reniform, never lobed but often with large rounded crenate or coarsely dentate or subentire; umbels usually several at each node; flowers 3–5, rarely 6 in an umbel, subsessile, petals deep red, ovate, acute or obtuse, imbricate, stamens red; fruit with very narrow commissure, didymous cocci.

Phenology : *Flowering* and *Fruiting* : generally in May but vary from place to place.

Distribution : Occurring in marshy places throughout India, Bangladesh and Pakistan, especially common and abundant along irrigation channels up to 3000 m.

Ecology and cultivation : In moist places, sides of the nullahs, bunds; wild and cultivated.

Chemical contents : *Herb* (extract)) : brahmic acid, thankunic acid, isothankuniside, thankunidide, polyacetylenes, asiatic acid, asiaticoside, madecassic acid, madegascaric acid, two saponin glycosides.

Centipeda minima (L.) A. Br. et Aschers (Asteraceae)

(2n = 14, 18, 20)

Syn : *Centipeda orbicularis* Lour., *Artemisia minima* L., *Dichrocephala schmidtii* Wight

Sanskrit names : Chikkani, Chhikika.

Vernacular names : Ben : Mechitta; **Hin :** Nakk chikni, Nagdowana; **Mar :** Nakashikani.

Trade name : Nakk chikni.

Traditional use : A snuff obtained from *Leaf* and *Seed* is used for colds in head; *Herb* : boiled and made into a thick paste applied to cure diseases of cheek and toothache; *Infusion of plant* : useful in ophthalmia; *Seed* : vermifuge.
UNANI : it is an ingredient of 'Nakchikni'.

Phytography : Common prostrate herb; branchlets thinly pilose, up to 15 cm; leaves alternate, subsessile, oblanceolate to spathulate, 4–7 cm by 2–3 mm, thick, base cuneate, decurrent, margin sinuate, apex acute; capitulum axillary, subsessile, yellow, globose, heterogamous, disciform, 0.3 mm across; receptacle convex, involucre campanulate; achene oblong-linear, 1 mm, faintly 4-ribbed.

Phenology : *Flowering* : May.

Distribution : Throughout the plains of India; Bangladesh and Pakistan.

Ecology and cultivation : Prefers moist places, especially by marshy bank of ponds.

Chemical contents : Contains an essential oil, myriogynic acid and a bitter principle, amorphous; *Plant* : contains an alkaloid, a glucoside, traces of saponin.

Centranthera indica (L.) Gamble (Scrophulariaceae)

Syn : *Centranthera procumbens* Benth.

Traditional use : *Herb* : used in fever and as external application for sore eyes and as febrifuge.

Phytography : A small herb; leaves opposite or upper alternate, oblong or linear, entire or toothed; flowers axillary or in bracteate spikes or racemes; bracteoles 2; capsules ovoid or sub-globose, loculicidal.

Phenology : *Flowering* : Winter.

Distribution : South India; Sri Lanka.

Cephaelis ipecacuanha (Bort.) A. Rich. (Rubiaceae)

(2n = 22)

Syn : *Psychotria ipecacuanha* Stokes.

English names : Ipecac, Ipecacuanha.

Vernacular names : Ben : Ipecac; **Nep :** Ipecac.

Trade name : Ipecac.

Traditional use : *UNANI* : Ingredient for 'Gurmar Buti'.

Modern use : Used as emetic and expectorant and in the treatment of dysentery and pyorrhoea; diaphoretic; it possesses cholagogue properties.

Phytography : A small straggling evergreen undershrub; roots much-branched, annulated or moniliform or beaded, 10–30 cm, grey or brown, when dried dark brown; stem 0.5 m high; leaves opposite, 5–10 cm long and 2.5–5.0 cm broad, eliptic, margin entire; flowers in cymes, white; fruits round or oval, black when ripe, containing two flat whitish seeds.

Phenology : Flowering starts from the 2nd year; production of fruits becomes maximum during 3–4 years; *Flowering* usually in very early Spring and *Fruiting* in Rains.

Distribution : Cultivated at Mungpoo near Darjeeling, at Rungbee Cinchona Plantation in Sikkim, Assam; and on the Nilgiris near Kollar.

Ecology and cultivation : The plant generally survives for 5–6 years. The plants can be raised from seeds as well as from leaf and stem cuttings; prefer artificial shade, soil rich in humus and acidic in nature. Alkaloids are present in leaves, stem and root, maximum being in the root. Harvesting at the roots mainly done at the end of 3 years. Darjeeling hills of West Bengal can luxuriantly grow ipecac at the base flat regions of the hills where the moisture is comparatively heavy during summer and temperature varies between a maximum of 30°–32°C and minimum of 20°–22°C; annual rainfall is about 400–500 cm.

Chemical contents : Annulated rhizomes are reported to contain cephaeline and emetine as active principles and also psychotrine; the Indian root contains 1.98% of total alkaloids and glucoside and 1.39% emetine; it also contains saponins; 3- to 4- year-old plants from Mungpoo gave highest yield of emetine. *Root* : cephaeline, δ-mannitol and a glucoside ipecoside in 1% yield.

Adulterants : East Indian Root, probably *Cryptocaryne* (Fam : Araceae); undulated ipecacuanha, derived from *Richardsonia scabra* (Rubiaceae); lesser striated ipecacuanha, derived from *Manettia ignita* (Rubiaceae), greater striated Ipecacuanha (root of *Psychotria emetica*); white ipecacuanha (root of *Hybanthus ipecacuanha* Fam. Violaceae) and Trinidad ipecacuanha (root of *Asclepias curassavica* Fam. Asclepiadaceae) can be used as adulterants.

Ceratonia siliqua L. (Caesalpiniaceae)

(2n = 24)

English names : Algaraba, Carab tree, St. John's bread, Karoub.

Vernacular name : Pun : Kharnub.

Trade name : Kharnub.

Traditional use : *Pod* : syrup and diabetic food are made from the pods, astringent, used in coughs and catarrhal infection; *Seed-husk* : purgative, astringent.
UNANI : an ingredient for 'Kharnobe-Shami'.

Phytography : Tree reaching a height of 14–18 m with thick crown; leaves alternate, pinnate, with 2–3 pairs of obovate or rotund shining leaflets, obtuse or emarginate, 40–100 mm long; racemes lateral; flowers red; pod 75–300 mm long.

Distribution : Native of eastern Mediterranean region; introduced in India; has become naturalised in Punjab and some other places.

Chemical contents : *Seed* : yields a fatty acid; *Carnob seed-gum* : consists of D-mannose and D-galactose.

Adulterant : *Seed* : used as a substitute for coffee.

Ceratophyllum demersum L. (Ceratophyllaceae)

(2n = 24, 38, 48)

Sanskrit name : Shivala.

Vernacular names : Ben : Sheoyala; **Hin :** Sivara; **Tel :** Nasu.

Trade name : Sivara.

Traditional use : *Plant* : antiperiodic, purgative, cooling, useful in biliousness, in scorpion bite.

Modern use : *Plant* (50% EtOH extract) : antibacterial.

Phytography : Aquatic herb; branchlets brittle, green, reddish, glabrescent; leaves 7–12 in close verticils, up to 2.5 cm, swollen near the middle, sometimes sparsely sericeous, denticulate, apex acute; flower solitary, mostly in the upper axils; nutlets ellipsoid, black when dry, 4.5 x 3.5 mm, furnished with 3-spurs; seed globose, testa stony.

Phenology : *Flowering* and *Fruiting* : with a peak during January–March.

Distribution : Cosmopolitan; all over India, Bangladesh and Pakistan.

Ecology and cultivation : Plains, in ponds, pools, marshes and swamps.

Chemical content : Hairs contain myrophyllin.

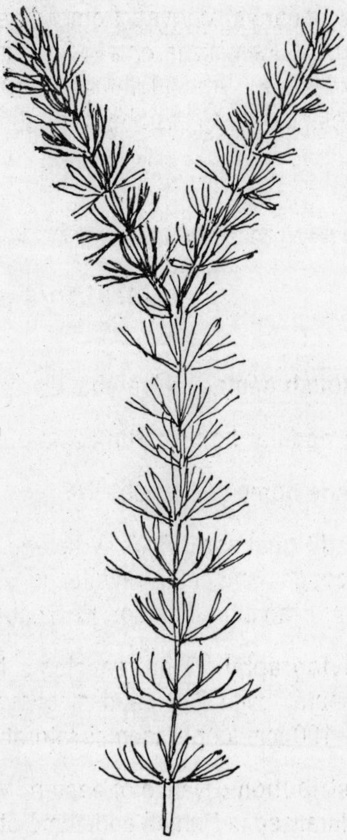

Ceratopteris Siliquosa (L.) Copel. (Ceratopteridaceae)

Syn : *Ceratopteris thallictroides* Brong., *Acrosticum thallictroides* L.

English names : Horned fern, Water fern.

Traditional use : *Frond* : used as poultice in skin complaints.

Phytography : Undershrub; sterile fronds narrowly deltoid, stipe slender, 50–250 mm long, laminae ± 50–250 mm long, 1- to 2-pinnatifid into lanceolate or oblong segments, not floating; fertile frond taller, 2–4 pinnate with ultimate segments linear.

Distribution : Found throughout India, Bangladesh and Pakistan up to 1000 m.

Ecology and cultivation : Common in tanks, ditches and swampy places or even dry ground during the rainy season; wild.

Remark : In China, plant is used as tonic and styptic.

Cerbera manghas L. (Apocynaceae)

(2n = 40)

Syn : *Cerbera oodollam* Gaertn.

English name : Dog bane.

Vernacular names : Ben : Dabur, Dhakur; **Kan :** Cande, Monde; **Mal :** Utalam, Chattankaya; **Mar :** Sukanu; **Tam :** Kodalma; Kattarali, Kottuma.

Trade names : Dog bane, Dabur.

Traditional use : *Root* : laxative; *Bark* : purgative; *Nut* : narcotic, poisonous, used in hydrophobia.

Phytography : Small tree with milky latex, glabrous; branches whorled, stout with leaf scars; leaves alternate at the end of the branches, coriaceous, black at maturity; flowers large, in terminal peduncle cymes, white with yellow throat, odorous fruits 5–15 cm long, solitary, subglobose.

Phenology : *Flowering* and *Fruiting* : June–August.

Distribution : Tidal forests and salt swamps along the coast.

Ecology and cultivation : Very rarely grows in the swamp-forests throughout salty swamps and sea coasts of India.

Chemical contents : *Root* : β-sitosterol; *Root-bark* and *Stem-bark* : gentiobiosyl thevetoside, glucosyl thevetoside, and thevetosides of tanghinigenine; *Leaf* : succinic acid,, nicotiflorin, rutin, *L*-(+)-bomesitol, manghaslin, clitorin, 17-βH-nerifolin, cardenolides; *Fruit* : theveside, theviridoside, iridiol glucosides; *Seed* : cerbertin, cerbertatin, Ac-cerbertin, thevetin B, cerbertin, di-Ac-neriifolin, Ac-thevetin B, mono-Ac-thevetin B, tanghinoside de-Ac-tanghinoside tanghinine, de-Ac- tanghinine.

Remarks : Plant used as fish poison. Fruit employed to kill dogs. Seeds poisonous. In Samoa, plant used in treatment of wounds and yaws.

Cereus grandiflorus Mill. (Cactaceae)

(2n = 22)

Syn : *Selenicereus grandiflorus* Britton and Rose.

English name : Night blooming cereus.

Trade name : Cereus.

Traditional use : *Fresh young shoot* : used as cardiac stimulant and as partial substitute for digitalis; *Liquid extract* or *tincture* : used in cases of dropsy and various cardiac affections.

Phytography : An ornamental cactus bearing large attractive sweet-scented flowers which open at night and wither before morning; stem about 25 mm thick with 7–8 low ribs; areoles with yellow white wool, spine needle-like, 6–12 mm long; flowers 1.75 cm or more long, tube with long brownish hairs; fruit ovoid, 75 cm long.

Distribution : Introduced into gardens.

Chemical contents : *Plant* : alkaloids; *Flowers* : cacticin, flavanoids, glycerides, narcissin.

Ceriops tagal (Perr.) C. B. Robins (Rizophoraceae)

Syn : *Ceriops candollena* Arn.

Vernacular names : Ben and **Hin :** Goran; **Tam :** Pandikutti; **Tel :** Godera.

Trade name : Goran.

Modern use : *Plant* : astringent; *Decoction of bark* : haemostatic.

Phytography : A small halophytic and mangrove plant; glabrous small shrub, 1–2 m high; branchlets marked with fallen leaf scars; leaves simple, 5–10 cm by 3–5 cm, coriaceous, obovate-oblong, apex rounded, glabrous; petioles 0.8–2.5 cm long, rugose, glabrous; flowers in axillary cymes, peduncled, white; fruits slightly conical.

Phenology : *Flowering* and *Fruiting* : June–January.

Distribution : Muddy shores and tidal creeks of Sundarbans and other coasts of India; Bangladesh.

Ecology and cultivation : Occasionally found on river slopes, the sides of the creeks and tidal forests of the Sundarbans; associated with other species.

Remark : In African coast, decoction of shoot is used as a substitute for quinine.

Ceropegia bulbosa Roxb. (Asclepiadaceae)

(2n = 22)

Vernacular names : Hin : Khapparkadu; **Kan :** Halliki; **Mar :** Khapparkadu, Patala kumbari; **Pun :** Galat; **Tel :** Palatige.

Trade name : Khapparkadu.

Traditional use : TRIBES OF RATAN MAHAL HILLS (Gujarat) : *Tuber* : as refrigerant; TRIBES OF MAHARASHTRA : *Tuber* : for vitality.

Phytography : A small herb; leaves linear-lanceolate to elliptic-obovate, 3–5 x 0.5–1.0 cm; peduncle 1–1.5 cm, bracts linear, 3 mm; pedicel up to 0.5 cm; flowers greenish or dark purple; follicles cylindric, tapering towards an acute apex, glabrous; seeds ovate-oblong.

Phenology : *Flowering* : October–November.

Distribution : Punjab, Upper Ganga plains, South India, Karnataka, Maharashtra; Pakistan.

Ecology and cultivation : Slender twiner in thickets.

Chemical content : Alkaloid ceropegine is the bitter principle.

Remark : Tribes of Rajasthan, Maharashtra and Goa eat the tubers and leaves as vegetables.

Ceropegia candelabrum L. (Asclepiadaceae)

Syn : *Ceropegia tuberosa* Roxb., *Riocreukia bubalina* Auct. non. W. Elliot.

Sanskrit name : Bhutumbi.

Vernaculars names : Hin : Patalatumbi; **Mar :** Khapparkadu; **Tel :** Bachalimanda.

Traditional use : *Tuber* : tonic in bowel complaints of the children, remedy for diarrhoea and dysentery.

Phytography : Herb; leaves elliptic to obovate, 2–3 x 0.8–1.0 cm, chartaceous, base acute to rounded, apex acutely apiculate, petiole up to 0.5 cm; peduncle up to 2 cm, bract linear, 3 mm, pedicel up to 0.5 cm, corolla tube greenish yellow, lobes 5, purplish green, shortly beaked.

Phenology : *Flowering* : October–November.

Distribution : Maharashtra, Karnataka and other parts of South India.

Ecology and cultivation : Slender twining on thickets.

Chasalia chartacea Craib. (Rubiaceae)

Syn : *Chasalia curviflora* Thw., *Psychotria curviflora* Wall.

Vernacular name : Tam : Vellakurinji.

Traditional use : *Decoction of root* : used in rheumatism, pneumonia, head disorders, ear and eye diseases and sore throat; *Root* : used against malaria; *Root* and *Leaf* : used as external applications for wounds, ulcers and headache.

Phytography : Small shrub, 60–90 cm high; branchlets glabrous, compressed; leaves elliptic, oblong-lanceolate, acute or acuminate, base tapering, main nerves 8–10 pairs, stipules short, 2-fid; flowers dimorphic, one form with exerted anther and included styles and another sessile or nearly so; berry the size of pea, dark purple.

Phenology : *Flowering* : Apri–July.

Distribution : Sikkim, Darjeeling, Assam, Meghalaya, ascending up to 1075 m, Maharashtra, Karnataka and the Andamans.

Ecology and cultivation : Common in some of the evergreen forests.

Cheilanthus tenuifolia Swartz (Polypodiaceae)

Vernacular name : San : Dodhari.

Trade name : Dodhari.

Traditional use : SANTAL : prescribe a preparation from the *Root* for sickness attributed to witchcraft or the evil eye.

Phytography : Small greenhouse or rock-loving, semi-hairy ferns, often hairy or woolly; fronds uniform, blades 1- to 4-pinnate or 1-pinnate and variously pinnatifid, segments often minute; sori borne at the enlarged tips of the veins, indusia formed by the revolute margins of the segments.

Distribution : Tamil Nadu, up to ± 1200 m, Bihar, Sikkim, West Bengal, plains of Assam, Meghalaya, Arunachal Pradesh; South Bangladesh and some parts of Pakistan.

Ecology and cultivation : Rocky areas.

Cheiranthus cheiri L. (Brassicaceae)

(2n = 12,14)

English name : Wall flower.

Vernacular names : Ben : Kheiri; **Hin :** Todrisurkh.

Trade name : Kheiri.

Traditional use : *Flower* : formerly used as emmenagogue, antispasmodic, resolvent; *Dried petals* : aromatic, stimulant; *Seed* : tonic, diuretic, expectorant, stomachic,

aphrodisiac, in dry bronchitis, fever and injuries to eyes.
UNANI : ingredient of 'Tudri or 'Surkh' or 'Zard'.

Phytography : Hardy perennial, erect herb, ± 30–75 cm high, thinly appressed, pubescent with a greying cast; leaves lanceolate to narrow lanceolate, acute, entire, 36–65 mm long, usually bunched beneath the flowers and at the end of the sterile shoots; flowers large, very fragrant, yellow or yellow-brown; fruits erect, angular, thick with short style tip.

Phenology : *Flowering* : early Spring.

Distribution : Native of southern Europe; in India, cultivated in gardens.

Chemical contents : *Plant* : cheiranthin, a glucoside, a heart poison, nerol, geraniol, benzylalcohol cheirinin and an alkaloid; *Flower* : source of essential oil; cheirotoxin, a glucoside, cheiroside A, a less active glucoside isolated from the seeds.

Chenopodium album L. (Chenopodiaceae)

(2n = 18, 36, 42, 54)

English name : Lamb's quarters.

Sanskrit names : Vastuka, Shakaraja, Yavashaka.

Vernacular names : Ben : Bethua shag; **Hin :** Bara bothus, Bethua sag, Khartua sag; **Man :** Monshaobi; **Mar :** Chakwit; **Mun :** Bhatua-ara; **Orn :** Bhatua-arxa; **Pun :** Bathua; **Sad :** Bhatua; **Tam :** Parukkirai; **Tel :** Pappukura.

Trade name : Bethua sag.

Traditional use : *ETHNIC COMMUNITIES OF HAZARIBAGH* (Bihar) : *Leaf* : to revive taste in tongue; *TRIBES OF VARANASI* (Uttar Pradesh) : *Seed* : in skin diseases; *ETHNIC COMMUNITIES OF DAHANU FOREST* (Maharashtra) : *Leaf* : in urinary troubles.
RIGVEDA : it can destroy all diseases; *ATHARVA VEDA* (*Vaidyakakalpa*) : it is beneficial in piles, worms and laxative; *CHARAKA SAMHITA* : it removes *tridosha* and enhances digestive power; *SUSHRUTA SAMHITA* : it is pungent, enhances memory, appetite, digestive power, strength of body, destroys all worms; *RAJANIGHANTU* : it is sweat, cooling, alkaline but slightly acidic, pacifies *tridosha*, appetizing, antipyretic, useful in piles, normalises urination and purgation; *AYURVEDA* : it is useful in anoraxia, cough, dysentery, diarrhoea, oedema, piles, small worms.

Modern use : *Plant* (50% EtOH extract) : Central Nervous System depressant.

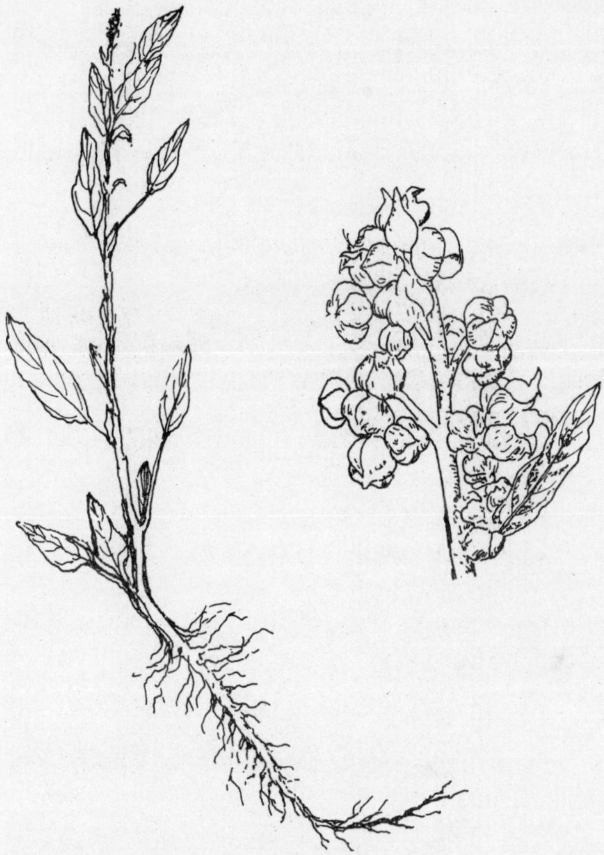

Phytography : Herb or subshrubby, up to 91 cm in height; branchlets with green and purple streaks, glabrous or thinly clothed with white vesicles; leaves simple, alternate, long petioled, ovate or lanceolate, 2.5–6.0 x 1.0–2.0 cm; flowers clustered at the nodes, whitish, 2 mm across; seeds horizontal, obscurely keeled, 1–2 mm across.

Phenology : *Flowering* and *Fruiting* : February–March.

Distribution : Throughout India, up to 4,260 m; Bangladesh.

Ecology and Cultivation : Mesophyte; wild and cultivated, occurs chiefly in the fields of mustard and wheat, on slopes; cultivated as a grain-crop and pot-herb in the western Himalaya.

Chemical contents : *Root* : ecdysteroids, β-ecdosyne, polypodine B; *Shoot* : ascorbic acid, β-carotene, β-sitosterol, caffeic acid, campesterol, catechins, ferulic acid, gallocatechin, imperatorin, n-triacontanol, p-coumaric acid, scopoletin, spinasterol, stigmasterol.

Remarks : Rural folks of Manipur, Assam, West Bengal, Bihar and Delhi consume the whole plant as vegetable.

Laves are consumed as vegetable by Bhoxas, Tharus, Kumaonis, ethnic communities of Varanasi, Mirzapur, eastern Rajasthan, Goa and Nilgiris.

Monpas (of Arunachal Pradesh) eat seeds.

In Hawaii, flowers and buds are used in weakness in children, stomach troubles and for fattening, while seeds for abortion.

Zulu tribes of Africa use infusion of leaves in intestinal ulcers, piles, throat troubles and eye complaints; and decoction of leaves as aphrodisiac, diuretic, tonic, in biliousness, hepatic disorders and spleen troubles.

Chenopodium ambrosioides L. (Chenopodiaceae)

(2n = 16, 22, 38, 64)

English names : American worm seed, Mexican tea.

Sanskrit name : Shvetachilli.

Vernacular names : Ben : Chandan beto; **Mal :** Katu ayamoddakam; **Man :** Monsaobi-manbi.

Traditional use : NAGA : *Leaf* : vermifuge, RURAL FOLK OF GOA : *Leaf* : in influenza, pneumonia, typhoid; RURAL FOLK OF DELHI : *Leaf* : anthelmintic.
RAJANIGHANTU : it is sweet, alkaline, cooling, pacifies *tridosha*, antipyretic, and used as diet. AYURVEDA : *Plant* : remedy against parasites; *Fruit-oil* : effective against hook-worms, round-worms.

Modern use : *Essential oil from leaves* : antimicrobial, antifungal against human pathogenic fungi.

Phytography : Strongly scented subshrubby, up to 121 cm; branchlets angular, ribbed glabrous or thinly pubescent; leaves oblong-lanceolate, 5–10 x 1–2.5 cm, lateral nerves 8–10 pairs, petioles up to 3 cm; panicles axillary or terminating the branchlets, 5–8 cm; flowers greenish, 2 mm across; fruit globular, compressed, with thin pericarp; seeds smooth, horizontal, rarely erect, margin entire.

Phenology : *Flowering* and *Fruiting* : throughout the year, peak in Winter.

Distribution : Found extensively in southern India, Maharashtra, Kashmir, West Bengal; Bangladesh and Pakistan; native of West Indies.

Ecology and cultivation : Hills, 800–1400 m, as a weed of cultivation; often gregarious near habitations; at lower elevations along river banks; ready coloniser.

Chemical contents : *Root* : magnesium, saponin, vitamin C; *Leaf* : ambroside, kaempferol-7-rhamnoside.

Adulterants : Used as a substitute for the American *C. ambrosioides* L. var. *anthelminticum* Gray (*C. anthelmintic* L.) which is the source of commercial worm seed oil or chenopodium oil; also a substitute for tea.

Chenopodium botrys L. (Chenopodiaceae)

(2n = 18)

Syn : *Ambrosia mexicana* Hort.

English names : Jerusalem oak, Feather.geranium.

Traditional use : AMCHI (of Ladakh) ; *Plant* : anthelmintic, in colds, dysuria, fever, gastritis, headache, liver complaints, stomachache, toothache, as stimulant and tonic.

Modern use : *Plant* : anti-asthmatic, antispasmotic, anticatarrh.

Phytography : Herb, much-branched, glandular, pubescent annual, 30–70 cm or more high, grown for its strong aromatic odour and long, profuse, fine panicles and knots; leaves oval or oblong, 15–50 mm long, sinuate toothed or pinnatifid; flowers many, minute in little cymes arranged in elongate terminal panicles.

Distribution : Himalaya— from Kashmir to Sikkim —up to an elevation of 1200–4750 m.

Ecology and cultivation : Prefers rocky areas, higher elevations, cold climate; wild.

Chemical contents : *Plant* : betanin in all parts; *Aerial parts* : chrysoeriol, chenopodic acid, essential oil, flavonoids, hispidulin, 7-Me-eupatulin, twelve polyoxygenated sesquiterpens, quercetin pyranosides, salvigenin, 5-salvigenin, sinensetin; *Fruit* : a new alkaloid.

Chloris virgata Sw. (Poaceae)

(2n = 14, 20, 26, 30)

Syn : *Chloris pubescens* Lagosea, *Rhadochloa virgata* (Sw.) P. Beauv., *Chloris tetrastachya* Hack.

Vernacular name : Raj : Gharaniagas.

Traditional use : *Decoction of plant* or *root* used by the Xosas of South Africa as an addition to baths for the treatment of colds and rheumatism.

Phytography : A characteristic grass of saline or desert tracts; rhizomatous, perennial, culms tufted, ascending, glabrous at nodes; leaf sheaths keeled, ligule membranous, ciliate, blades linear, flat or folded; spikes digitate, slender, spikelets subsessile, unilateral, biseriate; caryopsis ellipsoid or fusiform, compressed.

Distribution : Rajasthan, upper and lower Ganga plains, Central and South India; Bangladesh, Pakistan (Sind, Baluchistan).

Ecology and cultivation : Prefers hard soil, grows well in arid and semi-arid zones; wild.

Chlorophytum arundinaceum Baker (Liliaceae)

(2n = 28, 40, 42)

Vernacular names : Hin : Safed musli; **Mun :** Kulaebo, Piri jadu, Piri tar; **Orn :** Oro arxa.

Trade name : Safed musli.

Traditional use : *TRIBES OF ORISSA* : *Tuberous root* : in rheumatism; *TRIBES OF BASTAR* (Madhya Pradesh) : *Tuberous root* : in impotency.

Modern use : *Fruits* : as galactoglucan.

Phytography : Erect herb with tuberous roots; leaves radical, linear, oblong, flat or cananiculate, coriaceous, prominently nerved, base sheathing or narrowed into petiole, apex gradually tapering above; scape 1 or 2–5, leafless; flowers close or distant, bisexual; capsules globose, oblong, 3-winged/angled, loculicidal.

Distribution : Eastern Himalaya, Assam and Bihar.

Remarks : Garos eat leaves as vegetable; Tribes of Bastar (Madhya Pradesh) consume leaves and flowers as vegetables.
Mundas believe that a touch with this plant can bring an errant wife back to her husband; it is also given to newly bought cattle for bringing the animal under the new owner's influence.

Chloroxylon swietenia DC. (Rutaceae)

(2n = 20)

Syn : *Swietenia chloroxylon* Roxb.

English name : East Indian satinwood tree.

Vernacular names : Guj : Boeria, Billu, Halda, Hardi; **Hin :** Bhirra; **Kan :** Huragalii; **Mar :** Bheria; **Tam :** Porasu; **Tel :** Billu, Billydu.

Trade name : Bhirra.

Traditional use : *TRIBES OF ORISSA* : *Root* : in pain in neck; *TRIBES OF BASTAR* (Madhya Pradesh) : *Root* and *Bark* : as aphrodisiac; *TRIBES OF MAHARASHTRA* : *Leaf* : in wounds; *TRIBES OF DHARAMPURA FOREST* (Tamil Nadu) : *Leaf* : in inflammation.

Modern use : *EtOH* (50%) *extract of stem-bark* and *leaf* : spasmolytic.

Phytography : A small tree, unarmed, 4–18 m; branchlets pubescent; leaves even-pinnate, 1.5 x 2.5 cm, leaflets 10–15 pairs, alternate or (sub)opposite, thin coriaceous apex obtuse, petioles up to 2 cm; flowers pentamerous, bisexual, ± 1 cm across, cream-coloured; capsules oblong, 2.5 x 1.0 cm, persistent, woody; 3-valved, loculicidal; seeds 12–15, oblong, compressed, apically winged.

Phenology : *Flowering* : March–April.

Distribution : Dry, deciduous forests in the Indian peninsula, extending in the north to the Satpuras and Chotanagpur (Bihar).

Ecology and cultivation : Hills, 200–1200 m; very common on the lower slopes (dry deciduous); leaf fall in February–March; wild.

Chemical contents : *Bark* : *tert*-Bu-Ketone, swietonone, coumarin, rutamarin, 2 new alkaloids and 9 other coumarins, 3 swietenocoumarins G, H & I; *Heartwood* : 5 coumarins xanthoxyletin, 7-de-Me-ether, nodakenetin, skimonianine, r-fagarine, 2,4-di-OH-5-Ph-cinnamic acid, swietenol, xanthoxyletin, alloxanthoxyletin, 7-de-Me-suberosin, coumarin.

Remark : Garhwalis use leaf in wounds of cattle.

Chrozophora rottleri A. Juss. ex Spreng. (Euphorbiaceae)

(2n = 22)

Syn : *Chrozophora plicata* A. Juss., *C. prostrata* Dalz.

Sanskrit name : Suryavarta.

Vernacular names : Ben : Khudi okra; **Hin :** Shadevi, Subali; **Pun :** Nilakrai, Nilkanthi; **Tel :** Lingamiriyam.

Traditional use : *TRIBES OF SANTAL PARGANAS* (Bihar) : *Root* : in asthma, burns, cold, cough, cuts, wounds; *TRIBES OF MADHYA PRADESH* : *Leaf* : in dermatitis.

Modern use : *EtOH* (50%) *extract of plant* : spasmolytic.

Phytography : A spreading weed, up to 75 cm; branchlet fulvous, tomentose, 3-nerved from base; racemes (sub)terminal, monoecious, densely tomentose, up to 3 cm; male flowers (sub)sessile, crowded in upper axil, female ones pedicellate; capsules depressed, trilobed, stellate tomentose; seeds globose, testa smooth, shiny.

Phenology : *Flowering* and *Fruiting* : almost throughout the year.

Distribution : Found everywhere in the plains, especially in wasteland, on the sides of the meadows, ditches, etc., in India; Bangladesh and Pakistan.

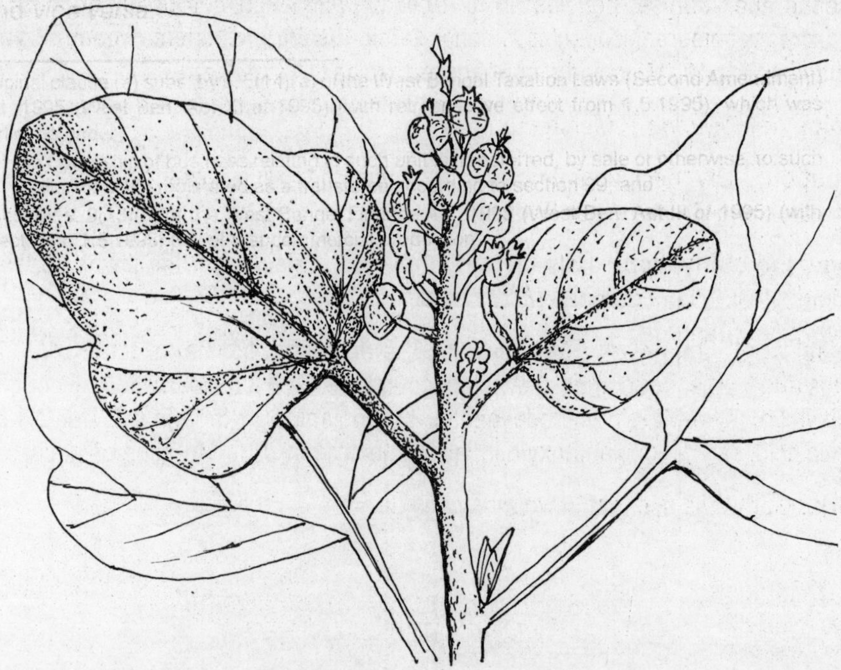

Ecology and cultivation : Plains from the coast, wasteland; wild.

Chemical contents : *Plant* : pyridinium salt oxidising enzyme; *Seed* : fatty acid compound, linoleate-rich oil.

Chrysanthemum coronarium L. (Asteraceae)

(2n = 10, 16, 18, 20, 28, 30)

Syn : *Anthemis coronaria* Hort.

English names : Garland chrysanthemum, Crown daisy.

Sanskrit names : Chandramallika, Shevantika.

Vernacular names : Ben and **Hin :** Chandramallika, Guldaudi; **Kan :** Shevanti; **Mar :** Scoti; **Pun :** Bagaur; **Tam :** Shamantippu; **Tel :** Chamanti; **Urd :** Gulechini.

Trade name : Chandramallika.

Traditional use : *Plant* (mixed with black pepper) : given in gonorrhoea; *Flower* : used as a substitute for chamomile, an aromatic, bitter and stomachic; *Bark* : purgative, used in syphilis.

Phytography : Often taller than *Chrysanthemum carinatum* L.; leaves not succulent, with broader and sharp toothed divisions, more closely placed, the main rachis about 4 mm broad, manifestly auricled on stamens; heads usually smaller and more likely to be fully double and globular, yellow or yellowish white, involucral bracts not keeled; achenes somewhat prismatic, strongly angled but not winged with minor ribs between.

Phenology : Cold season.

Distribution : Cultivated as an ornamental plant everywhere in gardens for its showy flowers.

Chemical contents : *Plant* : adenine, chlonine; *Leaf* : contains 7- glucosides of quercetin, quercetagetin and teolin.

Chrysanthemum indicum L. (Asteraceae)

(2n = 20, 30, 54)

Syn : *Chrysanthemum japonicum* Thunb., *Pyrethrum indicum* L.

English name : Japanese chrysanthemum.

Vernacular names : Ben : Chandramallika; **Hin :** Guldaudi; **Mar :** Akurkura, Sevati; **Pun :** Bagaur, Gendi; **Tam :** Akkarakaram; **Tel :** Chamundi; **Urd :** Guledawoodi.

Trade names : Chandramallika, Chrysanthemum.

Traditional use : *Plant* (mixed with black pepper) : in gonorrhoea; *Leaf* : used as a depurant and prescribed in migraine; *Flower* : bitter, tonic and aperient.

Modern use : *EtOH* (50%) *extract of stem* and *blossom* : anti-inflammatory; *Chrysantenone* : a good medicine for Parkinson's disease.

Phytography : An ornamental perennial, with branching more or less, slender looping stems 60–90 cm long, herbage grey pubescent; leaves thin, ovate in outline, small teethed acute or mucronate; heads many in small clusters on short peduncles; yellow rays, mid-nerve green.

Phenology : *Flowering* : Winter.

Distribution : : Cultivated in the gardens of all countries for ornamental multi-coloured flowers.

Chemical contents : *Plant* : essential oil, a new source of chrysanthemone, glucoside—chrysnathemin; yejuhua lactone, asteglasine-A; *Flower* : acacetin glyceride.

Remark : Causes contact dermatitis in man.

Chrysopogon fulva Springel (Poaceae)

(2n = 20, 80)

Syn : *Chrysopogon montanus* Trin., *C. coerulus* Duthie, *C. serrulatus* Trin., *Andropogon monticola* Schult., *Pollinia fulva* Sprengel.

Vernacular names : Hin : Goria, Gogar; **Mar :** Agiva; **Pus :** Lohigayab; **Tel :** Gogada gaddi.

Trade name : Gogar.

Traditional use : *Seed* : popular vermifuge.

Phytography : Culms 20–60 cm, densely tufted at base up to 1.5 mm wide; leaf sheaths 1.5–4.5 cm, stiff ciliate at collar, ligule up to 0.5 mm, blades 1.5–5 x 0.3–0.4 cm, flat, not distichus; panicles ovate, up to 7 x 3 cm, branches scabrid.

Distribution : Western Himalaya from Kashmir to Nepal up to 2000 m; hilly districts from Punjab to Bihar, South India; Pakistan and Bangladesh.

Ecology and cultivation : Plains as well as hills.

Chukrasia tabularis A. Juss. (Meliaceae)

(2n = 26)

English names : Bastard cedar, Chittagong wood, Red Indian wood, White cedar.

Vernacular names : Asm : Boga poma; **Ben :** Chikrassi; **Kan :** Dalmara; **Mal :** Akil; **Mar :** Pabha; **Tam :** Agil, Maleivembu, Vedivembu.

Trade name : Chikrassy.

Traditional use : *Bark* : astringent.

Modern use : *Stem-bark* (50% EtOH extract) : diuretic, hypotensive, spasmolytic; *Seed* (extract) : agglutinating.

Phytography : Tree, up to 25 m; banchlets lenticulate; leaves pinnate, 25 x 15 cm, leaflets ± 5 pairs, subopposite, ovate, lanceolate, 6–10.5 x 4–5.5 cm; panicles terminal; flowers fragrant, creamy, 4- to 5-merous, bisexual, pubescent, contorted; capsules globose, septifragal; massive, woody, persistent; seeds many, broadly winged at base.

Phenology : *Flowering* : April–September.

Distribution : Hill forests of Sikkim, western peninsula along the Western Ghats from Maharashtra towards South India and the Andamans; Bangladesh (Chittagong).

Ecology and cultivation : Hills 500 (1200) m; common.

Chemical contents : *Root* : cedrelone; *Stem-bark* : melianine, scopoletin, sitosterol, tabularin, 6,7-di-MeO-coumarin; *Wood* : chukrasins A, B, C, D & F; *Leaf* : a flavone, galloyel glucoside, quercetin galactoside, tannic acid; *Seed* : phragmalin esters, 12-OAc-phyramalin.

Remark : In Sri Lanka, plant is used in skeletal fractures.

Cicer arietinum L. (Fabaceae)

(2n = 14, 16, 24, 32)

English names : Gram, Carbanzos, Bengal gram, Chick pea.

Sanskrit name : Chanaka.

Vernacular names : Asm : Butmah; **Ben :** Chhola; **Guj :** Chana; **Hin :** Chana, But; **Kan :** Kadale; **Mar :** Harbara; **Mun :** Moroejang; **Ori :** Boot; **Orn :** But; **Pun :** Chola; **Sad :** Chana; **Tam :** Kadalai, Uppu kkadalai; **Tel** : Sanagalu.

Trade names : Chhola, Boot.

Traditional use : *RURAL FOLK OF KURUKSHETRA* (Haryana) : *Stem* : in jaundice; *Seed* : as tonic; *MUNDA* : the liquid obtained by macerating seeds as antibilous, cooling and tonic.

AYURVEDA : it quenches thirst.

CHARAKA SAMHITA : helps digestion, cooling, diuretic, effective in the diseases caused by bile (*pitta*) and phlegm (*sleshma*).

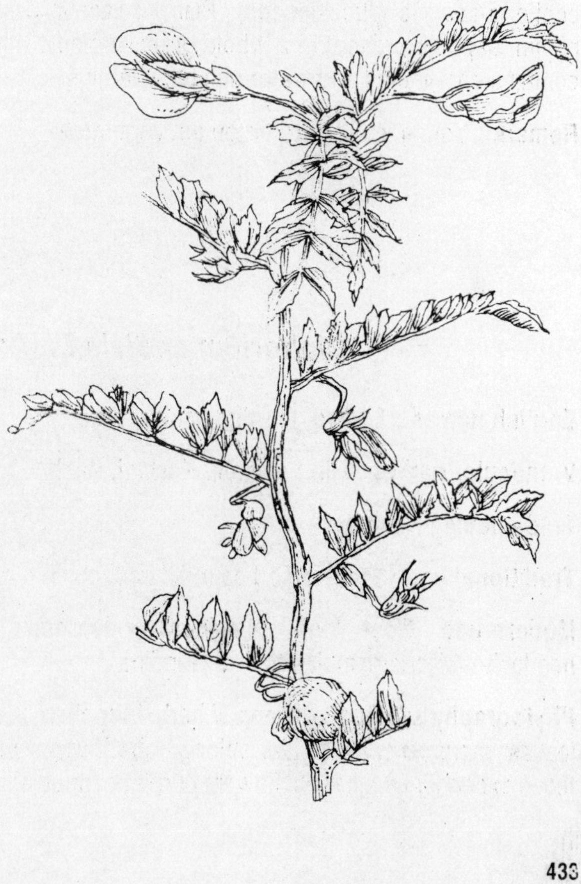

CHANAKAMLA : cures indigestion; *Powder of seed* : tonic.
UNANi : ingredient for the preparation of 'Nakhud'.

Modern use : *Seed* : hypocholesterolemic in human, germinated seeds more effective, reduce post-prandial plasma glucose in human; *EtOH* (50%) *extract of seed* : spasmolytic; *Seedcoat* : diuretic.

Phytography : A much-branched herb; branchlets glandular, pubescent; leaves odd-pinnate, 2.5–5.0 cm, leaflets up to 7 pairs, subopposite, ovate-elliptic, pubescent, base cuneate, margin sharply dentate, apex mucronate, flowers axillary, solitary, yellow; pods ovoid-oblong, 1.5 x 1 cm, glandular, pubescent; seeds 2.

Phenology : *Flowering* and *Fruiting* : Winter.

Distribution : Native to South West Asia; now cultivated extensively in most parts of India; Pakistan and Bangladesh.

Ecology and cultivation : Cultivated and found also as an escape. The crop is grown in heavy clay to light loam soils as a dry crop in *rabi* season.

Chemical contents : *Root* : medicagol, 12-0-Me-coumesterol; *Root* and *Stem* : coniferin-specific glucosidases; *Plant* : acetic acid, malic acid, oxalic acid, lecithin, phytin, saponin present in a whole gram, arginine, tyrosin, lycine, cystine, tryptophane; contains carotinoids and water-soluble vitamins.

Remark : Young leaves are eaten as vegetable.

Cichorium endivia L. (Asteraceae)

English names : Endive, Garden endive.

Vernacular names : Hin : Kasani, Kashini, Kashni.

Trade name : Kasani.

Traditional use : *Plant* : used as resolvent, cooling medicine and in bilious complaints.

Modern use : *Root* : tonic, demulcent, in dyspepsia and fever; *Fruit* : cooling, in fever, headache, bilious complaints and jaundice.

Phytography : Annual or biennial herb, branched, leafy stem, 60–90 cm high, glabrous; leaves many at base, brittle, oblong, lobed, upper passing into leafy bracts, commonly much exceeding the heads; flowers purple; pappus about 1/6 to 1/4 as long as the achene.

Distribution : Native of the Mediterranean region; cultivated mainly in North India and Pakistan.

Ecology and cultivation : Cultivated as salad plant.

Chemical contents : *Plant* : OH-flavone glycerides, dicaffeoyltartaric acid.

Adulterant : It is used as an adulterant of coffee.

Cichorium intibus L. (Asteraceae)

[sometimes spelt as *intybus*]

(2n = 18)

English names : Chikory, Succory, Wild endive.

Vernacular names : Ben, Hin and **Mar :** Kasni; **Tam :** Kashni; **Tel :** Kashini.

Trade name : Chikory.

Traditional use : *Cultivated plant* : used as tonic, in fever, vomiting, diarrhoea, enlargement of the spleen and its root as stomachic and diuretic; *Wild form of the plant* : considered as a tonic, emmenagogue, alexiteric.
HOMOEOPATHY : for liver and gall ailments.

Phytography : A hispid, stout, deep-rooted perennial, 90–180 cm high; upper branches practically leafless; leaves broadly oblong, oblanceolate or lanceolate, more or less clasping; flowers bright azure-blue, closing about night; pappus about 1/8–1/10th as long as the achene.

Distribution : Native to Europe but commonly found wild in Punjab, Andhra Pradesh; cultivated in Maharashtra and Gujarat; Pakistan (NW Frontier Province).

Chemical contents : *Plant* : Glucoside, cichorin, bitter substances lactucin, intylin, insulin.

Remark : After roasting and powdering, the dry roots are mixed with coffee (dandelion coffee).

Cinchona calisaya Wedd. (Rubiaceae)

(2n = 34, 68)

English names : Yellow bark cinchona, Peruvian bark.

Vernacular name : Tam : Kwainaamaram.

Trade name : Cinchona.

Traditional use : *Bark* : used as febrifuge, antiperiodic and tonic, in malaria and in the prophylaxis, whooping cough, hay-fever, enlargement of spleen, hemicornea and other neurologic affections, small pox, septic fevers, pneumonia, acute rheumatism, acute toncilitis, acute nasal catarrh, pyaemia, etc.; locally as an antiseptic; the drug is used in abscesses, cavities and ulcers; it is used as a wash and gargle in sore throat but indiscriminate use is harmful.

Modern use : *Bark* : source of quinine, specific for malaria.

Phytography : Evergreen tree, bark bitter, stipules deciduous; flowers pentamerous, scented, in terminal panicles; capsules 2-celled, 2-valved, dehiscing septicidally from the base; seeds numerous and peltate.

Distribution : Cultivated in India; at elevations of ± 500–1000 m at Sikkim and Moyar Valley in the Nilgiris.

Ecology and cultivation : Like *Cinchona ledgeriana.*

Chemical contents : *Bark* : contains a number of alkaloids, among which quinine, quinidine, cinchonine and cinchonidine are important ones from which different salts are derived.

Cinchona ledgeriana Moens ex Trimen (Rubiaceae)

(2n = 34, 68)

Syn : *Cinchona calisaya* Wedd. var. *ledgeriana* How.

English names : Quinine, Ledger bark cinchona.

Trade name : Yellow cinchona.

Traditional use : *Bark* : used for treating malarial fever; febrifuge, tonic and antiperiodic.

Modern use : *Leaf* (50% EtOH extract) : antiviral, analgesic.

Phytography : A hybrid between *C. calisaya* and *C. micrantha.*

Distribution : Darjeeling, Khasia Hills (Meghalaya) and the Nilgiris Hills (Tamil Nadu).

Ecology and cultivation : The cinchonas grow isolated or in small groups scattered among the trees of mixed forests, wellgrown trees reaching a height of 25–30 m and a circumference of about 2.1 m; the climate is moist and warm (12°–20°C).

Chemical contents : *Stem-bark* : quinine, quinidine; *Heartwood* : a glyceride and alkaloids; *Leaf* : cincophyllamine, quinone, quinchophylline, five monomeric and seven

quasi-dimeric indole alkaloids, bisindole monoterpen, phenolic acids, anthocyanins, flavonoids.
Plants after 5 years contain 5.14% total alkaloids in roots, 6.04% in trunks and 0.85% in branches; upper young leaves richer in alkaloid content than the older ones; in stem, alkaloid contents increases downwards.

Adulterants : Cuprea bark, *Cascarilla*, *Exostemma* and *Stenostemma*—these barks contain none of the cinchona alkaloids.

Cinchona officinalis L. f. (Rubiaceae)

Syn : *Cinchona coridaoninea* H.B.K., *C. lancifolia* Mut., *C. nitida* Ruiz and Pav.

English names : Quinine, Cascarilla.

Trade name : Cinchona.

Traditional use : *Bark* : chiefly used as bitter, stomachic, tonic and astringent.
UNANI : ingredient for 'Kanakana', useful for the treatment of malaria.
HOMOEOPATHY : for indigestion, general debility, stomach troubles, for excessive bleeding, hectic fever, headache, eye trouble, earache, toothache, stone and ascaris.

Modern use : *Bark* : source of quinine, yellow bark, specific for malarial fevers. Used medicinally as *C. calisaya*. As febrifuge, antiperiodic and tonic.

Phytography : An evergreen tree; bracts lanceolate; capsules septicidal.

Distribution : Cultivated in South India, especially in the Nilgiris; Peru, Ecuador, Bolivia, Columbia.

Ecology and cultivation : This species is suitable for higher altitude in South India, thrives well at higher elevations up to 1800–2600 m in the Nilgiris; grows up to 2200 m.

Chemical contents : *Bark* : yields 1–4% of which one-half to two-thirds is quinine; it also contains quinidine, cinchonine and cinchonidine, cinchotannic acid, quinetum, etc.

Adulterants : Cuprea bark, *Cascarilla*, *Exostemma* and *Stenostemma*.

Cinchona robusta How. (Rubiaceae)

English name : Cinchona.

Trade name : Cinchona.

Traditional use : *Branch* : used as febrifuge.

Modern use : *Bark* : source of quinine, specific for malaria.

Phytography : Like *Cinchona officinalis.*

Distribution : Grown at Naduvattam in Tamil Nadu at an altitude of ± 1200–2000 m.

Ecology and cultivation : It is a hybrid between *C. succirubra* x *C. officinalis*; grows up to 1500 m at Mungpoo and adjacent areas.

Chemical contents : *Branch* : yield quinine ± 2%; *Leaf* : avicularin.

Cinchona succirubra Pavon ex Kletzsch (Rubiaceae)

(2n = 34, 68)

English names : Red bark cinchona, Quinine drugist's bark.

Trade name : Red cinchona.

Traditional use : Used as a febrifuge for the treatment of malarial fever.

Modern use : *Red bark* : astringent, bitter stomachic and tonic; is used for pharmaceutical preparations.

Phytography : Although there are several varieties of cinchona, certain features are common to them. They vary in colour of the inner surface and in the fracture from twany yellow to brownish red and the inner surface is longitudinally striated. Externally, they are more or less rough and of a greyish colour. It is tree; flowers purplish.

Phenology : *Flowering* : round the year, especially in September–November.

Distribution : Hilly regions of Sikkim Himalaya, Nilgiris and Naduvattam plantations in South India.

Ecology and cultivation : Very widely introduced in the tropics during the colonial era, when extensive hybridisation has taken place. Grows well between an altitude of 620 and 1300 m; propagation is generally done by seeds; it is cross-fertilised and maintenance of genetic purity is a problem.

Chemical contents : Main alkaloid is cinchonine. Plants of 5 years' growth contain total alkaloids— in roots 7.13%, trunk 8.20%, branches 6% and quinine sulphate in roots 1.4%, trunk 2.98% and branches 1.89%.; leaves contain same alkaloids in the same proportions as bark; upper young leaves richer in alkaloid content than the older ones; in stem, alkaloid content increases downward.

Adulterant : Same as the other cinchonas.

Remark : Alkaloid contents are lower.

Cinnamomum tamala Nees & Eberm. (Lauraceae)

English name : Indian cassia lignea.

Sanskrit names : Tamalaka, Tejpatra.

Vernacular names : Asm & **Ben :** Tejpata; **Guj :** Tamalapatra; **Hin :** Tijpat; **Man :** Tezpat; **Tam :** Talishappattiri; **Tel :** Talishapatri.

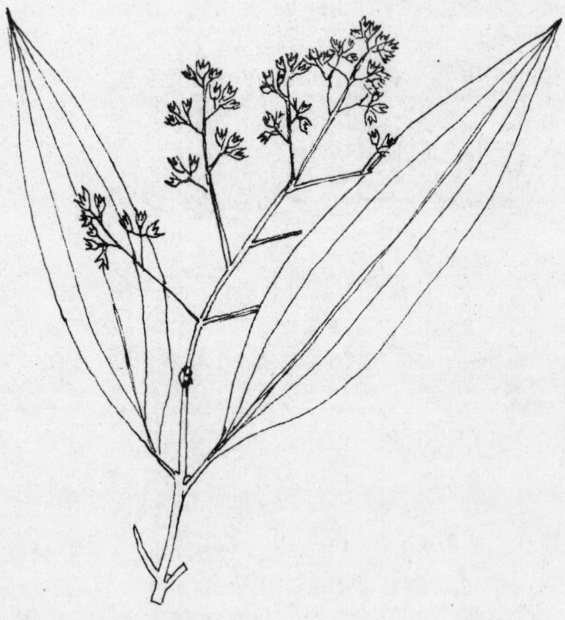

Trade name : Tejpat.

Traditional use : *Bark* : aromatic, used in gonorrhoea; *Leaf* : stimulant, carminative, used in rheumatism, colic, diarrhoea and in scorpion sting *UNANI* : it is the ingredient for preparation of 'Sazaj Hindi'/'Zarnab'(Tellispattar).

Phytography : A small tree with leaves usually ± 1–1.25 cm long, shining above, mostly oblong; flowers 2.5–5 mm long; deciduous tepals in fruit.

Phenology : *Flowering* : Winter.

Distribution : Distributed in the subtropical Himalaya ± 1000–3000 m, Meghalaya and Srihatta (Sylhet, Bangladesh) up to ± 1000–1200 m.

Ecology and cultivation : Cultivated for the leaves, but sometimes occur wild in the forests; leaves are collected every year when trees are young and on alternate years from old and weak ones.

Chemical contents : *Essential oil from leaf* : leaf yields an oil which resembles cinnamon leaf oil and contains-δ-phell andrene and 8% eugenol; *Essential oil from bark* : contains 70–80% cinnamic aldehyde.

Adulterant : Bark is a common adulterant of true cinnamon.

Cirsium arvense (L.) Scop. (Asteraceae)

(2n = 34, 36, 68)

Syn : *Cnicus arvensis* Hoffm.

English names : Canada thistle, Creeping thistle, Creeping plume thistle.

Traditional use : Emetic, tonic and diaphoretic.

Phytography : Handsome weed about 60 cm in height with a creeping root and very spinuous sessile and very slightly decurrent, oblong-lanceolate, wavy leaves; heads numerous, peduncled, ovoid, pale or lilac-purple, dioecious; involucral bracts broad-lanceolate, depressed with short spreading spine; cypsele linear-oblong, compressed.

Phenology : *Flowering* and *Fruiting*: Winter.

Distribution : West Bengal, Ganga plains from Sundarbans northeastwards up to Punjab, western Himalaya from Kashmir to Kumaon; Pakistan (NW Frontier Province).

Ecology and cultivation : A common English weed; not very common in India; ; prefers cold places.

Chemical contents : Glucosides, essential oil, alkaloids; *Seed* : contains 21.9% fatty oil; *Plant* : contains a volatile alkaloid and enicin, a crystalline bitter glucoside possessing emetic and emmenagogueic properties.

Cissampelos pareira L. (Menispermaceae)

(2n = 24)

Syn : *Cissampelos acuminata* Benth., *C. hirsuta* Buch. Ham., *C. convolvulacea* Willd.

English name : False pareira root.

Sanskrit name : Ambastha patha.

Vernacular names : Ben : Akandi; **Guj :** Venivel; **Hin :** Akanadi, Patat ki bel; **Kan :** Padavali; **Mal :** Kattuvali; **Mar :** Pathervel; **Tam** : Appatta; **Tel :** Adivi bankatige.

Trade name : Akandi.

Traditional use : *RURAL FOLK OF ASSAM* : *Stem* and *Leaf* : in infertility; *TRIBES OF JALPAIGURI* (West Bengal) : *Root* : in antifertility, bodyache, fever; *Whole plant* : in diarrhoea; *TRIBES OF PURULIA* (West Bengal) : *Whole plant* : in venereal diseases; *Root* : in cold, fever, stomachache; *Stem* : in fever; *SANTALS* : use different parts in dog bite, bronchitis, carbuncle, colic, diarrhoea, dysentery, epilepsy, stomachache, urinary complaints; *TRIBES OF CHOTANAGPUR* (Bihar) : *Root* : in fever, headache; *Leaf* : in antifertility; *RURAL FOLK OF MIRZAPUR* (Uttar Pradesh) : *Root* : in dysentery, stomachache *RURAL FOLK OF DEHRA DUN* (Uttar Pradesh) : *Root* : in catarrhal infection of bladder; *Leaf* : in sores; *KUMAONIS*: *Leaf*: in boil, burn, eye complaints, pimples; *GARHWALI* : *Root* : in diarrhoea, dysentery, fever; *FOLKS OF DELHI* : *Root*: in diuretic troubles, indigestion; *Leaf*: in itching sores; *TRIBES OF RAJASTHAN* : *Root* : in cold, cough; *Leaf* : in abscess, sores; *DANGS*

(Gujarat) : *Whole plant* : in dysentery; *Root* : in snake bite; *TRIBES OF SAURASHTRA* (Gujarat) : *Root* : in snake bite, urinary complaints; *Leaf* : in sores, wounds; *TRIBES OF SAGAR* (Madhya Pradesh) : *Root compound with leaf* : in diarrhoea, urinary complaints; *Leaf* : in skin diseases; *TRIBES OF BASTAR* (Madhya Pradesh) : *Root* : in enlarged spleen; *IRULAR* (of Tamil Nadu) : *Leaf* : in cold, fever.

Modern use : Expectorant, emmenagogue and febrifuge, muscular relaxant.

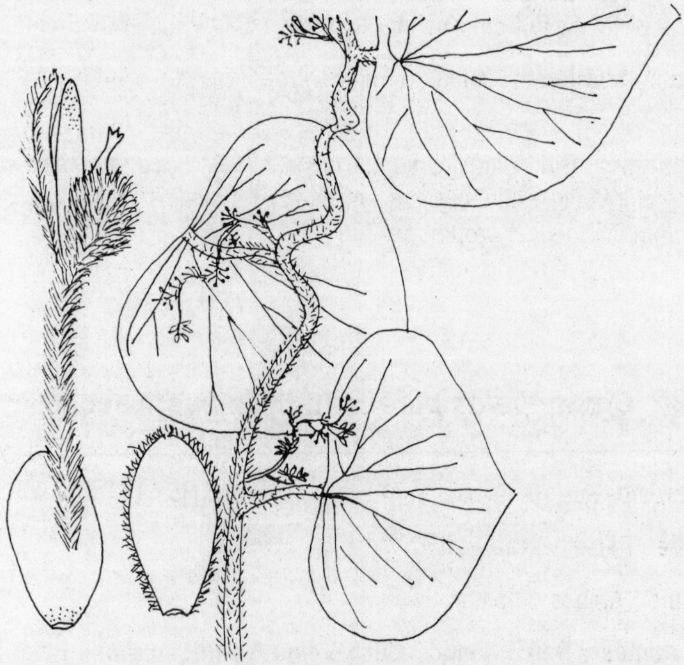

Phytography : Twining perennial shrub; leaves reniform, orbicular or cordate, 4.5 by 6.5 cm, peltate, chartaceous, 5- to 7-nerved, pubescent, base cordate or truncate; male inflorescence pendent type; flowers greenish; female inflorescence in thyroid clusters; fruits red when ripe, subtended by the conspicuous auricular bracts.

Phenology : *Flowering* : nearly throughout the year.

Distribution : Distributed in the tropical and subtropical tracts of India, Bangladesh and Pakistan.

Ecology and cultivation : Plains from the coast up to 1000 m, more abundant on the degraded slopes; extensive tomentose climber as compact mats on thickets.

Chemical contents : Alkaloid sepurine, bebeerine, cessampeline; pelosine or bebeerine present in the root; plant extract contains saponine and abundance of quartenary ammonium bases; alkaloid hyatin, hyatinin and a quercitol and sterol isolated from the roots; hyatin methodide (and methochloride) possesses almost equal degree of curari from activity as of δ-tubecurine.

Adulterant : Stems of *Chondodendron tomentosum* Ruiz. et Pavon are substituted for the genuine article.

Cisus pallida (Wt. & Arn.) Planch. (Vitaceae)

(2n = 20, 26)

Syn : *Vitis pallida* W. & A.

Vernacular names : Kan : Kondage; **Ori :** Takuanoi; **Tel :** Nallatige.

Traditional use : *Bruished root* : applied for rheumatic swellings.

Phytography : Erect shrub up to 1.5 m, branchlets glabrous; leaves simple, cordate, 7–18 x 8.5–19.0 cm, thin, coriaceous, nerves ± 6 pairs; petioles 4–15 cm, cymes up to 5 cm, panicled; peduncle up to 4 cm; flowers 8 mm across, petals red-tipped; berry pyriform, 1.0 x 0.8 mm; seed smooth.

Phenology : *Flowering* : January–August; *Fruiting* : throughout the year.

Distribution : Upper Ganga plains, Karnataka and other adjacent areas.

Ecology and cultivation : Hill above 750 m; common; initially erect, later on extensive, liane, stem with a white powdery indusium.

Cissus quadrangularis L. (Vitaceae)

(2n = 24, 44, 53)

Syn : *Vitis quadrangularis* Wall.

English names : Edible stemmed vine, Woody vine.

Sanskrit names : Asthisanhara, Vajravalli.

Vernacular names : Ben : Harjora, Harbhanga; **Guj :** Chodhari, Vedhari; **Hin :** Harjora, Harsankari; **Kan :** Mangaroli; **Mal :** Chaudhari, Kandavela; **Tam :** Piradai; **Tel :** Nalleru.

Trade name : Harjora.

Traditional use : ASURS (of Bihar) : *Root* : in cuts, wounds; *Root compounded with stem* : in bone fractures; *Stem compounded with leaf* : in labour pain; SANTALS : *Plant* :

in bone fracture, cuts, wounds; *TRIBES OF VARANASI* (Uttar Pradesh) : *Plant* : in bone fracture; *KUMAONI* : *Stem* : in bone fracture; *TRIBES OF DAHANU FOREST* (Maharashtra) : *Stem* : in bone fracture, pain in joints, swelling.

Modern use : *EtOH* (50%) *extract of aerial parts* : hypotensive; *Stem* (extract) : diuretic.

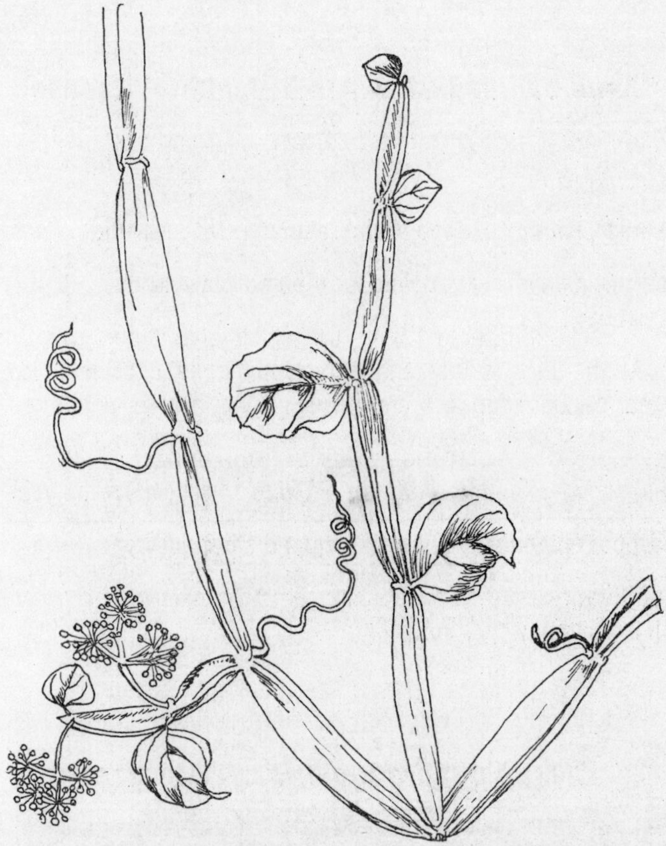

Phytography : A succulent twiner, rambling shrub, stem 4-angular, glabrous, winged or margined, contracted at the nodes; leaves simple, early caducous, ovate, suborbicular, subreniform, thick coriaceous; margin serrate; cymes umbellate; flowers 7 cm across, greenish yellow, red-tipped; berry globose, apiculate; seed smooth.

Phenology : *Flowering* : throughout the year.

Distribution : Throughout the hotter parts of India; Bangladesh and Pakistan.

Ecology and cultivation : Plains from the coast, in scrub jungles and wastelands up to 500 m on thickets, etc.

Chemical contents : Contains 3-ketosteroid, a steroidal principle; has acetylocholine-like action analoguous to muscarine and nicotine; in atropinised hearts, it shows cardiotonic activity; contains calcium oxalate, carotine, ascorbic acid.

Remarks : Tribes of hilly regions of Goa eat the stem and leaf; tribes of Saurashtra (Gujarat) consume the fruits.

Cissus repens Lam. (Vitaceae)

(2n = 96)

Syn : *Vitis repens* (Lam.) W. & A.

Vernacular names : Asm : Mei-hur-jarap; **Kan :** Elakombullaballi; **Mal :** Mrigampuli; **Nep :** Pureni; **Ori :** Diboria; **Tel :** Nelaboddu.

Traditional use : *TRIBES OF MEGHALAYA* : *Root compounded with leaf* : in wounds of dog bite; *NAGAS* : *Leaf* : in skin diseases.

Modern use : *EtOH* (50%) *extract of plant* : diuretic.

Phytography : A trailing plant; branchlets glabrescent; leaves simple, ovate-oblong, 6.5–10.0 cm x 4–6 cm, membranous, nerves ± 6 pairs, puberulous, base cordate, margin crenate-serrate, apex acuminate, petioles 2.5 cm; tendrils slender;cymes in compound umbels up to 4.5 cm; berry globose.

Phenology : *Flowering* : August; *Fruiting* : November.

Distribution : Assam, eastern tropical Himalaya, Maharashtra; Bangladesh (Chittagong).

Ecology and cultivation : Above 1500 m; occasional.

Remarks : Tribes of North-East region eat leaves as vegetables; people of Arunachal Pradesh eat the fruits.

Cissus setosa Roxb. (Vitaceae)

Syn : *Vitis setosa* Wall.

Vernacular names : Hin : Harmal; **Mar :** Khajdichavel; **Tam :** Pulippirandai; **Tel :** Puuabachali, Barubutsali.

Trade name : Harmal.

Traditional use : *Leaf* : stimulant, used in indolent tumours and applied externally to assist in the extraction of guinea worms.

Modern use : *EtOH* (50%) *extract of aerial parts* : antibiotic, hypotensive, spasmolytic.

Phytography : Vine up to 6 m; branchlets bristly-setose; leaves trifoliate; leaflets oblong-obovate, 5–9 x 3.5–5.5 cm, terminal one larger, laterals inequilateral, subsucculent margins, dentate-serrate, petiole 0; cymes up to 7 cm, glandular; flowers 6 mm across, green; berry red on ripening.

Phenology : *Flowering* : November–December; *Fruiting* : January.

Distribution : Gujarat, Maharashtra, Karnataka, Kerala, Tamil Nadu, Andhra Pradesh and Orissa.

Ecology and cultivation : Plains to foothills, scrub jungles, on thickets; occasional.

Cistanche tubulosa Wight (Orobanchaceae)

(2n = 40)

Syn : *Phelipaea calotropidis* Walp.

Vernacular name : Kal : Kasi.

Traditional use : *Plant* : used in sores, given in curd to prevent diarrhoea.

Phytography : A sprouted herb; general colour of the plant yellow with an occasional tinge of purple; stem 13–175 cm high, unbranched, furrowed, fleshy, glabrous or more or less pubescent, densely covered with triangular acute scales; flowers many in dense spikes; capsules 25 mm long, ovoid-oblong, laterally compressed; seeds α.

Phenology : *Flowering* : December.

Distribution : Rajasthan, Punjab; Pakistan (Sind and Baluchistan).

Ecology and cultivation : Arid zones; sprouted herb in the sand mainly.

Citrullus colocynthis Schrad. (Cucurbitaceae)

(2n = 24, 44)

Syn : *Cucumis colocynthis* L., *Colocynth vulgaris* Schrad.

English names : Colocynth, Bitter apple.

Sanskrit names : Aindri, Gavaksi, Indravalli, Indravaruni(ka), Visala.

Vernacular names : Asm : Koabhaturi; **Ben :** Makal; **Guj :** Indrak, Indronan-indravana; **Hin :** Indrayan; **Kan :** Paavamekkekayi, Tumtikayi; **Mar :** Kuruvran dawan, Paykimmuttii; **Pun :** Tumma; **Tam :** Pey kumotti; **Tel :** Etipuchcha, Patsakaya.

Trade name : Colosynth/Indrayan.

Traditional use : *BHATS* : *Root* : as abortifecient; *RURAL FOLK OF DELHI* : *Fruit* : as purgative; *TRIBES OF INDIAN DESERT* : *Fruit* : as galactogogue, purgative; *TRIBES OF SAURASHTRA* (Gujarat) : *Fruit* and *Seed* : as purgative.

AYURVEDA : useful in stomach troubles.

SIDDHA : *Juice of fruit* : an ingredient of *Arruttumatti camulam*; *Unripe fruit* : in preparation of *Arruttumatti* and *Arruttumatti caru*.

UNANI : used in the medicine of *Kala-azar*, also ingredient for the preparation of the drug 'Hanzal'.

HOMOEOPATHY : for colic pain, dysentery, diarrhoea, urinary troubles, sciatica and hip-joint diseases.

Modern use : Gastrointestinal stimulant; wellknown purgative acting as hydragogue cathartic.

Phytography : Perennial trailing herb, branchlets hirsute; leaves alternate, simple, petiolate, ovate or narrowly triangular, deeply trilobed, lobes pinnatifid, triangular, chartaceous, rigid; tendril simple, slender, petiole up to 3 cm; male flower 1 cm across, yellow; female flower ovary ovoid; fruits globose, striped green and white when young, yellow when ripe.

Phenology : *Flowering* and *Fruiting* : Winter.

Distribution : Native of Asia and Africa; occurs wild throughout India and Pakistan, particularly in the northewestern, central and southern India.

Ecology and cultivation : Xerophytic; desert and subdesert areas; also plains from the coast, along river banks, in wastelands; common.

Chemical contents : *Root* : α-elatarin, hentriacontane, saponins; *Fruit* : elaterin (cucurbitacin E), elatericin B (cucurbitacin I), dihydroelatericin B (cucurbitacin L); *Fruit-pulp* : δ-elaterin 2-δ-glucopyranoside (coloside A), hentriacontane, phytosterol, a mixture of fatty acids; *Peel-free flesh of ripe fruit* : a bitter oil 'citbittol'; *Fruit-peel* : hectacosane-1-ol, alkanes, ketone, waxes, essential oil; *Seed* : fixed oil, citrullonal, a OH-ketotetracyclictriterpene, a phytostuolin, 2-phytostarols, 2-hydrocarbons, a saponin, tannin.

Remarks : Together with *Ipomoea pescaprae* Sweet, it is cultivated in Gujarat and Karachi (Pakistan) as a sand-binder; grains are eaten by tribes of Rajasthan.

Citrus aurantifolia (Christm.) Swingle (Rutaceae)

(2n = 19, 27)

Syn : *Citrus medica* L. var. *acida* Watt., *Limmon aurantifolia* Christm.

English name : Lime.

Sanskrit name : Bijapura.

Vernacular names : Ben : Kaghzi lebu; **Hin :** Kaghzi nimbu; **Mal :** Erumichinarakam; **Tam :** Elumichai; **Tel :** Nimma.

Trade name : Kaghzi nimbu.

Traditional use : *TRIBES OF BASTAR* (Madhya Pradesh) : *Fruits* : in diarrhoea.
AYURVEDA : *Fruit* : as appetiser, antiseptic, antiscorbutic, refrigerant, stomachic, tonic and in bilious vomiting.
SIDDHA : *Fruit* : an ingredient of the drug *Elumiccam palam.*
UNANI : *Fruit* : an ingredient of 'Leemu Kaghzi'.

Modern use : *EtOH* (50%) *extract of aril parts* : depresses Central Nervous System, diuretic; *Fruits* : good source of vitamin C; *Extracts of fruit-skin* and *seed* : antibacterial, highly toxic to eggs and larvae of Caribbean fruit fly.

Phytography : A shrub or small tree with stiff, sharp spines; leaves elliptic-oblong, ± 50–75 mm long, blunt or sometimes rounded at apex, the short petiole manifestly or narrowly winged, articulation marked; flowers few in the axils mostly not exceeding ± 12 mm in length, white throughout; fruit round-oval to oval and small, ± 35–65 mm in diameter, sometimes with a small nipple, segments 10, shallowly roughing, exceedingly acid.

Phenology : *Flowering* and *Fruiting* : Summer and Winter.

Distribution : Cultivated in all parts of India; Bangladesh and Pakistan.

Ecology and cultivation : Plant of tropical humid climate.

Chemical contents : *Fruit-juice* : contains citric acid, volatile oil containing citral, limoneane, linaloul, linalyl acetate, terpineol and cymene; perfumed essential oil, composition of petilgrain oil; *Leaf* : contain coumarins, iso-pimpinellin, bergapten, citropten; *bark* : xanthyletin. The *peel* gives 6-glucoside of aureusidine.

Adulterant : The peel of the sweet orange is said to be mixed with that of the bitter orange.

Remarks : In Sri Lanka, the plant is used in fractures; in Samoa, fruits are used against skin irritation, nausea, sore throat and in eye complaints.

Citrus aurantium L. (Rutaceae)

(2n = 18)

Syn : *Citrus aurantium* var. *bigaradia.*

English names : Sour orange, Bigarade orange, Seville orange.

Vernacular names : Hin : Khatta; **Kan :** Heralay; **Mar :** Karna; **Tam :** Narattai; **Tel :** Mallikasangi.

Trade name : Khatta kamla.

Traditional use : *Leaf* : anthelmintic, stomachic; *Unripe fruit* : digestive, tonic, appetiser; *Ripe fruit* : aphrodisiac.

Modern use : *Plant* (extract) : antishock, effects cardiovascular receptors; *Leaf* (extract) : antibacterial; *Essential oil* : antifungal, antimicrobial; *Peel* : active against antiferitlity.

Phytography : A small tree, medium size, 6–9 m with long but not very sharp spines; leaves of medium size, ovate-oblong, shortly or bluntly acuminate, sinerate or crenate; flowers of medium size, single or several in axils, white, very fragrant; fruit globose, slightly flattened endwise, ± 75 mm in diameter, roughing, pulp acid and membranes bitter, core hollow at maturity.

Phenology : *Flowering* and *Fruiting* : Winter.

Distribution : Rarely met with, considered to be a sour orange or as a hybrid.

Ecology and cultivation : Tropical and subtropical parts; soil medium to light brown with or slightly heavier subsoil.

Chemical contents : *Root* : xanthyletin; *Leaf* : flavones tangeretin, nobiletin; *Flower* : glycerides, neohesperidin, naringin, β-sitosterol, *Peel* : cirantin, methoxylated flavones, neohesperidin, oil of neroli bigarade; *Dried peel* : nobiletin; *Seed* : limonin, de-Ac-nomilin, isolimonic acid and other compounds.

Remarks : Plant is claimed to have action against enlarged spleen, stomachache, menorrhagia and antispasmodic; in Sri Lanka, plant is used in fractures; boiled aqueous extract is carcinogenic.

Citrus limon (L.) Burm. f. (Rutaceae)

(2n = 18, 36)

Syn : *Citrus medica* L. var. *limonum* L.

English name : Lime/Lemon, Galgal or hill lemons.

Sanskrit name : Maha nimbu.

Vernacular names : Ben : Bara lebu, Gora lebu; **Guj :** Motulimbu; **Hin :** Bara nimbu; **Kan :** Bijapura, Bijori; **Tam :** Piriya yelumichai; **Tel :** Bijapuram.

Trade name : Bara nimbu.

Traditional use : *Fruits* : sour, thermogenic, digestive, carminative, stomachic, laxative, anthelmintic, antiscorbutic, stimulant, antiseptic and mosquito repellent and are useful in vitiated conditions of *kapha* and *vata*, dyspepsia, flatulence, colic, constipation, anoraxia, helminthiasis, scabies, fatigue, halitosis, vomiting, trembling of the limbs, hemicrania, psychopathy, pharygopathy, cough, bronchitis and heart burn.
UNANI : an ingredient for the preparation of 'Utraj'.

Phytography : Much branched, thorny shrub with spreading branches; leaves unifoliately compound, rachis winged, leaflets elliptic-oblong, alternate, coriaceous, entire or crenulate; flowers white in short racemes; fruits large, globose; berries with thick or thin rind, pulp pale, acidic; seeds numerous, horizontal testa coriaceous.

Distribution : Cultivated widely in the gardens of India, Bangladesh and Pakistan.

Chemical contents : *Oil of lemon-peel* : δ-limonene, δ-pinene, camphene, linalool, etc.; *Lemon-juice* : contains an antipneumonia factor; *Juice* : antibacterial; *Peel* : contains bitter principle, essential oil, hisperidin; *Juice* is very acidic with 5% citric acid, 5% sugar and small quantities of vitamin C, carotene and thiamine.

Citrus medica L. (Rutaceae)

(2n = 18, 28)

Syn : *Citrus medica* var. *medica* proper, *C. tuberosa* Mill.

English name : Citron.

Sanskrit names : Amlakishara, Begapura.

Vernacular names : Ben : Batapi; **Hin :** Bara nimbu; **Mal :** Gulam; **Tam :** Kadaranarathai; **Tel :** Lungamu.

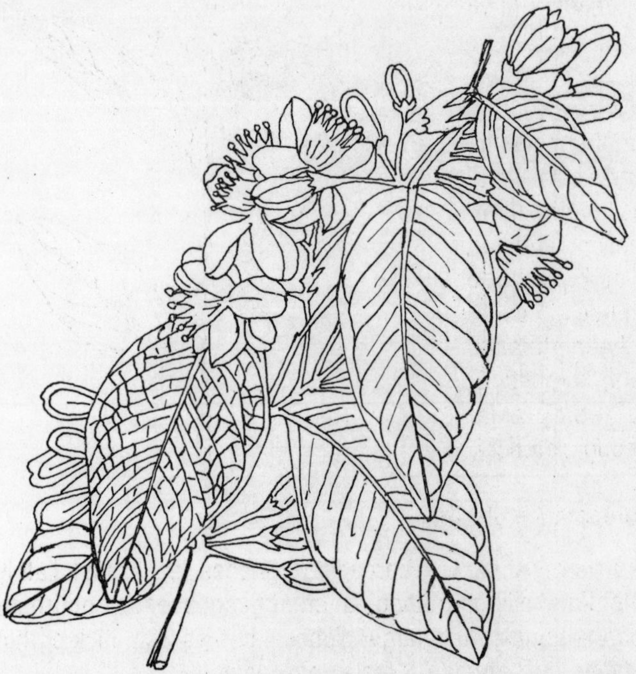

Trade names : Bara nimbu/Batapi lebu.

Traditional use : *Root* : laxative, anthelmintic, diuretic and beneficial in chronic constipation, colic, strangury, renal and vesical calculi, tumours, vomiting, helmianthiasis and dental caries; *Bud* and *Flower* : astringent, stimulant, anti-emetic, appetiser, abdominal pains, asthma, cough and cold; *Ripe fruit* : emollient, astringent, stimulant, digestive, carminative, cardiac stimulant, hepatopathy, flatulence, amenorrhoea, dysmenorrhoea, dysentery, leprosy and skin diseases; *Leaf* : anodyne; *Seed* : stimulant, anti-inflammatory and tonic.
UNANI : an ingredient for the preparation of 'Turanj'.

Phytography : An evergreen armed shrub with straggling thorny branches and smooth yellowish brown bark; leaves oblong or elliptic with acute or rounded apex, glabrous, coriaceous, pullucid-punctate, dull dark green on the upper surface; flowers white with a pinkish tinge, scented, in axillary cymes; fruits large berries, oblong or globose, fleshy, rind thick, rough, irregular or warted, yellow on ripening; seeds few and smooth.

Distribution : Sparsely cultivated throughout the warm climates of India, Bangladesh and Pakistan; in some places it is widely seen.

Ecology and cultivation : Prefers warm, moist climate.

Chemical contents : *Oil from peel* : limone, dipentene, citral.

Citrus paradisi Macf. (Rutaceae)

(2n = 18, 27, 36)

Syn : *Citrus decumana* var. *racemosa*, *C. decumana* var. *paradisi* Nichols, *C. racemosa* Marc.

English names : Grapefruit, Pomelo.

Vernacular names : Hin : Anjeer, Pahari nimbu.

Trade name : Pahari nimbu.

Traditional use : *Juice* : used for building up resistance to common colds and wound infections.

Phytography : Tender tree, more than 19 m high, twig pubescent, spine slender and not very sharp or locking; leaves glabrous, oval to broad ovate, 1–2 cm long, rounded or even retuse on young shoots, petals broadly winged; flowers solitary or usually in axillary clusters; fruits in clusters having grained flesh.

Phenology : *Flowering* and *Fruiting* : mostly in Autumn and Winter, respectively.

Distribution : Introduced in India recently and mainly cultivated in Punjab; Pakistan (Sind); origin unknown.

Ecology and cultivation : Prefers semiarid soil with dry moist climate.

Chemical contents : *Fruit* : source of vitamins C and B_1; *Peel* : rich in pectin; *Oil from peel* : limonene, oxygeneted volatile constituents, sesquiterpenes, aldehydes, geranials, cadinene, citral.

Citrus reticulata Blanco (Rutaceae)

(2n = 18, 36)

Syn : *Citrus nobilis* Andrews non Lour.

English name : King orange.

Sanskrit name : Airavata.

Vernacular names : Ben : Kamala lebu; **Hin :** Narangi, Santara.

Trade name : Kamala lebu.

Traditional use : *Flower* : fragrant, stimulant, diuretic, anodyne, antispasmodic, sedative, febrifuge and tonic, useful in catarrh, strangury, pectoral diseases, fever, general debility; *Rind of fruit* : anthelmintic, antidiarrhoeal, stomachic, tonic, useful in general debility, bilious disorders, dyspepsia and skin diseases; *Fruit* : sour, sweet, cooling, deodorant, appetiser, digestive, liver tonic, useful in vomiting, retching, vitiating conditions of *pitta* and *vata*, pectoral diseases, cardiac disorders.
UNANI : Ingredient of 'Gulbahar' (Baharnaranj)/'Sangtara'.

Phytography : Evergreen, bushy, moderate-sized tree with greenish white, glabrous branches; leaves tender, unifoliate, leaflets elliptic and ovate or acuminate, petioles short, almost wingless; flowers white, fragrant; fruits globose or subglobose with thin loose rind.

Distribution : Cultivated in most parts of India, especially in Assam, Sikkim, Madhya Pradesh, Punjab, Coorg, West Bengal; Bangladesh (Sylhet), Pakistan (Peshawar), at higher elevations.

Ecology and cultivation : Mostly dry places.

Chemical contents : *Oil* (from peel) : δ-limonene, carene, linalool, petingrain oil; *Juice of fruits* : contains sugars and free amino acids.

Citrus sinensis (L.) Osbeck (Rutaceae)

(2n = 18, 22, 36)

Syn : *Citrus aurantium* L. var. *sinensis* L., *C. aurantium* subsp. *decumane* var. *sinensis* Thell., *Aurantinum sinensis* Mill.

English name : Sweet orange.

Vernacular names : Ben : Mosambi; **Hin** and **Mar :** Malta, Mosambi; **Tam :** Santhagudi; **Tel :** Battavinarinja.

Trade names : Mosambi, Malta.

Traditional use : *Fruit* : purifies blood, allays thirotrin fevers, cures catarrh, improves appetite; *Juice* : useful in bilious affections and bilious diarrhoea; *Rind* : carminative, tonic; *Fresh rind* : rubbed on face as remedy for acne.

Phytography : Common orange extensively grown; not so hardy as *Citrus aurantium* L. and also distinguished by spines few or wanting; petiole narrowly winged or margined; fruits mostly smaller with pulp sweet and membranes not bitter and the core solid at maturity.

Distribution : Cultivated in many parts of India; Sylhet (Bangladesh) and Pakistan.

Chemical contents : *Fruit* : source of ascorbic acid, vitamin C; *Oil from peel* : -δ-limonen, decyclicaldehyde, linalool, di-terpenol, essential oil, nerali from and petitgrain oil from leaves and young shoots.

Cleoma viscosa L. (Capparaceae)

(2n = 20, 22, 48)

Syn : *Polanasia icosandra* W. & A.

English names : Wild mustard, Dog mustard, Stick cleomes.

Sanskrit name : Pasuganda.

Vernacular names : Ben : Hurhure; **Hin :** Hulhul, Hurhur; **Mal :** Atunarivela, Vela, Arya vela, Naivela; **Tam :** Naivelai, Naikaduku.

Trade name : Hurhur.

Traditional use : *Plant* : acrid, thermogenic, anthelmintic and sudorific; *Root* : stimulant, antiscorbutic and vermifuge; *Leaf-juice* : digestive and useful for otalgia; *Seed* : anthelmintic, carminative, constipating, febrifuge, cardiac stimulant and useful in fever, diarrhoea, worm infestations, cardiac disorders and dyspepsia.

Phytography : Annual, sticky herb with a strong pungent odour and clothed with glandular and simple hairs; leaves 3- to 5-foliate, gradually becoming shorter upwards; flowers yellow in lax racemes; fruits capsules, compressed, hairy; seeds brownish black on ripening.

Phenology : *Flowering* and *Fruiting* : throughout the year.

Distribution : Throughout the plains of India; Bangladesh; as an weed of cultivation in fallow fields; common by the wayside, dry river beds or even in very poor soil.

Ecology and cultivation : Wild; mainly wastelands.

Clerodendrum serratum (L.) Moon (Verbenaceae)

(2n = 48)

Syn : *Volkameria serrata* L.

Sanskrit name : Bharangi.

Vernacular names : **Asm :** Nangal banga; **Hin :** Bharangi; **Mal :** Cerutekku; **Mar :** Bharangi; **Tam** : Sirutikku; **Tel** : Gantubarangi.

Trade name : Bharangi.

Traditional use : *Root* : bitter, acrid, thermogenic, anti-inflammatory, digestive, carminative, stomachic, antispasmodic, anthelmintic, stimulant, febrifuge, and beneficial in vitiated conditions *kapha* and *vata*, inflammations, dyspepsia, anorexia, colic, asthma, bronchitis, hiccup tumours, dropsy, leucoderma, leprosy, fever, etc.; *Leaf* : for external application for cephalalgia and ophthalmia.
UNANI : ingredient of 'Baharangi' (Barangi).

Modern use : *Root* and *Leaf* : febrile and catarrhal.

Phytography : Slightly woody shrub with bluntly quadrangular stem and branches, usually three at a node, occasionally opposite, oblong-elliptic, coarsely and sharply serrate; flowers blue; fruits 4-lobed, purple drupe with a succulent pyrene in each lobe.

Phenology : *Flowering* : with peaks during December–February and July–September; *Fruiting* : throughout the year.

Distribution : Found east of Sutlej; very common in Assam, West Bengal and Bangladesh.

Ecology and cultivation : Hills above 1000 m, especially on cleared slopes; occasional.

Chemical contents : Block histamine response; contains a high active fraction about 100 times potent than the crude drug; alkaloid.

Clerodendrum viscosum Vent (Verbenaceae)

(2n = 48, 52)

Syn : *Clerodendrum infloretum* auct. non L.

Sanskrit name : Bhandirah.

Vernacular names : Hin : Tinabharmi, Bhates; **Mal :** Peruku; **Tam :** Perukilai.

Traditional use : *Leaf* : bitter, acrid, laxative, antiseptic, anti-inflammatory, cholagogue, demulcent, antipyretic and tonic; also useful in vitiated conditions of *kapha*, helmianthiasis, ascarides, abscesses, leprosy, tumours, bronchitis, cough, indolent ulcers, inflammations, intermittent fever, malarial fever, general debility, etc., also useful in scorpion sting, snake bite.
HOMOEOPATHY : excellent remedy in diarrhoea associated with nausea.

Modern use : Bitter tonic, antiperiodic, vermifuge, laxative, cholagogue and anthelmintic.

Phytography : Large, gregarious, villous shrub, up to 9 m in height with bluntly quadrangular branchlets; leaves large, 20–25 x 9–20 cm, ovate, acuminate, entire or denticulate, base cordate, hairy on both sides; flowers white, tinged with pink in terminal panicles; fruits more or less globose, drupes seated on the enlarged pink calyx, containing 1–4 pyrenes.

Phenology : *Flowering* and *Fruiting* : almost throughout the year.

Distribution : Throughout India, Bangladesh and Pakistan as an undergrowth in forests up to 1800 m.

Ecology and cultivation : Mainly in wastelands and waysides.

Adulterant : May be an excellent substitute for *Ipecac* and /or *Pulsatilla.*

Clitoria teratea L. (Fabaceae)

English name : Clitoria.

Sanskrit names : Aparajita, Girikarnika.

Vernacular names : Ben : Aparajita; **Hin :** Aparajit, Gokarni; **Mal :** Aral, Malyamukki, Samkhupuspam; **Tam :** Kannikoti, Girikanni, Kaakkattam, Sankupushpam.

Trade name : Aparajit.

Traditional use : *Root* : bitter, alexeteric, anthelmintic, depurative, aphrodisiac, refrigerant, laxative, memory sharpening, tonic, ophthalmic and beneficial to leprosy,

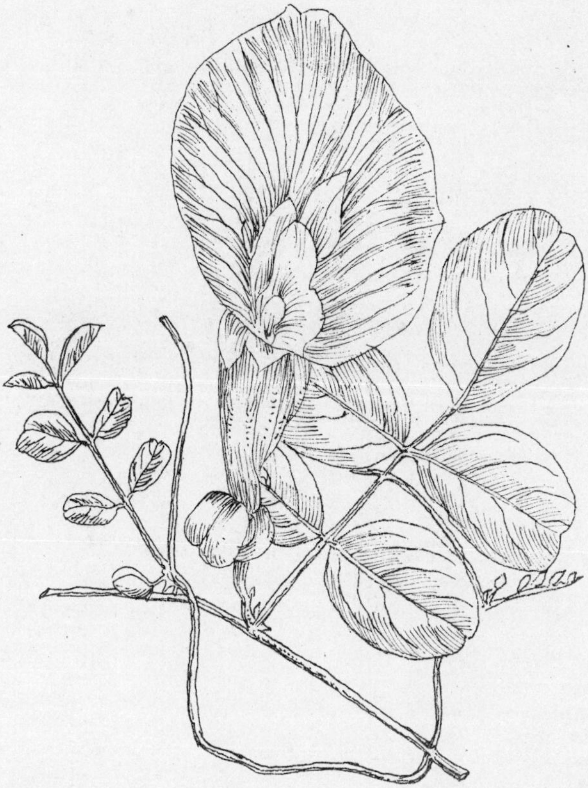

leucoderma and many other ailments; *Leaf* : in otalgia, hepatopathy and eruptions; *Seed* : cathartic and beneficial in visceralgia.

Phytography : Handsome twining perennial herb with terete stem and branches; leaves compound, imparipinate, leaflets 5–7, subcoriaceous, elliptic-oblong, obtuse; flowers blue or white, solitary, axillary or in fascicles; corolla papilionaceous; fruits more or less straight, flattened pods, sharply beaked; seeds 6–10, smooth, yellowish brown, reniform.

Phenology : *Flowering* : with a peak during March–May; *Fruiting* : throughout the year.

Distribution : India, Bangladesh and Pakistan; widely cultivated; sometimes found as ferals in hedges and on walls.

Ecology and cultivation : Common garden plant; occurs in hedges all over the tropical regions.

Chemical contents : *Leaf* : yields a 8-lactone, asparajitin; blue and white varieties contain a resin; *Seed* : gives the test of nucleoprotcia; contains fixed oil; *Root* and *Seed* : contain tannin.

Coccinia grandis (L.) Voight (Cucurbitaceae)

(2n = 24)

Syn : *Cephalandra indica* (W.A.) Naud., *C. cordifolia* Cogn., *Bryonia grandis* L.

English name : Ivy gourd.

Sanskrit names : Bimbi, Bimbika.

Vernacular names : Ben : Telakucha; **Hin :** Bimb, Kanturi; **Mal :** Kova, Koval; **Mar :** Bhimb; **Tam :** Kovai.

Trade name : Bimb /Telakucha.

Traditional use : *Root* : cooling, aphrodisiac and useful in vomiting, burning sensation and uterine discharges; *Leaf* : bitter, sweet, astringent and beneficial in *kapha* and *pitta*; *Fruit* : astringent, depurative, antipyretic, expectorant and useful in leprosy, skin diseases, intermittent fever, asthma, cough, bronchitis and jaundice; *Fruit and leaf of bitter variety* : bitter, acrid, thermogenic, emetic, purgative, digestive, liver tonic, sudorific and useful in vitiated conditions of *kapha* and *pitta*, wounds, ulcers, leprosy, jaundice, asthma, diabetes and many other ailments.

Phytography : Perennial, much branched, handsome tendril climber; roots sometimes tuberous; leaves deltoid or subrotund, angled or lobed, bight green above and pale green beneath, palmately 5-nerved; flower white, large, unisexual fruits ovoid or oblong or ellipsoid berries with white streaks, bright scarlet red on ripening; seeds ovoid, compressed, yellowish grey.

Phenology : *Flowering* : peak during December–March; *Fruiting* : throughout the year.

Distribution : Throughout the Indian subcontinent as cultivated and as wild.

Ecology and cultivation : Plains from the coast to 300 (500) m; semiwild; common in thickets and wastelands.

Chemical contents : Enzyme, hormones and traces of an alkaloid, pharmacologically inert; juice contains an amylas.

Cocculus hirsutus (L.) Diels (Menispermaceae)

(2n = 26)

Syn : *Cocculus villosus* Lam., *Menispermum hirsutum* L.

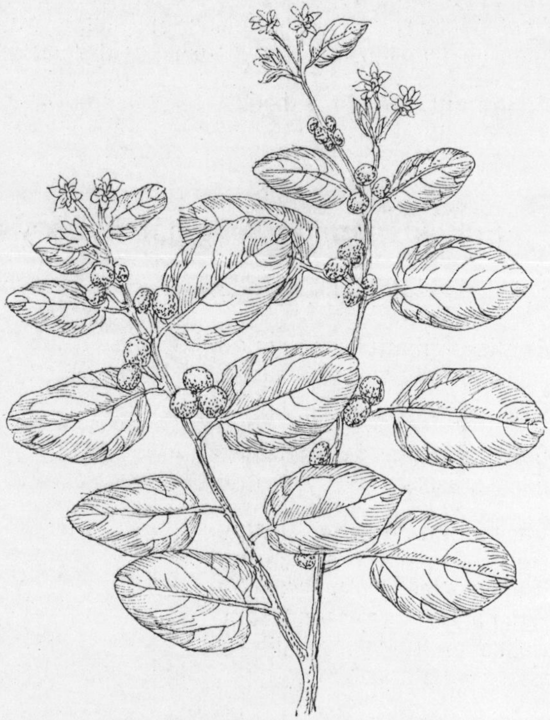

English names : Broom creeper, Ink Berry.

Sanskrit names : Patalagarudah, Chilihindah.

Vernacular names : Ben : Huyer; **Guj :** Vetavi; **Hin :** Patalagarudi, Jaljamni, Jamti ki bel, Khareta ki bel; **Kan :** Sagadi-balli, Dusari balli; **Mal :** Patalagarudakkoti; **Mar :** Vasanvel; **Tam :** Kattukkati, Siringattukoli.

Trade name : Huyur/Jamti ki bel.

Traditional use : *Root* : bitter, acrid, thermogenic, laxative, emollient, alternate, digestive, demulcent, alexeteric, carminative, diuretic, aphrodisiac, anodyne, expectorant, tonic and antipyretic, in vitiating conditions of *kapha* and *vata*, poisonous bites, leprosy, skin diseases, pruritus, dyspepsia, colic, bronchitis, gout, cephalalgia, intermittent fever, tubercular glands, fractures, spermatorrhoea, urethrorrhoea, hypertension, general debility, in acute rheumatism and venereal diseases.

HOMOEOPATHY : remedy for partial paralysis, general debility, female diseases, treatment for the children, occipital headache, for pregnant women, typhoid and cerebrospinal meningitis.

Modern use : *Leaf-juice* : when mixed with water, forms a jelly which is taken as a cooling medicine for gonorrhoea and used externally for eczema, prurigo, etc.

Phytography : Straggling, scandent shrub with soft, villous young parts; leaves simple, alternate, ovate-oblong, obtuse, apiculate, subcordate or truncate at the base, soft villous on both sides, petioles densely villous; male flowers small in axillary cymose panicles, female flowers 2–3 together in axillary clusters; fruits transversely rugose, purplish black drupes.

Phenology : *Flowering* and *Fruiting* : almost throughout the year.

Distribution : Tropical and subtropical regions of the Indian subcontinent.

Ecology and cultivation : Plains up to 700 m on thickets, particularly abundant in cleared localities.

Chemical contents : *Root*: show presence of D-trilobene and DL- coclaurine; *Stem* and *Root*: contain alkaloids, glucosides and sterols; water solution of the fraction of the extract sedative, hypotensive, bradycardiac, cardiotonic and spasmolytic.

Adulterant : *Root* : substitute for *Sarsaparilla.*

Cochlospermum religiosum (L.) Alston (Cochlospermaceae)

Syn : *Cochlospermum gossypium* DC.

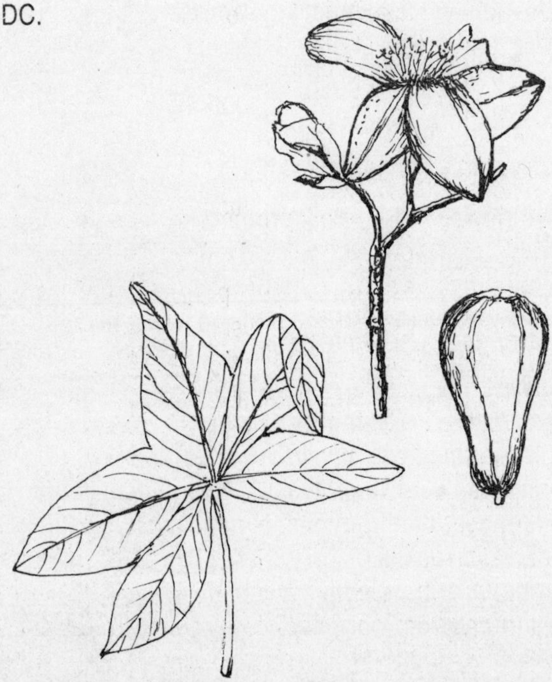

English names : Yellow-flowered cotton tree, White silk tree.

Sanskrit names : Girisalmalika, Silakarpasika.

Vernacular names : Hin : Gabdi, Galgal, Gjara, Pillikapas, Kumbi; **Mal :** Appakkutukka, Cempanni, Parappanni; **Tam :** Kattupparutti, Malapparutti; **Tel :** Tannakku, Kondagogu.

Trade name : Kumbi.

Traditional use : *Dried leaf* and *flower* : stimulant; *Young leaf* : cooling wash for hair; *Gum* (Katira) : thermogenic, anodyne, sedative and useful in cough, diarrhoea, dysentery, pharyngitis, gonorrhoea, syphilis and trachoma.
UNANI : an ingredient for the preparation of 'Katiera' (Katira).

Phytography : Small, deciduous tree, 2.4–5 m high with smooth, ash-coloured, fibrous, deeply furrowed bark containing gum; leaves palmately 5-veined, tomentose beneath; flowers large, golder yellow, very showy in terminal panicles appearing after leaf fall; fruits large, pea-shaped, 3- to 5-valved capsules; seeds many, covered with woolly hairs.

Phenology : *Flowering* : February–April; *Fruiting* : March–June.

Distribution : Uttar Pradesh, Bihar, Orissa, West Bengal, South India; less common in other parts of India and in Pakistan and Bangladesh.

Ecology and cultivation : Dry forests, especially on stony hills; leaf fall in January, new foliage in July.

Chemical contents : *Tree* : yields a gum.

Remark : Sometimes erroneously termed karaya gum.

Cocos nucifera L. (Arecaceae)

(2n = 32)

English names : Coco-palm, Coconut tree.

Sanskrit names : Narikelavriksah, Narikelah.

Vernacular names : Ben : Narikel; **Hin** : Nariyal; **Mal** : Narikelam; **Mar** : Narel; **Tam** : Tenkai-maram, Tennaimaram, Tenkai, Thennai; **Tel** : Narikelamu.

Trade name : Nariyal.

Traditional use : *Root* : astringent, diuretic, anthelmintic and beneficial in uterine disorders, pharyngodynia, blennorrhagia, bronchitis, hepatopathy, etc. ; *Juice of young spadix* : sweet, refrigerant, aperient, aphrodisiac, intoxicating, tonic, diuretic and useful in dyspepsia, diarrhoea, diabetes, leprosy etc.; *Oil* : disinfectant, insecticidal, appetiser, digestive and tonic and used as a medium for many medicated oils and ointments.

Phytography : Solitary, pinnate-leaved, monoecious palm; stem erect/inclined, curved, irregularly ringed, attains a height of about 30 m with a cylindrical, annulated stem bearing a crown of large leaves; leaves 2–5 m long, leaflet equidistant, narrow and tapering; inflorescence spadix with a hard, oblong, longitudinally splitting spathe enclosing many yellow/orange male flowers and few female flowers; fruits trigonously obovoid or subglobose or yellowish fibrous drupes; seed one, oval/spherical with a hard endocarp and oily white endosperm and sweet milky, watery fluid in the large cavity.

Phenology : *Flowering* and *Fruiting* : almost throughout the year.

Distribution : Cultivated throughout India, Bangladesh and Pakistan; mainly on the sea costs; cultivated in the tropics of Old and New World.

Ecology and cultivation : Prefers hot, damp locations with saline soils; plains and lower hills with irrigation facilities.

Chemical contents : From the kernel of coconut; water solution : glactomannan; *enzyme* : invertin, oxydase; *catalas milk* : histadine, orginine, lysine, tryosine, trytophan, proline, leucine, alanine; *Oil* : mixed glycerides, phytosterol and squalene.

Coffea arabica L. (Rubiaceae)

(2n = 22, 44, 66, 68)

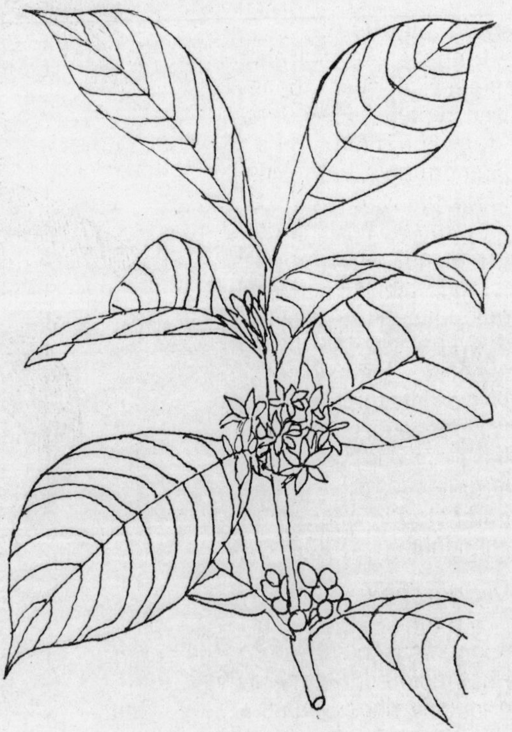

English names : Coffee, Arabian coffee.

Sanskrit name : Rajapiluh.

Vernacular names : Ben : Caphi; **Hin :** Caphi, Pilu; **Mal :** Kappi; **Ori :** Caphi; **Tel :** Kappi.

Trade name : Caphi.

Traditional use : *Seed* : bitter, stimulant, antipyretic, diuretic and useful as a medicinal beverage for the sick and convalescent; stimulates flow of digestive juices and intestinal peristalsis; *Infusion of unripe seed* : good for migraine, fever and gout; *Infusion of ripe toasted seed* : good for diarrhoea. HOMOEOPATHY : good medicine for neuralgia, insomnia, general weakness, general bodyache, headache, toothache, for labour pain, skin diseases and for stomach ailments.

Modern use : *Coffee* : palliative in spasmodic asthma, whooping cough, hysteria and cardioplegia; *Strong coffee* : remedy for poisoning by opium, alcohol and other narcotics.

Phytography : Glabrous large shrub with spreading branches; leaves simple, opposite with interpetiolar stipules, shiny, oblong, acuminate, main lateral nerves 6–10 pairs joined by intermarginal veins, flowers numerous, white, in axillary fascicles, corolla tube short; fruits obovoid, fleshy berries, purple when ripe; seeds concave or grooved ventrically.

Phenology : *Flowering* : white, fragrant, with a peak during March–May; *Fruiting* : ripen red by October.

Distribution : Cultivated in shade on hills, particularly in South India.

Ecology and cultivation : Grows well in medium rainfall with plenty of moisture in the subsoil; prefers cool, humid weather.

Chemical contents : Caffein content 1.0–1.2%, chlorogenic acid; *Seed* : alkaloid, caffeine; *Leaf* and *Fruit* : contain adenine, xanthine, etc.

Adulterants : Chicory, leaves of *Cassia tora*, *Tamarindus*, etc.

Coix lacryma-jobi L. (Poaceae)

(2n = 10, 12, 20, 24, 30)

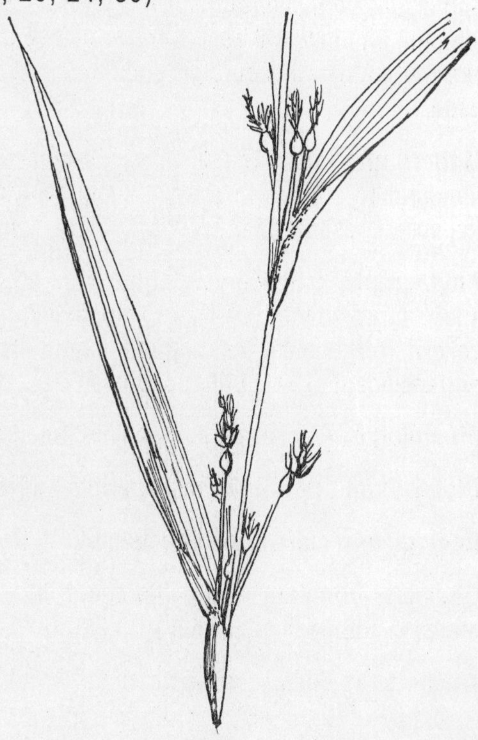

Syn : *Coix lacryma* L., *Lithagrostis lacryma-jobi* Gaertn.

English names : Job's tears, Adlay.

Sanskrit names : Gavednuka, Jargadi.

Vernacular names : Ben : Gurgur; **Hin :** Gurlu, Samkru; **Kan :** Sohrie, **Mal :** Kattgotampu, Kattappalunku; **Mar :** Ranmakkai; **Tam :** Kattukuntamani.

Traditional use : *Root* : menstrual disorders and strangury; *Seed* : bitter, sweet, tonic, diuretic, cathartic and depurative and used as a diet drink for invalids, also for pneumomycosis and pectoral diseases.

Phytography : Erect, perennial herb, upto 1.5 m high with basal rooting nodes; leaves linear lanceolate, midrib stout, ligule narrow, membranous, racemes nodding or drooping from long peduncles carrying male/female peduncles; fruits subglobose.

Phenology : *Flowering* and *Fruiting* : almost round the year.

Distribution : Throughout hotter and damper parts of India, Bangladesh and Pakistan, up to 1750 m; wild / cultivated.

Ecology and cultivation : Hills near water.

Chemical contents : Leucine, tyrocine, histidin, lysine, arginin and coicin.

Colchicum luteum Baker (Liliaceae)

(2n = 38)

English name : Golden collyrium.

Sanskrit name : Hiranyatutha.

Vernacular names : Hin : Hirantutiya; **Kas :** Virkum.

Traditional use : *Corm* : used for the treatment of rheumatism, gout and diseases of liver and spleen, laxative, aphrodisiac, alterative, aperient and external application to relieve pain and inflammation. HOMOEOPATHY : in neuralgic and rheumatic pain, it is extentisively used.

Modern use : used to relieve pain and inflammation and shorten the duration of acute gout and certain gouty affections but is liable to cause intestinal pain accompanied by vomiting and purging.

Phytography : Scape very short underground, flowers are very short, strap-shaped or linear, crocus-like, purple, white (rarely yellow), long tubular inserted in the perianth; corms are brownish or translucent or pale buff, opaque and cream-coloured.

Phenology : *Flowering* : generally in Autumn.

Distribution : Western Himalaya from Kashmir to Chamba, 1000–3000 m.

Ecology and cultivation : Prefers moist, rich meadows and limestone.

Chemical contents : Contains about 0.21–0.25% of colchicine and an abundance of starch in corms; in seeds 0.41–0.43% .

Adulterant : Used as a substitute for *Colchicum autumnale* L. (Meadow saffron).

Colocasia esculenta (L.) Schott (Araceae)

(2n = 22, 28, 42)

Syn : *Arun esculentum* L., *Colocasia antiquorum* Schott var. *esculenta* Schott

English names : Taro, Cocoyam.

Sanskrit name : Alupam.

Vernacular names : Ben : Kachu; **Hin :** Kachu, Kaccalu, Khuya, Arvi, Ghuiyan; **Kan :** Kachchi; **Mal :** Cempu; **Mar :** Allu; **Ori :** Saru; **Tam :** Sempu, Seppankizhangu; **Tel :** Chamadumpa.

Trade names : Kaccalu / Kachu.

Traditional use : *Leaf-juice* : styptic, stimulant and rubefacient and useful in internal haemorrhages, otalgia, otorrhoea, adenitis and bubose; *Corm-juice* : anodyne, demulcent, laxative and useful in somatalgia, alopecia areata, haemorrhoids, congestion of the portal system, as well as scorpion sting.

Phytography : Tuberous perennial with a group of underground farinaceous corms consisting of a central large one and surrounding ones with varying sizes; leaves with sheathing leaf-base and erect petiole $\pm$1.2 m long bearing thinly coriaceous peltate-ovate cordate lamina; spadix shorter than the petiole and much shorter than the spathe; spathe yellow; appendix also smaller than the spathe.

Phenology : *Flowering* : with a peak in November–December, *Fruiting* : throughout the year.

Distribution : Cultivated throughout the hotter parts of India, Bangladesh and Pakistan; as well as ferals.

Ecology and cultivation : Wild on the bank of streams, ponds, marshy and moist, shady situations, inside forests up to 2500 m.

Chemical contents : *Juice of tubers* : contains amylase; *Tubers* : contain sapotoxin *Leaf* and *Petiole* : good source of provitamin A and vitamin C.

Commiphora mukul (Hook. ex Stocks) Engl. (Burseraceae)

(2n = 26)

Syn : *Balsamodendron mukul* Hook. ex Stocks

English names : Mukul myrrh tree, Bolellium.

Sanskrit names : Gugguluh, Mahisaksah.

Vernacular names : Ben : Guggul; **Guj :** Guggul; **Hin :** Gugal, Guggul; **Kan :** Gugguli; **Mal :** Gulgulu, Mahisaksagulgulu; **Mar :** Gugguli; **Tam :** Gukkulu, Mahisaksi.

Trade name : Guggul.

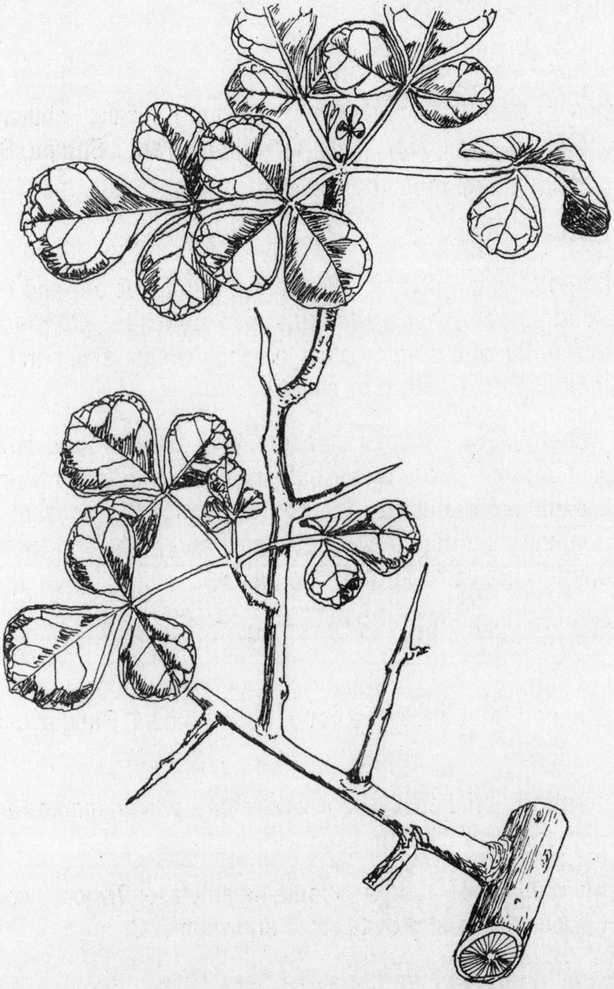

Traditional use : *Gum* : bitter, acrid, astringent, aromatic, anthelmintic, digestive, anodyne, antiseptic, nervine tonic, dumulcent, aphrodisiac, alterative, stimulant, liver tonic, detergent, depurative vulnerary, antispasmodic, emmenagogue, haematinic, diuretic, lithotriptic, rejuvenating and general tonic and useful in vitiated conditions of *vata,* gout, scrofula, sciatica, facial paralysis, diplegia, hemiplegia, leprosy, leucoderma, helmianthiasis, asthma, cough bronchitis, pectoral and hepatic disorders, epilepsy, fever, ulcers, cardiac disorders, coronary thrombosis, urinary calculus, diabetes, etc.
UNANI : Balsan (Habbe-Balsan); also for the preparation of 'Mugil'.

Phytography : Small, armed tree with spinescent branches and ash-coloured rough bark, peeling off in flakes, young parts glandular, pubescent; leaves alternate, 1- to 3-foliate, obovate, serrate toothed in upper parts; flowers small, brownish red; fruits ovoid drupes, red on ripening.
'Guggul' of commerce is the pale yellow, aromatic gum resin obtained from the bark.

Distribution : In arid, rocky tracts of Rajasthan, Karnataka; Sind, Baluchistan (Pakistan).

Ecology and cultivation : Xerophytic habit; prefers arid soil.

Chemical contents : 1.45% essential oil, containing myrcene, dimyrcene and polymyrcene; sugars in the *gum* and aldobiouronic acid structure.

Commiphora myrrha (Nees) Engl. (Burseraceae)

Syn : *Balsamodendron myrrha* T. Nees

English names : Common myrrh tree, Haroboi myrrh.

Sanskrit names : Bolah, Rasagandhah.

Vernacular names : Hin : Bol, Hirabol; **Mal :** Narumpasamaram, Narumpasa; **Tam :** Vellaippapolam.

Trade name : Bol.

Traditional use : *Gum* : bitter, acrid, astringent, anthelmintic, aphrodisiac, memory sharpening, expectorant, anti-inflammatory, diuretic, deodorant, ophthalmic, antiseptic, stimulant and tonic, and useful in vitiated conditions of *vata*, *pitta* and *kapha* of different ailments like amenorrhoea, menstrual disorders, asthma, phthisis, rheumatoid arthritis, etc.
UNANI : an ingredient for the preparation of 'Mur makki'. Also used in mouthwash, as an emmamagogue.

Modern use : Antiseptic mouthwash, as an emmenagogue, stimulant tonic.

Phytography : Small tree; a source of Hiraboi myrrh; gum resin exudates from wounds in the stem, pale yellow at first and solidifies later to brown-black.

Distribution : Indigenous to Northeast Africa, sold in Indian markets.

Ecology and cultivation : Prefers arid soil.

Chemical contents : *Essential oil* : bitter substance; *Mixture of resin* : volatile oil, etc.

Adulterants : Arabian myrrh (Fadhli), Yemen myrrh, Bdelliums, Bissabol, *Commiphora erythaea* var. *glabrecens,* Opague bdellium, African bdellium, Indian bdellium, gums, sand and vegetable fragments.

Commiphora wightii (Arn.) Bhandari (Burseraceae)

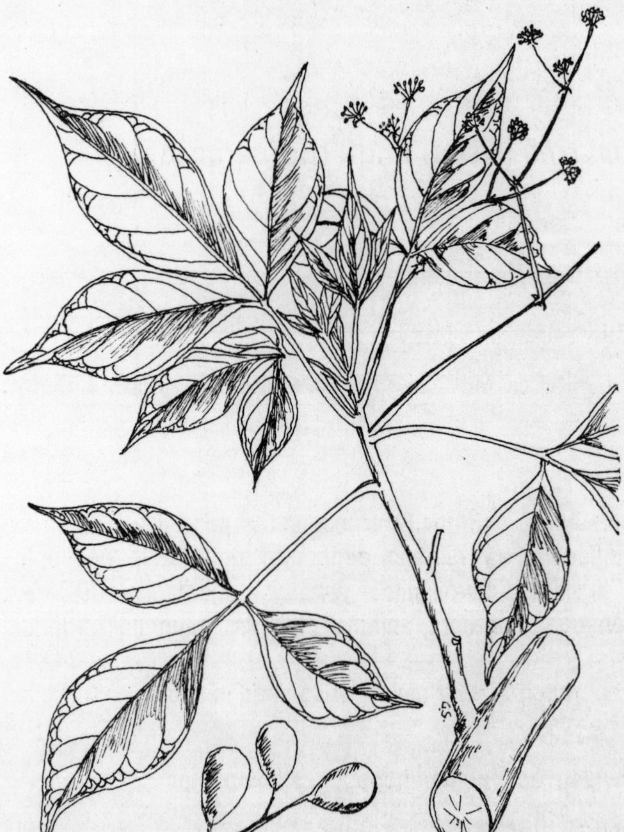

Syn : *Commiphora caudata* (W. & A.) Engl., *Protium caudatum* W. & A., *Balsamodendron wightii* Arnott, *B. roxburghii* Stocks

English name : Hill mango.

Sanskrit name : Ilkktah.

Vernacular names : Mal : Itinnil; **Tam :** Kiluvai.

Traditional use : *Root* : astringent, sweet, cooling, aphrodisiac and diuretic and useful in diabetes, strangury, fever and vitiated conditions of *vata* and *pitta*; *Leaf* : anodyne and used in rheumatalgia.

Phytography : Handsome deciduous tree with papery bark and greenish soft wood; leaves alternate, imparipinate, leaflets opposite,; flowers small in fascicles; fruits ovoid or subglobose, fleshy drupes as big as a pea.

Distribution : Dry forest of Andhra Pradesh, Karnataka, rarely in Kerala and Tamil Nadu.

Ecology and cultivation : Prefers semihigh / dry regions.

Coptis teeta Wall. (Ranunculaceae)

English names : Gold thread, Coptis.

Sanskrit names : Supita, Pita.

Vernacular names : Ben : Teeta; **Aru :** Mismi-teeta; **Hin :** Mamira, Mamiran; **Mal :** Pitarohini, Pitramanikkilannu; **Tam :** Pitarohini.

Trade name : Mismi-teeta.

Traditional use : *Rhizome* : bitter acid, cooling, ophthalmic, anthelmintic, diuretic, stomachic, laxative, carminative, depurative, antipyretic and tonic and beneficial in cough, bronchitis, atonic dyspepsia, colic, diabetes, leprosy, leucoderma, dropsy, etc.
UNANI : an ingredient for the preparation of 'Mameesa'. ('Mamira'/'Mamiran').

Phytography : Rhizomatous herbaceous, perennial with horizontally grown golden yellow woody underground rhizome bearing many fibrous roots; leaves ternatisect, leaflets 5.0–7.5 cm, ovate-lanceolate, pinnatifid, terminal largest, rachis 15–20 cm long; flowers small, white, on slender leafless scopes; fruits many-seeded follicles; seeds black, testa crustaceous.

Distribution : Temperate regions of Mishmi Hills at the northern frontier of Assam.

Ecology and cultivation : At high altitude.

Chemical contents : Berberine, coptine.

Corchorus aestuans L. (Tiliaceae)

Syn : *Corchorus acutangulus* Lam.

Vernacular name : Ben : Titapat.

Trade name : Titapat.

Traditional use : *Seed* : stomachic, in pneumonia.

Phytography : Subshrub, 50–75 cm, branchlets pubescent; leaves ovate, elliptic or oblong, 2–5 x 1-3 cm, chartaceous, basally 3- to 4-nerved, pubescent; cymes leaf-opposed, 2- to 3-flowered; flowers 7 mm across; capsules terete; seeds truncate.

Phenology : *Flowering* and *Fruiting* : almost throughout the year.

Distribution : Throughout hotter parts of India and Bangladesh.

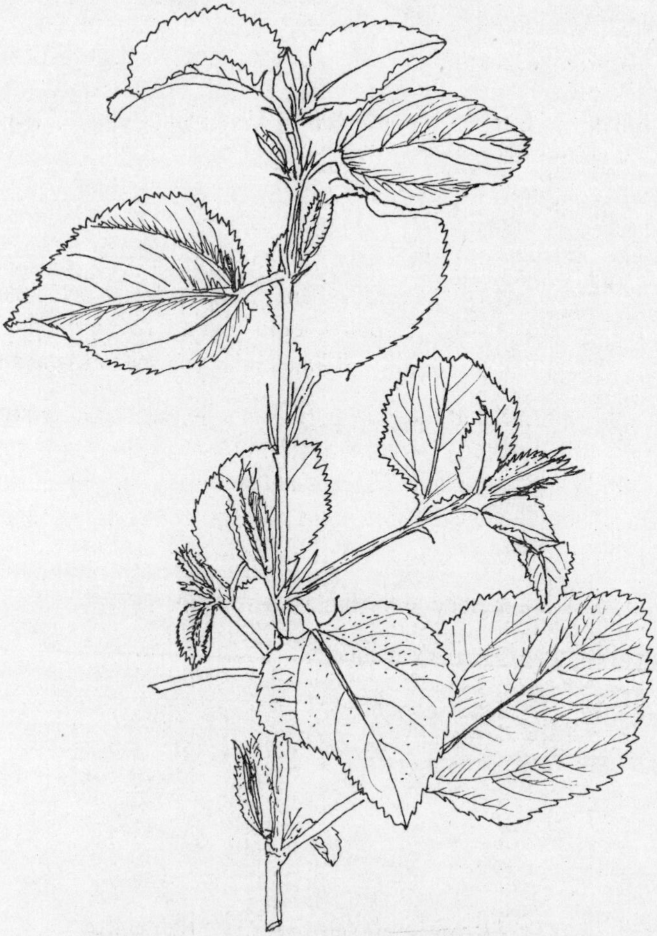

Ecology and cultivation : Plains from coast up to 750–1400 m; often gregarious by arable lands, fallow fields or as a weed of cultivation.

Corchorus capsularis L. (Tiliaceae)

(2n = 14)

English names : Jute, White jute.

Sanskrit name : Kalasaka.

Vernacular names : Asm : Titamara; **Ben :** Pat, Titapat; **Hin :** Narcha, Pat, Titapat.

Trade name : Pat.

Traditional use : *Leaf* : demulcent, bitter tonic, stomachic, laxative, carminative, antiperiodic anthelmintic, astringent and intestinal antiseptic; *Infusion of dried leaf* : bitter and commonly used as a stomachic tonic like chiretta, also used in fever, bilious trouble, dysentery, liver disorders, intestinal colic, gastric catarrh, skin diseases, atonic dyspepsia, mild jaundice and disorder of digestive system, for the treatment of dysentery, dried leaves are eaten with rice; *Decoction of dried root* and *unripe capsule* : good for diarrhoea; *Capsule* : used as a paste over swellings and abscesses, purgative and contains vitamin C.

Phytography : Annual herb, stem straight, smooth, cylindrical, very tall, upper portion branched; leaves light green, thin, oval, narrow pointed, toothed; flowers solitary or on 3 short, thick stalks, small, yellow; capsules short, globose, flattened at the top, wrinkled, ribbed, 5-valved; seeds few in each valve, without transverse partitions.

Phenology : *Flowering* and *Fruiting* : Rainy season.

Distribution : Widespread in parts of India and Pakistan; cultivated in West Bengal; Bangladesh, Sri Lanka, Myanmar.

Ecology and cultivation : It is an upland species, require warm and humid climate with medium rainfall and in a wide range of soils rich in K and P with pH between 5 and 7.

Chemical contents : *Leaf* : contains glucoside, capsularin which seems to be related to corchorin; corchorin and bitter substance corchoritin isolated from seeds; cardiac aglycon corchartoxin having heart action similar to digitalis group of genins but not as intensely obtained from seeds; corchoroside A and corchoroside B.

Corchorus olitorius L. (Tiliaceae)

(2n = 7, 14, 28)

English names : Tossa jute, Jute, Jew's mallow.

Vernacular names : Ben : Latikapat, Mithapat; **Hin :** Koshta, Mithapat; **Tam :** Peratti; **Tel :** Paritakura.

Trade name : Mithapat.

Traditional use : Same as allied species *Corchorus capsularis* L.

Phytography : Like that of *Corchoruos capsularis* L. to some extent excepting that the capsules are cylindric, 10-ribbed, erect, 50–75 mm long, 5-valved with 5 terminal blunt points; seeds many.

Phenology : *Flowering* and *Fruiting* : Rainy season.

Distribution : Like that of *Corchorus capsularis* L.

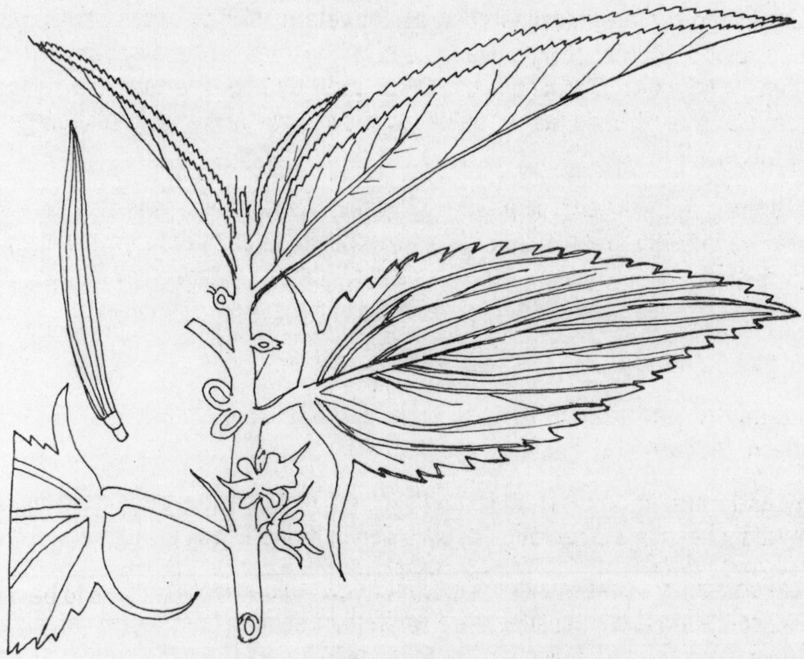

Ecology and cultivation : It is a lowland species; in other respects it is similar to *Corchorus capsularis* L.

Chemical contents : *Seed* : contains two digitalis glucosides, Corchoroside A and Corchoroside B.

Corchorus trilocularis L. (Tiliaceae)

Sanskrit name : Dirghachanchu.

Vernacular names: Guj : Kadvichhunchhadi; **Hin :** Kadukosta; **Mar :** Kaduhhunchh.

Traditional use : *Mucilage* : demulcent; *Seed* : in fever and obstruction of abdominal viscera.

Phytography : Erect subshrub, up to 1 m high; branchlets glabrescent; leaves oblong, ovate, lanceolate or elliptic, 3–7 x 2.0–2.5 cm, chartaceous, basally 3- to 4-nerved, glabrous, base cuneate, apex acute or obtuse; petioles 0.7–1.5 cm; stipules filiform, up to 7 mm; cymes leaf-opposed, 2- to 4-flowered; flowers 4 mm across, yellow; capsules 3 in a cluster, 3–6 cm, 3- to 4-angular, stellate hairy when young, scabrid when old, 3- or 4-valved, beak entire; seeds 1 mm, truncate.

Phenology : *Flowering* and *Fruiting* : throughout the year.

Distribution : India; Pakistan, Afghanistan, Bhutan and Sri Lanka.

Ecology and cultivation : Plains up to 1200 m; weed of cultivation.

Cordia dichotoma Forst. f. (Boraginaceae)

Syn : *Cordia obliqua* Willd., *C. myxa* non L.

English name : Sebesten plum.

Sanskrit names : Slesmataka, Bahuvarah.

Vernacular names : Ben : Buhul, Bohodari; **Hin :** Lasura, Lasora; **Mal :** Naruvari, Naruviri; **Mar :** Gondan; **Tam :** Naruvili; **Tel :** Cinna nakkeru.

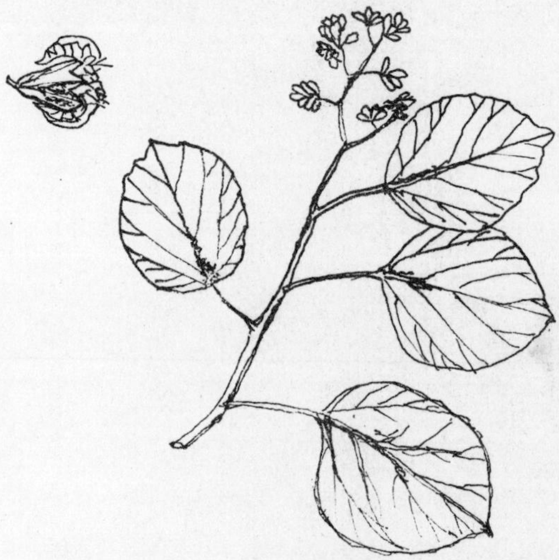

Trade name : Lasura.

Traditional use : *Bark* : bitter, acrid, astringet, anthelmintic, constipating, cooling, depurative and beneficial in diarrhoea, dyspepsia, fever, burning sensation, vitiated conditions of *kapha* and *pitta*, leprosy, helminthiasis and skin diseases; *Leaf* : aphrodisiac and useful in gonorrhoea and ophthalmodynia; *Fruit* : anthelmintic, sweet, cooling, diuretic, aphrodisiac, vulnerary, expectorant, febrifuge and in vitiated conditions of *vata* and *pitta*, ulcers, leprosy, skin diseases, chronic fever, dry cough and many other ailments like hyperdipsia, burning sensation, pectoral disease, etc.

Phytography : Medium-sized tree with a short trunk; leaves simple, entire or serrate, elliptic-lanceolate with a cordate base; flowers white, small in lax terminal or axillary cymes; fruits drupes, yellowish brown, pink or nearly black on ripening with a viscid sweetish, transparent pulp surrounding central stony part.

Phenology : *Flowering* : copious during January–April and September–October; *Fruiting* : throughout the year.

Distribution : Throughout India, Bangladesh and Pakistan, in warmer regions; wild and cultivated.

Ecology and cultivation : Plains from the coast up to 500–800 m.

Chemical content : *Bark* : contains 20% tannin.

Coriandrum sativum L. (Apiaceae)

(2n = 22, 24, 28, 44)

English name : Coriander.

Sanskrit names : Dhanyakam, Dhanaka.

Vernacular names : Ben : Dhane, Dhania; **Guj :** Konphir; **Hin :** Dhaniyam; **Kan :** Kothambri; **Mal :** Kattampala, Kottamalli, Malli, Kottampalari; **Mar :** Dhanna, Kothimber; **Tam :** Kottamalli; **Tel :** Dhaniyalu.

Trade name : Dhaniya.

Traditional use : *Leaf*: acrid, aromatic, analgesic, anti-inflammatory and styptic and are useful in halitosis, pharyngopathy, epitaxis, ulemorrhagia, chronic conjunctivitis, hiccup, suppuration, jaundice and odontalgia; *Fruit* : aromatic, sweet, acrid, bitter, astringent, emollient, anti-inflammatory, thermogenic, stomachic, carminative, appetiser, digestive, diuretic, constipating, antipyretic, stimulant and beneficial in vitiated conditions of *pitta*, burning sensation, cough, bronchitis, vomiting, dyspepsia, colic, diarrhoea, dysentery, anorexia, dropsy, scrofula, chronic conjunctivitis, gout, rheumatism, giddiness, intermittent fever, etc.

Phytography : Glabrous, aromatic, herbaceous annual, 30–90 cm high; leaves dicompound, lower ones long-petioled and upper ones short-petioled/subsessile, imparipinnatisect; flowers small, white/pinkish/purple in compound terminal umbels; fruits yellowish brown, globular and ribbed, separating into two halves containing a seed.

Phenology : *Flowering* : November–January; *Fruiting* : December–March.

Distribution : Cultivated in India, Bangladesh and Pakistan; indigenous from the Mediterranean region to Central Asia; cultivated in nearly all parts of Asia.

Ecology and cultivation : Plains/hills; cultivated.

Chemical contents : Essential oil, coriandrol, oxalic acid and Ca content.

Coscinium fenestratum (Gaertn.) Colebr. (Menispermaceae)

English names: Tree turmeric, False calumba.

Sanskrit names : Darvi, Daruharidra.

Vernacular names : Ben : Haldigach; **Hin :** Jhar-i-haldi; **Mal** : Maramannal; **Tam :** Maramaneal, Mancalkkoti; **Tel :** Monu pasupu.

Trade names : Jhar-i-haldi, Daruharidra.

Traditional use : *Stem* : bitter, anti-inflammatory, anodyne, ophthalmic, depurative, stomachic, antiseptic, sudorific, tonic, thermogenic and beneficial to ophthalmopathy, inflammation, wound, ulcer, abdominal disorder, jaundice, diabetes, fever and general debility in vitiated conditions of *kapha* and *vata*.

Phytography : Woody climbing shrub with cylindrical stem, yellow-brown externally, internally yellow and longitudinally flutted; leaves simple, oblong-deltoid, shiny and smooth above, minutely tomentose below; flowers yellow, divaricately branched inflorescence; fruits globose, drupes; seeds globose.

Distribution : Maharashtra, Karnataka, Tamil Nadu, Kerala and some other places.

Chemical contents : Berberin, saponin and some resinous materials.

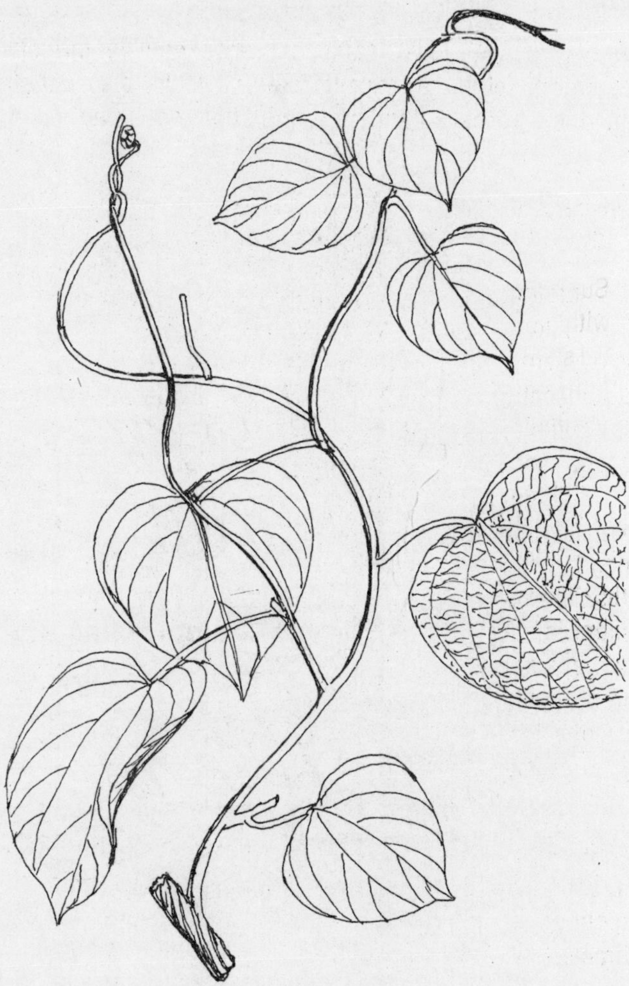

Costus speciosus (Koening ex Retz.) J.E. Smith (Zingiberaceae)

(2n = 18, 27, 36)

Syn : *Banksea speciosa* J. Koening

Sanskrit names : Canda, Kemuka.

Vernacular names : Ben : Keu; **Hin :** Kebu, Keyu, Kusi; **Kan :** Chengaivakoshtu; **Mal :** Cannakiiannu, Cannakkuvva, Narumeanna; **Mar :** Penva, Pushkarmula; **Tam :** Kostam, Kuiravam, Kottam; **Tel :** Chengaivakohtu.

Trade name : Keyu.

Traditional use : *Rhizome* : bitter, astringent, acrid, cooling, purgative, aphrodisiac, anthelmintic, depurative, febrifuge, expectorant and tonic, also beneficial in asthma, anaemia, bronchitis, leprosy, flatulence, constipation, fever, skin diseases and inflammation.

Modern use : in carcinogenic tumours.

Phytography : Succulent perennial herb with long leafy spirally twisted stems, 2–3 m high and horizontal rhizomes; leaves simple, spirally arranged, oblanceolate or oblong, glabrous above, silky pubescent beneath with broad leaf sheaths; flowers white, large, scented, in large terminal spikes; bracts bright red; fruits globose / ovoid capsules; seeds obovoid or semiglobose.

Phenology : *Flowering* : September–October; inflorescence globose of closely packed bracts.

Distribution : Throughout India; Pakistan and Bangladesh—in moist localities and in wastelands.

Ecology and cultivation : Hills above 800 m; moist places and wastelands also.

Chemical content : *Root* : rich in starch.

Crataeva magna (L.) DC. (Capparaceae)

(2n = 26)

Syn : *Crataeva nurvala* Buch. Ham., *C. religiosa* Hook. f. & Thoms. non Forst. f.

English names : Capertra, Three-leaved caper.

Sanskrit name : Varuna.

Vernacular names : Asm and **Ben** : Barun; **Hin** : Barna; **Kan** : Mavilinga; **Mal** : Varana; **Ori** : Baruna; **Pun** : Barna; **Sin** : Lunuwarana; **Tam** : Maralingam, Navala, Navilankai; **Tel** : Pedda-kulimidi.

Trade name : Varuna / Barun.

Traditional use : *MIKIRS* : *Bark* : as post-natal tonic, in urinary disorders; *TRIBES OF BASTAR* (Madhya Pradesh) : *Bark* : in antifertility, wounds; *NAYADIS* (of Tamil Nadu) : *Bark* : in fever, stomach troubles; *TRIBES OF BANSWARA* (Rajasthan) : *Bark* : in guinea-worms; *SANTALS* : various parts in baldness and hydrocele.

AYURVEDA : *Extract of root-bark* mixed with honey : in scrofulous enlargement of glands under lower jaw; *Powdered stem-bark* : as appetiser, in renal and urinary troubles, gastro-intestinal and urinary affections; *Decoction of stem-bark* compounded with root powder : in gravel; *Bark-collyrium* : in eye affections; *Stem-bark and fresh leaf compound* together as rubrifacient, in rheumatic pain; *Leaf-juice* : in rheumatism.

SIDDHA SYSTEM : *Root-bark* : ingredient of *Mavilankam*.

Modern use : *Root* (extract) : antibacterial against gram-positive and gram-negative bacteria; *Stem* (alcoholic extract) : antibacterial against *E. coli*; *Bark* (extract) : spasmolytic on animal and human uterii, cholinergic on smooth muscles of animals, anti-inflammatory.

Phytography : Small spreading tree; leaves trifoliate, aromatic when bruised; corymbs terminal, flowers large, white to yellow and purple, 2.5–5.0 cm in diameter, stamens purple, indefinite, adnate to base of gynophore, ovary on a slender stalk; berry fleshy, globose, 2.5–5.0 cm in diameter, many-seeded.

Phenology : *Flowering* : February–April; *Fruiting* : April–June.

Distribution : Near streams in Kannads, Malabar, Bihar and northeastern India, planted in gardens throughout India.

Ecology and cultivation : Mostly wild on low hills.

Chemical contents : *Root-bark* : β-sitosterol, lupen-3-one, lupeol, 3-epilupeol, quercetin, rutin, taraxasterol; *Bark* : betrulinic acid, ceryl alcohol, diosgenin, friedelin, lupeol; *Leaf* : stachydrine.

Remarks : In Sri Lanka, plant is used in fracture; fruits are eaten by tribes of northeastern India, Bihar and Orissa; wood is used to make combs, drums and turnery articles.

Cremanthodium decaisnei Cl. (Asteraceae)

English name : Rhend.

Traditional use : LADAKHIS : *Plant* : as sedative.

Phytography : Perennial acapigerous herb, robust, 15–25 cm high; leaves reniform, coriaceous, crenate, glabrous, petioles stout; heads 3.7–6.2 cm in diameter; achenes slender, pappus white.

Distribution : Alpine Himalaya from Kashmir to Sikkim between 3700 and 5000 m.

Ecology and cultivation : Alpine flora; wild.

Crepis flexuosa (DC.) Bth. & Hk. f. ex Cl. (Asteraceae)

Syn : *Crepis glauca* Benth., *Barkhausia flexuosa* DC.

Vernacular name : Lad : Samboo.

Traditional use : LADAKHIS : *Milky latex of plant* : in constipation.

Phytography : Most densely dichotomously corymbose branched herb, forming an obconical mass, 15–30 cm high; branches rigid, forking, spreading; leaves heterophyllous, alternate, lower petioled, pinnatifid or pinnate, lobes variously acutely cut, young leaves sessile, linear, entire; heads terminal and sublateral, peduncles slender, few-flowered, 0.8–1.2 cm in diameter, outer involucre of bracts minute; flowers yellow; achenes 0.5 cm long, very slender, ribbed to the tip, pale, obscurely scabrid, hardly beaked, pappus white.

Distribution : Lahul and Kumaon, between 3000 and 5800 m; western Tibet.

Ecology and cultivation : Plant of cold climate; wild.

Crinum amoenum Roxb. (Amaryllidaceae)

Vernacular name : Mir : Jah-jak.

Traditional use : MIRI : *Rhizome* : in stomach disorders.

Phytography : Bulb globose, 5.0 –7.5 cm in diameter; leaves 10–12, suberect, bright green, long, 40–60 cm by 2.5–3.7 cm, tapering from base to acuminate tip, margin cubcscabrous; scape 33–66 cm, subcylindric, rather slender; umbels 6- to 12-flowered, spathes 5 cm, lanceolate; flowers subsessile, perianth white, tube green.

Distribution : Sikkim, Meghalaya, ascending up to 2000 m; Nepal, Bangladesh (Sylhet), Myanmar.

Ecology and cultivation : Plant of tropical Himalaya.

Crinum defixum Ker. Gawl. (Amaryllidaceae)

(2n = 14, 20, 24)

Syn : *Crinum asiaticum* Roxb., *C. asiaticum* var. *angustifolium* Wall., *Amaryllis vivipara* Lamk.

Traditional use : TRIBES OF VARANASI (Uttar Pradesh) : *Tuber* : in bodyache; TRIBES OF JHABUA (Madhya Pradesh) : *Tuber* : in abdominal disorders; RURAL FOLK OF KURUKSHETRA (Haryana) : *Leaf* : in earache.

Phytography : Bulb with a fusiform stoloniferous base, neck cylindric; leaves few, erect, 60–100 cm by 1.8–2.5 cm, linear, concave; scapes 45–60 cm long, stout, umbels 6- to 15-flowered, spathes 3.7–5.0 cm; flowers sessile, perianth white, tinged with red, fragrant at night, filaments bright red; fruits subglobose, 2.5 cm in diameter, 1- to 2-seeded; seeds albuminous.

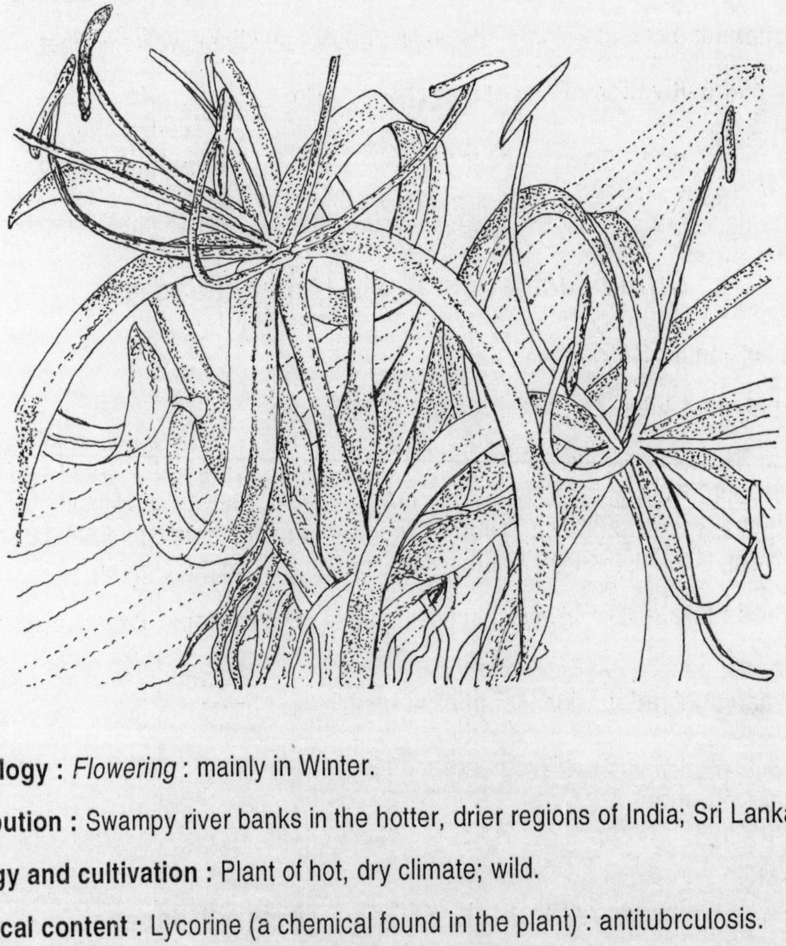

Phenology : *Flowering* : mainly in Winter.

Distribution : Swampy river banks in the hotter, drier regions of India; Sri Lanka.

Ecology and cultivation : Plant of hot, dry climate; wild.

Chemical content : Lycorine (a chemical found in the plant) : antitubrculosis.

Remarks : *EtOH* (50%) *extract of C. difixum* Ker. Gawl. var. *defixum* is anti-inflammatory, depresses Central Nervous System and diuretic.

Crinum latifolium L. (Amaryllidaceae)

(2n = 22, 24)

Syn : *Crinum zeylanicum* L., *C. speciosum* Herb., *Amaryllis latifolia* L.

Vernacular names : Ben : Sukhdarsan; **Bhi :** Barapunjaniya; **Hin :** Chindar; **Mar :** Gadambikand; **Tam :** Vishmungil.

Trade names : Sukhdarsan, Chindar.

Traditional use : *BHIL* (of Madhya Pradesh) : *Bulb* : in stomachache; *SANTAL* : *Bulb* : in tubercular fistula.

Phytography : Bulb globose, 12.5–15.0 cm in diameter, neck short; leaves many, 60–100 cm by 7.5–10.0 cm, bright green, keeled and waved; scape 60–100 cm, stout, green reddish or purple, spathes 7.5–10.0 cm, coloured like scape; umbels 10- to 20-flowered; flowers variable in size, white, more or less streaked or tinged with red towards centre, or red-purple nearly all over the back of lobes; fruits subglobose, 3.7–5.0 cm in diameter.

Phenology : *Flowering* : Mostly in Autumn; *Fruiting* : Winter.

Distribution : Throughout India; Sri Lanka, Bangladesh, Myanmar; wild or planted.

Crinum pratense Herb. (Amaryllidaceae)

Traditional use : *GARO* : *Bulb* and *Root* (together) : for diarrhoea; *KHASI* and *JAINTIA* : *Plant* : in bone fracture.

Phytography : Bulb ovoid or spherical, 10.0–12.5 cm in diameter, neck 5.0–6.2 cm; leaves 6 to 8, suberect or declinate, linear; scape 30 cm or more, decumbent, compressed; umbels 6- to 12-flowered; flowers subsessile, white, perianth tube 7.5–10.0 cm, fragrant, stamens very long, filaments red.

Distribution : India; Myanmar.

Crotalaria albida Heyne ex Roth (Fabaceae)

(2n = 16)

Syn : *Crotalaria montana* Roxb., *C. parva* Grah., *C. scoparia* Wall., *C. tenuis* Wall.

Vernacular names : Hin : Banmethi; **Mun :** Huring piri sokoe.

Traditional use : *RURAL FOLK OF SIWALIK* (Uttar Pradesh) *AND ORAON* (of Bihar) : *Root* : as purgative.

Phytography : Shrub, 30–60 cm high, branches many, slender, obscurely silky; stipules absent, leaves simple, short-stalked. firm, 2.5–5.0 cm, thinly silky beneath, glabrescent above; racemes mostly terminal, rarely lateral, 5–10 cm long, laxly 6- to 20-flowered; flowers pale yellow; pods oblong-cylindrical, 1.2–1.5 cm long, exserted from calyx, 6- to 12-seeded.

Phenology : *Flowering* and *Fruiting* : September–January.

Distribution : Tropical regions of India, ascending up to 1800 m in Kumaon and up to 2000 m in the western Himalaya, common in Chotanagpur (Bihar); Myanmar.

Ecology and cultivation : Tropical plant; wild.

Chemical contents : *Plant* : croalbidine, croalbinecine and pyrrolizidine.

Crotalaria bialata Schr. (Fabaceae)

(2n = 16)

Syn : *Crotalaria alata* Ham., *C. sagitticaulis* Wall.

Vernacular names : Mun : Marang gurgura; **Orn :** Xexel jhunka; **San :** Nagphani.

Traditional use : *ETHNIC COMMUNITY OF PURULIA* (West Bengal) : *Plant* : as tonic to pregnant women and to facilitate child birth; *TRIBES OF HAZARIBAGH* (Bihar) : *Root* : in urinary complaints; *ORAON* : *Plant* : against snake bite.

Phytography : Creeping herb or suberect undershrub, 30–60 cm high; stem and leaves softly haired; leaves subsessile, simple, thin, 5.0–7.5 cm long, obovate to ovate-oblong, stipules forming a broad wing; racemes lateral, leaf-opposed, 2- to 3-flowered; flowers bracteate, yellow, 1.25 cm, calyx densely silky; pods stalked, 3.0–4.3 cm long, linear-oblong, 30- to 40-seeded.

Phenology : *Flowering* : July–November; *Fruiting* : September–December.

Distribution : Kumaon to Meghalaya, ascending up to 1800 m, frequent in Chotanagpur plateau, available in Tamil Nadu up to 2500 m.

Ecology and cultivation : Tropical plant; wild.

Chemical content : *Seed* : usaramine.

Remark : Santals use the plant as fodder.

Crotalaria burhia Buch. Ham. ex Benth. (Fabaceae)

Vernacular name : Guj : Shinio.

Traditional use : *TRIBES OF SAURASHTRA* (Gujarat) *AND THE INDIAN DESERT* : *Plant* : as cooling agent, and on burns.

Modern use : *EtOH* (50%) *extract of root* : anticancer.

Phytography : Low undershrub; branches numerous, stiff erectopatent, clothed with fine pale momentum; leaves simple, scattered, subsessile, rigid, pale green, 0.6–3.7 cm long, exstipulate, deciduous; racemes terminal, elongated, 6- to 12- flowered, pedicels short; flowers yellow, small; pods oblong, 3- to 4-seeded.

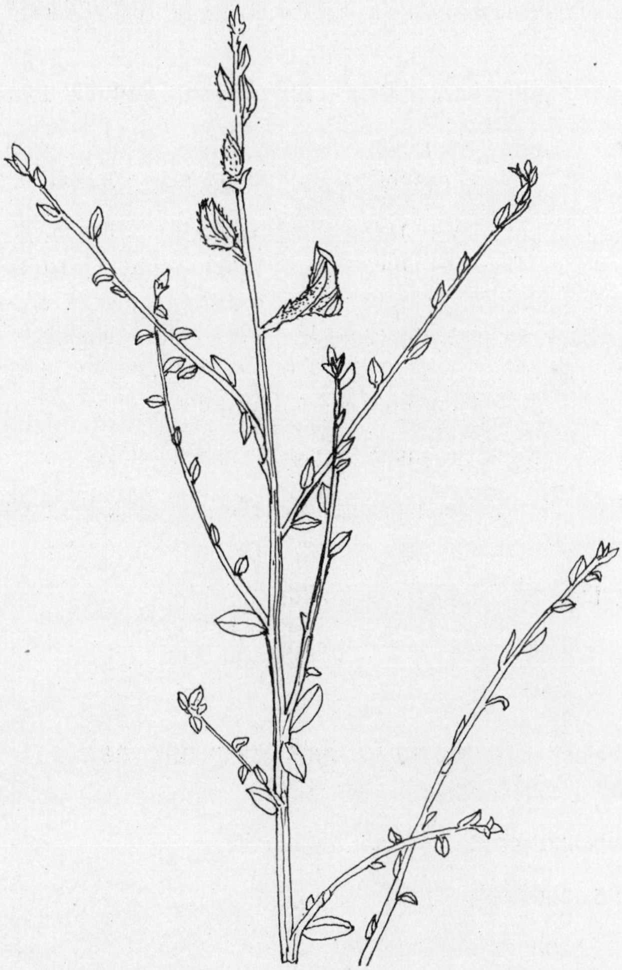

Distribution : Desert areas of Gujarat, Rajasthan and Punjab; Pakistan (Sind, Baluchistan).

Ecology and cultivation : Characteristic desert plant; wild,

Chemical contents : *Plant* : croburhine, crotalarine; *Stem* (young) and *Leaf* : hebtriacontane, β-sitosterol.

Remarks : It is claimed that the plant is useful in gout, hydrophobia and swelling, and contains antimicrobial principles.

Crotalaria calycina Shr. (Fabaceae)

Syn : *Crotalaria anthylloides* Don, *C. linearis* Herb., *C. roxburghiana* DC., *C. stricta* Roxb.

Vernacular names : Mun : Gara sokoe, Huring, Pirijiring; **Orn :** Ghugra.

Traditional use : *MUNDA* : *Pant* : in *adhkapar* (pain in one side of the forehead); *SANTAL* : in curbancle, cholera, dysentery, haematuria, pain, syphilis, venereal sores.

Phytography : Annual herb, 30–100 cm high, unbranched or with a few slender ascending branches, clothed with silky hairs; leaves linear or lanceolate, moderately firm, 5–10 cm long, glabrous above, obscurely silky beneath, stipules very minute, deciduous; racemes terminal, 2- to 12-flowered, rarely reaching 15 cm in length; flowers may also be solitary, flowers pale yellow, less than 2.5 cm in length, calyx densely covered with long silky brown hairs; pods sessile, 1.25–1.80 cm long, 20- to 30-seeded.

Phenology : *Flowering* : Rainy season; *Fruiting* : October–November.

Distribution : Tropical regions of India from Himalayan foothills to Karnataka, ascending up to 1800 m; Sri Lanka.

Ecology and cultivation : Tropical plant; wild.

Crotalaria juncea L. (Fabaceae)

(2n = 16)

Syn : *Crotalaria benghalensis* Lamk., *C. sericea* Willd.

English name : Sun hemp.

Sanskrit name : Shana.

Vernacular names : **Asm** : Shon; **Ben** : Shan; **Guj**, **Hin** and **Pun** : Sanai; **Kan** : Apsenabu, Sanna senabu; **Kon** : Khulkhuli; **Mal:** Thanthalakotti; **Mar** : Janalka nar, Khulkhula; **Mun** : Jirin; **Orn** : Sann; **Ori** : Channapata; **Sad** : Sanai; **Tam** : Canal, Sanappai; **Tel** : Jannamu.

Trade names : Shan, Shanai.

Traditional use : *TRIBES OF BIHAR* : *Leaf* : in antifertility; *TRIBES OF SIWALIK* (Uttar Pradesh): *Seed* : as blood purifier; *TRIBES OF MOUNT ABU* (Rajasthan): *Leaf* : in cough; *Seed* : for blood purification and skin diseases. *AYURVEDA* : it is a medicinal plant.

Phytography : Stiff shrub, 1.3–2.0 m high; branches slender, thinly silky; leaves simple, rather distant, firm, linear, 3.7–7.5 cm long, shining on both surfaces, silky brown hairs present; racemes terminal, loosely 12- to 20-flowered, nearly 30 cm long; flowers yellow, 5 cm long; pods 2.5 to 3.0 cm long, clothed with persistent silky hairs, 10- to 15-seeded.

Phenology : *Flowering* : August–January; *Fruiting* : December–February.

Distribution : From Himalayan foothills to Kanyakumari; Sri Lanka.

Ecology and cultivation : Wild; also cultivated for its fibre.

Chemical contents : *Seed* : phytoagglutinin; quercetin and alkaloids from tissue culture.

Remarks : The plant is known to cause hepatic veno-occlusive diseases originating in Equador. Cultivated in Saurashtra (Gujarat) for fodder. In Bihar, the top leaves are smoked for intoxication; prepared with milk, they are called *bhang*. Fibre obtained from the plant is used as cordage.

Crotalaria linifolia L. f. (Fabaceae)

Syn : *Crotalaria montana* Heyne, *C. sobolifera* Grah., *C. melanocarpa* Wall.

Vernacular name : Mun : Bir mindi.

Trade name : Bir mindi.

Traditional use : MUNDA : *Root* : against epilepsy.

Phytography : Annual erect herb, 20–60 cm high; branches clothed with short, adpressed, silky hairs; leaves simple, short-stalked, softly silky, exstipulate, linear; spikes terminal, 1.25–30 cm long; flowers pale yellow; pods black when mature, 8- to 10-seeded.

Phenology : *Flowering* : September–October; *Fruiting* : October–November.

Distribution : Sikkim, Bihar, Western Ghats; Sri Lanka, Myanmar.

Ecology and cultivation : Tropical plant; wild.

Remark : The plant is used to remove maggots from wounds of cattle.

Crotalaria pallida Dry. (Fabaceae)

(2n = 16)

Syn : *Crotalaria striata* DC.

Vernacular names : Mik : Runche; **Tot :** Thangkur.

Traditional use : MIKIR : *Root* : in swelling on joints; *Leaf* : as vermifuge.

Modern use : *EtOH* (50%) *extract of plant* : hypotensive; *Seed* : antitumour against Walker 256 and sarcoma 180 rat tumours.

Phytography : Erect herb, 60–120 cm high; branches robust, sulcate, thinly silky; leaves trifoliate, stipules minute, deciduous, petiole 5.0–7.5 cm, leaflets membranous, upper surface green and glabrous, lower surface pale and obscurely silky; racemes 15–30 cm long, 20- to 50-flowered; flowers yellow, striped with red; pods recurved, 3.7–5.0 cm long, 20- to 30-seeded.

Distribution : Throughout India; Sri Lanka.

Ecology and cultivation : Tropical plant; wild.

Chemical contents : *Stem-bark* and *Leaf* : flavonoids; *Seed* : mucronative, usaramine, nilgirine, crotastriatine.

Remark : Totos make cordage with stem fibre.

Crotalaria prostrata Rottl. ex Willd. (Fabaceae)
(2n = 16)

Vernacular names : Manb : Chotojhunjhuni; **Mun :** Nari sokoe; **Ori :** Vishnukarni.

Trade name : Vishnukarni.

Traditional use : TRIBES OF PURULIA (West Bengal) : *Whole plant* : after child birth; MUNDA : *Root* : against colic, toothache; TRIBES OF ORISSA : *Plant* : as antiemetic.

Phytography : A diffuse herb; stem slender, covered with silky yellow-brown hairs, 15–30 cm log; leaves simple, nearly sessile, 1.25–3.7 cm long, pale; racemes few-flowered; flowers 0.4 cm, yellow; pods nearly sessile, linear-oblong, 1.25–1.50 cm long, 12- to 15-seeded.

Phenology : *Flowering* and *Fruiting* : September–December.

Distribution : Plains of India, ascending up to 2000 m in the Chenab Valley; Sri Lanka.

Ecology and cultivation : Wild.

Crotalaria retusa L. (Fabaceae)
(2n = 16)

Syn : *Lupinus cochinchinensis* Lour.

English name : Rattle wort.

Vernacular names : Asm : Ghanta korna; **Ben :** Atasi; **Hin :** Jhunjhunia; **Kan :** Gijigiji gida; **Kon :** Ghagri; **Mal :** Thanthalakotti; **Mar :** Ghagri; **Ori :** Junka; **Pun :** Jhanjhanian; **Tel :** Pattigiligisha.

Modern use : *EtOH* (50%) *extract of aerial parts* : spasmolytic, depressor of Central Nervous System.

Phytography : Robust undershrub, 1.0–1.3 m high; branches stout, striated, glabrous; leaves short-stalked, stipules minute, lamina oblanceolate-oblong, 3.7–7.5 cm long, racemes laxly 10- to 12- flowered; flowers yellow, tinged with purple, nearly 2.5 cm; pods linear-oblong, 2.5–3.5 cm, 15- to 20-seeded.

Distribution : Tropical regions of India; Sri Lanka.

Ecology and cultivation : Tropical plant; usually wild, often grown as a garden plant.

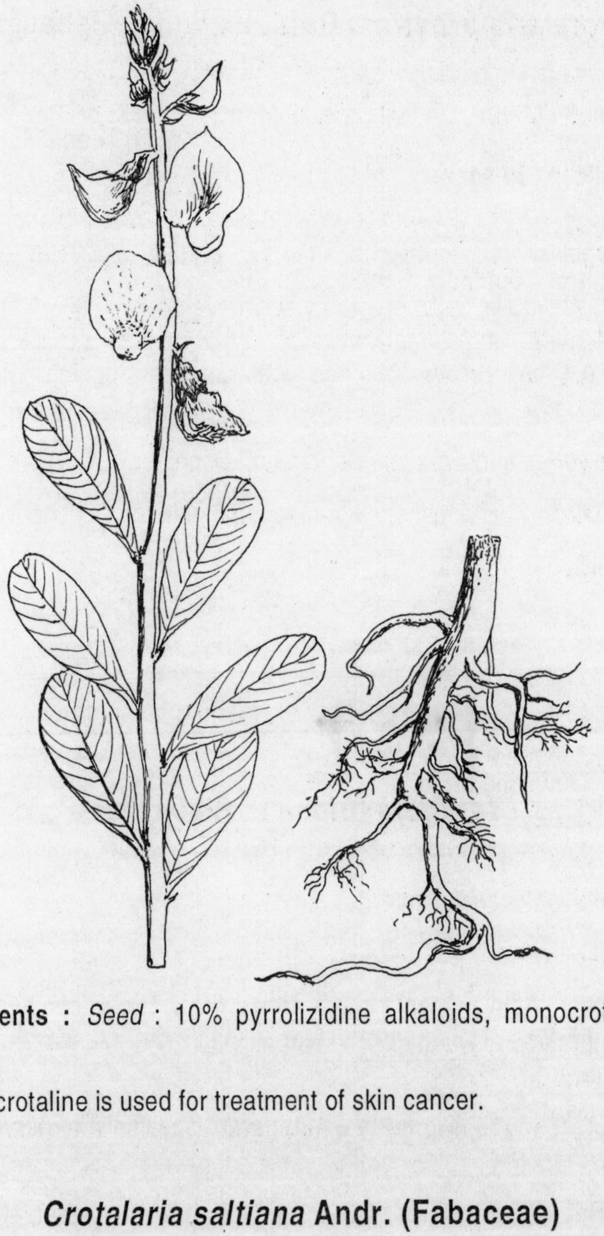

Chemical contents : *Seed* : 10% pyrrolizidine alkaloids, monocrotaline, retusine, spectbiline.

Remark : Monocrotaline is used for treatment of skin cancer.

Crotalaria saltiana Andr. (Fabaceae)

Syn : *Crotalaria striata* DC., *C. pisiformis* Guill. & Pers.

Vernacular names : Hin : Jhunjhunia; **Mun :** Korra.

Traditional use : *ETHNIC COMMUNITIES OF CHOTANAGPUR* (Bihar) : *Leaf* : in intestinal troubles; *MUNDA* (of Bihar) : *Root* : in leprosy and urine trouble.

Phytography : Erect low shrub, 1.0–1.6 m high; branches slender, silky; leaves alternate, petiolate, stipulate, imparipinnate, leaflets membranous, 7.5–10.0 cm long; green and glabrous above, pale and obscurely silky below; racemes 20- to 50-flowered, up to 33 cm long; flowers yellow; striped with red or brown, 1.6 cm; pods 3.7-5.0 cm long, rather reflexed, 20- to 30-seeded.

Phenology : *Flowering* and *Fruiting* : October–April.

Distribution : Throughout India; Bangladesh, Sri Lanka.

Crotalaria spectabilis Roth (Fabaceae)

(2n = 16)

Syn : *Crotalaria sericea* Retz., *C. macrophylla* Wei., *C. Cuneifolia* Sch.

Vernacular names : Mun : Haradijiling; **Orn :** Xar jhunka.

Traditional use : *TRIBES OF HAZARIBAGH* (Bihar) : *Plant* : in rheumatism; *MUNDA* : *Root* : given after childbirth; *TRIBES OF ALMORA AND SIWALIK* (Uttar Pradesh) : *Plant* : in scabies; *TRIBES OF BASTAR* (Madhya Pradesh) : *Stem* : anthelmintic.

Modern use : *EtOH* (50%) *extract of plant* : spasmolytic.

Phytography : Handsome bright green herb, 1.3–1.6 m high; leaves simple, oblanceolate-oblong, moderately firm, 7.5–15.0 cm long; inflorescence lax, 20- to 40-flowered, 30 cm long, bracts persistent; flowers 2.5 cm long, bright yellow or purplish; pods stalked, 2.5–5.0 cm long.

Phenology : *Flowering* : September–February; *Fruiting* : December–May.

Distribution : Tropical parts of India, ascending up to 1000 m in Kumaon.

Ecology and cultivation : Plant of tropical forests; wild.

Remarks : Tribes of West Bengal, Bihar and Orissa use the plant as antidote to poisoning of cattle, but the plant is known to cause outbreak of hepatic disease in cattle, horses, pigs and poultry. Rats fed on this species show pulmonary arterial changes.

Croton bonplandianum Bail. (Euphorbiaceae)

(2n = 16)

Syn : *Croton sparsiflorus* Morong.

Traditional use : *TRIBES OF EASTERN RAJASTHAN* : *Latex* : to heal cuts and wounds.

Modern use : *EtOH* (50%) *extract of plant* : hypotensive, spasmolytic; *Leaf* : antibacterial.

Phytography : Perennial herb, green, laticiferous; leaves simple, alternate, petiolate; racemes usually terminal, may be lateral, monoecious; flowers small, white, unisexual; fruits deciduous, 2-valved cocci; seeds small, smooth, albuminous.

Phenology : *Flowering* and *Fruiting* : almost throughout the year.

Distribution : Throughout the plains of India; Bangladesh and all other countries of South Asia.

Ecology and cultivation : : Mesophyte; wild, mainly in wastelands and cultivated fields.

Chemical contents : *Plant* and *Leaf* : sparsiflorine, crotonoflorine, crotsparinine, isoquinoline dienone, proaporphine, β-sitosterol, taraxerol, vomifoliol, ursolic acid, rutin (only in leaf); *Seed* : cocarcinogen, norsinoacutine, 3-OMe-4,6-di-OH-morphinandien 7-one, phorbol diesters, phorbol triesters.

Croton caudatus Geisel (Euphorbiaceae)

(2n = 20)

Syn : *Croton denticulatus* Blume, *C. drupaceous* Roxb., *Tiglium hispidium* Klotzsch.

Vernacular names : Aru : Sawaka; **Garo :** Sonaphula; **Tot :** Tiliker-rik.

Traditional use : TRIBES OF ARUNACHAL PRADESH : *Stem* and *Leaf* : in cold and cough; GARO : *Root-decoction* : in malaria; TOTO (of West Bengal) : *Leaf* : on sprain.

Modern use : *EtOH* (50%) *extract of aerial parts* : hypotensive; *EtOH* (50%) *extract of fruit* : depresses Central Nervous System, diuretic, hypotensive.

Phytography : Scandent shrub; branches scurfily stellately hairy; leaves alternate, simple, very variable in size and shape, upper surface smooth or scaberulous, lower scabrid or tomentose, glands present; racemes solitary, terminal, 10–25 cm long, slender, monoecious; flowers white, filaments silky hairy below, ovary stellately woolly; capsules large, globose or broadly oblong, woody, densely rusty, pubescent, 6-valved from top downwards, many-seeded; seeds very variable.

Distribution : Meghalaya, Arunachal Pradesh, Deccan; Bhutan, Bangladesh (Sylhet), Sri Lanka.

Ecology and cultivation : Plant of lower hills; wild.

Chemical contents : *Stem-bark* : dotriacotanol, β-amyrin, β-sitosterol, crotocandin, teucvidin, isocrotocaudin.

Remark : Totos use the seeds as masticatory.

Croton roxburghii Balak. (Euphorbiaceae)

Syn : *Croton oblongifolius* Roxb.

Vernacular names : Mun : Kuti; **Orn** and **Sad :** Putri.

Traditional use : *NAGA* : *Seed* : as purgative; *TRIBES OF PURULIA* (West Bengal) : *Root* : in rheumatism; *Latex* : in ringworm; *Leaf* : in wounds; *TRIBES OF MAYURBHANJ* (Orissa) : *Root* : in constipation, diarrhoea, dysentery; *TRIBES OF ORISSA* : *Root* : in sprain; *Bark* : in stomach troubles; *TRIBES OF SANTAL PARGANAS* (Bihar) : *Root* : in antifertility, snake bite, sore throat; *TRIBES OF HAZARIBAGH* (Bihar) : *Twigs* : in eye complaints, cataract, skin diseases; *TRIBES OF SINGHBHUM* (Bihar) : *Root* : in constipation, diarrhoea, dysentery; *MUNDA* : *Root* and *Bark* : as purgative; *SANTAL* : *Various parts* : in dog/jackal bite, cholera, epilepsy, fever, foul breath, headache, hydrocele, insanity, neuralgia, pleurisis, pneumonia, scabies; *BHIL* (of Madhya Pradesh) : *Root* : in stomach disorder, vomiting; *TRIBES OF MADHYA PRADESH* : *Root* : in snake bite, *Leaf* : in boils.

Modern use : *EtOH* (50%) *extract of Plant* : hypotensive.

Phytography : Small deciduous tree; leaves simple, alternate, petioles slender, 1.25–6.5 cm long, lamina 10–30 cm, leathery, very pale green when dry; racemes erect, often fascicled; flowers unisexual, 0.6 cm long; capsules 0.8 cm in diameter, globosely trilobed.

Phenology : *Flowering* : January–February; *Fruiting* : April.

Distribution : Nagaland, West Bengal, Bihar, Central India, the Deccan peninsula; Bangladesh, Myanmar, Sri Lanka.

Ecology and cultivation : Normally grows in hot, dry places; wild.

Chemical contents : *Bark* : β-sitosterol, oblongifoliol, deoxyoblongifoliol, oblongifolic acid.

Remarks : Mundas eat the young shoots. The wood is used to make gunpowder.

Croton tiglium L. (Euphorbiaceae)

(2n = 20)

Syn : *Tiglium officinale* Klotzsch.

English name : Croton seeds.

Sanskrit name : Jayapaala.

Vernacular names : Ben : Patabahar; **Garo :** Runibih; **Hin :** Patabahar, Jamalgota; **Kha :** Chicoc.

Trade names : Patabahar, Jamalgota.

Traditional use : *GARO AND KHASI* : *Plant* : in stomach disorders.
AYURVEDA : a medicial plant.
SIDDHA : *Kernal* : used in preparation of *Nervalam.*
UNANI : an ingredient of 'Ambe Mezhugu'–useful in rheumatism; as well as in preparation of 'Habbaus Salateen'.
HOMOEOPATHY : for cholera, skin diseases, urinary troubles and hiccup.

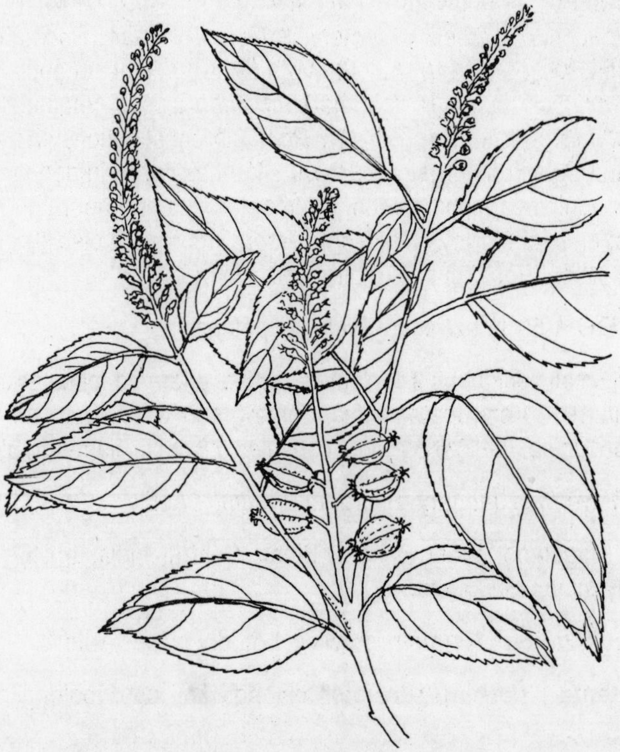

Phytography : Small evergreen tree, youngest shoots sparsely stellately hairy; petioles slender, 2.5–5.0 cm long, leaves 5–10 cm long, membranous, glandular beneath; racemes 5.0–7.5 cm long; flowers small, unisexual; capsules nearly 2.5 cm long, white, obtusely trigonous; seeds nearly 2 cm, oblong, pale, obtusely trigonous.

Distribution : Meghalaya, Assam, West Bengal.

Ecology and cultivation : Plant of tropical rainforest; wild.

Chemical contents : *Seed* : β-sitosterol, haemolytic protein, crotin-I; *Seed -oil* : a series of toxic irritating cocarcinogenic diesters of tigliane type of diterpene poliol, phorbol with acetic acid, 2-Me-butyric acid, tiglic acid and myristic acid.

Remarks : Plant-extract is inhibitory against P-388 lymphocytic leukemia in mice, increases intestinal motility and produces intestinal relaxant action in albino rat. Seed-oil and resin are insecticides.

Cryptolepis buchanani Roem. & Schult. (Asclepiadaceae)

Syn : *Cryptolepis reticulata* Wall., *Echites reticulata* Roth, *Nerium reticulatum* Roxb.

Vernacular names : Hin : Karanta; **Mun :** Bhains jambura nanri, Merom diring, Ṭonang utu; **Sad :** Bhains lakhan; **San :** Utri dudhi lota.

Trade name : Sariva (a chemical from the stem).

Traditional use : *TRIBES OF PURULIA* (West Bengal) : *Plant* : in venereal diseases; *MUNDA* : *Root* (mixed with oil of *Pongamia pinnata*) : rubbed on the body to cool it down; *SANTAL* : *Plant* : in body pain, dysentery, snake bite; *Root* : in anasarca, ascites, dropsy, sores; *Latex* : in cholera; *TRIBES OF CHOTANAGPUR* (Bihar) : *Root* : in intermittent fever, stomachache; *TRIBES OF ORISSA* : *Root* : for lactation, in rheumatism; *Latex* : in cuts, rheumatism; *TRIBES OF UPPER EAST GODAVARI* (Andhra Pradesh) : *Plant* : in rickets in children; *TRIBES OF DHARAMPURA FOREST* (Tamil Nadu) : *Latex* : on fissures of soles; *TRIBES OF MADHYA PRADESH* : *Root* : for lactation; *DANG* (of Gujarat) : *Root* : in abdominal pain; *TRIBES OF SIWALIK* (Uttar Pradesh) : *Plant* : in rickets in children.

Modern use : *EtOH* (50%) *extract of root* : antiamphetaminic, Central Nervous System depressor, hypotensive.

Phytography : Twining, evergreen, glabrous shrub; leaves simple, opposite, petiole 1.25 cm, lamina 7.5–15.0 cm by 2.5–6.2 cm, leathery, shining above, acuminate; cymes very shortly peduncled, paniculate; flowers white, tubular, 1.25 cm in diameter; follicles 5–10 cm long, straight, rigid; seeds compressed.

Phenology : *FLOWERING* : May–June; *Fruiting* : December–February.

Distribution : Kashmir to Assam, ascending the Himalaya up to 1300 m, Karnataka; Sri Lanka.

Ecology and cultivation : Wild.

Chemical contents : *Plant* : buchanamine; *Stem* : nicotinoyl glucoside; *Leaf* : α- & β-amyrin.

Remark : In Sri Lanka, the plant is used in fractures.

Cucumis callosus (Rottl.) Cogn. (Cucurbitaceae)

Syn : *Cucumis trigonous* Roxb., *Bryonia callosa* Herb.

Vernacular names : Mun : Jaiti dimbu; **Orn :** Dimbo.

Traditional use : *TRIBES OF KOLHAPUR* (Maharashtra) : *Fruit* : in wounds; *TRIBES OF MOUNT ABU* (Rajasthan) : *Fruit* : in skin diseases.

Modern use : *EtOH extract of fruit* : analgesic, anti-inflammatory.

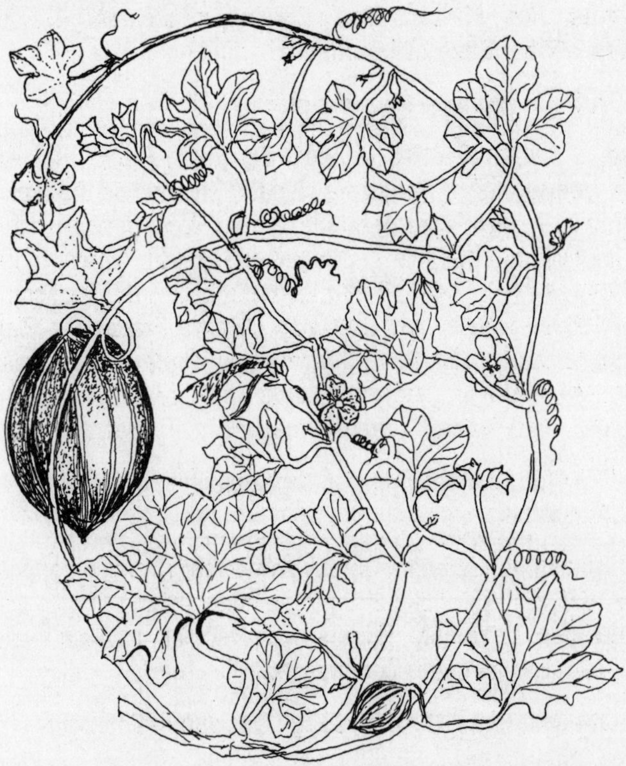

Phytography : Climbing herb, may be procumbent and trailing; stem scabrid; tendril simple; leaves simple, alternate, petiolate, palmately 5-angled, may be 3- to 7-lobed, 2.5–5.0 cm in diameter; flowers yellow, unisexual, monoecious, males clustered in the axils, females solitary; fruit fleshy, indehiscent, spherical, green with white stripes; seeds many, compressed.

Phenology : *Flowering* and *Fruiting* : August–October.

Distribution : Throughout India.

Ecology and cultivation : Grows on cultivated high ground, especially among the cotton plants; wild.

Chemical contents : *Fruit* : proteolytic enzyme, steroid, triterpenoid compounds, cucurbitacin B.

Remarks : Tribes of Mount Abu (Rajasthan) use leaves to destroy lice on head. Tribes of Salsette Island (near Mumbai) eat fruits as a vegetable.

Cucumis melo L. (Cucurbitaceae)

(2n = 20, 22, 24)

Syn : *Cucumis utilissimus* Roxb., *C. flexuous* L., *C. chata* Wall.

English names : Muskmelon, Sweet melon.

Vernacular names : Asm : Bangi; **Ben :** Phuti, Bangi; **Guj :** Tarbuch; **Hin** and **Pun :** Kharbuza, Kakri, Phuti; **Kan :** Balli, Ibbudlu, Kaarabuja, Kekkarike; **Kon :** Chibud, Chiboll, Melanv; **Mal :** Thannimathai; **Mar :** Chibunda, Kharbuj; **Mun :** Birataear; **Orn :** Phutkhira; **Ori :** Kharbuja; **Sad :** Kanghera; **Tam :** Mulam, Velapandu; **Tel :** Vellari verai, Velopatam.

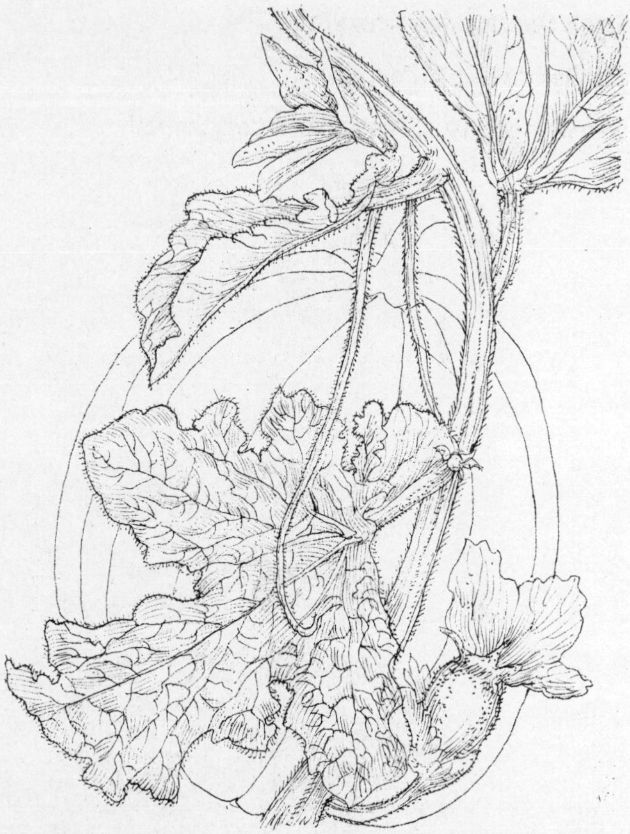

Trade names : Kharbuja, Kakri.

Traditional use : RURAL FOLK OF KASHMIR : *Seed* : as cooling agent.
UNANI : *Fruit-pulp* and *juice* : demulcent, useful in tan freekles.

Modern use : *Root* : emetic and purgative; *Seed-extract* : diuretic; *Seed* ($NaHCO_3$ and HCL extracts) : anticancer, against ehrlich ascites tumour cells; *Seed-oil* : useful in urinary infections.

Phytography : Usually a procumbent annual creeper, may be climber; stem green, scabrous, branched; tendril simple; leaves simple, alternate, petiolate, pentangular or slightly lobed, scabrid on both surfaces; flowers yellow, unisexual, monoecious, males clustered in axils, females solitary; fruits fleshy, indehiscent, spherical, sometimes with sweet smell; seeds many, compressed.

Phenology : *Flowering* and *Fruiting* : Summer.

Distribution : Throughout India; Bangladesh, Pakistan and South Asia.

Ecology and cultivation : Tropical plant, prefers sandy loam and sunny hot days; cultivated in most hot countries.

Chemical contents : *Fruitstalk* : cucurbitacin B and E; *Fruit* : caffeic acid, chlorogenic acid, ferulic acid, oil; in Korean plants, β-sitosterol.

Cucumis sativus L. (Cucurbitaceae)

(2n = 14, 24, 28)

Syn : *Cucumis muricatus* Wall.

English name : Cucumber.

Sanskrit names : Sukasa, Trapusha, Trapushi.

Vernacular names : Asm : Tianh; **Ben :** Shasha; **Guj :** Kakri; **Hin :** Khira; **Kan :** Mullusante, Southe kayi; **Kon :** Touxin; **Mal :** Mullenvellari, Mulluvelarikak; **Mar :** Kakri; **Mun :** Taear; **Orn :** Pal-xanja; **Ori :** Kakudi; **Pun :** Khira; **Sin :** Pipina; **Tam :** Mullivelarika, Pipini; **Tel :** Dozakaia.

Trade names : Khira, Shasha, Kakri.

Traditional use : ANGAMI NAGAS : *Fruit* : in whooping cough; BHOXA (of Uttar Pradesh) : *Leaf* : in fever; KUMAONI : *Fruit* : in pneumonia.
CHARAKA SAMHITA and SUSHRUTA SAMHITA : *Fruit* and *Seed* : useful in urinary complaints, in diseases caused by *kapha* (phlegm) and *vata* (wind), in stones.
SIDDHA : *Seed* : used to prepare *Vellari.*
UNANI : an ingredient of 'Khiyar'.

Modern use : *Fruit* : in blood sugar and indigestion.

Phytography : Annual climber; stem scabrous; tendrils simple; leaves alternate, petiolate, simple, ovate, pentangular or slightly lobed, hispidulous on both surfaces, 7.5 –12.5 cm in diameter; flowers yellow, unisexual, monoecious, 2.5 cm in diameter; fruits green, commonly cylindric, grows up to 30 cm in length; seeds white, many, compressed.

Phenology : *Flowering* : Rainy season; *Fruiting* : almost throughout the year.

Distribution : Throughout India; Bangladesh, Pakistan and South Asia.

Ecology and cultivation : Grows in warm and warm temperate climates, prefers sandy soil; cultivated and wild.

Chemical contents : *Leaf* : free cucurbitasides B and C, ferredoxin, stigmast -en-3β-ol, α-spinosterol; *Flowers* (Male and Female): *sterols; Fruit* : rutin, cucurbitin, volatile compounds; *Seed* : glucosides including cucurbitaside; *Seedling* : free and bound sterols.

Remark : Fruits are eaten as vegetable and salad.

Cucurbita maxima Duch. ex Lam. (Cucurbitaceae)

(2n = 24, 40, 44, 48)

English names : Pumpkin, Sweet gourd.

Vernacular names : Asm : Ronga alu; **Ben :** Mitha kumra; **Hin :** Mitha kaddu; **Kan :** Seegumbala; **Kon :** Dudin; **Mal :** Mathangai; **Mar :** Kala bhopala; **Ori :** Mitha kokharu; **Pun :** Halwa kaddu; **Tam :** Pushanikai; **Tel :** Gummadi.

Trade names : Mitha kaddu, Misti kumra.

Traditional use : KORKU (of Maharashtra) : *Seed* : in intestinal worms.

Modern use : *Fruit* : emollient, refrigenant, sedative, useful in haemoptysis, pulmonary haemorrhage; *Seed* (extract) : anthelmintic *in vivo* and *in vitro*; *Seed-oil* : in migraine and neuralgia.

Phytography : Annual, large, climbing or trailing herb; stem green, pentangular, ribbed and furrowed, covered with white hairs, branched; tendril divided into five unequal parts at the top; leaves alternate, simple, petiolate, 5-angled, 10–15 cm in diameter, petiole and leaf blade covered with soft white hairs; flowers unisexual, monoecious, yellow, 7.5–10.0 cm long; fruiting peduncle stout, corky, striated, pepos large, fleshy, green when young, reddish brown when matured, many-seeded; seeds white, large, exalbuminous, compressed.

Phenology : *Flowering* and *Fruiting* : late Winter.

Distribution : Throughout India; Bangladesh.

Ecology and cultivation : Cultivated in warm and temperate areas.

Chemical contents : *Fruit* : vitamin A; *Seed* : oil, sterols, triterpenoids.

Remarks : Stem, leaf, flower and fruit are eaten as vegetables. Shell of the ripened fruit used to make utensils by tribes of Santal Parganas, and also used to make musical instrument.

Cucurbita moschata (Dich.) Poir. (Cucurbitaceae)
(2n = 24, 40, 48)

Syn : *Cucurbita maxima* W.A., *C. melopepo* Roxb., *C. camolenga* Wall.

English name : Muskmelon.

Traditional use : NAGA : *Fruit* : to stop bleeding; SANTAL : *Fruit* : in asthma, bronchitis, headache, phthisis; TRIBES OF EASTERN RAJASTHAN : *Peduncle* and *thalamus of mature fruit* : as antidote to scorpion sting.

Modern use : *Fruit* (pulp) : for boils, ulcers; *Fruit* (dried pulp) : in haemoptysis, haemorrhage from pulmonary organs; *Seed* : anthelmintic against tapeworm, diuretic, inhibits growth of young worms of *Sehistosoma japonica,* useful in gonorrhoea and urinary diseases.

Phytography : Annual, large, climbing herb; stem hispid or hairy; tendrils multifid; leaves simple, alternate, petioled, large, covered with hairs; flowers unisexual, monoecious, yellow; pepos large, many-seeded; seeds compressed, exalbuminous.

Phenology : *Flowering* and *Fruiting* : mostly in Summer.

Distribution : Throughout India; Bangladesh, Pakistan and South Asia.

Ecology and cultivation : Widely cultivated in tropical and subtropical regions.

Chemical content : *Pollen* and *Stigma* : flavonoids.

Cucurbita pepo DC. (Cucurbitaceae)
(2n = 22, 24, 28, 40, 42, 46, 80)

English name : Hungarian pumpkin.

Vernacular names : Ben : Kumra; **Hin :** Konrar, Kumra, Safed kaddu; **Mun :** Kakaru; **Kas :** Gol aal; **Orn** and **Sad :** Konhra.

Trade names : Safed kaddu, Kumra.

Traditional use : RURAL FOLK OF KASHMIR : *Flower* : in diarrhoea.

Modern use : *Seedcoat* : anthelmintic; *Seed-extract* : fungitoxic against *Helminthosporium oryzae*, minor insecticde.

Phytography : Annual, heavy creeper; stem hispid; leaves simple, alternate, petiolate, leaf blade 5-lobed, 10–15 cm in diameter, soft hairs present on petiole and leaf blade;

flowers yellow, unisexual, monoecious, 7.5–10.0 cm long, peduncle of female flower shorter than that of the male flower; fruiting peduncle woody, strongly grooved; pepo large, many-seeded; seeds exalbuminous, compressed.

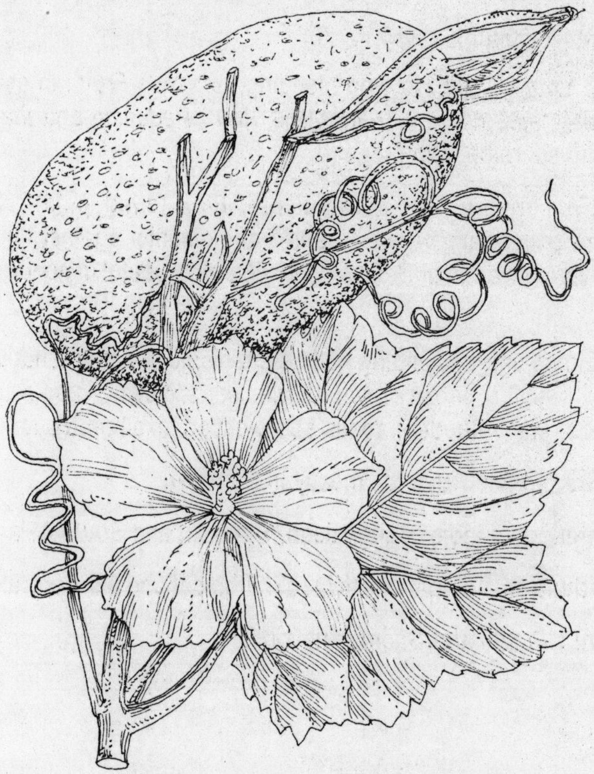

Phenology : *Flowering* and *Fruiting* : mainly in Rainy season.

Distribution : Throughout India; Bangladesh and Pakistan.

Ecology and cultivation : Cultivated in all warm and temperate regions.

Chemical contents : *Stem* and *Leaf* : alkaloids, ulcer preventing vitamin U; *Flower* : flavonol glycerides; *Pollen* and *Stigma* : flavonoids; *Seed :* cucurbitine, stigmasta-7, 24 (28)-dien-3β-ol.

Remarks : Leaves and flowers eaten as vegetables; utensils made with fruits.

Cuminum cyminum L. (Apiaceae)

(2n = 14)

English name : Point caraway.

Sanskrit names : Ajaji, Jiraka, Svetajiraka.

Vernacular names : Ben : Jira; **Guj :** Sufed jirum; **Hin :** Safed jira; **Mar :** Jira; **Tam :** Shiragam; **Tel :** Jiraka, Jirana.

Trade name : Jira.

Traditional use : *TRIBES OF SAGAR DISTRICT* (Madhya Pradesh) : *Fruit* : in diarrhoea, dysentery, inflammation of breasts and testicles; *NAYADI* (of Kerala) : *Fruit* : in stomachache; *IRULAR, KOTA AND TODA* (of Tamil Nadu) : *Seed* : to reduce labour pain. *BRAHMAVAIVARTA PURANA* : helps in removing phlegm; *AGNI PURANA* : plant of great medicinal value.
SIDDHA : *Dried fruit* : ingredient of *Chitrakam.*

Modern use : *EtOH* (50%) *extract of fruit* : spasmolytic, hypotensive; *Fruit extract* : inhibits *Clostridum botulinum*, toxin production by *Aspergillus*; *Essential oil* (from fruit) : antibacterial, larvicidal; *Essential oil* (from seed) : antifungal, antimicrobial.

Phytography : Slender annual herb; stem green, furrowed, usually with hollow internodes, glabrous; leaves green, alternate, petiolate, twice or thrice tripartite, ultimate segments filiform; umbels compound, many bracts and bracteoles, rigid, linear; flowers white, small, bisexual; fruits cremocarps; seeds many, albuminous.

Phenology : *Flowering* and *Fruiting* : Winter.

Distribution : Punjab, Decan, widely cultivated in the plains of Indian subcontinent.

Ecology and cultivation : Plains and hills; cultivatd for condiment.

Chemical contents : *Fruit* : cuminin, volatile oil, fixed oil, proteins, flavonoids, glucopyranosides of luteolin and apigenin; *Seed* : fat, 5,6-octadeconic acid.

Remark : Fruits are used as spice.

Curculigo capitulata (Lour.) Kuntze (Hypoxidaceae)

Syn : *Curculigo recurrata* Dry.

Vernacular names : Nag : Koritong, Stitamat.

Traditional use : *NAGA* : *Rhizome* : in eye trouble, vaginal diseases; *ANGAMI NAGAS* : *Rhizome* : in eye trouble, haematuria; *WAR JAINTIA* : *Shoot* : in sprain.

Phytography : Rootstock tuberous; leaves very variable, 60–90 cm by 7.5–15.0 cm, recurved, narrowed into a channelled petiole, 30–60 cm long, villous below, glabrous above; scapes 7.5–22.0 cm long, stout or slender; raceme decurved, subcapitate; flowers 5–10 cm in diameter; berries 0.4-0.8 cm in diameter; seeds black, grooved.

Phenology : *Flowering* and *Fruiting* : in Spring.

Distribution : Meghalaya, Nagaland; Bangladesh (Chittagong), Nepal, Myanmar, Sri Lanka.

Ecology and cultivation : Plant of tropical Himalaya; wild.

Curculigo orchioides Gaertn. (Hypoxidaceae)

(2n = 18)

Syn : *Curculigo brevifolia* Dry., *C. malabarica* Wight, *Hypoxis orchioides* Kurz

Sanskrit names : Mushali, Talamuli, Bhumitala.

Vernacular names : Guj and **Hin :** Kali musli; **Kan :** Nelatale; **Mal :** Nelappanakighanna; **Mun :** Sereng jadu; **Orn :** Dhakira, Kinda; **Ori :** Talamuli; **Sad :** Dinda kita; **Sin :** Heen-bin-tal; **Tam :** Nilapani; **Tel :** Nilaptali guddalu.

Trade name : Kali musli.

Traditional use : *TRIBES OF PURULIA* (West Benal) : *Root* : in fever, filaria, venereal diseases and as tonic; *TRIBES OF ORISSA* : *Root* : in ear complaints, bleeding through nose; *Seed* : as galactogogue; *TRIBES OF SANTAL PARGANAS* : *Root* : as antidote to scorpion sting, snake bite, etc.; *TRIBES OF HAZARIBAGH* : *Root* : in epilepsy, leucorrhoea; *TRIBES OF BASTAR* (Madhya Pradesh) : *Rhizome* : in unconsciousness of children; *BHIL* : *Root* : in impotency; *TRIBES OF ARAKA VALLEY* : *Rhizome* : in eye complaints; *TRIBES OF DAHSAN VALLEY* : *root* : in dysentery, scorpion sting; *TRIBES OF KHED TALUKA* (Maharashtra) : *Rhizome* : in itching; *TRIBES OF EASTERN RAJASTHAN* : *Root* in impotency and as tonic.
AYURVEDA : medicinal plant.
SIDDHA : *Tuber* is an ingredient of *Nilappanai*.
UNANI : ingredient of 'Musli siyat'.

Modern use : *EtOH* (50%) *extract of plant* : anticancer and hypoglycaemic.

Phytography : Herbaceous plant, up to 45 cm high; rootstock tuberous or elongate, sometimes 30 cm long, stout; leaves long- or short-petioled, lanceolate, 15–40 cm by 1.25–2.0 cm, plicate, glabrous, scape 2.5 cm, clavate, flattened, hidden by leaf-sheath; bracts lanceolate, membranous; flowers subsessile, lowest bisexual, rest male, bright yellow, 1.25 cm broad; capsules 1.25 cm, oblong, beaked, 1- to 4-seeded; seeds black, shining, 0.3 cm long, deeply grooved in wavy lines.

Phenology : *Flowering* : May–August.

Distribution : Subtropical Himalaya from Kumaon eastwards, ascending up to 2000 m, Khasia Hills, Manipur, Western Ghats.

Ecology and cultivation : Subtropical plant; wild.

Chemical contents : *Root* : a new aliphatic MeO-ketone, 3-MeO, 5-Ac, 31-tritriacontane; *Rhizome* : glyceride, yuccagenin, lycorine.

Remarks : Tribes of eastern India use rhizome in eye troubles of cattle.
Ethnic communities of hilly regions of Maharashtra, Goa and Santal Parganas eat rhizome and root as vegetables.

Curcuma amada Roxb. (Zingiberaceae)

(2n = 42, 64)

English name : Mango ginger.

Sanskrit name : Karpura haridra.

Vernacular names : Asm and **Ben :** Amada; **Hin :** Am haldi; **Mar :** Amba halada; **Mun :** Dundid-ba, Kead sanga, Keo sanga; **Orn :** Bondo balka; **Ori :** Ambakassia ada; **Sad :** Ban haldi; **Tel :** Miami diallam.

Trade name : Am haldi.

Traditional use : ORAON : *Tikur* (prepared with crushed root mixed with rice flour) : as a sedative in influenza.
UNANI : ingredient of 'Amba haldi'.

Phytography : Herb; rooststock ovoid, sessile tubers thick, cylindric, pale yellow inside; leafy tuft 60–90 cm high, petiole as long as the lamina, leaf blade 30–45 cm by 15 cm or more, pale green, tapering to base and apex; scape from the centre of the leafy tuft, spike 7.5–15.0 cm by 3.7 cm, bracts pale green, those of the coma tinted with pink; flowers as long as the bracts, whitish with pale yellow lip.

Phenology : *Flowering* : August–September.

Distribution : Native of West Bengal, Bihar, Konkan.

Ecology and cultivation : Common in tropical jungles; widely cultivated.

Remarks : The freshly cut rootstock has the smell of green mango, as such, used as aromatic agent in food by the Bengalis.
The tribes of Medinipur (West Bengal) consume the rootstock as vegetable.

Curcuma angustifolia Roxb. (Zingiberaceae)

(2n = 32, 62, 64)

Vernacular names : Hin : Banhaldi; **San :** Paroda.

Trade name : Banhaldi.

Traditional use : NAGA : *Rhizome* : in bone fracture, swelling of body; SANTAL : *Tuber* : in diarrhoea, gravel, fever, subnormal temperature after fever; BHIL : *Rhizome* : in fever.
UNANI : an ingradient for 'Zard chob', 'Zarambad'.

Phytography : Rootstock small, globose; tubers many, oblong, at the end of long fibres; leaves with petiole 30–45 cm, blade lanceolate, green, 15–30 cm by 5.0–7.5 cm; spike with peduncle 15–30 cm, bracts green, ovate, 2.5 cm; flowers 1.25 cm long, yellow.

Phenology : *Flowering* : in Autumn; *Fruiting* : in Winter.

Distribution : Tropical Himalaya from Kumaon eastwards, between 1000 and 1300 m, hills of Bihar; Nepal.

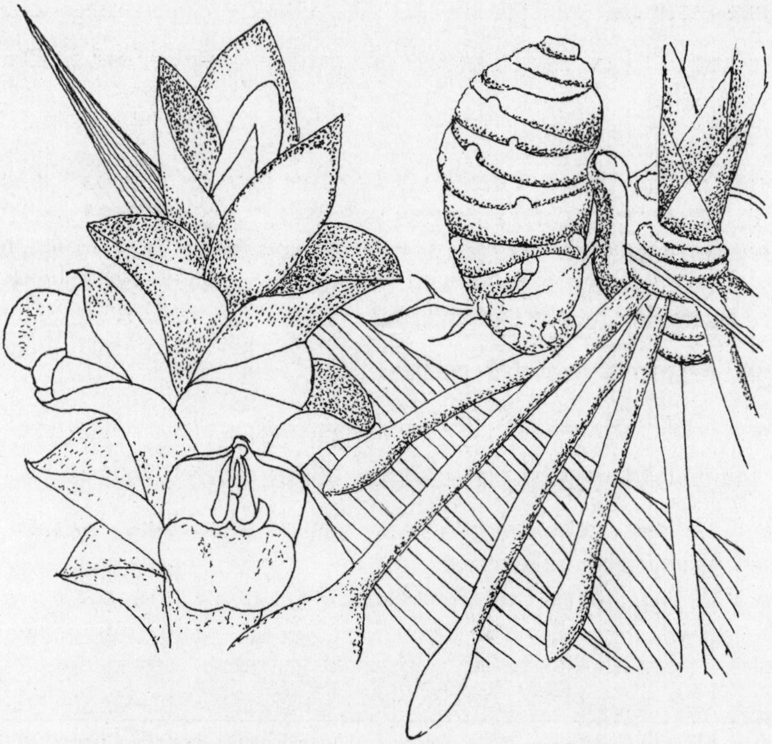

Ecology and cultivation : Plant of tropical hills; wild.

Remark : Tribes of Bastar (Madhya Pradesh) consider it a sacred food.

Curcuma aromatica Salisb. (Zingiberaceae)

(2n = 42)

Syn : *Curcuma zedoaria* Roxb.

English name : Wild turmeric.

Vernacular names : Hin : Bonholdi; **Mun :** Bir sasang, Hatu bundu sasang; **Meg :** Jalam-dike, Jia-lekechiang.

Trade name : Bonhaldi.

Traditional use : *ETHNIC COMMUNITIES OF MEGHALAYA* : *Rhizome* : in child birth, indigestion, intestinal worms, rheumatism and in removing dead foetus from womb; *ETHNIC COMMUNITIES OF CHOTANAGPUR* : *Rhizome* : in fever, pain.
SIDDHA : *Rhizome* : an ingredient of *Kasturi maneal.*
UNANI : an ingredient for preparation of 'Zarambad'.

Modern use : *EtOH* (50%) *extract of rhizome* : spasmolytic; *Rhizome-powder* : anthelmintic; *Essential oil from Rhizome* : antifungal, anthelmintic, antimicrobial, used for treatment of early stages of cervical cancer.

Phytography : Herb; rootstock large, 2.5 cm in diameter, tubers sessile, yellow, aromatic inside; petiole long, green, leaf blades 30–60 cm by 10–20 cm, plain green above or variegated with lighter and darker green, lower surface clothed with persistent soft hairs, may not mature at the time of flowering; spikes with peduncle 30 cm, bracts ovate, pale green, 3.7–5.0 cm, those of coma, tinged with pink; flowers 2.5 cm, upper half funnel-shaped, yellowish.

Phenology : *Flowering* : April–June.

Distribution : Throuoghout India; Sri Lanka and other South Asian countries.

Ecology and cultivation : Tropical plant; wild, cultivated in gardens

Chemical contents : *Essential oil from Rhizome* : curcumol and curdion.

Remarks : War Jaintias consume the shoot as vegetable.

Curcuma leucorhiza Roxb. (Zingiberaceae)

Vernacular name : Hin : Parored.

Trade name : Parored.

Traditional use : *ETHNIC COMMUNITIES OF SINGBHUM* (Bihar) : *Rhizome* : in enlarged liver and spleen, ulcer in stomach.

Phytography : Herb; rootstock large, ovoid; sessile tubers cylindric, sometimes 30 cm long; petiole long, green, leaf blade 30–60 cm by 12.5–15 cm, broad at middle, green and glabrous on both sides; spike vernal, bracts green, 3.7 cm, obtuse; flowers white or pale yellow, shorter than bracts.

Distribution : Bihar.

Ecology and cultivation : Wild; in scrub jungles and wasteland.

Curcuma longa L. (Zingiberaceae)

(2n = 32, 42, 62, 64)

Syn : *Curcuma domestica* Val.

English name : Turmeric.

Sanskrit name : Haridra.

Vernacular names : Asm : Holodhi; **Ben :** Halud; **Guj :** Halada; **Hin :** Haldi; **Kan :** Arisina; **Kon :** Holldi, Ghor hollad; **Mal :** Manjella-kua; **Mar :** Halede; **Mun :** Hatu sasang, Sasang; **Orn :** Balka; **Ori :** Haladi; **Sad :** Haldi; **San :** Oerel sasan; **Tam :** Manjal; **Tel :** Pasupu.

Trade name : Haldi.

Traditional use : *ETHNIC COMMUNITIES OF ARUNACHAL PRADESH AND ASSAM* : *Rhizome* : in migrain; *SANTALS* : (i) *Rhizome* : in hazy vision, inflammation of eye, night blindness, subnormal temperature after fever, spleen consumption, *Basli rog* (pain similar to rheumatism), rheumatism due to draught, lock-jaw, *Rosbi* (stealth convulsions with indistinct speech), chronic scabies, sores and curbuncles, infantile atrophy, indigestion, prolapsus ani and fistula ani, bronchitis, cough and cold, puerperal fever; (ii) *Extract of Rhizome* : in rhagades; (iii) *Bulb* : in drying up of lactation; (iv) *Flower* : in cholera, sores in throat, syphilis; (v) *Dried Flower* : in icterus; *ETHNIC COMMUNITIES OF RANCHI* and *HAZARIBAGH* (Bihar) : (i) *Leaf* : in cold, fever, pneumonia; *GARHWALI* : *Rhizome* : in pimples and feckles on face, wounds, leprosy; *KUMAONI* : *Rhizome* : in cough, insect stings, wounds; *ETHNIC COMMUNITIES OF KURUKSHETRA* (Haryana) : *Rhizome* : in body pain, headache; *TRIBES OF ARAKU VALLEY* (Andhra Pradesh) : *Rhizome* : as anthelmintic.
RIGVEDA : *Rhizome* : improves body complexion and apetite; *YAJURVEDA* : *Rhizome* : is a blood-purifier, improves body complexion; *CHARAKA SAMHITA* : laxative, useful in leprosy

and against contaminuos microbes; *SUSHRUTA SAMHITA* : digestive; *AYURVEDA* : *Rhizome* : effective against bacterial infection, skin diseases, intestinal worms, liver complaints, stammering, filaria, asthma, sprain, boils, wounds, conjuctivitis, thirst due to phlegm, allergic reactions, against leeches, minor ingredient of a drug for malarial fever.

SIDDHA : *Rhizome* : ingredient of *Kappu mancal, Manchal.*

UNANI : ingredient of 'Majnoor-e-falsfa', useful in gastrointestinal complaints; *Powder of rhizome* : used as antifertility agent.

Modern use : *Rhizome* : ingredient of 'Geriforte'—effective in senile pruritis, *Vitafix* —useful in premature ejaculation, insect repellent against houseflies, insecticide, antifungal; *EtOH* (50%) *extract of rhizome* : antiprotozoal, spasmolytic, hypotensive, Central Nervous System depressent; daily consumption of 1 gm raw rhizome helps to fight decaying metabolism and thus prevents cancer; *Essential oil from rhizome* : anti-arthritic, antifungal, anti-inflammatory, antibacterial.

Phytography : Herb with large ovoid rootstock, sessile tubers thick, cylindric, bright yellow inside; petiole 60 cm long, leaves green, 30–45 cm by 10–20 cm; peduncle 15 cm or more long, hidden by sheathing petiole, spikes 10–15 cm, about 5 cm in diameter; bracts pale green, ovate, about 3.7 cm long, those of the coma pale pink; flowers yellow, as long as the bracts.

Phenology : *Flowering* : Autumn.

Distribution : Indigenous to Paresnath (Bihar); widely cultivated in West Bengal and other parts of India; Bangladesh, Sri Lanka.

Ecology and cultivation : Tropical plant; cultivated throughout the tropics.

Chemical contents : *Essential oil from rhizome* : curcumin.

Remarks : In Hawaii, rhizome is used against growth of nostrils, for cleaning blood and as gargle; green rizome is given for whooping and other coughs. In Sri Lanka, Rhizome paste is used in skeletal fracture. Extensive research is being carried out on the nutritional and medicinal value of this plant.

Cuscuta capitata Roxb. (Cuscutaceae)

(2n = 28)

Vernacular names : Lad : Deelasazin, Handathapa.

Traditional use : AMEHI (of Ladakh) : *Plant* : in fever, melancholia and as antiseptic on cuts, wounds and ulcers.
UNANI : material for 'Aftimoon' and 'Kassoos'.

Modern use : *Plant* : in anginal and renal pain.

Phytography : Slender, leafless, twinning, red-coloured parasite; flowers capitate, scarcely 0.3 cm, covered on the exterior with minute papillae; capsules 0.2 cm, thin, fragile, enclosed by unchanged corolla; seeds 4, ellipsoid, brown.

Distribution : From Kashmir to Himachal Pradesh between 2000 and 4000 m.

Ecology and cultivation : Parasite on angiosperms in temperate western Himalay; wild.

Cuscuta chinensis Lamk. (Cuscutaceae)

Syn : *Cuscuta carinata* Br., *C. ciliaris* Hohen., *Grammica aphylla* Lour.

Vernacular names : Mun : Maga ronara, Ronara; **Orn :** Magha banda; **Sad :** Sarguja banda; **San :** Alagjari.

Traditional use : SANTAL : (i) *Root* : in anascara, ascites, dropsy, *gorob sitka* (a peculiar trouble during pregnancy); (ii) *Leaf* : in cancer in face.

Phytography : Delicate, thread-like, pale yellow twiner; pedicel 0.4 cm or less; flowers 0.25 cm or less, green, tubular; capsules 0.25 cm or less, fragile, irregularly breaking, normally 4-seeded.

Phenology : *Flowering* : July–November.

Distribution : Kashmir, Chamba, between 1600 and 2300 m, Bihar, Deccan peninsula; Bangladesh, Sri Lanka.

Ecology and cultivation : A common parasite on *Guizotia abyssinica*; wild.

Remarks : Santals use the plant to treat rinderpest and bone fracture, leaves in dropsy in cattle. In China, the plant is used as diuretic.

Cuscuta reflexa Roxb. (Cuscutaceae)

Syn : *Cuscuta grandiflora* Wall., *C. verrucosa* Sweet, *C. macrantha* G. Don

English name : Dodder.

Sanskrit name : Amaravela.

Vernacular names : Asm : Akashi lota, Ravanar nari; **Ben :** Alokelata, Swarnalata; **Guj :** Akaswel; **Hin :** Akashbel; **Kan :** Badanike, Bandalike, Muditali; **Kon :** Amarval; **Mar :** Amar vel; **Mun :** Ara roanar; **Orn :** Xoxro banda; **Ori :** Nirmuli; **Pun :** Amarbel; **Sad :** Sindur banda; **Tam :** Kodiyagumdal, Poonai-kkali; **Tel :** Lanjasa var-amu, Paniyanakau.

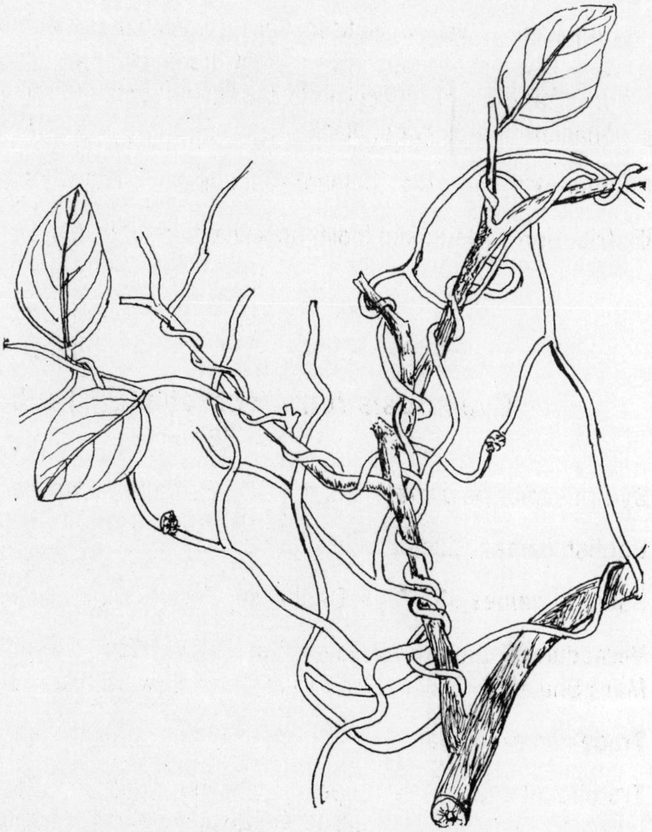

Trade names : Dodder, Swarnalata.

Traditional use : *ETHNIC COMMUNITIES OF NORTH-EAST INDIA* : *Seed* : as purgative; *ETHNIC COMMUNITIES OF MEGHALAYA* : *Plant* : in bodyache; *ETHNIC COMMUNITIES OF SIKKIM* : *Plant* : in bone fracture; *ETHNIC COMMUNITIES OF PURULIA* (West Bengal) : *Stem* : in eye troubles, wounds; *ETHNIC COMMUNITIES OF SANTAL PARGANAS* (Bihar) : *Plant* : in burns, cuts, itching, nervine weakness, skin diseases, wounds; *BHOXA* (of Uttar Pradesh) : *Plant* : in headache; *THARU* (of Uttar Pradesh) : *Plant* : in headache, swelling of body, legs, testicles; *ETHNIC COMMUNITIES OF MIRZAPUR* (Uttar Pradesh) : *Plant* : in cough; *ETHNIC COMMUNITIES OF PITHORAGARH* (Uttar Pradesh) : as anthelmintic, in diptheria, fall of hair; *GARHWALI* : *Plant* : in fever, indigestion, liver complaints; *KUMAONI* : *Seed* : in stomachache; *RURAL FOLK OF KURUKSHETRA* : *Plant* : in oedema; *ETHNIC COMMUNITIES OF RAJASTHAN* : *Plant* : in displaced

kneecap, *Seed* : as carminative; *BHIL* : *Stem* : in jaundice; *ETHNIC COMMUNITIES OF CHANDRAPURA* (Maharashtra) : *Seed* : as purgative.
UNANI : one of the constituents of the drug 'Itrifal Ustakhudus' used for purging viscid matters from stomach and brain.

Modern use : *Plant* (water-soluble portion of fresh alcoholic extract) : relaxant and spasmolytic in guineapig and rabbit; *EtOH* (50%) *extract of aerial parts* : antiviral.

Phytography : A yellow, leafless, fleshy, twining parasite, branches stout, fleshy, forming dense yellow masses on shrubs and low trees; racemes lax, 2.5–10.0 cm long, clustered or subspicate; flowers subcylindric, 0.8 cm long, scales prominent; capsules tough, fleshy, subquadrate; seeds large, black.

Phenology : *Flowering* : October–December; *Fruiting* : December–January.

Distribution : Thoughout India, ascending up to 2600 m.; Bangladesh, Sri Lanka.

Cyamopsis tetragonoloba (L.) Taub. (Fabaceae)

(2n = 14)

Syn : *Cyamopsis psoralioides* (L.) DC., *Psoralla tetragonalobe* L.

English name : Cluster bean.

Sanskrit names : Bakuchi, Dridhabija, Gorani, Ksudrasimbi, Goraksaphalini.

Vernacular names : Hin : Govar; **Guj :** Guwar; **Kan :** Gorikay; **Mal :** Kottavara, Kottamara; **Mar :** Bhavachi, Gowar; **Pun :** Kulti, Guar, Kuwara; **Tam :** Kottavarai.

Trade name : Govar.

Traditional use : *Pods* : appetiser, cooling, digestive, galactogogue, laxative, sweet and tonic and useful in constipation, anorexia, agalactic, nyctalopia and vitiated condition of *pitta*.

Phytography : Erect, robust annual, 30–60 cm high, clothed with appressed greying hairs; stem grooved; leaves trifoliate, leaflets elliptic, acute, sharply dented, pubescent, main nerves prominent; flowers small, purplish in axillary racemes; fruits thick fleshy pods, in clusters, subtetragonal, more or less hairy; seeds 5–12 in each pod, compressed to some extent.

Distribution : Cultivated in many parts of India, Bangladesh and Pakistan.

Ecology and cultivation : Plains, up to 500 m; not known in the wild condition; widely introduced in the drier tropics.

Cyathocline purpurea (Don) Kuntze (Asteraceae)

Syn : *Cyathocline lyrata* Cass., *Tanacetum purpureum* Don

Vernacular names : Mun : Bir buduba, Kuru toropuru.

Traditional use : *AMACHI* (of Ladakh) : *Leaf* plus *Stem* : in intermittent fever; *ETHNIC COMMUNITIES OF MADHYA PRADESH* : *Root* : in stomachache; *ETHNIC COMMUNITIES OF KHEDBRAHMA* (Gujarat) : *Leaf* : in headache.

Modern use : *Plant* : *EtOH* (50%) *extract of Plant* : spasmolytic; *Essential oil* : antimicrobial, anthelmintic, against nodular worms, tapeworms, hookworms, spasmolytic, oxytocic.

Phytography : Erect, odorous, annual herb, 30–60 cm high; stem soft; leaves alternate, pinnatisect, 1.5 cm long, segments teethed; heads 0.4 cm in diameter, yellow usually, may be variously coloured; achenes minute, pappus absent.

Distribution : Kashmir up to 1300 m, Kumaon between 1300 and 1600 m, Uttar Pradesh, Bihar, West Bengal, Assam, Manipur, western peninsula; Nepal, Bhutan, Bangladesh (Chittagong), Myanmar.

Chemical contents : *Essential oil* : lyratol, lyratic acid, isovangustin, a guaianolide.

Cycas rumphii Miq. (Cycadaceae)

Syn : *Cycas circinalis* var. *angustifolia* Miq., *C. macrocarpa* Griff., *Zamia corsoniana* G . Don, *Olus calappoides* Rumph.

English name : Cycas.

Vernacular names : Ben : Bileti khejur; **Kan :** Godduyishalu; **Mal :** Intimpote; **Sin :** Madugaha; **Tam :** Kamappu, Gavala; **Tel :** Kamakshi.

Trade name : Cycas.

Traditional use : *ETHNIC COMMUNITIES OF ANDAMAN AND NICOBAR ISLANDS* : (i) *Young Leaf* : as galactogogue; (ii) *Seed* : in stomachache.

Phytography : An evergreen palm-like tree, 5 m or more high, rarely branched when old; leaves in terminal crown, linear-oblong, pinnate, 1.6–3.0 m long, petiole 45–60 cm long, leaflets linear, 25–30 cm long, about 1.25 cm wide; male cones shortly peduncled, often 45 cm long; carpophylls numerous, crowded round the stem apex, densely woolly; ovules 1–5 in notches on either side of the stalk of carpophyll; seeds elllipsoid or globose.

Distribution : Malabar coast, dry hills in West Tamil Nadu, Andaman and Nicobar Islands; Sri Lanka.

Chemical contents : *Leaf* : biflavones; *Seed* : cycasin, β-glycosidase.

Remarks : Ethnic communities of the Andaman and Nicobar Islands eat the young leaves, fruits and seeds as vegetables.

Cymbopogon citratus Stapf (Poaceae)

(2n = 40, 60)

Syn : *Andropogon citratus* DC., *A. citriodorum* Desf., *A. roxburghii* Nees

English name : West Indian lemon grass.

Sanskrit name : Bhustrina.

Vernacular names : Hin : Gandhatrin, Sugandh rohisha, Aghyaghas; **Mal :** Cennanampullu, Incippullu, Vasanappullu; **Tam :** Vassanapillu.

Trade name : Lemon grass.

Traditional use : *Plant* : aromatic, acrid, anthelmintic, appetizer, alexipharmic, anaphrodisiac, bitter, thermogenic and useful in anorexia, bronchitis, cholera, fever, flatulence, neuralgia, sprains, and vitiated conditions of *kapha* and *vata*.

Phytography : Tall, tufted, perennial with a short rhizome and fascicles of leaves; leaf blades linear, tapering upward to a long setaceous point, glabrous, margin rough, midrib stout below, whitish upward, ligules very short, sheath terete, flowers in spikelets, glumes distinctly concave at the lower half.

Distribution : Cultivated throughout India.

Ecology and cultivation : Cultivated in barren and poor soil along the hilly slopes; a well-drained gravelly soil/sandy loam is suitable for cultivation.

Chemical contents : *Plant* : essential oil, citral.

Cymbopogon jwarancusa Schult. (Poaceae)

Syn : *Andropogon jwarancusa* Jones

Sanskrit name : Lamajjaka.

Vernacular names : Ben : Karankusa; **Hin :** Lamjak; **Mar :** Izkhir; **Pun :** Lamjak.

Traditional use : *Grass* : used to purify blood, in cough, chronic rheumatism and cholera; *Flower* : styptic.
UNANI : an ingredient for preparation of 'Azakhan'.

Modern use : *Plant-extract* : Aromatic tonic in dyspepsia, sudorific in gout, fever and rheumatism.

Phytography : Grass; usually pale, often glaucous green, 15 cm to 2 m in height; stem tufted on an annulate creeping rootstock; leaves flat or convolute, 15–60 cm long, narrow, linear, tips capillary, rigid or flexuous, glabrous or scaberulous, lower surface glaucous or pruinose, midrib slender, sheaths glabrous, lower portion dilated, ligule short, oblong, membranous, ciliate; panicles very variable, densely or loosly branched, spathes longer than the peduncles or the spikes, narrow, lanceolate, spikes 0.6–1.8 cm, spikelets sessile and pedicelled, sessile spikelets pale or reddish, pedicelled spikelets usually longer than the sessile ones, purple, narrow, lanceolate.

Phenology : *Flowering* : April.

Distribution : Western Himalaya from Kashmir to Kumaon, ascending up to 2200 m; Pakistan (West Punjab and Sind).

Ecology and cultivation : Xerophytic.

Chemical content : *Plant* : essential oil.

Cymbopogon martinii (Roxb.) Wats. (Poaceae)

Syn : *Andropogon martinii* Roxb., *A. schoenanthus* L. var. martinii Hk. f., *Cymbopogon martinianus* Schult, *C. pachnodis* (Trin.) Wats.

English names : Rosha grass, Palma rose, Rusa grass.

Sanskrit names : Dhyamakah, Rohisha.

Vernacular names : Hin : Gandhabel, Rusaghas, Gandhijghas; **Mal :** Sambharapullu; **Tam :** Kavattampillu, Munkilpul, Curaippul.

Trade names : Rusaghas, Palmarosa.

Traditional use : *Plant* : acrid, appetiser, bitter, carminative, cardiotonic, digestive, depuratve, diuretic, galactogogue, febrifuge, sudorific, thermogenic, beneficial in vitiated

conditions of *kapha* and *vata*, agalactia, bronchitis, cough, catarrh, dyspepsia, leprosy, neuralgia, epileptic fits for children, strangury, pharyngopathy and fever.

Phytography : Perennial, sweet-scented grass, 1.5–2.5 m high; leaves linear-lanceolate, subcordate, amplexicaul and finely cordate; flowers in panicled spikelets, turning reddish brown when mature, glumes channelled corresponding to a midrib on the inner side, awned.

Distribution : Cultivated in dry localities in India; Pakistan and adjacent countries.

Ecology and cultivation : Prefers drier soils.

Chemical contents : There are two varieties of the grass, *Motia* and *Sofia*; the former yields palmarosa oil with 90–95% gerran oil content and the later ginger grass oil, essential oil (Rosha grass oil).

Cymbopogon nardus (L.) Rendle (Poaceae)

Syn : *Andropogon nardus* L.

English name : Citronella grass.

Vernacular name : Hin : Ganjni.

Trade names : Citronella, Citronella grass.

Modern use : *Essential oil* : antimicrobial, antibacterial and antifungal.

Phytography : Perennial grass with stout rootstock; stem tall, stout, paniculately branched; leaves simple, alternate, distichous, long, narrow, flat; panicles large, profusely branched, oblong or subpyramidal, branches loosely or closely packed, erect, often drooping; spathes laxly or closely imbricate, lanceolate, 0.8–1.8 cm, spikes with 4–5 pairs of spikelets, 0.6 cm, lanceolate, awned, glume 1-winged.

Distribution : Eastern Himalaya, West Bengal; Sri Lanka.

Ecology and cultivation : Prefers hot and moist climate; mainly wild.

Chemical contents : *Leaf* : oil of citronella containg r-cadinene and elemicin.

Adulterants : Kerosene, turpentine, alcohol, resin and other essential oils.

Cynodon dactylon (L.) Pers. (Poaceae)

Syn : *Cynodon filiformis* Voigl., *C. linearis* Willd., *C. maritimus* H.B. & K., *C. repens* Dulac.

English names : Dog grass, Fodder grass.

Sanskrit name : Durva.

Vernacular names : Asm : Duborubon; **Ben :** Durba; **Guj :** Durva; **Hin :** Doob; **Kan :** Garike, Hullu, Kudigarike; **Kon :** Dhurva, Hariali; **Mal :** Arugampullu; **Mar :** Durva, Harali; **Mun :** Dubila tasad; **Ori :** Duba ghasa; **Pun :** Doob; **Sad :** Dubla ghans; **San :** Dhubi ghas; **Tam :** Aruganapullu; **Tel :** Gericha gaddi, Gerike.

Trade names : Durva, Doob.

Traditional use : *MIRI* (of Assam) : *Plant* : in gastric ulcer; *ETHNIC COMMUNITIES OF MEGHALAYA* and *ASSAM* : *Plant* : in menstrual complaints; *TOTO* (of West Bengal) : *Plant :* in bleeding, wounds; *SANTAL* : (i) *Plant* : in convulsions, cramps, headache, wounds from bear's bite; (ii) *Bulb* : in carbuncle, sores; *ETHNIC COMMUNITIES OF CHHINDWARA* (Madhya Pradesh) : *Stem* : as tonic; *ETHNIC COMMUNITIES OF SAGAR* (Madhya Pradeeh) : *Leaf* : in bleeding, dropsy, piles, urinary complaints, vomiting; *GARHWALI* : *Root* : as cooling agent, in dysentery, hysteria, eye troubles.
AYURVEDA : it cures irregularities of bile, antiviral, blood purifier, cures sterility of women, useful in leucorrhoea, in cuts and wounds, pyorrhoea, urinary troubles, dysentery, vomiting, itching, nose-bleeding, checks loss of hair.

Modern use : *Plant* (i) *Aqueous-extract* : antifungal, antilithic, diuretic, hypoglycaemic; (ii) *EtOH* (50%) *extract* : antiviral, (iii) *Juice* : applied to stop nose bleeding, in skin diseases; *Rhizome* : useful in genito-urinary disorders; *Pollen-extract* : beneficial in rhinitis and asthma.

Phytography : Perennial, creeping grass; stem prostrate, often widely creeping and forming matted tufts with short ascending branches; leaves distichous, short, narrow, flat, glaucous, ligulate; spikes up to 12.5 cm, branched, radiating, green or purplish, rachis very slender, spikelets not more than 0.25 cm; grains oblong, compressed.

Phenology : *Flowering* and *Fruiting* : Throughout the year.

Distribution : Throughout the plains of India, ascending up to 1600 m; Bangladesh, Myanmar, Sri Lanka.

Ecology and cultivation : Plant of warm tropical climate; grows on all types of soils.

Chemical contents : *Plant* : 28 terpenes, sterols, fattyol.

Remarks : A sacred plant to the Hindus. According to Hindu mythology, Brahma resides in the root, Vishnu in the stem and Shiva in the leaf; an essential item in their every religious rites.
In Mexico, leaves wetted with morning dew are rubbed on the warts to cure them.

Cynoglossum lanceolatum Forssk. (Boraginaceae)

Syn : *Cynoglossum racemosum* Roxb., *C. micranthum* Dalz. & Gib.

Traditional use : *ETHNIC COMMUNITIES OF ORISSA* : *Plant* : in muscular pain; *ETHNIC COMMUNITIES OF MADHYA PRADESH* : (i) *Root* : as aphrodisiac; (ii) *Plant* : in muscular pain.

Modern use : *EtOH* (50%) *extract of plant* : antiaccholine.

Phytography : Erect herb; leaves alternate, simple, petiolate, broadly lanceolate, hispid; racemes long; flowers ultimately distant, bluish or purple; nutlets shortly ovoid, 0.3 cm in diameter.

Distribution : Plains from West Bengal to Punjab; Pakistan, Bangladesh.

Ecology and cultivation : Mesophyte.

Chemical contents : *Plant* : cynaustraline and cynaustine.

Cynoglossum zeylanicum Thunb. ex Lehm. (Boraginaceae)

Syn : *Cynaglossum meeboldii* Brand., *C. denticulatum* var. *zey* Cl.

Vernacular names : Mun : Huring jikipota; **Orn** and **Sad :** Latenga.

Traditional use : *DANG* (of Gujarat) : *Root-paste* : in diarrhoea, vomiting; *ETHNIC COMMUNITIES OF RAJASTHAN* : *Leaf* : in asthma, cough; *ORAONS* (of Bihar) : *Root* (cooked with Jatingi oil—*Guizotia abyssinica*—and sugar) : in wounds.

Phytography : Erect herb; stem soflty hairy or strigose; leaves simple, alternate, petiolate, ovate or oblong denticulate; racemes large; flowers small, bluish; nutlets vey strongly margined, sometimes obliquely attached to the carpophore.

Phenology : *Flowering* : Winter.

Distribution : Hills of Bihar, Deccan peninsula, southern part of Rajasthan.

Ecology and cultivation : Can withstand xeric conditions.

Chemical contents : *Plant* : hepatotoxic pyrrolizidine alkaloids.

Cyperus distans L. f. (Cyperaceae)

Syn : *Cyperus elata* Rottb., *C. graminicola* Steud.

Vernacular names : Kha : Ganechi; **Mun :** Piri jimtu.

Trade name : Ganechi.

Traditional use : *KHASI AND GARO* : *Plant* : in jaundice.

Phytography : A robust leafy sedge; stolons up to 5 cm, clothed by dark-brown eliptic tristichous leaves, often as long as stem, 0.6–0.8 cm broad; umbels 20–30 cm in diameter, from a single head, copiously 3–4 times compound, bracts leaflike, rather longer than umbel; nuts dusty black.

Phenology : *Flowering* and *Fruiting* : Rainy season.

Distribution : All warm regions of India, ascending up to 1000 m; Sri Lanka.

Cyperus rotundus L. (Cyperaceae)

Syn : *Cyperus hexastachyus* Rottb., *C. albidus* Herb., *C. tenuiflorus* Royle

English name : Common sedge.

Sanskrit names : Mustagundra, Musta, Aambuda, Ghana, Jalada, Varivaha.

Vernacular names : Asm : Motha; **Ben :** Mutha; **Guj :** Motha; **Hin :** Mutha; **Kan :** Tungahullu; **Mal :** Kora; **Mar :** Bimbal; **Mun :** Piri, Jimtu; **Ori :** Mutha ghasa; **Sin :** Kalanduru; **Tam :** Korai; **Tel :** Pura gaddi.

Trade name : Mutha.

Traditional use : *ETHNIC COMMUNITIES OF MEGHALAYA* : *Bulb* : in delirium, dysentery, jaundice; *ETHNIC COMMUNITIES OF ASSAM* : *Root* : in jaundice; *BHOXA (*of Uttar Pradesh) : (i) *Root* : in bowl complaints; (ii) *Plant* : in heat stroke; *GARHWALI* : *Root* : as astringent, diuretic, in sores and wounds; *KOLE* (of Uttar Pradesh) : *Plant* : in urinary complaints; *ETHNIC COMMUNITIES OF BANSWARA* (Rajasthan) : *Root* : in snake bite; *ETHNIC COMMUNITIES OF CHANDRAPURA* (Maharashtra) : *Root* : in sores, wounds.
AYURVEDA : ingredient of drugs named *Aryamuttanga,* and *Bhadramustaka.*
UNANI : a minor ingredient of the drug 'Jawarish Jalinoos', used in diseases of intestine, stomach and urinary bladder.

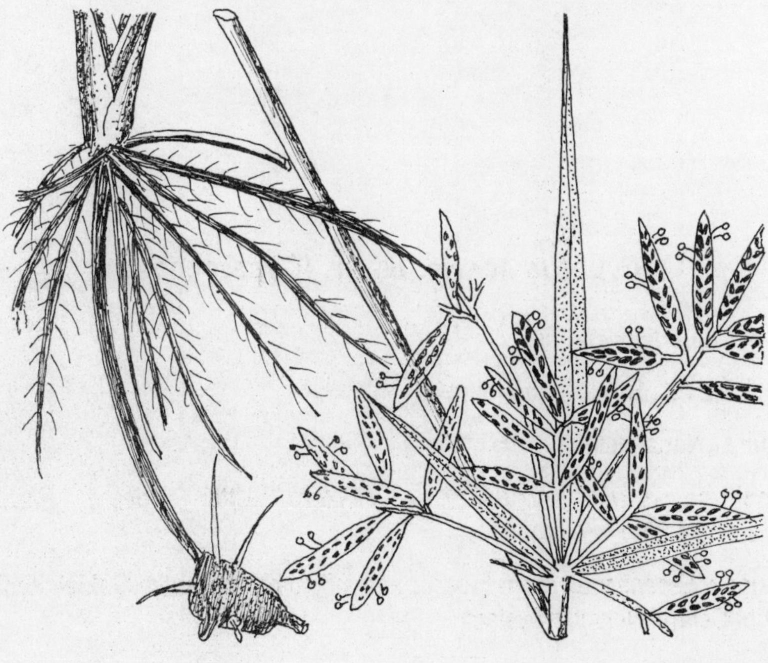

Modern use : *Tuber* (aqueous/alcoholic extract) : hypotensive in cats, causes systolic heart arrest in frogs, potentially anti-inflammatory, antipyrretic, analgesic, useful in dysentery; *Essential oil* : anti-insect, antimicrobial, anthelmintic and shows juvenile hormone mimicking activity.

Phytography : A small sedge, 10–60 cm in height; stolon slender, 10–20 cm, thickened into black woody ovoid tubers; aerial stem triquetrous at top; leaves tristichous, long, often overtopping the stem, 0.4–0.8 cm broad; umbels frequently compound, primary rays 5–20 cm.

Phenology : *Flowering* and *Fruiting* : July–December.

Distribution : All warm regions of India, ascending up to 2000 m, as a pestiferous weed; Bangladesh.

Ecology and cultivation : Wild in fallow land, wasteland, cultivated land, on roadside, bunds, nullahs, etc.

Chemical contents : *Tuber* : a triterpenoid, oleanolic acid and its glyceride, sitosterol; *Essential oil* : sesquiterpenes.

Adulterants : *Cyperus scariosus* R. Br., *C. arundinacoum* Baker.

Remarks : Ethnic communities of Sikkim and Hazaribagh District of Bihar eat the roots, while people of the Indian desert eat the bulbs. The stem is used for making mats.
In Bahama Island, tea prepared with this plant is consumed in cold condition to cure jaundice.

Cyperus scariosus Br. (Cyperaceae)

Syn : *Cyperus pertenuis* Roxb.

Vernacular name : Hin : Nagarmotha.

Trade name : Nagarmotha.

Traditional use : ETHNIC COMMUNITIES OF MIRZAPUR (Uttar Pradesh) : *Rhizome* : as antidote to snake bite.

Modern use : *Essential oil* (from tuber) : anti-inflammatory, antimicrobial, hypotensive, stimulant of Central Nervous System.

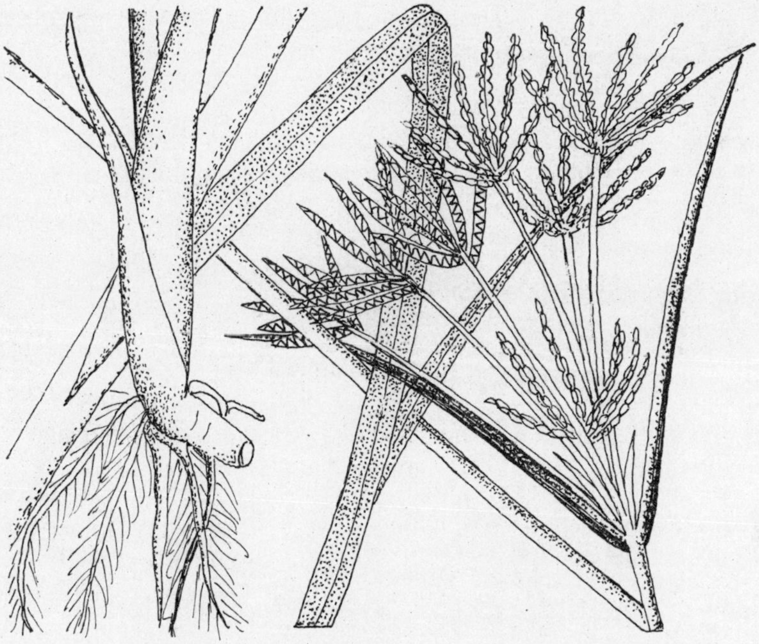

Phytography : A glabrous herb; stolon slender, 0.8–5.0 cm by 0.25 cm, clothed by elliptic, acute, lax, concolorous scales; stem 40–90 cm long, slender, triquetrous at top; leaves variable, usually short, narrow, weak; umbels slender, thin, rays slender.

Distribution : Common in the Sundarbans; available in the plains of India.

Ecology and cultivation : Prefers halophytic and swampy condition.

Chemical contents : *Essential oil* (from tuber) : sesquiterpene, cyperenone, cyperenol, patchoulenol, isopatchoulenone, β-selinene, isopatchoula-3,5-diene, retundene, rotundenol; *Leaf* : leptosidin.

GLOSSARY OF SOME SELECTED MEDICAL TERMS

Abscess : Localised collection of pus in a tissue.

Acidity : Sourness, a sign of indigestion (under digestive disorders).

Acne vulgaris : Skin eruptions on inflammatory base, occurs usually during puberty and adolescence.

Adenitis : Inflammation of gland.

Alterative : A drug which alters course of diseases and re-establishes function of an organ.

Amenorrhoea : Abnormal suppression or absence of menstruation.

Anaemia : A condition caused by lowering of number of red blood corpuscles.

Anaesthetic : An agent to produce partial or total loss of the sense of pain, touch, temperature, etc.

Analgesic : A drug that reduces or relieves pain.

Anasarca : Oedema in subcutaneous tissues all over the body.

Angina pectoris : A disease of heart—signalled by pain in chest, shoulder, etc.

Anorexia : Loss of appetite.

Anthelmintic : Agent which expels or destroys intestinal worms.

Antiallergic : Agent that counters allergic reactions.

Antibacterial : An agent capable of reducing or stopping the action and proliferation of bacteria.

Antidote : An agent that counteracts the action or effect of poisons.

Antiemetic : Agent which prevents or lessens tendency to vomit.

Antifertility : A drug that inhibits formation of ova or sperm or interferes with the process of fertilisation.

Antihelmintic : A drug that acts against helminths.

Antilithic : Which prevents the formation of calculii or promotes their dissolution.

Antipyretic : Agent that reduces body temperature, i.e., fever.

Antiphlogistic : Prevents or relieves inflammation.

Antiscorbutic : A drug that cures scurvy.

Antiseptic : Agent that arrests development or reproduction of bacteria and other microorganisms.

Antispasmodic : Allays, relaxes or controls spasms.

Antiviral : Agent to treat or stop viral infection.

Aperient : Mild purgative.

Aphrodisiac : Stimulates sex desire or power.

Aphthae : *A boil in mucus membrane in mouth.*

Apoplexy : A general term for cerebral haemorrhage, thrombosis, vasospasm.

Appendicitis : Inflammation of the vermiform appendix of the ceacum.

Appetiser : An agent which stimulates appetite.

Aromatic : Substance having pleasent smell or flavour.

Arthritis : Inflammation of joint(s).

Ascaris : A genus of parasitic worms infecting small intestine.

Ascites : .Abnormal accumulation of serous fluid in peritoneal cavity of abdomen.

Asthma : A lung disease characterised by wheezing, difficult breathing and feeling of suffocation.

Astringent : An agent which contracts tissues and checks secretions.

Atrophy : Washing or emaciation with loss of strength.

Auge : Periodic fever with shivering.

Beri-beri : An acute form of dropsy and oedema.

Bile complaints : Bile is a thick bitter fluid secreted by liver—irregularity in its secretion.

Biliousness : General condition of indigestion, constipation, headache, etc., caused by derangement of the bile.

Bite : Cut into or nip with the teeth by dog, jackle and other wild animals.

Bleeding from nose : (see nasal problem).

Blister : Vesicle on skin filled with serum caused by friction, burning, etc.

Blood pressure : The pressure caused on the walls of blood vessels due to the flow and pumping of the blood by heart.

Blood purifier : Agents believed to remove impurities from or deficiencies of blood.

Blood sugar : Sugar in blood, if the level of its concentration goes beyond certain limit, causes a disease called diabetes.

Bodyache : (see pain).

Boil : A localised pyrogenic infection originating in a hair follicle; hard inflamed suppurating tumour.

Bone fracture : Breaking of a bone.

Breast pain : (see mammary diseases).

Bronchitis : Inflammation of bronchi.

Burn : Sore or mark caused by fire on body of a being.

Burning sensation inside body : Includes feeling hot inside the body or in palms, soles, etc.

Calculus : Deposit of solid matter, like salts, in any part of the body, such as kidney.

Cancer : A malignant new growth or tumour in any part of the body.

Carbuncle : An extensive boil which discharges pus from many points.

Cardiac complaint : Ailments relating to heart.

Caries : Decay or rottenness or ulceration of bone —it is related to teeth.

Carminative : The medicine which cures flatulence, promotes expulsion from or reduces formation of gas in stomach.

Cataract : Opacity or clouding of lens of eye causing partial blindness.

Catarrh : Inflammation of mucous membranes.

Cerebral affections : Ailment of cerebral region.

Chancre : Sore caused by syphilis.

Chicken pox : A contagious disease resulting in eruption of papules.

Chilblain : Localised inflammation of skin in cold weather.

Cholera : An acute infectious disease caused by *vibro* resulting in diarrhoea, vomiting, cramps and suppression of urine.

Cirrhosis : Diminution and deformity of liver.

Cold : Exposure to cold or sudden change of temperature causing inflammed state of mucous membrane with hoarseness of voice, running nose, sore throat, etc.

Colic : Severe gripping pains due to contraction of muscles in walls of intestines.

Conjunctivitis : Inflammation of conjunctiva.

Constipation : Costiveness; condition of bowels when defection of faeces is irregular or difficult.

Consumption : Wasting of tissues; tuberculosis.

Contraceptive : A drug or device to prevent uterine conception.

Contusion : Injury to tissues by blow without causing break in skin.

Convulsion : Violent irregular motion of limb or body due to involuntary contraction of muscles.

Cooling : Reducing temperature or thirst.

Corn : Horny induration of cuticle with hard centre and root, often penetrating deep into subjacent tissue, caused by pressure of shoe or feet.

Cough : A violent effort of the lungs to throw off injurious matter accompanied by harsh sound from throat.

Cramp : Sudden painful involuntary contraction of muscles.

Cut : Breaking of skin due to external means.

Dandruff : Scurf; dead skin in smalll scales among hairs.

Debility : Feebleness; relaxation of solid or muscular fibre.

Dehydration : Removal of constitutional water.

Delirium : Derangement of mental functions accompanied with incoherent speech.

Demulcent : Substance to soothen surface(s).

Dermatitis : Inflammation of skin causing discomforts, like eczema.

Diabetes : A metabolic disorder resulting in elevated blood glucose.

Diaphoretic : Drug or treatment causing sweating or perspiration.

Diarrhoea : A common symptom of gastrointestinal diseases resulting in excessive looseness of bowels accompanied with frequent discharge of watery stool.

Diphtheria : An acute infectious disease with inflammation of mucous membrane of throat in which air passages become covered and impeded by leathery membrane.

Disinfectant : A substance which destroys harmful germs, cleanses infections.

Diuretic : An agent that induces discharge of urine.

Dropsy : A leakage of the watery part of the blood into any of the tissue or cavity of the body.

Dysentery : A disease caused by bacteria or protozoa causing inflammation of mucous membrane and glands of large intestine.

Dyspepsia : Indigestion.

Dysuria : Difficult or painful urination.

Ear complaints : All ailments of ear.

Ecbolic : An agent which hastens delivery or produces abortion.

Eczema : Acute or chronic noncontagious inflammation of skin often accompanied by itching.

Elephantiasis : Enlargement of cutaneous and subcutaneous tissues caused by obstruction of lymphatics—the affected part resembles elepahant's hide.

Emetic : Causes or produces vomiting.

Emmenagogue : A substance which restores normal menstrual function.

Emollient : A substance applied externally to soften the skin of living animal.

Epilepsy : A nervous disorder causing temporary loss of consciousness with or without convulsions.

Eruption : Acne, breaking out of rash, pimples, blister, etc.

Erysipelas : Local febrile disease, producing deep red colour on skin and subcutaneous tissue.

Expectorant : Which promotes ejection of phlegm, etc., from chest or lungs by regulating bronchial secretions.

Eye diseases : Ailments of eye.

Epistaxis : Nasal haemorrhage.

Excoriations : Linear break in the skin surface, usually covered with blood or crusts.

Fertility : Capability to conceive or cause conception.

Fever : Abnormal body heat and quickening of pulse.

Filaria : A term for a group of embryo worms of *Filarioidea* family.

Fistula : A pipe-like sore with a narrow opening.

Fit : A sudden attack by convulsions.

Flatulence : Excessive gas in stomach and intestine.

Galactagogue : A substance which induces or stimulates secretion of milk.

Gangrene : Necrosis, usually with loss of vitality or decomposition in some part of the body.

Gastric complaints : Disorders of stomach.

Germicide : Agent to destroy germ.

Giddiness : Dizziness, feeling of things circling round the person causing instability.

Goitre : Morbid enlargement of thyroid gland, often showing as large pendulous swelling in neck.

Gonorrhoea : A veneral disease—inflammatory discharge of mucus from urethra or vagina.

Gout : A constitutional hereditary condition of uric acid metabolism, resulting painful inflammation of smaller joints.

Guinea worm : A thread-like tropical parasitic nematode.

Gyneacological problems : Ailments women related to menstruation, conception, delivery, etc.

Haematoma : Collection of blood at some part of body.

Haematuria : Presence of red blood corpuscles in urine.

Haemoptysis : Spitting of blood from bronchi, larynx, lungs or trachea.

Haemorrhage : Discharge of blood from blood vessels.

Haemorrhoids : Piles with painful swelling at the anus.

Haemostat : Agent that arrests bleeding.

Hallucinogenic : A drug or substance that produces hallucinations, imaginary things apparently seen or heard.

Headache : Pain in head.

Hemicrania : Severe pain affecting one side of head.

Hemiplegia : Paralysis of one side of body.

Hepatic complaints : Diseases of liver.

Hepatitis : Inflammation of liver caused by viral infection.

Hernia : Abnormal protrusion of an organ through its containing wall, usually associated with abdominal cavity.

Hydrocele : Pathological accumulation of serum fluid in body cavity, especially in the testes.

Hydrophobia : Dread of water with convulsions, generally caused by bite of a rabid animal.

Hypertension : High blood pressure.

Hypnotic : Sleep inducing.

Hypoglycemic : Low level of sugar in blood.

Hypotensive : Agent to reduce blood pressure.

Hysteria : A functional disturbance of nervous system of psychoneurotic origin, especially in women, characterised by convulsions, and usually attended with disturbance of intellectual faculty.

Impotence : Inability to engage in sexual intercourse.

Indigestion : Digestive disorder.

Inflammation : Morbid process affecting any part or organ of the body with swelling, heat, pain and redness.

Infertility : Incapability to conceive or cause conception.

Influenza : A severe epidemic febrile disease with rapid prostration; severe catarrh and weakening fever.

Injury : Contusion, cut, wound, bruise caused by external hit.

Insect-bite : Bites/stings of insects.

Insomnia : Sleeplessness.

Intestinal diseases : Diseases of intestine.

Intoxication : State of being drunk, high excitement beyond self control.

Itch : Irritation in skin.

Jaundice : Morbid state caused by obstruction of bile, and marked by yellowness of skin, tissues and secretions, constipation and loss of appetite.

Joint diseases : Variety of painful diseases causing swelling, immobility and even deformation of joints.

Kidney diseases : Kidney stone, renal complaints, calculus, etc.

Labour : Pains of child birth.

Lactagogue : Which promotes secretion and flow of milk.

Lactation problems : Disfunctions of mammary glands including galactagogue, etc.

Laxative : Medicine tending to loosen bowels.

Leprosy : Chronic contagious diseases due to infection with *Lepra bacillus*, varying in manifestation as the skin, nerve or other tissues are affected.

Leucoderma : A skin condition characterised by defective whitish pigmentation.

Leucorrhoea : A white or yellowish mucopurulent discharge from vagina.

Lithotriptic : The agent which causes dissolution of a calculus.

Liver complaints : Disorder in the function of liver; including hepatic diseases.

Loose motion : Watery motion.

Lumbago : Rheumatic affection in loins.

Lung diseases : Pulmonary diseases, like cough, pleurisy, pneumonia, respiratory complaints.

Malaria : An acute or chronic disease, usually accompanied with intermittent and remittent fever, caused by protozoa (*Plasmodium*) and transmitted by Anopheles mosquito.

Masticatory : Substance chewed for increased salivation.

Measles : An acute infectious disease, caused by virus infection, and marked by red pustules.

Melancholia : A disease of mind marked by sadness, depression and illgrounded fears.

Menorrhagia : Diseases associated with menstruation, including premenstrual tension.

Menstrual complaints : Irregularity in the monthly discharge from uterus.

Mental diseases : Relating to functions of mind.

Migraine : A recurring and very painful headache on one side of the head, often accompanied by giddiness, vomiting.

Miscarriage : Premature involuntary expulsion of foetus.

Mumps : Virus infection which causes acute inflammation of parotid and salivary glands.

Nail diseases : Diseases of nail.

Narcotic : Substance which tends to induce drowsiness, sleep, stupor, even paralysis of nervous system.

Nasal diseases : Sinus, bleeding from nose, etc.

Nausea : Sensation of sickness of stomach, which may cause vomiting.

Nephritis : Inflammation of kidney(s).

Nervine diseases : Nervous disorder, neuralgia, etc.

Neuralgia : Sudden severe pain radiating along the course of a nerve.

Neurosis : A disease of nerve(s).

Night blindness : Inability to see in dim light.

Oedema : Flow of excessive clear fluid from blood into other tissue(s).

Oxytocic : Any drug that hastens child birth.

Pain : Distress of body, like bodyache, headache, muscular pain, etc.

Paralysis : Loss of power of motion.

Parturition : The act of child birth.

Phthisis : Pulmonary tuberculosis.

Piles : Enlarged or dilated blood vessels; disease marked by tumours of veins of lower rectum.

Pimple : Small solid round outgrowth on skin.

Pleurisy : Inflammation of the pleura membrane enveloping lung, usually caused by chill and marked by pain in chest or side, fever, etc.

Pneumonia : Inflammation or congestion in lungs.

Poison : Toxic substance.

Polyuria : Excessive micturition.

Postnatal : After child birth.

Prenatal : Before child birth—generally related to gynaeocological or other problems related to pregnancy.

Prickly heat : Heat rash, irritation of the skin caused by heat.

Pruritus : Itching.

Psoriasis : A chronic inflammatory skin disease with reddish patches covered with white scales.

Puerperal fever : Fever during child birth.

Pulmonary complaints : Lung and chest disorder.

Purgative : The agent which causes evacuation of bowels.

Pyorrhoea : A purulent discharge with pus.

Rabies : Canine madness, an infectious viral disease caused in man by bite of infected dogs and some other canine animals.

Refrigerant : Having cooling properties on surface of body.

Renal disease : Disorders in function of kidney(s).

Respiratiory diseases: Problems related to respiration.

Revive taste : To rectify tastelessness.

Rheumatism : A general term; refractory disease of muscle, tendon, joint, bone, nerve, resulting in discomfort and disability.

Rhinitis : Inflammation of the mucous membrane of the nose.

Rickets : A vitamin D deficiency disease of the young, affecting the developemnt of bone.

Ringworm : A common parasitic fungus contagious disese on skin.

Rubefacient : An agent that stimulates skin to redness and increase flow of blood in that part.

Scabies : A contagious parasitic disease of skin causing itching.

Sciatica : A pain in sciatic nerve, neuralgia of hip and thigh.

Scorpion-bite : Sting by scorpion, usually causing severe irritation and pain.

Scrofula : Tuberculosis of the lymphatic glands, morbid constitutional condition with glandular swellings and tendency to consumption.

Scurvy : A nutritional disorder with swollen gums, livid spots, etc., caused by deficiency of ascorbic acid.

Sedative : Drug which soothes or allays irritability.

Sex disorders : Absence of sexual vitality, vigour, also includes night pollution.

Sinus : From sinusitis, the inflammation of sinus.

Skin diseases : Diseases like scabies, ringworm,. eczema, fissures and emollients.

Skin outgrowth : Lavandula.

Small pox : A severe and fatal disease with eruptions on skin and fever due to pox virus infection.

Snake bite : Cut by snake.

Soporific : Agent that produces sleep.

Sore : A general term for variety of morbidly tender tissues, boils and wounds.

Spasmolytic : A drug that prevents spasm.

Spermatopoietic : Spermatogenic.

Spermatorrhoea : Involuntary semen-discharge.

Spleen complaints : Disorder in spleen-function.

Splenomegaly : Enlargement of spleen.

Sprain : The wrench or twist of a ligament or muscle or a joint causing pain and swelling.

Sprue : Intestinal malabsorption with steatorrhoea.

Steriliser : An agent that destroys unwanted bacteria or other pathogens.

Sterility : Incapability to conceive or cause conception.

Stimulant : An agent that increases functional activity of an organ.

Stomachache : Pain or unspecified disorder in stomach.

Stupor : Lethergy.

Stye : A small inflammated tumour on edge of eyelid.

Sudorific : That which causes perspiration.

Sun stroke : Exposure to excessive heat causing severe perspiration.

Swelling : Inflammation.

Syphilis : A contagious venereal disease caused by *Treponema pallidium.*

Tetanus : An infectious disease caused by bacillus *Clostridium tetani* and marked by spasm of many or all muscles of voluntary motion.

Throat complaints : Sore throat, catarrh, tonsilitis, diphtheria, etc.

Tonic : A drug that invigorates or improves normal tone of the body of a patient.

Tonsilitis : Inflammation of the tonsil(s).

Tuberculosis : An infectious disease caused by tubercle bacillus.

Tumour : Generally an abnormal swelling of body.

Typhoid : Infectious disease caused by some waterborne bacteria.

Ulcer : An open sore on external or internal surface of body, discharging pus and other materials.

Unconsciousness : The state of absence of consciousness.

Urethritis : Inflammation of urethra.

Urinary complaints : Disorders in urinary system, e.g., blood in urine, dysuria, night wetting, haematuria, urethritis, etc.

Uterine disorders : Irregularity in the normal function of the uterus or womb.

Vaginal complaints : Diseases of vagina.

Venereal diseases : Diseases communicated by sexual intercourse, viz., gonorrhoea, syphilis, etc.

Vertigo : Giddiness, dizziness in which patient feels that surrounding objects are whirling.

Vesicant : Any agent that causes blisters.

Veterinary : Ailments of domestic animals and birds.

Vomiting : Throwing up the contents of the stomach by mouth.

Weakness : A feeling of loss or reduction in normal strength of body.

Whitlow : An old term for any suppurative inflammation on finger.

Whooping cough : An infectious disease found among children.

Worms : A variety of limbless or apparently limbless creeping-creatures which are parasites on human body, mainly in the intestinal tract.

Wound : Incision in soft tissues or skin caused by external injury.

GLOSSARY OF SOME SELECTED BOTANICAL TERMS

Accrescent : Enlarging with age.

Achene : Small, dry, hard, one-chambered, one-seeded fruit.

Acicular : Needle-shaped.

Actinomorphic : Flower having radial symmetry.

Acuminate : Suddenly tapering to a sharp point.

Acute : Gradually tapering to a point.

Acyclic : Arranged in spirals and not in whorls.

Adnate : Adherent.

Albuminous : Provided with albumen or endosperm.

Alternate : One leaf at a node, and leaves are so arranged that a line connecting the consecutive leafbases will take a spiral course on the stem.

Amplexicaul : Leafbase embracing the stem completely.

Androgynous : Bearing staminate and carpellate flowers in the same inflorescence.

Annual : Plants which complete their life-cycle within a year.

Anther : Pollen containing part of a stamen.

Apetalous : Lacking petals.

Apocarpous : When the carpels of a flower are free from each other.

Arborescent : Resembling a tree.

Aril : A small appendage from or around the hilum of a seed.

Articulate : Jointed.

Asepalous : Without sepals.

Attenuate : Drawn out into a long, slender extension.

Auricle : An earshaped appendage, normally from the top of the leafsheath in grasses.

Awn : Bristle-like appendage.

Axil : The angular space between stem and leaf-stalk.

Baccate : Having the nature of berry, pulpy or fleshy.

Basifixed : Filament attached to the base of the anther.

Balausta : Fruit with firm rind, berried within and crowned with lobes of adnate calyx.

Barbed : Equipped with retrorse projections.

Barbellate : Finely barbed.

Berry : Fleshy, indehiscent, few- to many-seeded fruit derived from a gynoecium of fused carpels.

Bicarpellary : Pistil with two carpels.

Biennial : The plant which completes its life-cycle in two years.

Bifid : Two-cleft.

Bifoliate : A compound leaf having two leaflets.

Bilabiate : Two-lipped, referring especially to zygomorphic corolla.

Bilocular : Two-chambered.

Bipinnate : Twice pinnate, rachis (modified mid-rib) producing secondary axes which bear leaflets.

Bisexual : When androecium and gynoecium present in the same flower.

Blade : Expanded flat part of a leaf.

Bract : Modified leaf at the base of a flower.

Bractlet : Bract borne on a secondary axis.

Bracteate : Bract present.

Bracteole : Secondary bract.

Bracteolate : With bracteoles.

Bulb : Modified underground stem, most of the leaves of which are thickened at the base with stored reserve food.

Bulbil : Modified axillary vegetative or floral bud.

Caducous : Falling off early.

Calyculus : Simulated calyx composed of bracts and bracteoles but not of sepals.

Calyx : The outermost, generally green, floral envelope.

Calyx tube : The tube formed by calyx.

Campanulate : Bell-shaped.

Campylotropous : The ovule is so bent as to bring the hilum, micropyle and chalaza close together.

Canescent : Bearing greyish hairs.

Capitate : Headlike.

Capsule : Dry, dehiscent fruit, developed from a compound gynoecium (constituted of two or more carpels), splitting along two or more lines of suture.

Carcerulus : Dry, schizocarpic fruit, developed from a bicarpellary ovary, and ultimately breaking into four nutlets.

Cardiopetalous : With heart-shaped petals.

Carpel : The ovuliferous organ of the flower; often regarded as a single, modified mega-spore-bearing leaf.

Caryopsis : Small, dry, indehiscent, one-seeded fruit in which the seedcoat is adherent to the wall of fruit.

Caryophyllaceous : Five petals with comparatively long claws, and limbs placed at right angles to the claws.

Catkin : An elongate spike with comparatively soft axis, which, failing to bear the weight of flowers, bends down.

Caudate : Bearing a tail-like appendage—normally the term is applied to leaf-apex.

Caudex : The type of stem found in palms.

Cauline : Borne on the stem only.

Centric : A leaf which is more or less cylindrical.

Circinate : A prefoliate state of a leaf where the leaf remains coiled or rolled from apex to base like the tail of a dog.

Cirrhose : A slender coiled apex.

Cladode : Cylindrical or flattened green branch of one internode or of limited growth.

Clavate : Club-shaped, gradually thickened upwards.

Claw : Narrow basal part of a petal.

Cleistogamous : Flowers which remain closed to ensure self-pollination.

Climber : A plant which attaches itself to a nearby support by means of some special device for growing upwards.

Coccus : Spherical structure; a part of a schizocarpic fruit.

Cochleate : Shaped like a spiral shell.

Coherent : Fused or united with the organs of same kind.

Cohesion : The phenomenon of union of similar organs, viz., union of sepals or petals or stamens.

Coleoptile : The sheath surrounding the young stem of grasses.

Coleorhiza : The sheath surrounding the young root of grasses.

Collateral : Placed side by side.

Comose : Leaf surface bearing a tuft or tufts of hairs.

Complete : A flower having all four sets of floral organs.

Complicate : When the leafblade is folded upon itself.

Conduplicate : The pattern of a leaf in which the blade is folded lengthwise with the upper surface within.

Confluent : Blended into one.

Conical : Cone-like—broad at the base and gradually tapering towards the apex.

Connate : When the bases of some organs appear to have fused.

Connective : The part of a stamen which joins two anther-sacs.

Connivent : Coming close together or touching similar organs but not united.

Convolute : Longitudinally rolled up from one margin to other like the rolling up of a sheet of paper.

Cordate : Heart-shaped with a basal notch.

Coriaceous : Leathery.

Corm : A stout, solid, fleshy, underground stem growing in the vertical direction.

Corneous : Horny texture.

Corolla : Second outer whorl of a flower, consists of a number of petals.

Corona : Collar-like or tubular appendage of corolla, forming more of less a crown-like structure.

Corrugate : Wrinkled or folded having prominent ridges.

Corymb : A type of racemose inflorescence where the main axis is comparatively short, and the lower flowers have longer stalks or pedicels than upper ones so that all the flowers are brought to more or less same level.

Corymbose : Corymb-like.

Cotyloid : A concave receptacle, sometimes to the extent of surrounding the carpels.

Creeper : A trailing shoot that produces roots from nodes, almost throughout its length.

Cremocarp : Dry, dehiscent, two-seeded fruit; each half a mericarp, borne on a thread-like carpophore.

Crenate : Shallowly rounded or obtusely toothed margin.

Crenulate : Finely crenate.

Crispate : An extreme form of undulate.

Cruciform : Cross-shaped.

Crustaceous : Hard and brittle texture.

Culm : Jointed stem with solid nodes and hollow internodes.

Cuneate : Wedge-shaped with broad apical end.

Cupule : A small cup-shaped structure formed usually by fusion of bracts together.

Cuspidate : Ending in a sharp rigid point.

Cyamium : A kind of follicle.

Cyathium : A special type of inflorescence, where the involucre is cup-shaped; it encloses a single reduced female flower, surrounded by a number of reduced male flowers.

Cyme : A determinate inflorescence, where the apical flower opens first.

Cynarrhodion : A fleshy hollow fruit enclosing achenes.

Cypsela : An achene developed from an inferior ovary and invested by an adnate calyx.

Deciduous : Falling off soon after maturity; not persistent.

Decompound : Applied to a compound leaf when the leaf is more than thrice pinnate.

Decumbent : Reclining or growing flat along the ground, but with the tip ascending.

Decurrent : The leafblade appearing to extend down the sides of the stem and adnate to the stem.

Decussate : Applied to leaves which occur in pairs alternately at right angles at adjacent nodes.

Deltoid : Like a delta; triangular.

Dentate : Usually applied to leaf margins, where the margin bears sharp teeth directed outwards at right angles to the margin.

Denticulate : Minutely toothed leaf margin.

Diadelphous : Said of stamens, when the stamens are grouped into two bundles.

Diandrous : Applied to flowers which bear two stamens only.

Dichasium : A type of cymose inflorescence, where the main axis end in a flower and at the same time it produces two lateral younger flowers; the process is repeated by each set of branches.

Dichlamydous : Applied to flowers having both whorls of perianth, i.e., calyx and corolla.

Dichogamous : Said of flowers in which stamens and carpels mature at different times.

Dichotomy : A pattern of branching where the axes divides into two equal halves.

Diclinous : Applied to plants/flowers when androecium and gynoecium are borne on separate flowers.

Didynamous : Androecium having four stamens in two pairs of different heights.

Diffuse : Plants having stem spread over the ground.

Digitate : A palmately compound leaf in which the petiole bears terminally a number of leaflets which appear to be radiating from a common point like fingers from the palm.

Dimerous : Applied to flowers when each whorl consists of two units.

Dimorphic : Those have two forms.

Dioecious : Plants having male and female flowers borne on different individuals.

Disk : Outgrowth of the receptacle, may be epigynous or hypogynous.

Dissected : Deeply divided or cut into many slender segments.

Disk florets : Tubular flowers in the centre of a capitulum.

Distichous : Arranged in two vertical lines, as the leaves of grasses.

Divided : Separated to the very near the base.

Dorsifixed : Filament attached to the back of anther.

Dorsiventral : Flattened structure with distinct dorsal (upper) and ventral (lower) surfaces.

Drupaceous : Resembling a drupe.

Drupe : The fleshy, one-seeded, indehiscent fruit with the seed enclosed in a stony endocarp, while the mesocarp remains thick and fleshy, and the epicarp thin.

Echinate : Beset with prickles.

Emarginate : With a shallow notch at the apex.

Emergence : An outgrowth developing from superficial layers.

Endocarp : Innermost zone of pericarp.

Endosperm : The food-storage tissue of grain, which normally stores proteinaceous food; this nutritive tissue normally nourishes the embryo in seed plants.

Ensiform : Sword-shaped.

Entire : Even and smooth without any marginal serrations.

Entomophillous : Insect loving; the term usually refers to insect pollinated flowers.

Epicalyx : The whorl of sepaloid green structures just outside and below the calyx.

Epicarp : The outermost zone of pericarp.

Epigynous : Situated upon or on top of the ovary.

Epipetalous : Borne upon petals—normally referred to the stamens occurring on petals.

Epiphyte : A plant growing upon another plant but using it as support only and not drawing nutrition from the latter.

Equitant : Conduplicate leaves standing face to face and one completely overlapping the other by both the margins.

Erect : A rigid and firm stem holding itself in an upright position.

Erose : Irregularly toothed.

Exasperate : Rough with hard projecting points.

Exserted : Projecting beyond the summit, as stamens protruding from the corolla.

Exstipulate : Stipules are absent.

Extra-axillary : Situated outside the axil.

Extrorse : Facing outwards.

Falcate : Sickle-shaped or scythe-shaped.

False fruit : A fruit not developed from a fertilised ovary, but from some other part(s) of a flower.

Farinaceous : Either containing starch or starch-like materials in nature.

Fasciculated : Clustered or bundled, normally applied to root.

Filiform : Like a filament or thread.

Fimbriate : Fringed along the edge.

Fistular : Stem or leaf having a hollow interior throughout its length.

Flabellate : Fan-shaped.

Flaccid : Flabby, withered and limp.

Flocoose : With tufts of silky or woolly hairs that rub off easily.

Floret : Reduced flower.

Foliaceous : Leaf-like.

Follicle : A dry one-chambered fruit developed from single carpel, and splits along one suture only.

Free-lateral : Free and situated on sides—-usually applied in respect of stipules.

Fruticose : Shrub-like.

Fugacious : Withers as soon as formed.

Funicle or *Funiculus* : The slender stalk of the ovule, by which it remains attached to the placenta.

Fusiform : Spindle-shaped—narrowed both ways from a swollen middle.

Gamopetalous : With united petals.

Gamophyllous : Perianth with united segments.

Gamosepalous : With united sepals.

Gibbous : Tubular structure having the base or one side slightly dilated to a pouch; this term is usually applied to corolla.

Glabrous : A structure having its surface smooth and free from hairs or outgrowths of any kind.

Glabrate : Nearly glabrous or becoming glabrous with maturity.

Glandular : Possessing gland(s).

Gland-dotted : When glands are scattered like dots on a surface.

Glaucous : The surface is covered by a waxy coating causing a shiny bluish or whitish tinge.

Globose : Spherical like a globe.

Glochidium : A barbed bristle.

Glochidate : Equipped with barbed bristles.

Glumaceous : Glume-like.

Glume : Chaff-like scaly bracts, usually found in the grasses and similar plants.

Glutinous : Covered with a sticky exudation.

Gregarious : Growing in clusters.

Gynandrophore : The internode of the thalamus between corolla and androecium.

Gynandrous : Usually refers to the stamens which are adnate to carpels as a whole or by their anthers only.

Gynobasic : Applied to the style which arises from the depressed centre of a syncarpous ovary, actually the style adheres by its base to the prolongation of the torus upwards between the carpels.

Gynophore : The elongated internode of the thalamus on which pistil is situated—this looks like a stalk of the pistil.

Gynostegium : A sheath or covering of the gynoecium.

Gynostemium : The column in an orchid flower, formed by adnation of stamens and styles.

Habit : Characteristic appearance of a plant.

Habitat : Natural home of a plant.

Halophyte : A plant which grows in saline substratum.

Hastate : Arrow-shaped leaf with two basal lobes directed outwards.

Head : Capitulum inflorescence.

Helicoid : Coiled like a helix.

Herb : Small plant with soft stem, may vary from a few millimetres to a metre or so in height.

Herbaceous : Non-woody, like a herb having a green colour and a leafy texture.

Herkogamy : The phenomenon of having some sort of barrier between the stamens and the pistil of the same flower.

Hermaphrodite : Bisexual.

Hesperidium : A superior, many-celled, fleshy fruit where the endocarp projects inwards forming distinct chambers, and the mesocarp and epicarp, fused together, form a separable skin, e.g., orange and other citrus fruits.

Heteroanthy : The phenomenon of having stamens of different lengths.

Heterogamous : Bearing two kinds of flowers—male and female.

Heteromerous : Refers to a flower where the number of members of different whorls is irregular.

Heteromorphic : Plants bearing two or more different forms of flowers with anthers and stigmas at different levels.

Heterophyllous : The same individual plant bearing different types of leaves.

Heterostyly : When one plant bears some flowers with long styles and others with short styles.

Hilum : The point of attachment of the of the body of the ovule to its stalk. On a developed seed, it is represented by a scar, indicating the point where the seed breaks from the stalk.

Hirsute : Having moderately stiff separate hairs.

Hispid : Having stiff or bristly hairs.

Hispidulous : Minutely hispid.

Hyaline : Colourless and translucent or transparent.

Hypanthium : The tube of the receptacle upon which the calyx, corolla and stems are borne; perianth tube.

Hypanthodium : It is the typical inflorescence of banyan, fig, etc., where the fleshy receptacle forms a hollow cavity with an apical opening guarded by scales, and the flowers are borne on the inner wall of the cavity.

Hypocotyl : The axis of the embryo lying immediately below the cotyledons.

Hypogynous : Borne beneath or below the gynoecium; usually this term refers to stamens, petals and sepals.

Imbricate : This is a type of aestivation (mode of arrangement of the sepals or of the petals in a floral bud with respect to the members of the same whorl), where one of the sepals or petals is overlapped on both the margins, and one of them is external, and each of the remaining one is overlapped on one margin and overlaps the next one on the other margin.

Imparipinnate : A unipinnate leaf with an odd terminal leaflet.

Imperfect flower : A flower missing either stamens or carpels; a unisexual flower.

Incised leaf : A leaf with prominently indented margin.

Indehiscent : Applied to fruits which do not open.

Indeterminate : Of indefinite growth.

Indigenous : Native to a region.

Infinite : Very great in number; applied to inflorescences indicating indefinite growth.

Inflexed : Leaf with margins turned inwards.

Inflorescence : Arrangement of flowers of a plant in relation to axis and to each other.

Infundibuliform : Funnel-shaped.

Innate : Natural—applied to stamen when the filament is attached to the base of the anther.

Insectivorous : A plant that supplements its nitrogen supply by digesting small insects that become trapped on leaves.

Inserted : Applied to stamens which remain within the flower.

Integument : The envelope of an ovule—which develops to seed-coat in seeds.

Interpetiolar : Between two petioles—usually applied to the stipules.

Intrapetiolar : Applied to the stipules which remain between the petiole and the axis of stem.

Introrse : Turned inwards—normally applied to the stamens when the grooved ventral side of anthers faces the centre of the flower.

Involucel : Smaller involucre round the base of each branch of an inflorescence.

Involucre : A whorl of bracts surrounding the base of an inflorescence or several flowers.

Involute : Applied to a leaf in which the two sides are rolled lengthwise over the upper surface towards the midrib.

Irregular : Asymmetrical—applied to flowers or floral parts.

Keel : A prominent dorsal rib or ridge, as in glumes of grasses or in the lower petal of papilionaceous flower or in some carpels.

Kernel : The nucellus of an ovule or of a seed, i.e., the body of seed within the hard shell.

Labiate : Lipped—usually applied to corolla.

Lacerate : Torn.

Laciniate : Cut into deep irregular segments; fringed.

Lacunae : Cavities.

Lanceolate : Lance-shaped—much longer than width, widening above the base and tapering to the apex.

Legume : Dry dehiscent fruit formed from monocarpellary ovary, and splits along both sutures.

Lenticel : Some aerating pores formed in the bark and look like lens-shaped raised spots on the surface of the axis.

Lianas or *lianes* : Very thick and woody perennial climbers, which twine themselves round tall trees.

Lichen : An extraordinary group of plants; a lichen is a peculiar symbiotic combination of an alga and a fungus.

Ligule : Thin, scarious, often hairy projection from the junction of the sheathing leaf-base and lamina.

Ligulate : Furnished with a ligule.

Limb : The broader or expanded part of a gamo-sepalous calyx or gamopetalous corolla as distinct from a tube or tubular structure.

Linear : Like a line—long and narrow with sides parallel or nearly so.

Parachute mechanism : Some fruits and seeds are equipped with some appendages which help the fruits and seeds to float in the air for some time—this mechanism is so named.

Paripinnate : A unipinnate compound leaf bearing even number of leaflets.

Partite : Cleft nearly to the base.

Parietal : Belonging to or borne on wall—usually applied to a type of placentation.

Pectin : A water-soluble chemical found in the cell-walls and cells of some plants.

Pedate : Palmately lobed leaves where the lobes are arranged like the claw of bird.

Pedicel : The short stalk of a flower borne in an inflorescence.

Pedicellate : Equipped with pedicel.

Peduncle : The main axis of an inflorescence together with lateral axes; when a flower is borne singly, the stalk of the same is also denoted by this term.

Pedunculate : Equipped with a peduncle.

Peltate : Usually referred to such a leaf which is shield-shaped with petiole attached to the centre of the underside.

Pendulous : Hanging like a pendulum.

Pentamerous : Flowers having five (or multiples of five) members in each whorl.

Pentandrous : Flowers having five stamens.

Pepo : One-chambered, many-seeded, fleshy fruit developed from inferior ovary with parietal placentation.

Perennial : Living through many, at least more than two years.

Perfect flower : The flower containing both the essential whorls, i.e., androecium and gynoecium.

Perfoliate : When the base lobes of a leaf fuse together after completely clasping the stem so that it appears that the stem has perforated the leaf, this term is applied.

Perianth : Collective name of the accessory whorls, i.e., calyx and corolla, of a flower.

Pericarp : Wall of the fruit.

Perigynous : Literally means standing around the gynoecium; in some flowers, the thalamus grows upward around the ovary in the form of a cup, carrying on its rim the sepals, petals and stamens—such floral parts and the flower as a whole are called perigynous.

Persistent : Leaves and floral parts which live for long duration; sepals which remain even on the fruits.

Personate : When the two lips of a bilabiate corolla are so placed that the mouth of the corolla remains closed.

Petal : Each constituent of corolla.

Petaloid : Like a petal.

Petiole : The stalk of a leaf.

Phanerogams : Plants with manifest flowers.

Phanerophytes : This term includes trees and shrubs—the plants which are easily visible.

Phylloclade : Green, leaf-like flattened or swollen stem.

Phyllode : A green, flattened petiole which looks like and performs the functions of a lamina.

Phyllotaxy : The various patterns in which the leaves are arranged on the stem.

Pilose : Surface with distinct, soft, long and scattered hairs.

Pinna : A primary division of a pinnately compound leaf.

Pinnate : Feather-like—a compound leaf in which leaflets are arranged in two rows along the common axis.

Pinnatifid : Pinnately cleft.

Pinnatipartite : Pinnately parted.

Pinnatisect : Pinnately incised nearly to the rachis.

Pinnule : Secondary pinna.

Pistil : Gynoecium—the whorl consisting of carpels.

Pistillate : Applied to the flower which contains only pistil and not androecium.

Pistillode : A sterile pistil devoid of fertile ovules.

Placentation : Arrangement of placentae (ovule holding tissues) in an ovary.

Plicate : When a leaf, in bud condition, remains folded longitudinally several times along the prominent ribs.

Plumose : Feathered.

Plumule : The shoot in the embryo.

Pneumathodes : Numerous minute pores of special lenticels found on the exposed parts of breathing roots of halophytes.

Pneumatophores : Negatively geotropic vertical branches bearing pneumathodes.

Pollination : Transference of pollen grains from the anther to the stigma.

Pollinium : A mass of pollen grains adhering together, and shed as a unit.

Polyadelphous : Stamens disposed into several groups by connation of their filaments.

Polyandrous : With indefinite number of stamens.

Polycarpellary: With indefinite number of carpels.

Polygamous : Same plant bearing both unisexual and bisexual flowers.

Polymorphic : Having several forms.

Polypetalous : The corolla in which all the petals remain free from each other.

Polyphyllous : When all the members of a perianth remain free from each other.

Polysepalous : The calyx in which all the sepals remain free from each other.

Pome : A type of fleshy fruit which develops from a compound gynoecium with the seeds encased within a cartilaginous wall, and the fruit remains surrounded by fleshy edible thalamus.

Prefoliation : Arrangement of the young leaves in the bud prior to opening.

Prickle : Sharp pointed or curved and unbranched short and superficial outgrowth.

Prop root : Roots developing from the main stem or from the branches, grow downwards and act as pillars supporting the main stem or branches.

Prostrate : Lying flat on the ground.

Protrandous flower: A flower in which the androecium matures before the gynoecium.

Protogynous flower : A flower in which the gynoecium matures earlier than the androecium.

Pseudopodium : False axis formed by joining up successive branches.

Puberulent : With very fine downy hairs.

Puberulous : Slightly hairy.

Pubescent : Covered with short, soft hairs.

Pulverulent : Dusted with fine powder.

Pulvinus : The swollen cushion-like base of petiole.

Punctate : Dotted with depressions or with resinous glands.

Quincuncial : A type of aestivation, where out of five parts two are exterior, two interior, and the fifth one having one margin exterior and the other interior.

Raceme : An indefinite inflorescence in which the flowers are stalked and arranged on the long axis with the oldest flower near the base and the youngest near the top.

Racemose : Raceme-like.

Rachis : The axis of an inflorescence or of a compound leaf.

Radical : Arising from the root.

Radicle : The root of the embryo.

Ray floret : The floret of capitulum situated near the periphery.

Receptacle : The part of the axis of a flower which supports other parts.

Reduplicate : A type of valvate aestivation where margins of each constituent project outwards.

Regma : A dry, schizocarpic fruit with elastically opening segments or cocci.

Regular flower : A flower which can be divided into two exactly equal halves by any vertical section passing through the centre.

Reniform : Kidney-shaped.

Repand : Undulate or wavy.

Replum : False partition wall developed by the extended placentae.

Reticulate : Net-like.

Retinaculum : The hook-like processes to which seeds remain attached.

Retuse : An obtuse leaf apex with a slight notch in the middle.

Revolute : A leaf with margins rolled lengthwise towards the lower side.

Rhachilla : The secondary axis in the inflorescence of grasses.

Rhizome : A thickened prostrate underground stem; usually horizontal.

Rootstock : Vertically growing rhizome.

Rostellum : A beak-like projection from the upper edge of the stigma in orchids.

Rotate : Wheel-shaped, usually referred to corolla.

Rotund : Somewhat circular in outline; usually applied to describe the shape of a leaf of waterlily, garden nasturtium, etc.

Rugose : Wrinkled.

Rugulose : Slightly wrinkled.

Runcinate : Coarsely serrate to sharply incised with the teeth pointing towards the base.

Runner : A slender prostrate stem with long internodes, creeping on the ground, and rooting at the nodes.

Saccate : Gibbous.

Sagittate : Usually applied to the arrow-shaped leaf having basal lobes directed downwards.

Salver-shaped : Usually applied to corolla having a slender tube and an abruptly expanding limb; hypocrateriform.

Samara : A dry, indehiscent, 1- or 2-seeded, winged fruit, developing from a superior, bi- or tri-carpellary ovary.

Samaroid : Winged fruits where the wings develop from dry, persistent sepals.

Saprophyte : The plant which lives on dead organic matter.

Scabrous : Pubescent.

Scaly : Scale-like.

Scape : The flowering shoot developed from underground stem.

Scapigerous : Scape-producing plant.

Scapose : Like a scape.

Scarious : Dry, thin, scale-like.

Sciophyte : Shade-loving plant.

Scorpioid : Scorpion-like; when the lateral branches develop on alternate sides forming a zigzag.

Scrambler : Weak plant which actually does not climb, but slowly grows over other bushes.

Scrub : A plant community dominated by shrubs or bushes.

Secund : Parts or organs directed to one side only.

Semiparasite : Partial parasite.

Sepal : The constituent of calyx.

Sepaloid : Like a sepal.

Septicidal : Usually applied to fruits where splitting takes place along the septa thus separating each loculus.

Septifragal : Usually applied to fruits where dehiscence is loculicidal or septicidal, but the valves fall off leaving the seeds attached to the central axis.

Septum : Partition.

Serrate : Margin with sharp teeth pointing forward.

Serrulate : Minutely serrate.

Sessile : Stalkless.

Setacious : Bristle-like.

Setose : Bristly.

Setulose : With minute bristles.

Sheath : The tubular basal portion of a leaf of a grass or sedge, which encircles the stem.

Sheathing : Usually referred to prominent leafbase which clasps and invests the stem.

Shrub : A plant which usually gives rise to a number of woody branches from the level of the soil forming a bush, in which the main trunk gets lost.

Sigmoid : Curved like "S".

Silicula : A short, broad and flat siliqua.

Siliqua : A dry, dehiscent, long, narrow, two-chambered fruit developed from a bicarpellary pistil with two parietal placentae.

Simple fruit : A single fruit developed from the ovary of a flower with or without accessory parts.

Sinuate : With a deep wavy margin.

Sorosis : A multiple fruit developed from a spike or spadix.

Spadix : A spike with a fleshy axis and enclosed by one (sometimes more) large brightly coloured bract.

Spathe : Flat, broad, usually brightly coloured bract enveloping a spadix.

Spathellae : A special name given to the spathes which separately cover the branches of a compound spadix.

Spathulate or *spatulate* : Spatula-shaped leaf apex.

Spicate : Like a spike.

Spike : An indefinite, elongate inflorescence bearing sessile flowers.

Spikelets : A small spike; the unit of inflorescence of grasses and sedges.

Spine : Leaf or part of a leaf modified to a small, sharp, pointed, woody or hard body.

Spinose : Leaf-surface covered with prickles.

Spiral : Winding continually—usually applied to arrangement of leaves or floral organs.

Sporophyll : Spore-bearing leaf.

Sporophyte : Spore producing plant.

Spur : A tubular or sac-like projection, usually from the calyx or corolla, commonly producing nectar in its tip.

Squamose : Scaly or scale-like.

Squarrose : With minute scales.

Stamen : The pollen producing organ of a flower consisting of filament and anther; the unit of androecium.

Staminate : Bearing stamens; flowers which bear only stamens and no carpel is described by this term.

Staminode (Staminodium) : A sterile or abortive stamen.

Standard : Vexillum—the broad outer petal in a papilionaceous flower.

Stellate : Star-shaped.

Stigma : The terminal end of the style, i.e., the part that receives the pollen grains.

Stilt roots : Roots that grow obliquely downwards from near the base of the stem and provide additional support and anchorage.

Stipe : A stalk.

Stipel : Stipule-like appendages found on the two sides of the base of a leaflet.

Stipule : A pair of appendages present on two sides of a petiole at the base of some leaves.

Stolon : A slender lateral branch which originates from the base of the stem, bends down on or into the ground, and grows horizontally outwards for a shorter or longer distance.

Stoloniferous : Which produces or bears stolon.

Strap-shaped : Ligulate.

Style : The slender elongated projection of ovary, usually from top, and acting as the stalk of the stigma.

Subulate : Awl-shaped.

Sucker : A lateral branch of underground stem, but it grows obliquely upwards and gives rise to a leafy shoot or a new plant.

Sulcate : Grooved or furrowed.

Superior : Usually applied to an ovary which is situated above the point of attachment of perianth and stamens, or surrounded by a hypanthium that is not adnate to it.

Superficial : This term usually indicates origin from upper layers of tissues; also applied to the placentation when the placentae in a multilocular ovary is distributed all round the inner surface of the partition wall.

Suture : The line of union of two margins of one or more carpels.

Syconus : A multiple fruit developed from a hollow, pear-shaped, fleshy receptacle which encloses a number of minute, reduced, male and female flowers.

Synandrous : This is a condition when two or more stamens are completely united by their filaments as well as anthers.

Syncarpous : Having carpels united.

Syngenesious : Stamens united by their anthers —often forming a cylinder around style.

Syngynia : When the ovaries of two adjacent flowers fuse while other parts remain free.

Taproot : The permanent root developed from the radicle via the stage of primary root.

Tassel : Apical panicles of maize on which the male flowers are borne.

Taxonomy : Systematic botany, primarily concerned with the nomenclature and classification of plants.

Tegmen : Inner seed-coat.

Tendril : A slender, leafless, spirally coiling, usually green structure which helps the plant to cling upon a support.

Tentacles : Glandular hairs which secrete sticky fluid.

Ternate : In threes; trifoliate.

Testa : The outer and usually hard coat of a seed.

Tetradynamous : Having four long and two short stamens.

Tetramerous : When the number of floral members is four or multiple of four.

Tetrandrous : When the number of stamens in a flower is four.

Thalamus : The suppressed swollen end of a flower-axis on which the floral parts, viz., sepals, petals, stamens and carpels are inserted.

Therophyte : Annual herb.

Thorn : A hard, often straight, pointed structure developed usually through modification of axillary bud.

Thyrsus : A special type of inflorescence having an indefinite main axis while secondary and other axes are of cymose nature.

Tomentose : Densely woolly with matted hairs.

Tomentulose : Delicately tomentose.

Trifoliate : A palmately compound leaf with three leaflets.

Trigonous : Three angled.

Truncate : A type of leaf apex, which appears as abruptly cut across or nearly so.

Tuber: The swollen end of a special underground branch of a stem, bearing scale-like leaves with buds (eyes) in their axils.

Tuberous : Tuber-like.

Twisted : Convolute—with parts rolled up in such a way that the outer part of each covers the inner part of the one in front of it, while in turn a part is covered by the one behind it.

Twiner : A plant that climbs by means of spirally coiling stems.

Umbel : An umbrella-shaped racemose inflorescence with shortened internodes, consequently the pedicellate flowers appear to have originated from a common point.

Umbellate : Of the form of an umbel.

Uncinate : Hooked.

Undulate : Wavy.

Unicostate : Having one prominent vein.

Unifoliate : A palmately compound leaf having only one leaflet.

Unipinnate : A pinnately compound leaf where the leaflets are directly borne on the midrib.

Urceolate : Urn-shaped hollow and contracted at the mouth.

Urticle : A small bladdery structure.

Valvate : A type of aestivation where the parts meet by their abrupt edges but not overlapping.

Valvular dehiscence : The whole or portions of the wall of anther open out like trap-doors releasing the pollens.

Veins : Rigid linear structures found in the leaf-blade, these form the fibro-vascular tissue system of the lamina.

Velutinous : Surface of a leaf is velvety, due to a coating of fine, soft hairs.

Venation : The arrangement of veins of a leaf.

Vernation : Prefoliation—the mode or arrangement of leaves in the bud.

Vernicose : Shiny, as though varnished, leaf surface.

Verrucose : Surface covered with wart-like elevations.

Versatile anther : When filament is attached to merely a point on the back of the anther, so that the anther can swing freely.

Versicolor : Changing colour.

Verticillaster : A special type of inflorescence composed of a pair of opposed bichasial cymes bearing sessile or almost sessile flowers.

Verticillate phyllotaxy : Whorled phyllotaxy—more than two leaves at a node in a circle around it.

Vexillary : Typical aestivation of the papilionaceous corolla where of the five petals, the posterior one almost covers two lateral petals, and the latter in turn nearly overlap the two anterior ones.

Vexillum : The posterior and usually the largest petal in the papilionaceous corolla.

Villose : Surface covered with long, soft and closely arranged hairs.

Villous : Surface covered with long, soft, shaggy hairs that are not matted.

Viscid : Sticky.

Vitta : Tube in pericarp containing aromatic oil.

Vivipary : A kind of germination when seeds germinate inside the fruit while it is still on the parent tree and is nourished by it.

Viviparous : The plant in which vivipary takes place.

Whorl : A circle or cycle or ring of organs arranged around an axis—as the leaves in some stems, the organs of a flower.

Whorled phyllotaxy : Verticillate phyllotaxy.

Xeromorphic : Morphological character developed in xerophytes to facilitate them to adjust in the dry environment.

Xerophyte : The plant which can grow and subsist normally in dry localities like desert.

Zoophilous : Plants pollinated by animals.

Zygomorphic : A flower which can be divided into two similar halves by vertical section passing through one particular plane.